D1514692

# Cell Culture Models of Biological Barriers

# Cell Culture Models of Biological Barriers

*In vitro* test systems for drug absorption and delivery

Edited by

# Claus-Michael Lehr

Saarland University, Saarbrücken, Germany

London and New York

First published 2002
by Taylor & Francis
11 New Fetter Lane, London EC4P 4EE

Simultaneously published in the USA and Canada
by Taylor & Francis Inc,
29 West 35th Street, New York, NY 10001

*Taylor & Francis is an imprint of the Taylor & Francis Group*

© 2002 Taylor & Francis

Typeset in Baskerville by
Integra Software Services Pvt. Ltd, Pondicherry, India
Printed and bound in Great Britain by
TJ International Ltd, Padstow, Cornwall

Every effort has been made to ensure that the advice and information in this book
is true and accurate at the time of going to press. However, neither the publisher
nor the authors can accept any legal responsibility or liability for any errors or
omissions that may be made. In the case of drug administration, any medical procedure
or the use of technical equipment mentioned within this book, you are strongly
advised to consult the manufacturer's guidelines.

*British Library Cataloguing in Publication Data*
A catalogue record for this book is available
from the British Library

*Library of Congress Cataloging in Publication Data*
A catalog record for this book has been requested

ISBN 0–415–27724–8

# Contents

# Color plates

All color artworks are reproduced in black and white in the maintext. The color plate section appears between pages 326 and 327.

# Figures

# Tables

# Contributors

**Dr Kenneth L. Audus**   Department of Pharmaceutical Chemistry, The University of Kansas, 2095 Constant Avenue, Lawrence, KS 66047-2504, USA

**Dr Udo Bakowsky**   Department of Biopharmaceutics and Pharmaceutical Technology, Saarland University, Im Stadtwald 8.1, 66123 Saarbrücken, Germany

**Dr H. Bakowsky**   Martin-Luther University Halle/Wittenberg, Department of Physiological Chemistry, 06096 Halle/Saale, Hollystraße 1, Germany

**Dr Michael B. Bolger**   University of Sourthern California, Department of Pharmaceutical Science, USC School of Pharmacy, 985 Zonal Avenue, PSC 700, Los Angeles CA 90089, USA

**Dr Gerrit Borchard**   LACDR, Department of Pharmaceutical Technology, PO BOX 9502, NL-2300, RA Leiden, The Netherlands

**Dr Ronald T. Borchardt**   Pharmaceutical Chemistry, The University of Kansas, 2095 Constant Avenue, Lawrence, KS 66047, USA

**Dr Lee Campbell**   Cardiff University, Welsh School of Pharmacy, Redwood Building, Kind Edward VII Avenue, Cardiff CF1 3XF, UK

**Dr Weiqing Chen**   Pharmaceutical Chemistry, The University of Kansas, 2095 Constant Avenue, Lawrence, KS 66047, USA

**Dr Yusuf K. Durlu**   World Eye Hospital, Retina Unit, Dünya Göz Hastanesi, Nispetiye Caddesi, Aydin Sokak No. 1, 1. Levent, Besiktas, 80620 Istanbul, Turkey

**Dr Carsten Ehrhardt**   Department of Biopharmaceutics and Pharmaceutical Technology, Saarland University, Im Stadtwald 8.1, 66123 Saarbrücken, Germany

**Dr Tanja Eisenblätter**   Institut für Biochemie, WWU Münster, Wilhelm Klemm Str. 2, 48149 Münster, Germany

**Dr Bora Eldem**   Hacettepe University, Medical School, Department of Ophthalmology, Retina Unit, 06100 Sihhiye, Ankara, Turkey

**Dr Türkan Eldem**   Hacettepe University, Faculty of Pharmacy, Department of Pharmaceutical Biotechnology, 06100 Sihhiye, Ankara, Turkey

**Dr Bogdan I. Florea**   LACDR, Department of Pharmaceutical Technology, PO BOX 9502, NL-2300, RA Leiden, The Netherlands

**Dr Robert Fraczkiewicz**   Simulations Plus Inc., 1220 W. Avenue J, Lancaster, CA 93534-2902, USA

**Dr Helmut Franke**   Institut für Biochemie, WWU Münster, Wilhelm Klemm Str. 2, 48149 Münster, Germany

**Dr Geat Fricker**   Institut für Pharmazeutische Technologie und Biopharmazie, INF 366, 69120, Heidelberg, Germany

**Dr Lawrence X. Fu**   FDA Food and Drug Administration, Center for Drug Evaluation and Research, Rockville MD 20857, USA

**Sabine Fuchs** Department of Biopharmaceutics and Pharmaceutical Technology, Saarland University, Im Stadtwald 8.1, 66123 Saarbrücken, Germany

**Dr Akira Fukuhara**   Department of Pharmaceutical Chemistry, The University Of Kansas, 2095 Constant Avenue, Lawrence, KS 66047-2504, USA

**Dr Franz Gabor**   Institute of Pharmaceutical Technology and Biopharmaceutics, University of Vienna, Althanstraße 14, A-1090 Vienna, Austria

**Dr Hans-Joachim Galla**   Institut für Biochemie, WWU Münster, Wilhelm Klemm Str. 2, 48149 Münster, Germany

**Dr Thomas M. Gilman**   Simulations Plus Inc., 1220 W. Avenue J, Lancaster, CA 93534-2902, USA

**Dr Gerhard Gstraunthaler**   Department of Physiology, University of Innsbruck, Fritz-Pregl-Strasse 3, A-6010 Innsbruck, Austria

**Dr Mark Gumbleton**   Cardiff University, Welsh School of Pharmacy, Redwood Building, Kind Edward VII Avenue, Cardiff CF1 3XF, UK

**Dr Hans Häberlein**   Universität Marburg, Institut für Pharmazeutische Biologie Deutschhausstraße 17, A-35032 Marburg, Germany

**Dr Thomas Hartung**   Biochemical Pharmacology, University of Konstanz, D-78457 Konstanz, Germany

**Dr D. Hoekstra**   University of Groningen, Department of Membrane Cell Biology, 9700 Av. Groningen, A. Deusinglaan 1, The Netherlands

**Dr Kazutoshi Horie**   Pharmaceutical Chemistry, The University of Kansas, 2095 Constant Avenue, Lawrence, KS 66047, USA

**Dr Hans E. Junginger**   LACDR, Department of Pharmaceutical Technology, PO BOX 9502, NL-2300, RA Leiden, The Netherlands

**Dr Kwang-Jin Kim**   Associate Professor of Medicine, Room HMR-914, USC, Keck school of Medicine, 2011 Zonal Avenue, Los Angeles, CA 90033, USA

**Dr Carsten Kneuer**   Department of Biopharmaceutics and Pharmaceutical Technology, Saarland University, Im Stadtwald, Building 8.1, D-66123 Saarbrücken, Germany

**Dr Annette M. Koch**   Swiss Federal Institute of Technology Zürich (ETH), Department of Applied Biosciences, Institute of Pharmaceutical Sciences, Irchel Campus, Winterthurer Straße 190, Ch-8057 Zürich, Switzerland

**Dr Alf Lamprecht**   Department of Biopharmaceutics and Pharmaceutical Technology, Saarland University, Im Stadtwald 8.1, 66123 Saarbrücken, Germany

**Dr Vincent H. L. Lee**   University of Southern California, School of Pharmacy, 1985 Zonal Avenue, PSC 704, Los Angeles, CA 90089-9121, USA

**Dr Claus-Michael Lehr**   Department of Biopharmaceutics and Pharmaceutical Technology, Saarland University, Im Stadwald 8.1, 66123 Saarbrücken, Germany

**Dr Manfred Liebsch**   Bundesinstitut für gesundheitlichen, Verbraucherschutz und Veterinärmedizin, BgVV-Zebet, Diedersdorfweg 1, 12277, Berlin, Germany

**Dr Claire Meaney**   LACDR, Department of Pharmaceutical Technology, PO BOX 9502, NL-2300, RA Leiden, The Netherlands

**Dr Hans P. Merkle**   ETH, Eidgenössische Technische Hochschule Zürich, Dpt Pharmazie, Winterthurer Straße 190, CH-8057 Zürich, Switzerland

**Dr Helga Möller**   Zentrallaboratorium, Deutscher Apotheker, Carl-Mannich Str. 20, 65760 Eschborn, Germany

**Dr Hanne Mørck Nielsen**   Department of Pharmaceutics, The Royal Danish School of Pharmacy, 2 Universitetsparken, 2100 Copenhagen, Denmark

**Dr Thorsten Nitz**   Institut für Biochemie, WWU Münster, Wilhelm Klemm Str. 2, 48149 Münster, Germany

**Dr V. Oberle**   University of Groningen, Department of Membrane Cell Biology, 9700 Av., Groningen, A. Deusinglaan 1, The Netherlands

**Dr Meral Özgüç**   Tübitak (The Scientific and Technical Research Council of Turkey), DNA/Cell Bank and Gene Research Laboratory, Hacettepe University, Medical School, Institute of Child Health, 06100 Sihhiye, Ankara, Turkey

**Dr W. Pfaller**, Institut für Physiologie und Balneologie der Universität Innsbruck, Fritz-Preglstrasse 3, A-6010 Innsbruck, Austria

**Dr Katherina Psathaki**   Institut für Biochemie, WWU Münster, Wilhelm Klemm Str. 2, 48149 Münster, Germany

**Dr U. Rothe**   Martin-Luther University Halle/Wittenberg, Department of Physiological Chemistry, 06096 Halle/Saale, Hollystraße 1, Germany

**Dr Ulrich F. Schäfer**   Department of Biopharmaceutics and Pharmaceutical Technology, Saarland University, Im Stadtwald 8.1, 66123 Saarbrüecken, Germany

**Dr M. Christiane Schmidt**   ETH, Eidgenössische Technische Hochschule Zürich, Dpt Pharmazie, Winterthurer Straße 190, CH-8057 Zürich, Switzerland

**Dr Horst Spielmann**   Bundesinstitut für gesundheitlichen, Verbraucherschutz und Veterinärmedizin, BgVV-Zebet, Diedersdorfweg 1, 12277, Berlin, Germany

**Dr Jennifer Sporty**   University of Southern California, School of Pharmacy, 1985 Zonal Avenue, PSC 704 Los Angeles, CA 90089-9121, USA

**Dr Boyd Steere**   Simulations Plus Inc., 1220 W. Avenue J, Lancaster, CA 93534-2902, USA

**Dr Pekka Suhonen**   Department of Pharmaceutics, University of Kuopio, POB 1627, 70211 Kuopio, Finland

**Dr Fuxing Tang**   Pharmaceutical Chemistry, The University of Kansas, 2095 Constant Avenue, Lawrence, KS 66047, USA

**Dr Joseph J. Tukker**  Utrecht Institute of Pharmaceutical Sciences, University of Utrecht, PO BOX 80082, NL-3508, Utrecht, The Netherlands

**Dr Anna-Lena Ungell**  DMPK and Bioanalytical Chemistry, AstraZeneca R&D Mölndal, S-431 83 Mölndal, Sweden

**Dr Arto Urtti**  Department of Pharmaceutics, University of Kuopio, POB 1627, 70211 Kuopio, Finland

**Dr Donna A. Volpe**  Division of Product Quality Research, Food and Drug Administration, HFD-941, NLRC-2400B, 5600 Fishers Lane, Rockville, MD 20857, USA

**Dr H. Wagner**  Department of Biopharmaceutics and Pharmaceutical Technology, Saarland University, Im Stadtwald 8.1, 66123 Saarbrücken, Germany

**Dr Joachim Wegener**  Institut für Biochemie, WWU Münster. Wilhelm-Klemm Str. 2, 48149 Münster, Germany

**Dr Michael Wirth**  Institute of Pharmaceutical Technology and Biopharmaceutics, University of Vienna, Althanstraße 14, A-1090 Vienna, Austria

**Dr Laurie Withington**  BD Biosciences, Bedford, MA 01730, USA

**Dr Walter S. Woltosz**  Simulations Plus Inc., 1220 W. Avenue J, Lancaster, CA 93534-2902, USA

**Dr Amber Young**  Department of Pharmaceutical Chemistry, The University Of Kansas, 2095 Constant Avenue, Lawrence, KS 66047-2504, USA

**Dr Lawrence X. Yu**  Division of Product Quality Research, Food and Drug Administration, HFD-941, NLRC-2400B, Rockville, 5600 Fishers Lane, MD 20857, USA

**Dr N. Zghoul**  Department of Biopharmaceutics and Pharmaceutical Technology, Saarland University, Im Stadtwald 8.1, 66123 Saarbrücken, Germany

**Dr I. Zuhorn**  University of Groningen, Department of Membrane Cell Biology, 9700 Av. Groningen, A. Deusinglaan 1, The Netherlands

# Preface

This book aims to provide a practical approach to contemporary cell culture-based *in vitro* techniques for drug transport studies at biological absorption barriers. It is not primarily written for experts, but rather for the novices in the field.

Traditionally, pharmaceutical scientists have a strong background in chemistry and physicochemistry, while often their biological education has been more elementary. Many pharmaceutical scientists are therefore not familiar with laboratory methods of cell biology. In contrast, over the past ten years, several sophisticated cell or tissue-based test models have been introduced in the field of drug delivery. These models have been found to be very useful in characterizing the permeability of epithelial tissues, and in studying formulations or carrier systems for improved drug delivery and enhanced absorption. Compared with *in vivo* trials on animals or humans, cell culture models are more convenient and faster, and offer ethical and cost advantages. Most importantly, they can be more easily standardized and validated. By appropriate automatization, even high-throughput screening can be realized. In the drug development process, epithelial cell culture systems therefore allow savings of money and time.

This book is based on our experience of three 10-day intensive courses, encompassing both laboratory courses and classes, which were held between 1998 and 2000 at the Department of Biopharmaceutics and Pharmaceutical Technology at Saarland University, Saarbrücken. The primary target group of readers are pharmaceutical scientists working in industry or in academia, who are planning to implement cell culture or *ex vivo* models in their current projects. The book strives to provide all the necessary information for somebody considering setting up such models in his own research environment. In addition, it will provide useful information to those pharmaceutical scientists who are not actually planning – or do not have the facilities – to use cell culture systems in their own laboratory, but instead are considering having such experiments carried out in collaboration with external contractors. Finally, the book may be useful to students attending graduate courses on this subject.

The first part of the book is dedicated to general aspects of epithelial cell culture systems, also including issues of validation and quality control. In the second part, methods used to model particular epithelial barriers of pharmaceutical interest are reviewed. Each chapter is intended to provide a brief overview on the existing methods. Furthermore, we have asked each of the authors to describe his or her model of choice in more detail, providing the reader with point-by-point protocol to be used at the bench side. The third part of the book discusses a selection of novel or just-emerging research methods in visualizing or quantifying the inter-actions between drugs or drug carrier systems and epithelial cells.

# Part 1

# General aspects of epithelial cell and tissue culture

# Chapter 1

# Basic aspects of cell growth and cell cycle in culture

*Lee Campbell and Mark Gumbleton*

## INTRODUCTION

The use of cell culture as part of pharmaceutical science investigations has dramatically increased over the last 20 years to the point where it has become almost a non-specialist technique. In the following chapters of this textbook the reader will be introduced to various cell culture protocols from the isolation of cells from tissue and their subsequent use in primary cultures, e.g. isolation and culture of alveolar type-II cells (Chapter 12), or the isolation of brain microvascular cells and their use, following passage or subculture, as finite cell lines (Chapter 18), through to the extensive exploitation of a wide range of continuous cell lines whose growth in culture is not limited and does not require the repeated need to undertake demanding tissue and cell isolation procedures. To provide a framework for the understanding of the following chapters, and with consideration that the text is aimed primarily at the early postgraduate student level, it is the aim of this contribution to inform the reader of the varied nature of cell growth characteristics in culture. Concurrently, it is anticipated that many aspects of the basic terminology used in this discipline will be clarified. Furthermore, a basic overview of tight junctional physiology and regulation will be provided, as this is an implicit feature of all transepithelial or transendothelial transport investigations.

## MAMMALIAN CELL MEMBRANE COMPOSITION

It is assumed that the reader has a general knowledge of eukaryotic cell structure and the subcellular organelles. If this is not the case, any basic text in biochemistry can provide this background. However, consideration of the composition and structure of biological membranes is fundamental to understanding the relationship between a drug's physicochemical properties and its membrane transport. As such, an overview of the main features of biological membranes is provided below.

A plasma membrane encloses every cell, defining the cell's extent and maintaining the essential differences between the cell's interior and its environment. Membranes of eukaryotic cells, including the plasma membrane and internal membranes such as the endoplasmic reticulum, have a common general structure; they are assemblies of lipid and protein molecules held together mainly by non-covalent interactions. Lipid molecules are arranged as a continuous bilayer approximately

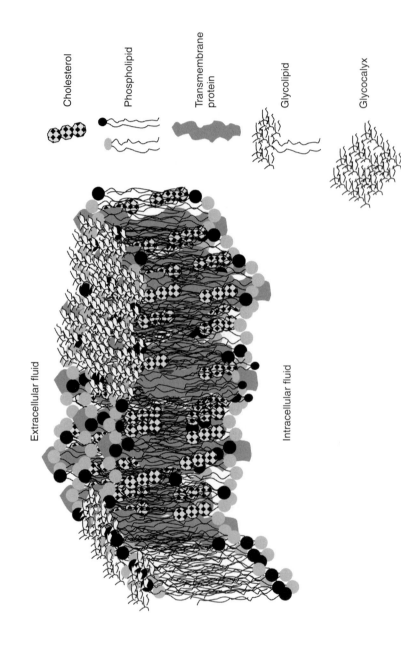

Cholesterol

Phospholipid

Transmembrane protein

Glycolipid

Glycocalyx

Extracellular fluid

Intracellular fluid

*Figure 1.1* Schematic of cell membrane comprising lipid bilayers, transmembrane proteins and an outer negatively charged glycocalyx.

5 nm thick (Figure 1.1). Protein molecules embedded within the lipid bilayer mediate most of the functions of the membrane, e.g. the transport of specific solutes, the catalysis of membrane-associated reactions as receptors for receiving and transducing chemical signals, and the provision of structural links connecting plasma membrane to cytoskeleton, or cell–extracellular matrix, and cell–cell, adhesion etc.

Lipid molecules constitute about 50 per cent of the mass of most mammalian plasma membranes, nearly all the remainder being protein. The three major types of lipid in cell membranes are phospholipids, cholesterol and glycolipids. All these three are amphipathic. The fluidity (viscosity) of the lipid bilayer depends on its composition. Perturbation of membrane fluidity is the mechanism of action of many membrane-penetration enhancers. A phospholipid molecule, e.g. phosphatidylcholine, possesses a polar head group, e.g. choline-phosphate-glycerol, and two hydrophobic fatty-acid tails or chains. The shorter the hydrophobic tails and the higher degree of unsaturated *cis*-double bonds in the fatty acid chain, then the tendency for them to pack together is reduced, e.g. the membrane is more fluid. The plasma membranes of most eukaryotic cells contain a variety of phospholipids. For example, those based on glycerol include phosphatidylcholine, phosphatidylserine (possesses net negative charge at physiological pH 7.4) and phosphatidylethanolamine. Sphingomyelin is a membrane phospholipid based on ceramide.

Eukaryotic plasma membranes contain large amounts of cholesterol, with up to one molecule for every phospholipid molecule. Cholesterol molecules orient themselves in the bilayer with their hydroxyl groups close to the polar head groups of the phospholipid molecules. Their rigid, plate-like steroid rings interact with – and partly immobilize – those regions of the phospholipid hydrocarbon chain that are closest to the polar head groups, leaving the remainder of the chain flexible. Cholesterol decreases the permeability of the membrane to small hydrophilic molecules and enhances both the flexibility and mechanical stability of the bilayer.

Glycolipids are oligosaccharide-containing lipid molecules which are located exclusively in the outer leaflet of the membrane bilayer, e.g. the leaflet exposed to interstitial (extracellular) fluid that bathes all cells. The polar sugar groups are exposed at the cell surface contributing to the cell's glycocalyx (see below). Glycolipids differ remarkably from one animal species to another, and even among tissues in the same species. In bacteria and plants, almost all glycolipids are derived from glycerol-based lipids, e.g. phosphatidylcholine. But in animal cells glycolipids are almost all based upon ceramide (e.g. sphingomyelin). These glycosphingolipids have a general structure comprising a polar head group and two hydrophobic fatty-acid chains. Glycolipids are distinguished from one another by their polar head group, which consists of one or more sugar residues. The most widely distributed glycolipids in the plasma membranes of eukaryotic cells are the neutral glycolipids whose polar head groups consist of 1–15 or more uncharged sugar residues.

The varying lipid compositions of different membranes may well reflect the specific environments required for the optimal functioning of the respective membrane proteins. A good example of this relates to the 'raft hypothesis' (Simons and Ikonen, 1997), which proposes that sphingolipids and cholesterol cluster in biomembranes to form 'raft-like' microdomains floating in a glycerophospholipid-rich environment, creating a mosaic of raft and non-raft membrane domains. The lipid rafts serve as platforms for the lateral sorting of proteins, selectively sequestering certain proteins and excluding others, and, as such, the rafts orchestrate numerous cellular functions including the vesicular sorting and transport of biosynthetic molecules, cellular differentiation and cell signaling.

Most of the specific functions of biological membranes are carried out by proteins. The amounts and types of protein in a membrane are highly variable, e.g. in the myelin membrane, which serves mainly to insulate nerve cell axons. Less than 25 per cent of the membrane mass is protein, whereas protein constitutes close to 75 per cent of the internal membrane of mitochondria and approximately 50 per cent of the mass of the average plasma membrane. Membrane proteins associate with a lipid bilayer in many different ways. Transmembrane proteins extend across the bilayer as a single α-helix, or as multiple α-helices. These transmembrane proteins are amphipathic: they have hydrophobic regions that pass through the membrane and interact with the hydrophobic tails of the lipid molecules in the membrane interior, and hydrophilic regions that are exposed to the aqueous environment on both sides of the membrane – the interstitial fluid and cell cystosol. Transmembrane proteins always have a unique orientation in the membrane which reflects both the asymmetrical manner in which they are inserted into the bilayer and the different functions of their cytoplasmic and extracellular domains. The great majority of transmembrane proteins are glycosylated with the oligosaccharide chains exposed to the extracellular environment. This exposure contributes to the cell's glycocalyx. Some membrane proteins are attached covalently to the bilayer only by means of a fatty-acid chain, while some cell-surface proteins are attached covalently via a specific oligosaccharide to phosphatidylinositol (a major phospholipid) in the outer leaflet of the plasma membrane. The proteins linked to the lipid bilayer by these first four methods are referred to as integral membrane proteins – they can be released only by disrupting the bilayer with detergents or organic solvents. Finally, proteins can be attached to the membrane by noncovalent interactions with other membrane proteins. Proteins linked by this last mode are referred to as peripheral membrane proteins – they can be released by gentle extraction procedures such as exposure to solutions of high or low ionic strength or pH.

Cells that perform a barrier function, such as epithelial, endothelial or mesothelial cells, confine proteins (and lipids of the outer leaflet) to specific domains within a membrane. The confinement of specific proteins and lipids to specific membranes leads to the polarization of the cell, e.g. the cell possesses two distinctively different membrane domains and, by inference, different capabilities for active and facilitated processes. The polarized nature of epithelial cells is maintained by the 'fence' function of the tight-junctional complex.

All eukaryotic cells have carbohydrate on their surface, both as polysaccharide chains covalently bound to integral membrane proteins (glycoproteins) and as oligosaccharide chains covalently bound to lipids (glycolipids). The total carbohydrate in plasma membranes constitutes 2–10 per cent of the membrane's total weight. Most plasma membrane protein molecules exposed at the cell surface carry sugar residues. As with proteins and lipids, the distribution of carbohydrate in biological membranes is asymmetrical. Carbohydrate chains of glycoproteins and glycolipids of both internal and plasma membranes are located exclusively on the noncytosolic surface. The term glycocalyx is used to describe the carbohydrate-rich peripheral zone on the outer surface of the cell plasma membranes. The glycocalyx is characterized by a net negative charge which results from the presence of sialic acid (a family of 9-carbon carboxylated sugars) or sulfate groups at the non-reducing termini of the glycosylated molecules. This 'blanket' of negative charge functions to protect the underlying membranes. It has important influences on the functioning of the cell, e.g. cell–cell recognition, and in xenobiotic–membrane interactions.

## GROWTH OF THE CULTURED CELL

The first aspect of terminology that should be clarified is that of *primary culture*. Freshly isolated cells taken from animal or human tissue which are then grown in an *in vitro* environment are termed primary cultures until they are passaged or subcultured for the first time. The process of *passage* or *subculture* is where cells from an *adherent culture* (e.g. a culture where growth is based upon adherence to an extracellular matrix laid down by the cells onto tissue culture plastic) or from a *suspension culture* (e.g. cells grown in a suspension of culture medium) are transferred from one vessel to a second vessel. In the case of adherent cell cultures the passage or subculture is achieved by, firstly, dissociating the cell–matrix and cell–cell, interactions by briefly exposing the cell cultures to a solution of the protease trypsin and/or the divalent ion ($Ca^{++}$) chelator, ethylenediamine tetraacetic acid (EDTA). The latter binds $Ca^{++}$ and, therefore, disrupts key $Ca^{++}$-dependent adhesion forces between cells, and between cells and the matrix. In the case of suspension cultures, the passage or subculturing process simply involves transfer of an aliquot of the cell suspension to fresh medium. In both cases, the process of passage involves a diluting of the cell density (either cells per $cm^2$ area of plastic or cells per ml volume of suspension) to allow regrowth of the cells under conditions that are not limiting with respect to nutrient supply, growth area or volume.

The process of obtaining cells from tissues generally involves an enzymatic tissue disaggregation process, e.g. the use of elastase or trypsin for the isolation of lung alveolar type-II epithelial cells (see Chapter 12), or of collagenase and dispase for the isolation of brain microvascular cells (see Chapter 18). The key to establishing a primary culture is to achieve a high level of purity of the cell type in which one is interested. The complexity of the tissue will thus tend to dictate the complexity of the isolation procedure. For example, obtaining a highly pure isolate of hepatocytes from the liver (a tissue comprising only a few cell types with the predominant one being the hepatocyte itself) is more straightforward than obtaining type-II alveolar epithelial cells from the lung (a tissue which comprises many cell types – approximately 40 – and within which the type-II cell comprises approximately 16 per cent of the lung cell population).

Clearly, any primary isolate from a tissue will be a mixture of different cell types, although, as stated above, the aim is always to achieve a highly pure primary culture. Nevertheless, in primary culture cell selection processes will be occurring and will be based upon cell growth rates. Some cells will be capable of proliferation and will increase in number in the culture, some cells will survive but not replicate, others will not be able to survive under the particular conditions of culture used. Hence, as the primary culture proceeds the distribution of cell types will change.

After the first passage or subculture the cells become a cell line and may be propagated or subcultured several times depending upon their nature and growth properties. Isolation of cells from normal tissue usually gives rise to a culture with a finite and short lifespan. Most normal cells have a finite lifespan of 20–100 population doublings or divisions.

The relationship between population doubling and passage or subculture may warrant some clarification. Within each passage or subculture the cells will undergo a number of *population doublings* or divisions which will depend upon the density at which the cells are originally placed in the culture vessel (i.e. *cell seeding density*), the growth rate of the cells and the time the cells are left within the culture vessel. For example, if adherent cells are seeded into a culture vessel at a density of 50,000 cells per/cm$^2$ and allowed to grow to fill the area of the culture vessel (e.g. to reach *confluence*) and to achieve at confluence a final cell count of 1,600,000 cells/cm$^2$, then the culture can be considered to have undergone five population doublings during this particular passage or subculture phase; e.g. one doubling 50,000–100,000; two doublings 100,000–200,000; three doublings 200,000–400,000 etc. If the same is repeated upon the next passage or subculture, then the cells would have undergone ten population doublings in two passages. In this example, it is assumed that 100 per cent of the cells that are seeded into the culture vessel (50,000 cells/cm$^2$) would have adhered and proliferated. This is never the case and only a certain per cent of the cells seeded into a culture vessel will attach and undergo cell proliferation. This per cent of the cells that adhere and are able to grow is termed *plating efficiency* and is used as a measure of proliferative or survival capacity of the cells following, for example, the thawing of cells from liquid nitrogen storage or in the cell seeding of primary or low passage cultures. All aspects of a cell's environment (e.g. growth matrix and media or cell seeding density) will affect plating efficiency of a culture, including prior treatments, such as the extent of tissue digestion or cell disaggregation during subculturing etc.

With each successive passage the cells within the population that show the highest ability to proliferate or divide will eventually predominate and the non-proliferating or slowly-proliferating cells will be diluted out. During these initial passages from the primary isolate additional features within the culture environment can be modified to select for specific cells. For example, the nature and concentrations of the basement membrane laid down exogenously onto the plastic by the investigator will modify selection (e.g. coating of the tissue culture-treated plastic with commercially available laminin, collagens or fibronectin), as will the addition of growth supplements to the media, e.g. epidermal growth factor (EGF). Such selective media or substrates will be of critical importance in the selection process, and especially in cases where the purity of the primary isolates is compromised.

After a few passages from the primary cell, the cell lines will either undergo senescence (blockage of cell proliferation) leading to cell death, or undergo a transformation to become a continuous cell line. The process of a cell line becoming a continuous cell line is termed *transformation*. Transformation involves a genotypic change and can occur spontaneously, or can be chemically- or virally-induced. The process of experimental cell immortalization is a distinct process, whereby scientists seek to generate a continuous cell line using exogenous genetic material, e.g. transfection with SV40 large T-antigen, and most often by exploiting viral vectors. The spontaneous genotypic alteration within cells in culture and leading to cell transformation probably arises because of a change occurring within the culture itself of one or more cells, rather than a very small number of 'transformed cells' being present within the original primary isolate. Transformation describes the morphological and kinetic alterations that occur in the cells in culture. The appearance of the transformed cells usually undergoes marked alteration in cytomorphology, e.g. decrease in cell size, with an increase in the ratio of nuclear volume to (serum- and) cytoplasmic volume. The cells display a greater growth rate and are less serum and growth-factor dependent. As the cells have transformed to become a continuous cell line where proliferation is the dominant feature, they loose many of the features of the differentiated phenotype present within the original donor cell. Primary cells or low passage cell lines are generally considered more representative of the respective cell within the original tissue than the 'equivalent' continuous cell line. This includes preservation in the expression of tissue and cell-specific properties and functions. It is this phenotypic divergence of the transformed cell from the original donor cell that may represent a significant disadvantage to experimentation – a feature that needs to be considered when investigators are defining the scientific objectives of their cell culture studies. Isolation of cells from malignant tumor tissue can also give rise to continuous cell lines which can be propagated over decades. Continuous cell lines derived from spontaneous transformation of normal cells or from isolation of malignant cells show many commonalities, although cells transformed from the normal phenotype can become a continuous cell line without becoming malignant.

Figure 1.2 shows a typical cell growth profile for an homogeneous cell line in culture, where the y-axis represents cumulative cell number (log scale) and the x-axis represents time in culture following cell seeding. The figure displays a number of notable features. Firstly, the nominal cell seeding density at the time of initial cell plating (time zero) is 50,000 cells per culture vessel. Cell counts performed within the first 24–36 hours show a decrease in cell number. This decrease is associated with a plating efficiency that is less than 100 per cent (e.g. actual adherence and survival of the cells is lower than the nominal predicted seeding density) and also where the cells in the culture have yet to begin proliferating; this phase is called the *lag phase*. Once the cells have adjusted to their culture environment (including the need to adhere to the substratum in the case of adherent cell lines) then net cell proliferation can be observed. In the example shown in Figure 1.2 the cells enter this net proliferation phase from about 36 hours postseeding, and between this time and 110 hours postseeding the cumulative cell number (logarithmic scale) in the culture vessel can be seen to increase linearly with time. This phase is termed the *exponential phase* of cell growth reflecting a constant rate of cell division. It is from this phase that a doubling time (e.g. the time for the cell population to double in number) for the culture can be determined. This doubling time is a feature of the cell type and also the culture environment. For example, in Figure 1.2 the doubling time approximates 12–14 hours during the exponential phase of growth. After the exponential phase the cell growth rate can be seen to decline, occasionally

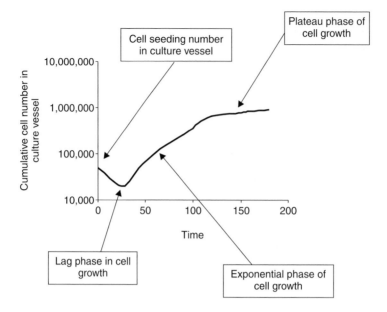

*Figure 1.2* Characteristic growth profile for a cell line seeded into a culture vessel. The plot shows the cumulative cell number per vessel (log scale) as a function of time. Distinct phases of growth can be seen: a lag phase, an exponential phase, and a plateau phase.

reaching a plateau level where there is not net growth of cells in the culture. This phase represents a point where the cell density in the culture limits either the available area or volume for growth or becomes limiting in respect to nutrient supply. In the case of adherent cell cultures that grow as a monolayer, such as the common continuous cell lines Caco-2 (intestinal model), MDCK (renal epithelial model), Calu-3 (bronchial model), A549 (alveolar epithelial type-II model) to name but a few, it represents a point of complete coverage of the substratum growth area by the cells (e.g. the cells have reached confluence), a point where they contact each other and their growth is inhibited (e.g. contact inhibition). Even for continuous cell lines the plateau phase or confluent phase of cell growth can represent a phase where differentiated characteristics of the cells become expressed. For example, Caco-2 cells reach confluency within 5–8 days at 250,000 cells/cm$^2$ seeding density on a 0.33 cm$^2$ area of culture vessel (e.g. Transwell 6 mm diameter insert). To obtain a more differentiated phenotype, however, the cultures are most often left until 19–21 days postseeding before any transport investigations are undertaken (see Chapter 10).

## THE CELL CYCLE

Throughout life there is a need to retain the ability to regenerate aged and damaged cells of specialized organ systems in order to sustain normal homeostasis. However,

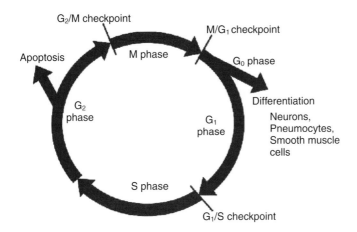

*Figure 1.3* The cell cycle of a eukaryotic cell showing: a $G_1/G_0$ (resting) phase; an S phase where DNA is synthesized; a $G_2$ phase that serves as an intermediate stage prior to mitosis; and mitosis itself (M phase) where nuclear division occurs.

such replicative processes need to be stringently controlled in order to prevent abnormal growth and tumor development (e.g. cancer). The lifespan of eukaryotic cells can be divided into four sequential but distinct phases, collectively termed the cell cycle (see Figure 1.3) which is described below.

1   $G_1/G_0$ *Phase (resting)* – $G_1/G_0$ is the period of the cell cycle that generally has the longest duration. During $G_1$ cells may actually withdraw from the cell cycle and undergo quiescence, otherwise known as $G_0$. It is during $G_0$ that cells (e.g. neurons and pneumonocytes) no longer have the capacity to proliferate but undergo the process of terminal differentiation in order to undertake their specialized function. Commitment to $G_0$ may occur when cells reach maximal confluence (contact inhibition) or when nutrients are in short supply. Failure of cells to terminally differentiate triggers their progression through the cell cycle and it is during the late stage of $G_1$ that the cells prepare themselves for entry into S phase and another round of replication.

2   *S Phase (DNA synthesis)* – S phase is the only period of the cell cycle in which DNA synthesis can occur and has a typical duration of between 6 and 8 hours within the cultured cell. It is during the first half of S phase that DNA and histone synthesis is maximal.

3   $G_2$ *Phase* – $G_2$ phase is an additional gap phase that separates both the S and M phases of the cell cycle. This phase is an important regulatory period of the cell cycle where progression to mitosis may be delayed if DNA is damaged or external environmental factors are unfavorable.

4   *M Phase* – M phase is the period of the cell cycle where nuclear division occurs and can be further subdivided into two distinct phases where cells first undergo chromosomal duplication (mitosis) followed by cytoplasmic division (cytokinesis) for the generation of daughter cells.

*How is the cell cycle regulated?* Strict control of proliferation and differentiation are two fundamental cellular processes, and are central to both normal tissue development and homeostasis. Identification of a whole plethora of key cell cycle regulatory elements, namely the cyclin dependent kinases (Cdk) and their respective inhibitors, has greatly enhanced our understanding of the regulatory growth pathways that exist in the cell. It is now understood that perturbations to such pathways are prerequisites for tumorigenesis and malignant disease. Commitment to cell division is essentially regulated by extracellular biochemical cues such as growth factors. Upon cellular stimulation Cdk bind to their regulatory subunits known as cyclins. This binding results in the cyclins' activation, which then elicits a series of orchestrated events needed for cells to transverse important checkpoints in the cell cycle in order for one round of cell division to take place. One important checkpoint is located at $G_1/S$ since it is at this stage that cells are either committed to division or exit from the cell cycle and undergo differentiation. During $G_1$ phase of the cell cycle Cdks are held in an inactive state; however, appropriate proliferation signals result in their activation and the molecular brake at the $G_1/S$ restriction site is released, prompting cells to irreversibly enter S phase. Entry into S phase correlates with the hyperphosphorylation of the retinoblastoma tumor suppresser protein (Rb) resulting in its inactivation. This key step in cell cycle progression is mediated by cyclin D-directed Cdk4 and Cdk6 phosphorylation. Inactivation of Rb results in the transcription of E2F responsive genes that are required for progression through S phase.

During each phase of the cell cycle, cells periodically synthesize different cyclin components which in turn positively regulate their respective Cdk. For example, during S phase the cells accumulate cyclin E resulting in the activation of Cdk2, a step critical for the initiation of DNA synthesis. Likewise, during late S phase $G_2$ cells accumulate cyclins A and B, respectively, where upon activation of their respective Cdks changes in the phosphorylation state of key proteins occur initiating spindle assembly, nuclear envelope structure and chromosome condensation. The activity of Cdks at the various stages of the cell cycle needs to be tightly regulated if uncontrolled cell proliferation is to be avoided. This is primarily achieved through the expression of specific Cdk inhibitors which target the cyclin D/Cdk4 complex, ultimately arresting cells in $G_1$. p21 and p16$^{INK4A}$ represent two such CDK inhibitors that physically disrupt the association between cyclin D and Cdk4, thereby preventing the hyperphosphorylation of the Rb. Not surprisingly, the expression of p21 appears to be synonymous with cell differentiation. In Rb competent cells the induction of p21 synthesis is dependant upon p53, whereas cells deficient in Rb, bypass the $G_1$ checkpoint and undergo p53-mediated apoptosis during $G_2/M$. Expression of p53 and subsequent cell cycle arrest can be induced by extracellular stresses such as irradiation, thus preventing the replication of damaged or mutated DNA. Indeed, mutations in the p53 gene have been identified in approximately 50 per cent of all human cancers. The p53 growth control pathway is in turn regulated by the opposing actions of the stimulatory p19$^{ARF}$ Cdk inhibitor and the targeted downregulation of p53 by the oncogene product MDM2. The principal components of the p53 and Rb regulatory growth pathways and their functional interaction is summarized in the flow diagram depicted in Figure 1.4. Other cell cycle checkpoints also exist at the $G_2/M$ and $M/G_1$ interfaces where cells are prevented

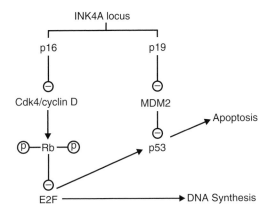

*Figure 1.4* The principal components of the p53 and retinoblastoma tumor suppresser protein. (Rb) regulatory growth pathways and their functional interaction.

from undergoing mitosis if DNA is damaged or cell division arrested if chromosomal duplication, spindle formation and cell size are perturbed in any manner.

While the above two paragraphs describe an overview of the cell cycle and its regulation, more in-depth reviews are available to the reader (Chin *et al.*, 1998; Gao and Zelenka, 1997; Roberts, 1999).

*What are the strategies for cell synchronization?* Cells in the exponential phase of growth are distributed through the four phases of the cell cycle and, therefore, such populations are termed asynchronous. However, for investigators wishing to determine if the action or permeability of a given drug is cell cycle-dependent then synchronized populations of cells are necessary. For example, optimal expression of reporter gene constructs using DNA–liposome complexes only occurs during mitosis, while anthracycline cytotoxics exert their chemotherapeutic effect when cells are in $G_1$ phase of the cell cycle. Several procedures exist that give rise to synchronized cell cultures and can be crudely divided into those that utilize physical methodologies or those that use chemical agents. Both strategies have inherent advantages and disadvantages, some of which will be briefly discussed below.

One of the most commonly used methods for obtaining cells that are synchronized in M phase is mitotic detachment. This procedure relies on the fact that during mitosis the attachment of cells to the cell culture substratum is reduced and as a consequence cells in M phase may be harvested by gentle shaking of the culture plates. The recovered mitotic cells are then reseeded into a culture vessel and such cell populations are ideal for studying events in the $G_1$ and S phases as they progress through the cell cycle. Whilst the degree of synchronization using this method is high (over 90 per cent of the recovered cells are mitotic) typical yields can be as low as 5–6 per cent of the total cell population. Additionally, this method is limited to the use of cells that grow as adherent monolayers. The issue of low cell yield may be partly overcome by exposing the cells to a microtubule depolymerizing agent such as colchicine or nocodazole in order to arrest cells in M phase prior to mitotic 'shake off'.

Another reliable physical method for obtaining cells in different phases of the cell cycle is centrifugal elutriation. The procedure takes advantage of the fact that cells in different phases of the cell cycle vary in sizes and density (generally cells in $G_1$ tend to be smaller than cells in other phases of the cell cycle) and as such will differentially sediment when subjected to a gravitational force. Centrifugal elutriation can be performed using an elutriator, specifically, a Beckman J2-21M centrifuge equipped with a JE-6B rotor. Cells suspended in an elutriation medium (phosphate buffered saline (PBS) with 0.9 mM $CaCl_2$ and 0.5 mM $MgCl_2$ containing 5 per cent fetal bonine serum (FBS)) are introduced into a separation chamber and spun at a constant rotor speed of approximately 2000 rpm. The final separation of the cells (elutriation) is then achieved by raising the pump speed in a stepwise fashion (Nguyen et al., 1999). The elutriated fractions are then harvested and an aliquot from each fraction is stained with ethidium bromide. The stained fractions are then analyzed in a flow cytometer to determine their precise position in the cell cycle. One distinct advantage of adopting centrifugal elutriation for synchronization is that cells in different phases of the cell cycle may be obtained simultaneously from one cell population.

The clear advantages of using physical methods for obtaining synchronized cells are that the issues of toxicity that exist when using alternative chemical methods are circumvented. However, limitations do exist in that mitotic detachment results in low cell yield, whilst the fractions harvested by centrifugal elutriation may display some degree of overlap, e.g. a fraction of cells enriched with $G_1$ cells may also contain a percentage of cells in $G_2$ and S phases of the cell cycle. Such redundancy has given rise to strategies that rely upon chemical reagents for cell synchronization.

Over many years various methods that use different chemical reagents have evolved as potent but reversible blockers of the cell cycle. Some of the methods that are currently used will be briefly described below.

One of the simplest and most extensively used methods for obtaining synchronized cultures is serum deprivation. Typically, the culture media is removed and the monolayers are washed thoroughly with prewarmed (37 °C) PBS in order to remove any residual serum. Cells are then cultured for up to 48 hours in their normal culture medium containing low concentrations of serum (0.5–1 per cent) or no serum at all. Low serum concentration will allow cells to accumulate in the $G_1$ phase of the cell cycle. After 48 hours the cells may be exposed to serum conditions again, resulting in a burst of DNA synthesis (usually between five and seven hours after the re-addition of serum) in a synchronized manner as cells proceed through the S phase of the cell cycle. The degree of synchrony achieved using this method may be as high as 80 per cent, although this will vary depending upon the cell type. Also, consideration must be given to the medium used during serum-deprivation experiments since different media will vary in their content of essential nutrients, such as amino acids, glucose and vitamins. It may delay the time taken for cells to accumulate in $G_1$ if they are cultured in a 'rich' medium, e.g. Dulbecco's modified Eagle's medium (DMEM) compared with RPMI 1640. If S phase cells are required for experimentation then hydroxyurea (final working concentration 1.5 mM) may be supplemented to the cultures six hours after the re-addition of serum. The ability of hydroxyurea to block cells in S phase results from its inhibition of ribonucleotide diphosphate reductase, thus limiting the supply of deoxyribonucleotides for DNA polymerization. Exposure to hydroxy-urea is typically for 14 hours after which the medium is removed and the mono-layers washed and replenished with new media without containing hydroxyurea. As with all chemical reagents, the concentration of hydroxyurea effective for cell synchronization needs to be carefully determined for each cell line, since high concentrations can elicit toxicity whilst low concentrations may prove inadequate in halting the cell's progression through S phase.

Prolonged exposure of cells to low concentrations of methotrexate has also been proposed and successfully used to investigate S phase-related events. Methotrexate blocks cells in early S phase by depleting intracellular stores of reduced folate via the inhibition of dihydrofolate reductase. The optimum dose and exposure time to methotrexate in order to achieve satisfactory synchrony again may vary depending upon the cell line adopted for experimentation. However, exposure of cells to concentrations in the region of 0.04–0.08 µM for between 16 and 24 hours are commonly reported to give good results (Sen *et al.*, 1990). More recently, cell treatment with aphidicolin, a reversible inhibitor of DNA-polymerase has been used as another way of trapping cells in S phase of the cell cycle (Matherly *et al.*, 1989). Cells in the exponential phase of growth are exposed to aphidicolin (4 µg/ml) for 18 hours, after which the inhibitor is removed and cells are cultured for a further 8 hours. The reintroduction of aphidicolin at a higher concentration (8 µg/ml) for a further 18 hours is then undertaken in order to improve the degree of synchrony. The inhibitor is then finally removed and the cells are cultured in normal media for up to 30 hours at 37 °C (Sourlingas and Sekeri-Pataryas, 1995; Fielding *et al.*, 1999). This technique has proved successful for the synchronization of both primary cultures and continuous cell lines.

Whilst mitotic detachment is commonly used for obtaining cells synchronized in M phase, protocols do exist whereby chemical reagents are used to arrest cells during mitosis. Instances when the use of chemicals for M phase synchronization is more desirable than mitiotic detachment occur when cells retain a significant degree of attachment to the culture substratum during mitosis. For example, epithelial cells are generally regarded to be more resistant to 'mitotic shake' off than fibroblasts. N-acetyl-leucyl-leucyl-norleucinal (ALLN), a neutral cystein protease inhibitor, has been effectively used to block the exit of cells from M phase via the inhibition of cyclin B degradation. Protocols do exist whereby cells are synchronized in M phase using a double block strategy (Uzbekov *et al.*, 1998). Initially, cells are cultured in the absence of serum and then incubated with aphidicolin (2 µg/ml) for a total of 30 hours. Following this incubation the media is removed and the cells are washed with PBS and subsequently incubated for 8 hours in medium containing ten per cent FBS. After 8 hours ALLN is then added to the culture medium at a final concentration of 40 µg/ml. The period of exposure to ALLN is typically 6 hours. Alternatively, following release from aphidicolin and incubation with medium containing ten per cent FBS, nocodazole (0.5 µg/ml for 7 hours) may be used instead of ALLN.

While the techniques used for cell synchronization are varied and none are perfect, there are important points the researcher must keep in mind before choosing a particular procedure. The method used must synchronize cells at a specific point of the cell cycle and the reagents used should not be toxic to the cell or interfere with its metabolic processess in any way. Both of these requirements must be met whilst giving sufficient cell yields for analytical purposes.

## STRUCTURAL ELEMENTS OF TIGHT-JUNCTIONAL COMPLEXES

Epithelial and endothelial cells form cellular barriers separating compartments of different composition. In forming such barriers the epithelial or endothelial cells polarize and form intercellular junctions. Tight-junctions (TJ) (or zonula occludens) are the most apical intercellular junctions and represent a boundary between two distinct cell membrane domains, i.e. the apical and basolateral domains. This 'fence' function of TJ complexes restricts the inter-mixing of apical and basolateral membrane lipids and proteins. Furthermore, the formation of TJ complexes serves to maintain a selective barrier for the transepithelial or transendothelial transport of solutes or ions. The restrictive intercellular diffusion pathway ('paracellular' pathway) generated by the TJ complex serves to minimize the transfer of potentially harmful solutes while maximizing the functional significance of active transport systems within the cell's membranes. For some recent reviews on TJ biology and regulation the reader is directed to the following citations: Anderson and Van Itallie (1999); Balda and Matter (2000); Cereijido *et al.* (2000); Lapierre (2000); Tsukita and Furuse (2000a,b); Tsukita *et al.* (1999); Turner (2000).

By using the technique of freeze fracture electron microscopy the TJs appear as a set of continuous, anastomosing intramembraneous strands which contact similar strands on the adjacent cells and thus seal the intercellular space. Figure 1.5 represents a schematic of the proteins involved in the transmembrane generation of extracellular TJ strands. Occludin (a protein of molecular weight approximately 60 kDa and comprising four transmembrane domains) is one of the major constituents of the TJ strands or fibrils, where it is exclusively localized. There is only one occludin gene, although two isoforms can be generated by alternate splicing of the primary RNA transcript. Occludin was the first candidate protein for fulfilling the functional restrictive properties of the TJ fibril network. However, recent observations suggest that occludin alone cannot fulfill this role. With only a single gene product it is difficult to envisage how occludin alone could account for the varied differences in paracellular permeability between different tissues. Also, recombinant cell lines transfected with mutated occludin appear to maintain an unaffected network of TJ strands. Furthermore certain cell types lack occludin but still form TJ strands. Occludin deficient embryonic stem cells can differentiate into polarized epithelial cells bearing TJs (Saitou *et al.*, 1998).

In 1998 Tsukita's group (Furuse *et al.*, 1998a) identified claudin-1 and claudin-2 as novel integral membrane proteins, bearing no sequence similarity to occludin, and localizing to TJ strands. Furthermore, work from the same laboratory (Furuse *et al.*, 1998b) showed that TJ strands could be reconstituted within cultured fibroblasts (cells that do not normally express claudin or occludin or generate TJ complexes) by introduction of claudin-1 or -2, and that occludin could be recruited to, and integrated with, the reconstituted claudin-based backbone of TJ strands.

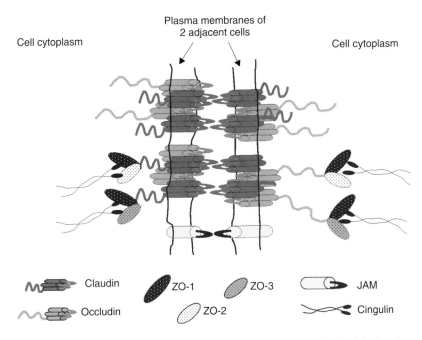

*Figure 1.5* Proposed interactions of the major proteins associated with tight junctional strands. Occludin and claudins appear to be major constituents of the tight junctional strands or fibrils. Claudins bind homotypically to the tight junctional strands of adjacent cells. A junction-associated membrane protein (JAM) has also been localized to tight junctional complexes. Accessory proteins localized to the cytoplasmic surface of tight junctional complexes include, among others, the zonula occluden (ZO) proteins, ZO-1, ZO-2 and ZO-3 and cingulin.

Claudins bind homotypically to the TJ strands of adjacent cells. To date, 20 members of the claudin gene family have been identified (Tsukita and Furuse, 2000a); their molecular weight is approximately 22 kDa and they comprise four transmembrane domains. The members of this family appear to show a differential pattern of tissue distribution, e.g. claudin-5 has been found only in endothelial cells. Current opinion is that the claudin family of proteins serve as the primary, but not the sole, molecule that *fulfills* the functional barrier properties of the TJ strands. In addition, a junction-associated membrane protein (JAM) (molecular weight 40 kDa), a member of the immunoglobulin (Ig) superfamily, has been localized to TJ complexes (Martin-Padura *et al.*, 1998), although very little is known about the functional role of this protein.

In addition there are several accessory proteins localized to the cytoplasmic surface of TJs. The zonula occluden (ZO) proteins, ZO-1, ZO-2 and ZO-3 form heterodimers potentially serving as the major molecular scaffold for the TJ network. The ZO complexes (through ZO-1 or ZO-2) form crosslinks between the TJ strands and actin filaments. Cingulin is a double stranded myosin-like protein that associates with the cytoplasmic face of the TJ complex and apparently directly with

ZO proteins. Cingulin may function in linking the TJ strands with the actomyosin cytoskeleton.

The ion $Ca^{++}$ is often used in pharmaceutical studies either to enhance TJ restrictiveness or as a target for EDTA divalent ion chelation. This latter function opens up TJs to make the cellular barrier 'leakier'. The $Ca^{++}$ acts primarily on the extracellular side of the cell to interact with the extracellular part of E-cadherin, a critical cell adhesion molecule in the zonula adheren's junctions that lies underneath the ZOs. This molecule fulfills a key role in cell–cell interactions.

Extracellular $Ca^{++}$ activates E-cadherin, which is then able to aggregate with other E-cadherin molecules on the same cell, an arrangement that favors binding to E-cadherin of an adjacent cell. In the absence of $Ca^{++}$ E-cadherin are inactive and functional TJ complexes will not form.

The $Ca^{++}$ ions also promote binding of E-cadherin with intracellular-located catenins which in turn bind to vinculin, actinin and, indirectly, to the cytoskeleton of actin. The cytoskeleton appears to fulfill a key role in delivering signals from the adherens junctions to the TJ; inhibitors of microfilaments and microtubules disrupt TJ junction formation. The intercellular interactions between E-cadherin also activates phospholipase C, which then cleaves phosphatidyl 4,5-bisphosphate into inositol 1,4,5-trisphosphate and diacylglycerol activates protein kinase C (PKC). There follows mulitiple phosphorylation steps leading to the synthesis and assembly, and even in the disassembly, of the TJ strands and a decrease in paracellular transport of markers such as mannitol, sucrose and dextrans.

## CONCLUSION

It is hoped that this chapter has provided the non-specialist with a background knowledge and understanding that will allow greater appreciation of the chapters to follow. Much of the information is available in textbooks and review articles. However, it is only with experimentation that much of the information in this chapter, and indeed the following chapters, can be truely valued.

## REFERENCES

Anderson, J. M. and Van Itallie, C. M. (1999) Tight Junctions: closing in on the seal. *Curr. Biol.*, **9**, R922–R924.

Balda, M. S. and Matter, K. (2000) Transmembrane proteins of tight junctions. *Sem. Cell Dev. Biol.*, **11**, 281–289.

Cereijido, M., Shoshani, L. and Contreras, R. G. (2000) Molecular physiology and pathophysiology of tight junctions I: biogenesis of tight junctions and epithelial polarity. *Am. J. Physiol.*, **279**, G477–G482.

Chin, L., Pomerantz, J. and DePinho, R. A. (1998) The INK4a/ARF tumor suppressor: one gene – two products – two pathways. *TIBS*, **August**, 291–296.

Fielding, C. J., Bist, A. and Fielding, P. E. (1999) Intracellular cholesterol transport in synchronised human skin fibroblasts. *Biochem.*, **38**, 2506–2513.

Furuse, M., Fujita, K., Hiragi, T. and Tsukita, S. (1998a) Claudin-1 and -2: novel integral membrane proteins localising to tight junctions with no sequence similarity to occludin. *J. Cell Biol.*, **141**, 1539–1550.

Furuse, M., Sasaki, H., Fugimoto, K. and Tsukita, S. (1998b) A signle gene product claudin-1 or -2 reconstitutes tight junction strands and recruits occludin in fibroblasts. *J. Cell Biol.*, **143**, 391–401.

Gao, C. Y. and Zelenka, P. (1997) Cyclins, cyclin-dependent kinases and differentiation. *BioEssays*, **19**, 307–315.

Lapierre, L. A. (2000) The molecular structure of the tight junction. *Adv. Drug Deliv. Rev.*, **41**, 255–264.

Martin-Padura, I., Lostaglio, S., Schneemann, M., Williams, L., Romano, M., Fruscella, P., *et al.* (1998) Junctional adhesion molecule, a novel member of the immunoglobulin superfamily that distributes at intercellular junctions and modulates monocyte transmigration. *J. Cell Biol.*, **142**, 117–127.

Matherly, L. H., Schuetz, J. D., Westion, E. and Goldman, I. D. (1989) A method for the synchronisation of cultured cells with aphidicolin: application to the large-scale synchronisation of L12210 cells and the study of the cell cycle regulation of thymidylate synthase and dihydrofolate reductase. *Anal. Biochem.*, **182**, 338–345.

Nguyen, P., Broussas, M., Cornillet-Lefebvre, P. and Potron, G. (1999) Coexpression of tissue factor and tissue factor pathway inhibitor by human monocytes purifies by leukapheresis and elutriation. Response of nonadherent cells to lipopolysaccharide. *Transfusion*, **39**, 975–982.

Roberts, J. M. (1999) Evolving ideas about cyclins. *Cell*, **98**, 129–132.

Saitou, M., Fujimoto, K., Doi, Y., Itoh, M., Fujimoto, T., Furuse, M., *et al.* (1998) Occludin-deficient embryonic stem cells can differentiate into polarised epithelial cells bearing tight junctions. *J. Cell Biol.*, **141**, 397–408.

Sen, S., Erba, E. and D'Incalci, M. (1990) Synchronisation of cancer cell lines of human origin using methotrexate. *Cytometry*, **11**, 595–602.

Simons, K. and Ikonen, E. (1997) Functional rafts in cell membranes. *Nature*, **387**, 569–572.

Sourlingas, T. G. and Sekeri-Pataryas, K. E. (1995) Aphidicolin large scale synchronisation of rapidly dividing cell monolayers and the analysis of total histone and histone variant biosynthesis during the S and $G_2$ phases of the Hep-2 cell cycle. *Anal. Biochem.*, **234**, 107–110.

Tsukita, S. and Furuse, M. (2000a) Pores in the wall: claudins constitute tight junction strands containing aqueous pores *J. Cell Biol.*, **149**, 13–16.

Tsukita, S. and Furuse, M. (2000b) The structure and function of claudins, cell adhesion molecules and tight junctions. *Ann. N. Y. Acad. Sci.*, **915**, 129–135.

Tsukita, S., Furuse, M. and Itoh, M. (1999) Structural and signalling molecules come together at tight junctions. *Curr. Opin. Cell Biol.*, **11**, 628–633.

Turner, J. R. (2000) 'Putting the squeeze' on the tight junction: understanding cytoskeletal regulation. *Sem. Cell Dev. Biol.*, **11**, 301–308.

Uzbekov, R., Chartrain, I., Philippe, M. and Arlot-Bonnemains, Y. (1998) Cell cycle analysis and synchronization of the Xenopus cell line XL2. *Exp. Cell Res.*, **242**, 60–68.

Chapter 2

# Cell culture media: selection and standardization

*Tanja Eisenblätter, Katherina Psathaki, Thorsten Nitz, Hans-Joachim Galla and Joachim Wegener*

## INTRODUCTION

The phenotype of cultured cells often differs from the characteristics that predominate in the tissue from which it was derived. This difference is due to the loss of the three-dimensional architecture of the surrounding tissue, reduced cell–cell and cell–matrix interactions and an altered hormonal and nutritional milieu. These and other factors that regulate geometry, growth and function *in vivo* are absent *in vitro*, creating an environment that favors spreading, migration and proliferation of unspecialized cells, rather than the expression of differentiated functions. Therefore, providing cells *in vitro* the appropriate culture conditions is a fundamental prerequisite to allow them to express their specialized functions.

Most cell types grow as adherent monolayers and they need to attach and spread out on the substrate before they can start to proliferate and differentiate (anchorage dependence). Cell–substrate adhesion is mediated by a group of adhesive proteins that naturally form the extracellular matrix (ECM) – an extracellular network of various proteins and proteoglycans that often serves as a scaffold for the organization of a particular tissue. These ECM proteins are recognized by specific cell surface receptors whose cytoplasmic domains are associated with the cytoskeleton to ensure mechanical stability. The most prominent group among these *anchor proteins* are the so-called integrins that bind matrix molecules like fibronectin (FN), laminin (LAM), vitronectin (VN) and collagen. Another group known as transmembrane proteoglycans bind to matrix proteoglycans, various collagens and also growth factors. Apart from cell–substrate contact, many cells like epithelial cells also need to make cell–cell contact for optimum survival and growth. Cell–cell adhesion is mainly provided by transmembrane proteins of the (CAM) ($Ca^{2+}$-independent adhesion molecules) and Cadherin-family ($Ca^{2+}$-dependent adhesion molecules) which directly interconnect two adjacent cells. These adhesion molecules are also linked to the cytoskeleton which transmits changes in cell shape and signaling between the cell surface and the nucleus.

Once the cells have correctly attached and spread out and culture conditions are otherwise favorable they start to proliferate. The term *cell proliferation* actually means that one particular cell progresses through another division cycle to form two daughter cells. This cell cycle is divided into the four phases: $G_1$ (Gap 1), S (DNA synthesis), $G_2$ (Gap 2) and M (Mitosis). In the $G_1$ phase the cell can either progress towards DNA synthesis and another division cycle or exit the cell cycle to rest ($G_0$ phase) or to differentiate. Therefore, it is during $G_1$ that the fate of the cell is either set to proliferation or differentiation (or apoptosis). Entry into the cell cycle, equivalent with further cell proliferation, is regulated by environmental as well as intracellular signals. Among the former is, for example, the cell density in the culture dish. Low cell density promotes the cell's entry into the cell cycle provided that mitogenic growth factors such as epidermal growth factor (EGF), fibroblast growth factor (FGF) and/or platelet-derived growth factor (PDGF) are present. Cyclins, which interact with cell division cycle (CDC) kinases, are the most prominent among intracellular regulators. However, cell-proliferation may also be inhibited by extracellular factors like transforming growth factor (TGF-$\beta$) or high cell densities, as well as by inhibitory intracellular proteins like retino blastoma tumor suppressor protein (Rb) and p53. The link between extracellular control elements (positive, PDGF; negative, TGF-$\beta$) and intracellular regulators (positive, cyclins; negative, p53) of cell proliferation is made by cell membrane receptors and signal transduction pathways which often involve protein phosphorylation and second messengers such as cAMP and $Ca^{2+}$.

Cell proliferation is generally incompatible with the expression of a differentiated phenotype. Conditions required to induce differentiation are high cell density, enhanced cell–cell and cell–matrix interactions, and the presence of various differentiation factors, such as certain paracrinic growth factors (TGF-$\beta$), hormones (hydrocortisone, HC), vitamins and inorganic ions (especially $Ca^{2+}$). Once the cell has reached its full differentiation, this status has to be preserved by appropriate culture conditions, otherwise dedifferentiation may occur. Dedifferentiation has originally been used to describe the loss of differentiated properties of a particular tissue when it is grown in culture. But dedifferentiation must be distinguished from de-adaptation and selection. De-adaptation implies that expression of a certain functional phenotype which is under regulatory control by hormones, cell–cell or cell–matrix interaction, would be re-induced as soon as the correct conditions have been re-established. Selection, on the other hand, implies that an undifferentiated progenitor cell has succeeded to dominate the culture because of its greater proliferative potential compared to that of a terminally differentiated cell. This potential does not preclude the possibility that the progenitor cell may be induced to a fully differentiated cell if the correct environmental conditions are established. However, once the wrong lineage has been selected, no amount of induction can bring back the required phenotype.

The brief background on the biology of cultured cells given above was intended to emphasize that proper selection of cell culture conditions is of immense importance in order to induce animal cells to proliferate, differentiate or to keep them from dedifferentiation. As appropriate culture conditions and, most notably, the culture medium are always individually different from cell type to cell type, it is impossible to give an exhaustive survey within this chapter. Instead, we will try to summarize certain principles of medium composition and illustrate them by examples from our laboratory and our own experience. These examples basically comprise two cell culture models that we have established to mimic the blood–brain and the blood–cerebrospinal fluid barrier *in vitro*, respectively. *In vivo*, both these physiological barriers separate the circulating blood flow from the central nervous system (CNS) and thereby provide this organ with a constant chemical environment that is necessary for correct signal processing. The blood–brain barrier (see Chapter 18) is built up by the endothelial cells that form the vascular wall in brain capillaries. Effective intercellular contacts between adjacent endothelial cells make the vascular wall highly impermeable for hydrophilic substrates and ions. The blood–cerebrospinal fluid barrier, located in the *choroid plexus* (CP), is formed by a sheet of epithelial cells that underlies the endothelial cells of the blood vessels in this tissue. Since endothelial cells are fenestrated and thus highly permeable within the CP tissue, the epithelial cells have to provide the diffusion barrier between blood and CNS. For many of the general ideas about medium composition, we will learn strongly on Freshney's well-known monograph 'The culture of animal cells' (Freshney, 2000). For more detailed information the interested reader is referred to this source.

## MEDIUM COMPOSITION – CONCEPTS AND EXAMPLES

### General remarks on basal medium composition

As briefly mentioned in the introduction, culture conditions have to be carefully adjusted in order to successfully culture cells *in vitro* apart form their natural environment. Besides a well-balanced medium composition, there are certain physico-chemical requirements that the medium and the environment must fulfill.

#### *Physicochemical properties*

The optimal temperature for a particular cell culture depends on the organ temperature of the animal from which the cells were obtained. Therefore, most human and warm-blooded animal cell lines are cultivated at 37 °C, while cold-blooded animals that do not regulate their blood temperature within narrow limits tolerate a wide temperature range, between 15 and 27 °C. In general, overheating is a more serious problem than underheating.

Most cell lines grow well at pH 7.4. However, since atmospheric $CO_2$ causes the pH to decrease in open culture dishes and overproduction of $CO_2$ or lactic acid in metabolically active cell lines will lower the pH even further, culture medium must be buffered. In spite of its poor buffering capacity at physiological pH, bicarbonate ($NaHCO_3$) is the most frequently used buffering compound because of its low toxicity, low cost and nutritional benefit to the culture. Sometimes bicarbonate is applied in combination with HEPES, a small organic substance with strong buffering capacity at pH 7.2–7.6. Bicarbonate counterbalances the effect of atmospheric $CO_2$ tension in a way that the chemical equilibrium represented by equation (2.1) is pushed back to the left and the pH is increased accordingly.

$$H_2O + CO_2 \leftrightarrow H_2CO_3 \leftrightarrow H^+ + HCO_3^- \tag{2.1}$$

Thus, the resulting pH in a bicarbonate-buffered medium can be precisely controlled by the amount of bicarbonate added to the culture medium and by the $CO_2$ tension in the incubator atmosphere. With the introduction of HEPES into tissue culture, there was some speculation that, since $CO_2$ was no longer necessary to adjust the pH, it could be omitted. This speculation proved to be untrue (Kimura, 1974), at least for a large number of cell types, particularly at low cell concentrations. Although 20 mM HEPES can control medium pH within the physiological range, the absence of atmospheric $CO_2$ allows equilibrium (2.1) to move to the left, eventually eliminating dissolved $CO_2$ and ultimately $HCO_3^-$ from the medium. This chain of events appears to limit cell growth, although it has not yet been clarified whether the cells require the dissolved $CO_2$ or $HCO_3^-$ or both. Taken together, cultures kept in open dishes need to be incubated in an atmosphere of $CO_2$, while cells at moderately high concentrations ($\geq 1 \times 10^5$ cells/ml) and grown in sealed flasks do not need to have $CO_2$ added to the gas phase, provided that the bicarbonate concentration is kept low and HEPES is added. At low cell concentrations, however, and for some primary cultured cells, it is necessary to incubate in an atmosphere of $CO_2$ (Freshney, 2000).

Less important in cell and tissue culture is the amount of oxygen dissolved in the culture medium, as cultured cells often rely on anaerobic glycolysis. When the adherent cells are covered with a medium layer of 2–5 mm height (0.2–0.5 ml/$cm^2$), a sufficiently high rate of oxygen diffusion to the cells is guaranteed. One problem associated with higher amounts of dissolved oxygen is its toxicity due to free radical formation. This problem is more likely to cause difficulties in serum-free media as serum contains various antioxidants capable of eliminating reactive oxygen species within certain limits.

### Concepts of medium composition

Apart from these physicochemical properties, the right medium composition is decisive for cell growth and proliferation. Although there is a large variety of different media available, the basic ingredients and their individual concentrations are always rather similar: an isotonic, buffered basal medium with inorganic salts, nutrients, essential amino acids and vitamins. All components, their range of concentration and their function in cell culture – as far as they have been clarified – are summarized in the following chapter.

### Amino acids ($10^{-5}–10^{-3}$ M)

All essential amino acids, which can not be synthesized by the cells, and almost all nonessential amino acids, are generally required for metabolism. They are added to the culture medium either in order to compensate for the cellular inability to make them or because they are made but lost by leakage into the medium. Since glutamine has a very limited shelf-life, it is commonly added to the medium immediately before use in a final concentration of 2 mM. It serves as the major carbon source and as a precursor in the synthesis of proteins and other intermediates. Glutamine is also an important energy source because it can enter the energy metabolism via the citric acid cycle.

### Vitamins ($10^{-8}–10^{-6}$ M)

The water-soluble vitamins (e.g. B-group, choline, inositol, nicotinamide, folic acid) are always present in media, while the fat-soluble vitamins A, D, E, K are only present in some very complex media like M199, RPMI 1640. Vitamin limitation is usually expressed in terms of reduced cell survival and growth rates.

### Inorganic salts ($10^{-6}$–$10^{-1}$ M)

The major ionic components in medium are $Ca^{2+}$, $Mg^{2+}$, $Na^+$, $K^+$, $Cl^-$, $SO_4^{2-}$, $PO_4^{3-}$ and $HCO_3^-$. Calcium ($10^{-3}$ M) is required for cell adhesion and is therefore reduced in suspension culture. It acts as an intermediary in signal transduction and its concentration can determine whether cells will proliferate or differentiate. Magnesium ($10^{-4}$ M) is essential for many enzymes to express their full functionality as well as for membrane stability. Sodium ($10^{-1}$ M), potassium ($10^{-3}$ M) and chloride ($10^{-1}$ M) are responsible for the membrane resting potential, while sulfate ($10^{-6}$ M), phosphate ($10^{-3}$ M) and hydrocarbonate ($10^{-2}$ M) have individual roles as anions required by the ECM and as regulators of intracellular charge.

### Glucose ($10^{-3}$–$10^{-2}$ M)

Beside glutamine, glucose is included as the major source of energy. It is metabolized by glycolysis to form pyruvate, which may be converted either to lactate or acetoacetate or it may enter the citric acid cycle to be decomposed to $CO_2$.

### Antibiotics

Antibiotics are used to reduce the frequency of contamination, especially with primary cultured cells. Common antibiotics are penicillin (100 U/ml; inhibition of bacterial cell wall synthesis), streptomycin (100 µg/ml; influence on protein synthesis) and gentamicin (50 µg/ml; inhibition of bacterial protein synthesis) (Darling, 1993). However, the use of laminar-flow hoods, coupled with strict aseptic techniques, makes antibiotics basically unnecessary as they have a number of significant disadvantages: they encourage the development of antibiotic-resistant organisms; they may hide mycoplasma infections; they have antimetabolic effects that can cross-react with mammalian cells; and finally, they encourage poor aseptic technique. For all these reasons, it is often recommended that routine culturing should be performed in the absence of antibiotics.

A variety of other compounds, including certain proteins, peptides, nucleosides, citric acid cycle intermediates, pyruvate and lipids are present in complex media. These constituents are more important when serum concentration is reduced. Even in the presence of serum they may help in maintaining certain specialized cells.

## Complete medium – supplementation with serum

In the early days of cell and tissue culture, mammalian cells were routinely kept in blood or blood cloths since researchers were unable to establish a cell culture medium capable of promoting cell growth and proliferation (Alberts *et al.*, 1995). It took many years to recognize what was missing in those highly nutrient media that were already supplemented with glucose, amino acids, vitamins and even some hormones. Nowadays, we are aware of the necessity of so-called growth factors and mitogens as unconditional medium constituents that keep cultured cells from being arrested in the $G_0$ phase of the cell cycle. Since proliferation is a major objective of culturing cells *in vitro* in order to multiply and expand a cell population isolated from an intact organism, growth factor supplementation of the culture medium attracts considerable interest. However, growth factors are not only involved in cell proliferation but also in their differentiation, survival, protein biosynthesis and motility. A complex mixture of growth factors and mitogens is commonly introduced into the basal medium by the addition of serum, a fluid derived from animal blood after the clothing and removal of all blood cells. Many of them have been identified throughout the last years, such as PDGF, FGF, EGF or vascular endothelial growth factor (VEGF). PDGF is regarded to be the major growth factor in serum (Antoniades *et al.*, 1979), whereas all others are only present in small amounts (Freshney, 2000).

## Proliferation

Growth factors usually interact with highly specific cell surface receptors that transmit the information about their binding into the cytoplasm. It is important to note that cell growth and cell proliferation are very different phenomena. Whereas the former is regarded as the extension of cell dimensions and protein content of one cell, often described by the cytoplasm to DNA ratio, the latter is defined as the duplication of nucleus (mitosis) and cell body (cytokinesis) of one progenitor cell into two daughter cells. Thus, growth and proliferation of one cell species may be regulated individually by different signal molecules. With respect to cell-type, specific functions in a multi-cellular organism (epithelial cells, muscle cells, neurons) and the necessity for an independent regulation of proliferation, it is not at all surprising that each cell species responds individually to a certain growth factor or a certain growth factor combination. For culturing cells *in vitro* this individualized response may have two consequences: first, not every serum is equally well-suited to culture one particular cell type, and second, properly selected serum may induce proliferation of one cell type but retard that of another and may thus allow investigators to obtain pure cultures of one cell species.

Whereas the potency to purify cell cultures just by the choice of serum is rather limited, it is common practice to add certain supplements to the culture medium to make it selective. Figure 2.1 compares phase contrast micrographs of primary cultured epithelial cells from porcine CP, after two cell populations derived from the same preparation have been grown in serum-containing medium for ten days (=10 days *in vitro*; 10 DIV). For both cultures the same basal medium Dulbecco's Modified Eagle DME/Ham's F12 supplemented with ten per cent fetal calf serum (FCS) was used, but only the culture shown in Figure 2.1B was subjected to 20 µM cytosine arabinoside (Ara-C) within the medium. The cells in Figure 2.1A show an irregular morphology and the white arrows point towards some elongated fibro-blasts that have also been released from the tissue during the preparation. Due to shorter proliferation times these fibroblasts eventually overgrow the epithelial cell culture (Gath *et al.*, 1997). Contaminating fibroblasts are entirely absent from the culture treated with Ara-C (Figure 2.1B), in which the cells exhibit a regular cobble-stone-like morphology. Ara-C, a nucleoside with arabinose as a sugar component,

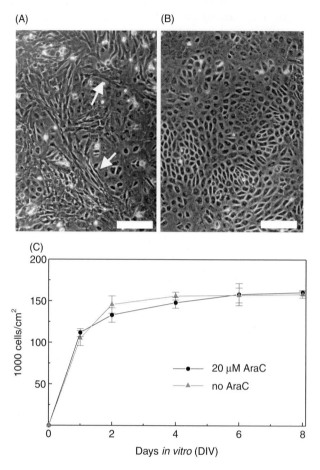

*Figure 2.1* Phase contrast micrographs of *choroid plexus* (CP) epithelial cells cultured in DME/ Ham's F12 supplemented with ten per cent fetal calf serum (FCS) (A) and additionally with 20 μM cytosine arabinoside (Ara-C) (B). The arrows in Figure A point towards fibroblasts that contaminate this epithelial cell culture and are absent from the culture shown in Figure B. The scale bar corresponds to 20 μm. Figure C compares the number of CP epithelial cells as a function of culture time for complete medium with and without Ara-C.

is a well-known inhibitor of DNA synthesis (Brewer and Scott, 1983) and it is, therefore, cytostatic. Ara-C is, however, not a substrate of the highly specific nucleoside transport systems in CP epithelial cells (Spector, 1982) and remains extracellular. Nucleoside transport systems in fibroblasts do not distinguish between ribose and arabinose as sugar residues, which results in Ara-C uptake and erasure of the fibroblast population. Figure 2.1C compares the number of cells in CP primary cultures for conditions with and without Ara-C in the medium. The curves show that proliferation of the cells is not affected by the presence of Ara-C, which suggests that Ara-C supplementation makes the medium selective for CP epithelial cells (Gath *et al.*, 1997). Use of Ara-C is just one method among others that make culture conditions selective. In order to obtain pure cultures of the same epithelial cells, Crook *et al.* (1981) report adding the proline-analog *cis*-hydroxy-proline to the culture medium to retard fibroblast proliferation.

## Attachment and spreading

Another very important effect of serum is to promote attachment of the cells to *in vitro*-surfaces, since many mammalian cell types are anchorage-dependent (ad), i.e. they need to attach and spread on a substrate before they can proliferate or differentiate (Freshney, 2000). It is nowadays generally accepted that in these cells the interactions with a culture substrate trigger certain signaling pathways that, in concert with other regulatory mechanisms, take control over the cell cycle. The interaction of ad cells with the culture substrate is mediated by a group of adhesive proteins that form the ECM *in vivo*. The most prominent members among those adhesive proteins are the various collagens, LAM, FN (Yamada, 1997), VN (Tomasini and Mosher, 1991) and fetuin (Fisher, 1958). Thus, it seems that tissue culture surfaces have to be pre-coated with one of these proteins in order to make the substrate compatible with attachment and spreading. However, when complete (serum-containing) medium is used to seed suspended cells in tissue culture dishes pre-coating was found to be an unnecessary measure. Some of the above mentioned adhesive proteins – most notably VN and FN – are present in serum in significant concentrations (Tomasini and Mosher, 1991) and adsorb instantaneously to the *in vitro*-surfaces (Vogler and Bussian, 1987). These proteins provide the molecular surface composition that is required for specific cell–substrate interactions. This phenomenon can be nicely demonstrated when the time course of cell attachment and spreading is compared for serum-containing and serum-free medium. A novel experimental means to perform these kind of studies, referred to as electric cell–substrate impedance sensing (ECIS), has been recently described (Wegener *et al.*, 2000). ECIS is based on using small gold-film electrodes (d = 250 µm) as a culture substrate for ad cells. The electrical capacitance of such an electrode changes in predictable ways when cells attach and spread on the electrode surface so that capacitance readings as a function of time allow investigators to monitor the establishment of cell–substrate contacts. Figure 2.2 compares the time course of cell attachment and spreading when suspended Madin–Darby Canine Kidney (MDCK)-cells are seeded on the electrode, either in serum-containing or serum-free medium. The parameters of the measurement are adjusted in a way that the time course of the capacitance directly mirrors the increasing surface coverage of the electrode due to cell spreading. It is apparent from the data that MDCK cells attach and spread immediately on the electrode surface when serum-containing medium is used, whereas it takes a lag-phase of roughly eight to ten hours before the cells start to spread in serum-free medium. Spreading in serum-free medium eventually occurs since the cells are capable of synthesizing and secreting adhesive proteins themselves (Wegener *et al.*, 2000). Only after the adhesive proteins have been deposited on the culture substrate can the cells mechanically establish stable substrate contacts that are required for the active spreading process.

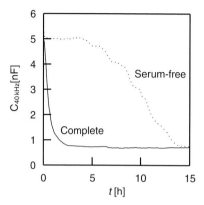

*Figure 2.2* Time course of the electrode capacitance during attachment and spreading of initially suspended Madin–Darby Canine Kidney (MDCK) cells (strain-II) after they have been seeded in complete or serum-free medium. The shift in capacitance measured with alternating current of 40 kHz frequency linearly mirrors the increasing surface coverage on the electrode surface.

## Differentiation

At a certain point of the cell cycle, the fate of a mammalian cell is either set to proliferation, differentiation or apoptosis (programmed cell death). Which of these opposing pathways is selected critically depends on the presence of growth factors and mitogens in the surrounding medium (Alberts *et al.*, 1995). Thus, by addition of serum to the basal medium the cell may be triggered to proliferate, but it may never reach its differentiated phenotype. The most frequently used sera are calf serum and fetal calf serum (FCS). But sera from many other species (e.g. horse, pig, man) are commercially available and the selection depends on the particular cell line. Most times the appropriate serum is found empirically. Very often, fetal serum is the first choice since it contains larger amounts of growth factors which promote cell proliferation and fetuin that mediates cell attachment. An easily available parameter to determine the differentiation of barrier-forming epithelial and endothelial cell layers is the electrical resistance of the cell layer (TEER: transepithelial/transendothelial electrical resistance) indicative of the effectiveness of the barrier towards ion permeation (Lo *et al.*, 1999). Figure 2.3 shows TEER values of primary cultured epithelial cells from porcine CP that were grown to confluence in basal medium supplemented with ten per cent FCS. The established cell monolayers were subsequently exposed to basal medium supplemented with ten per cent FCS (A), ten per cent heat-inactivated FCS (30 minutes at 60 °C) (B), ten per cent ox serum (C) or serum-free medium (D) for another four days before TEER values were recorded. As shown in Figure 2.3, TEER values were minimal when either FCS or heat-inactivated FCS was added to the culture medium, indicating the least degree of differentiation of the cells into a barrier-forming epithelial phenotype. Using ox serum (OS) instead more than doubled the resistance values but did not reach the same degree of barrier efficiency that was determined under serum-free

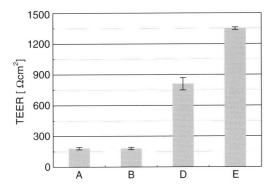

*Figure 2.3* Transepithelial electrical resistance (TEER) of confluent *choroid plexus* (CP) epithelial cells grown on permeable filter substrates after they have been incubated with DME/Ham's F12 supplemented with (A) ten per cent fetal calf serum (FCS), (B) ten per cent heat-inactivated calf serum and (C) ten per cent ox serum. Cells represented by column (D) have been incubated for the same time in serum-free medium.

conditions. These observations demonstrate that the concentration of growth factors, which is higher in fetal serum compared with adult OS, may determine the degree of differentiation of this cell type. Components of the complement system that can be inactivated by a moderate heating protocol are apparently not responsible for the incomplete barrier function of the CP cells cultured in FCS-containing medium. On the other hand, using OS or serum-free medium from the beginning of the culture immediately after isolation did not result in sufficient proliferation to establish confluent cell layers. In accordance with this interpretation, medium supplementation with FCS induced the highest proliferation rates and was thus used to grow the cells to confluence (Hakvoort *et al.*, 1998).

Other functions of serum that cannot be discussed in detail in this article are trypsin-inhibition by α2-macroglobulin, binding iron for bioavailability by transferrin, and inhibition of cell growth and promotion of differentiation by the growth factor TGF-$\beta$ (Freshney, 2000). Apart from the minerals, iron, copper and zinc, serum contains trace elements like selenium which probably helps to detoxify free radicals as a cofactor for glutathione synthetase (McKeehan *et al.*, 1976). Lipids like linoleic acid, oleic acid and ethanolamine are present in small amounts and are usually bound to proteins such as albumin. Hormones like insulin promote uptake of glucose and amino acids (Stryer, 1995) and may owe its mitogenic effect to this property. Growth hormones are particularly present in fetal serum and they may also have a mitogenic effect in conjunction with so-called insulin-like growth factors (IGFs).

## Serum-free medium

Although serum contains a potent mixture of growth factors and mitogens that have proven to be capable of promoting proliferation of many different cell types, there are a number of disadvantages that are associated with the presence of serum in culture medium. Below, we will list a number of those aspects and in doing so we basically followed the arguments given in Freshney's outstanding book (Freshney, 2000).

1   Apart from the major constituents albumin and transferrin, serum contains a wide range of minor components like nutrients, peptide growth factors, hormones, minerals and lipids that may have considerable effects on cell growth. However, serum concentrations and the precise mechanism of action have not been fully determined for all those individual components.

2   Serum composition may vary from batch to batch, and changing serum batches requires extensive and time-consuming testing. When different cell lines are used, each type may require a different batch of serum.

3   Standardization of experimental protocols that involve serum-containing medium is difficult among different laboratories owing to batch-to-batch variations of the serum.

4   To anyone interested in isolating and characterizing cell products, the presence of serum creates a major obstacle to purification. Characterizing the permeation rate of a given compound or its active transport across epithelial and endothelial cell layers often requires the use of label-free substrates. Quantification of these compounds by means of chromatography or spectroscopy is often much easier in a matrix of serum-free than in serum-containing medium. Even certain measurements may be less reproducible when serum is present.

5   Periodically, the supply of serum is restricted owing to the spread of diseases. Moreover, serum is occasionally contaminated with viruses which represent an additional unknown factor outside the operator's control.

6   Cost is often cited as a disadvantage of serum supplementation. But if serum is replaced by many defined constituents, the cost may be as high as that of the serum. Therefore, it depends on the size of supplementation.

For these reasons an increasing number of laboratories try to adapt their cultures to serum-free conditions, which have three further advantages, as follows: the ability to select for a specific cell type by control over growth-promoting activity; the possibility to regulate proliferation and differentiation processes by switching from a growth factor, after necessary amplification, to a differentiation factor; in serum-free medium many cell types show an enhanced differentiation owing to a reduced proliferation.

The latter aspect can be demonstrated by studying our primary cultures of porcine CP epithelial cells under serum-free conditions (Hakvoort *et al.*, 1998). As already mentioned in the preceding paragraph, the barrier function of CP epithelial cells towards ion permeation is increased manifold when the cells are cultured in serum-free medium. This TEER increase is independently confirmed when the permeation of a membrane-impermeable probe like fluorescein-labeled dextran (molecular weight = 4 kDa) is used to probe barrier efficiency (Figure 2.4). Unlike ions, dextran molecules of this molecular weight can only traverse the epithelium via paracellular pathways, indicating that barrier-forming cell–cell contacts (tight junctions) must have been strengthened by serum withdrawal. Immunocytochemical detection of the tight junction associated protein ZO-1 confirms this conclusion

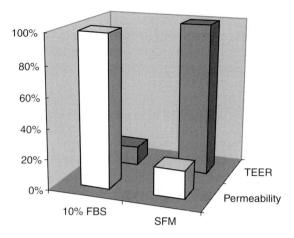

*Figure 2.4* Transepithelial electrical resistances (TEER) and permeation rates of 4 kDa-dextran for *choroid plexus* (CP) epithelial cells incubated in complete medium (ten per cent fetal bovine serum, FBS) or serum-free medium (SFM) for four days after a confluent cell layer had been established. For easy comparison, TEER readings under serum-free conditions and permeability readings under serum-containing conditions were set to 100 per cent, respectively.

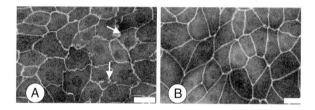

*Figure 2.5* Immunocytochemical detection of the tight junction-associated protein zonula occludens (ZO)-1 in *choroid plexus* (CP) epithelial cells grown on permeable membranes in the presence of ten per cent fetal bovine serum (FBS) (A) and after incubation in serum-free medium (SFM) for four days *in vitro* (DIV) (B). The arrows in Figure A point towards a fuzzy staining pattern of ZO-1 in serum-containing medium. Scale bars correspond to 20 μm.

(Figure 2.5). Immunostainings of CP epithelial cells incubated with serum-containing medium show a pattern which is characterized by numerous extrusions and invaginations with significant numbers of branching points (arrows in Figure 2.5A). After incubation in serum-free medium for four days, the ZO-1 staining outlines the cells without any fuzzy appearance. Similar effects on epithelial barrier function and molecular changes at the tight junctions have been previously observed for other epithelia (Chang *et al.*, 1997; Mortell *et al.*, 1993).

The improved differentiation of CP epithelial cells under serum-free conditions is, however, not only apparent from their improved barrier function but also from other structural and functional properties. Scanning electron micrographs of

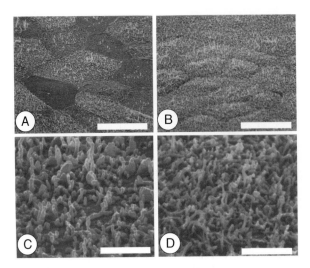

*Figure 2.6* Scanning electron micrographs of *choroid plexus* (CP) epithelial cells after they have
been incubated in complete medium (A, C) or serum-free medium (SFM) (B, D) for
four days. Scale bars in images A and B correspond to 20 μm, those in C and D
correspond to 2 μm.

these cells after they have been kept either in serum-containing or serum-free
medium are compared in Figure 2.6. The images show that *plexus* cells incubated
in serum-free medium express extended microvilli which are homogeneously
distributed across the cell layer. Microvilli trimming on the surface of cells grown
in serum-containing medium is, however, rather heterogeneous and significantly
less extended. Since these membrane protrusions are a morphological indicator of
transporting epithelia that enlarge their surface area in order to increase transport
efficiency, the conclusion applies that only in serum-free medium the epithelial
phenotype is properly expressed.

An increased transport activity of CP epithelial cells under serum-free conditions
is easily documented from the pH-gradient that the cells establish between the
apical and basolateral compartment when they are cultured on a permeable filter
substrate. Figure 2.7 demonstrates that the cells generate a pH difference between
both fluid compartments in the order of 0.5 pH units when cultured in the absence
of serum, whereas pH differences in medium supplemented with ten per cent FCS
are barely significant. Although establishment of a pH-gradient has no direct
physiological equivalent, various transport systems in the epithelial membrane are
involved in its establishment (Hakvoort *et al.*, 1998). One of the most essential
transport proteins for all mammalian cells is the $Na^+,K^+$-ATPase that actively
removes $Na^+$ from the cytoplasm and introduces $K^+$. These processes are not only
necessary in order to establish the normal resting potential across the cellular
membrane, but is also important to keep the cytoplasm osmotically balanced. The
expression of the $Na^+,K^+$-ATPase can be studied by means of immunocytochemistry
and Figures 2.8A,B compare the results for CP epithelial cells cultured in

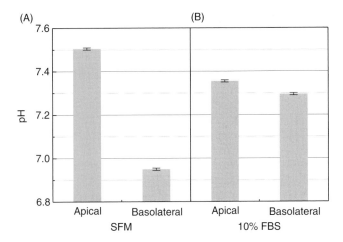

*Figure 2.7* pH readings in the apical and basolateral chamber of an experimental set-up in which a confluent cell layer of *choroid plexus* (CP) epithelial cells adherently grown on a porous filter separates the two fluid compartments. When the cells are incubated in serum-free medium (A) for 24 hours, they establish a pH gradient between the compartments in the order of 0.5 pH units. The pH difference between apical and basolateral chamber remains insignificant when cells from the same preparation are cultured in complete medium (B).

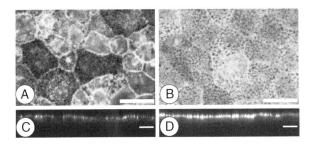

*Figure 2.8* Immunocytochemical detection of $Na^+,K^+$-ATPase in *choroid plexus* (CP) epithelial cells that were either incubated with serum-containing medium (A, C) or serum-free medium (SFM) (B, D) four days prior to the staining. Images were recorded with a confocal laser scanning microscope, and in A and B the focal plane was parallel to the cell surface, whereas in C and D we recorded side scans with the focal plane perpendicular to the cell surface. All scale bars correspond to 20 μm.

serum-containing (Figure 2.8A) or serum-free medium (Figure 2.8B). In serum-containing medium fluorescence intensity of the labeled secondary antibody is not at all homogeneously distributed and is sometimes almost absent from individual cells, indicating heterogeneous expression of the ATPase. In contrast, cells incubated with serum-free medium for four days are brightly and homogeneously fluorescent,

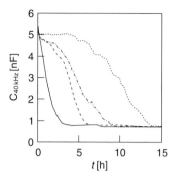

*Figure 2.9* Time course of the capacitance of an electric cell–substrate impedance sensing (ECIS) electrode when suspended Madin–Darby canine kidney (MDCK) (strain II) cells are seeded on the electrode surfaces that are precoated with different proteins prior to inoculation: (–) fibronectin, FN; (---) vitronectin, VN; (-.-.-) laminin, LAM; (....) bovine serum albumin (BSA).

indicative of significant amounts of the transporter in all cells of the cell layer. This result was independently confirmed from measurements of the $Na^+,K^+$-ATPase activity. Figures 2.8C,D compare the side view for both cell populations, with the image plane being perpendicular to the cell surface. Whereas, again, an irregular fluorescence is detected for cells kept in complete medium, the apical membranes of cells grown in serum-free medium are brightly stained. These images reveal that the $Na^+,K^+$-ATPase is predominantly located in the apical membrane of CP epithelial cells.

Using serum-free medium during the establishment of a confluent cell monolayer from suspended cells requires a precoat of the substrate with an adhesive protein in order to provide the molecular requirements for attachment and spreading. Under these conditions it is possible to study the interaction of the cells with a well-defined surface composition, and the importance of a particular ECM-protein for cellular functions. It is well-known that certain cell types show a clear preference for one or more ECM proteins in terms of attachment and spreading efficiency. Figure 2.9 demonstrates spreading of MDCK cells on different protein coatings as a function of time when suspended cells are seeded in serum-free medium (Wegener *et al.*, 2000). Similar to the data shown in Figure 2.2, the electrical capacitance of an ECIS electrode is used to monitor the increasing surface coverage of the electrode by the spreading cells. It is apparent from the recorded raw data that MDCK cells attach and spread fastest on a fibronectin (FN) coated substrate, followed by laminin (LAM), vitronectin (VN) and bovine serum albumin (BSA). Spreading on BSA is only possible after the cells have modified the substrate by secreting adhesive proteins themselves. Readings of the transepithelial electrical resistance when the cell layers are completely established does not, however, result in any significant differences, indicating that the protein coatings do not have any impact on the formation of the epithelial barrier and thus differentiation.

## Serum-free medium plus individual supplementation

Use of chemically defined, serum-free medium allows one to test individual additives for their role in proliferation and differentiation of one particular cell type. This way the individual role of certain growth factors and hormones can be experimentally addressed. A rather striking example for the role of an individual supplement is the activity of hydrocortisone (HC) on primary cultured endothelial cells from porcine brain microvessels (PBCEC). Similar to epithelial cells from porcine CP, these cells respond to serum withdrawal with a pronounced strengthening of endothelial barrier function, indicative of improved differentiation. Barrier function may be further enhanced when the medium is supplemented with physiological concentrations of the glucocorticoid HC (Hoheisel *et al.*, 1998). HC is known to promote cell attachment (Fredin *et al.*, 1979) and cell proliferation (McLean *et al.*, 1986), however, under certain conditions it can also be cytostatic (Freshney, 2000) and induce cell differentiation (McCormick, 2000). Figure 2.10 compares TEERs and transendothelial permeation rates of $^{14}$C-sucrose for confluent PBCEC monolayers incubated either in complete medium containing ten per cent OS but no HC (+OS/–HC), complete medium supplemented with 550 nM HC (+OS/+HC), serum-free medium without HC (–OS/–HC) or serum-free medium supplemented with HC (–OS/+HC). Both addition of HC and withdrawal of serum induce significant alterations in endothelial barrier function that appear to be additive. Thus, we conclude that serum-free medium supplemented with HC provides the best conditions for proper endothelial differentiation.

On a molecular level the expression of proteins that are known to be associated with the tight junctions (ZO-1, occludin, claudin-1, claudin-5) was surprisingly not enhanced when HC was added to the culture medium. However, immunocytochemical

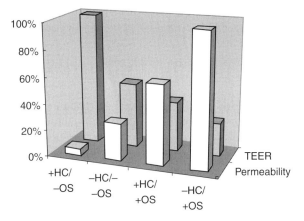

*Figure 2.10* Transendothelial electrical resistances (TEER) and sucrose permeation rates for porcine brain microvessel endothelial cells (PBCEC) that were incubated in DME/Ham's F12 medium supplemented with ten per cent ox serum (OS) (+OS/–HC), supplemented with ten per cent OS and 550 nM hydrocortisone (HC) (+OS/+HC), without HC and OS (–OS/–HC) and HC only (–OS/+HC). For easy comparison the maximum TEER readings under (–OS/+HC) conditions as well as maximum sucrose permeation rate for (+OS/–HC) conditions were set to 100 per cent.

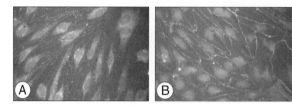

*Figure 2.11* Immunocytochemical detection of the tight junction-associated protein claudin-5 in porcine brain microvessel endothelial cells (PBCEC) when the medium contained either 550 nM hydrocortisone (HC) (B) or no HC (A). The micrographs show a clear re-orientation of claudin-5 from the rough endoplasmic reticulum to the site of cell–cell contacts when HC is present in the medium.

studies revealed that some of these proteins were dislocated from cell–cell contact sites in the absence of HC but were found in the rough endoplasmic reticulum instead. As a typical example for this kind of protein disarrangement Figure 2.11 compares the localization of claudin-5 in PBCEC when these cells were either cultured in serum-free medium, with HC (Fig. 2.11B) or without (Fig. 2.11A).

HC also induced rather striking morphological alterations within PBCEC which become apparent from electron micrographs of thin frozen sections. With the presence of HC the cells become very flat – except within the nuclear region – and elongated, as is generally expected for differentiated brain microvessel endothelial cells. Without HC in the medium, they even tend to form in multilayers. The cells also strengthen their substrate contacts and become much more resistant to mechanical challenges which they may have to face in form of shear flow in the vascular wall. HC shares its pro-differentiating activity in PBCEC with other gluco-corticoids like the synthetic analog dexamethasone, but not with chemically related mineralo-corticoids (Engelbertz *et al.*, 2000).

### Serum replacements

Since it is often desirable, on the one hand, to exactly know and control the com-position of the culture medium, yet is indispensable, on the other, to add a certain amount of growth factors, addition of serum is often bypassed by using so-called serum-replacements. An investigation into the interaction between PBCEC and immune competent cells facilitates the study of inflammatory processes at the blood-brain barrier *in vitro*. Since serum always contains trace amounts of HC, and HC is known to have an anti-inflammatory activity, serum-containing medium may interfere with the outcome of the experiments. On the other hand, when grown in serum-free medium PBCEC have shown to be rather sensitive to mechanical medium perturbation, as they occur when immune cells are added to the culture. Thus, medium supplemented with a carefully selected serum replace-ment seems to be advantageous. These mixtures contain known quantities of chemically defined growth factors, hormones and mitogens. Figure 2.12A com-pares TEERs of PBCEC that were grown to confluence and then incubated in

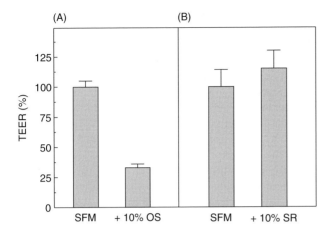

*Figure 2.12*  Transendothelial electrical resistances (TEER) of confluent monolayers of porcine brain microvessel endothelial cells (PBCEC) after the cells had been incubated in serum-free basal medium (SFM) or basal medium supplemented with (A) ten per cent ox serum (OS) or (B) ten per cent serum replacement (SR). Serum replacement does not hamper the expression of a barrier-forming phenotype as it is observed with serum supplementation.

basal medium only, or in the same medium but supplemented with ten per cent ox serum. As shown previously, the presence of ox serum significantly reduces barrier efficiency to less than 33 per cent of the serum-free values which were set to 100 per cent. Figure 2.12B shows the same comparison for PBCEC cultured either in basal medium or basal medium supplemented with ten per cent serum replacement. Serum replacement does not impair the electrical resistances of PBCEC monolayers, indicating that expression of the barrier-forming phenotype is not hampered under these conditions. Apparently, a certain factor or a combination of factors present in serum responsible for barrier reduction is omitted from the serum replacement used here. It is noteworthy, however, that unless designed and optimized for one particular cell type, serum replacements often lag behind fetal serum where its ability to promote growth and proliferation is concerned. Additionally, some uncertainty remains as to whether undefined impurities from biotechnologically produced supplements may have been introduced into the medium.

## Conditioned medium

Although it is often a major objective to obtain pure cultures of one particular cell type, these cells may not reach their *in vivo*-phenotype without certain signals from another cell species that they communicate with within their natural environment. This interaction can either be mediated by physical contact or by secretion of signal molecules into the medium that diffusively reach the target cells. The interaction via physical contact can only be modeled *in vitro* by so-called co-cultures in which both cell species are cultured in one dish (sometimes) separated by a permeable membrane that still allows physical contact. Co-culture systems will, however, not be discussed in this chapter. Diffusive signal molecules, on the other hand, secreted by one cell species and processed by the other can be simulated *in vitro* by means of conditioned media. In this instance, the donor cell species is incubated with serum-containing but preferably serum-free medium for a certain time-span. During that time these cells are expected to secrete the signal molecules into the medium, which is then termed 'conditioned'. It may contain substrate-modifying matrix constituents like collagen, FN and proteoglycans as well as growth factors like FGF, IGF-1 and -2, and PDGF and intermediary metabolites. The acceptor cell species is then incubated in the conditioned medium. Sometimes nutrient consumptions of the donor cells are compensated by adding a certain amount of fresh medium. Since the signal molecules absent in ordinary cell culture medium are now present, the cell may have all induction factors necessary to express their *in vivo*-phenotype.

An example taken from the field of modeling the blood–brain barrier *in vitro* is the treatment of cerebral microvessel endothelial cells with medium that has been conditioned by astrocytes. *In vivo* cerebral microvessels are entirely covered by astrocyte endfeet so that astrocytes and endothelial cells are more or less in direct contact. The question then arises whether physical contact or secretion of a soluble astrocytic factor is necessary to induce microvessel endothelial cells to fully express the blood–brain barrier phenotype *in vitro*. When brain capillary endothelial cells were cultured in astrocyte conditioned medium (ACM), Rubin *et al.* (1991) found a significant increase in TEER and a reduction of the sucrose permeation rate. Consistently, the barrier-forming network of tight junction became more complex upon treatment of endothelial cells with ACM (Wolburg *et al.*, 1994). Interestingly, even human umbilical vein endothelial cells – usually devoid of tight junctions – expressed these special cell–cell contacts when they were incubated in medium containing astrocyte derived factors (Tio *et al.*, 1990). This observation underlines that paracrine signal molecules may be decisive for proper cell differentiation.

## ACKNOWLEDGMENT

This work has been financially supported by the *Deutsche Forschungsgemeinschaft* (*DFG*) within *SFB 293* and *492* as well as within *Graduiertenkolleg GRK 233/1-97*.

## REFERENCES

Alberts, B., Bray, D., Lewis, J., Raff, M., Roberts, K. and Watson, J. D. (1995) Molecular biology of the cell. Verlag Chemie Weinheim.

Antoniades, H. N., Scher, C. D. and Stiles, C. D. (1979) Purification of human platelet-derived growth factor. *Proc. Natl. Acad. Sci. USA*, **76**, 1809.

Brewer, M. and Scott, T. (1983) Concise encyclopedia of biochemistry. Walter de Gruyter Verlag Berlin.

Chang, C., Wang, X. and Caldwell, P. B. (1997) Serum opens tight junctions and reduces ZO-1 protein in retinal epithelial cells. *J. Neurochem.*, **69**, 859–867.

Crook, R. B., Kasagami, H. and Prusinger, S. B. (1981) Culture and characterisation of epithelial cell from bovine choroid plexus. *J. Neurochem.*, **37**, 845–854.

Darling, S. J. M. a. D. C. (1993) Animal Cell Culture. BIOS Scientific Publishers Limited, Oxford.

Eagle, H. (1973) The effect of environmental pH on the growth of normal and malignant cells. *J. Cell Physiol.*, **82**, 1–8.

Engelbertz, C., Korte, D., Nitz, T., Franke, H., Haselbach, M., Wegener, J. *et al.* (2000) The development of *in vitro* models of the blood–brain and blood–CSF barriers. In Begley, Bradbury and Kreuter (eds), the blood–brain barrier and drug delivery to the CNS, Marcel Dekker, Inc. New York, pp. 33–63.

Fisher, H. W., Puck, T. T. and Sato, G. (1958) Molecular growth requirements of single mammalian cells: The action of fetuin in promoting cell attachment of glass. *Proc. Natl. Acad. Sci. USA.*, **44**, 4–10.

Fredin, B. L., Seiffert, S. C. and Gelehrter, T. D. (1979) Dexamethasone-induced adhesion in hepatoma cells: The role of plasminogen activator. *Nature*, **277**, 312–313.

Freshney, R. I. (2000) Culture of animal cells: a manual of basic techniques. Wiley-Liss, New York.

Gath, U., Hakvoort, A., Wegener, J., Decker, S. and Galla, H.-J. (1997) Porcine choroid plexus cells in culture: expression of polarized phenotype, maintenance of barrier properties and apical secretion of CSF-components. *Eur. J. Cell Biol.*, **74**, 68–78.

Hakvoort, A., Haselbach, M., Wegener, J., Hoheisel, D. and Galla, H.-J. (1998) The polarity of choroid plexus epithelial cells *in vitro* is improved in serum-free medium. *J. Neurochem.*, **71**, 1141–1150.

Hoheisel, D., Nitz, T., Franke, H., Wegener, J., Hakvoort, A., Tilling, T. *et al.* (1998) Hydrocortisone reinforces the blood–brain barrier properties in a serum free cell culture system. *Biochem. Biophys. Res. Comm.*, **244**(1), 312–316.

Kimura, A. I. a. G. (1974) TES and HEPES buffers in mammalian cell cultures and viral studies: Problems of carbon dioxide requirements. *Exp. Cell Res.*, **83**, 351–360.

Lo, C.-M., Keese, C. R. and Giaever, I. (1999) Cell–substrate contact: another factor may influence transepithelial electrical resistances of cell layers cultured on permeable filters. *Exp. Cell Res.*, **250**, 576–580.

McCormick, C. and Freshney, R.I. (2000) Activity of growth factors in the IL-6 group in the differentiation of human lung adenocarcinoma. *Brit. J. Cancer.*, **82**, 881–890.

McKeehan, W. L., Hamilton, W. G. and Ham, R. G. (1976) Selenium is an essential trace nutrient for growth of WI-38 diploid human fibroblasts. *Proc. Natl. Acad. Sci. USA.*, **73**, 2023–2027.

McLean, J. S., Frame, M. C., Freshney, R. I., Vaughan, P. F. T. and Mackie, A. E. (1986) Phenotypic modification of human glioma and non-small cell lung carcinoma by glucocorticoids and other agents. *Anticancer Res.*, **6**, 1101–1106.

Mortell, K. H., Marmorstein, A. and Cramer, E. B. (1993) Fetal bovine serum and other sera used in tissue culture increase epithelial permeability. *In vitro Cell Dev. Biol.*, **29A** (3pt l), 235–238.

Rubin, L. L., Hall, D. E., Porter S., Barbu, K., Cannon, C., Horner, H. C. *et al.* (1991) A cell culture model of the blood–brain barrier. *J. Cell Biol.*, **115**(6), 1725–1735.

Spector, R. (1982) Pharmacokinetics and metabolism of cytosine arabinoside in the central nervous system. *J. Pharm. Exp. Ther.*, **222**, 1–6.

Stryer, L. (1995) *Biochemistry*, W. H. Freeman, New York.

Tio, S., Deenen, M. and Marani, E. (1990) Astrocyte-mediated induction of alkaline phosphatase activity in human umbilical cord vein endothelium: an *in vitro* model. *Eur. J. Morphol.*, **28**, 289–300.

Tomasini, B. R. and Mosher, D. F. (1991) Vitronektin. *Prog. Hemost. Thromb.*, **10**, 269–305.

Vogler, E. A. and Bussian, R. W. (1987) Short-term cell attachment rates: a surface sensitive test of cell–substrate compatibility. *J. Biomed. Mater. Res.*, **21**, 1197–1211.

Wegener, J., Keese, C. R. and Giaever, I. (2000) Electric cell–substrate impedance sensing (ECIS) as a noninvasive means to monitor the kinetics of cell spreading to artificial surfaces. *Exp. Cell Res.*, **259**, 158–166.

Wolburg, H., Neuhaus, J. and Kniesel, U. (1994) Modulation of tight junction structure in blood–brain barrier endothelial cells. *J. Cell Sci.*, **107**, 1347–1357.

Yamada, K. M. a. G., B. (1997) Molecular interactions in cell adhesion complexes. *Curr. Opin. Cell Biol.*, **9**, 76–85.

# Chapter 3

# Bioelectrical characterization of cultured epithelial cell (mono)layers and excised tissues

*Kwang-Jin Kim*

## INTRODUCTION

Epithelial cells line mucosal surfaces of various organs in the body. These specialized cells participate in the exchange of substances between the outside world and internal body fluid compartments (e.g. interstitial and vascular fluids). An epithelium consists of highly polarized cells adjoined by cell membrane proteins from neighboring cells. These cell–cell associating proteins include gap junctions (GJ), cell adhesion molecules (CAM), junctional adhesion molecules (JAM), and tight junction (TJ)-associated proteins (e.g. claudins and occludins) (Liu *et al.*, 2000; Mitic *et al.*, 2000; Tsukita *et al.*, 2000). This unique array of cell–cell association is further assisted by many intracellular proteins, including zonulae occludentes (ZO) proteins (ZO-1, -2, and -3), catenins ($\alpha$-, $\beta$- and $\gamma$-catenins), cytoskeletal proteins (e.g. microtubules and microfilaments) and other cellular proteins (e.g. p120) whose functions remain ill defined (Liu *et al.*, 2000; Mitic *et al.*, 2000; Tsukita *et al.*, 2000). Epithelial cells are endowed with some common (e.g. facilitative glucose transporter, Na,K-ATPase) transport processes that are ubiquitously present for homeostatic control and cell survival as well as for unique (proton-dependent peptide transporter, specific type(s) of ion and water channels) processes that dictate the characteristic function of the epithelium.

The primary function of an epithelium is to limit the uncontrolled leak of various solutes across the barrier, which allows selective absorption/secretion of molecules across the barrier with the aid of various transcellular processes. Each epithelium specializes in selective and controlled translocation of specific ions (e.g. Na, Cl, $HCO_3$, K, Ca) and various hydrophilic (e.g. nucleosides, nucleotides, sugars, amino acids, peptides, and proteins) and lipophilic (e.g. substrates for multidrug resistance transporters) solutes in an orchestrated and unique fashion under elaborate regulation by neurohumoral factors. Epithelial water transport is also thought to be tightly controlled in a similar fashion.

Structure–function relation of a given epithelial barrier embodies an important field of transport physiology. This chapter is organized to introduce some basic electrical means of investigating epithelial barrier properties that may be useful for students and investigators who are interested in studying drug delivery/transport across various epithelial barriers. Those who wish to pursue more in-depth investigations in a particular area of epithelial transport physiology/biology should consult a number of excellent treatise and review articles on epithelial transport (Fleisher and Fleisher, 1990; Schultz, 1979; 1980; 1998; Schultz *et al.*, 1996; Schultz and Frizzell, 1976; Wills *et al.*, 1996). It should also be pointed out that the focus of this chapter lies in the characterization of epithelial cell (mono)layers cultured on permeable supports (that are to be used for the study of transepithelial drug transport), but many of the approaches described herein can be applied to excised tissues with little modification.

## METHODS FOR BIOELECTRICAL CHARACTERIZATION OF CULTURED EPITHELIAL CELL (MONO)LAYERS

Regardless of the *in vitro* model types (e.g. excised epithelial tissue or a cultured cell (mono)layer model) utilized in transport studies, the first priority is to ascertain the viability/integrity of the model. As a first step towards fulfilling such assurance, one may begin with the determination of transepithelial electrical resistance (TEER) of a given model. Transport characteristics of an epithelial barrier are largely categorized into two parts: one dealing with passive properties and the other with active properties. Passive properties can be embodied by the leakiness or tightness of the epithelium, whereas active properties include activities of Na-pump and ion channels.

### Screening of cultures for their viability and suitability for transport studies

With the advent of the wide use of *in vitro* culture models of polarized epithelial cells grown on permeable substrata, the need for quickly assessing the viability/ integrity of such epithelial cell layers has led to a particular screening device. This screening device is based on reusable electrodes and the self-contained unit of their associated electronic amplifier. These devices are made by a number of manufacturers that include World Precision Instruments, Warner Instruments, and Millipore. All these units are based on the Ohm's law such that the TEER is estimated from the ratio of $dV$ over $dI$, where $dV$ is the observed electrical potential difference in response to the electrical current ($dI$) of a sinusoidal (10 Hz) shape and small amplitude ($\sim 10\,\mu A$). The sinusoid is generated by the device and passed across the epithelial barrier through a pair of electrodes comprised of platinum wire or Ag/AgCl pellet. The potential difference across the epithelium is measured via a pair of Ag/AgCl electrodes composed in pellet form. Ag/AgCl electrodes have relatively small offset potential compared with most other electrodes. Fabrication of such electrodes in a small format allows the electrodes to be placed onto a small surface area (e.g. $\sim$ one-third of a square centimeter) available for cells cultured on commercially available permeable supports.

The specific description of how to use these devices to measure TEER and spontaneous potential difference across a given cultured cell model is not discussed in this chapter, since each device comes with a detailed instruction. Instead, some practical points and important caveats are illustrated. The most formidable task in use of these devices may be the maintenance of reliable electrodes and the issue of sterility. Culture fluids usually contain various amounts of proteinaceous materials that tend to stick to the electrode surface. Proteins adherent to the electrodes are the leading cause of unstable electrode readings. Treatment of electrodes by dipping them overnight in the dilute (~0.1–0.5 N) hydrochloric acid detaches the proteins from the electrodes, although this procedure can not be repeated once the electrode surface erodes beyond its useful lifespan (~ a year or so depending on frequency of usage). Sometimes gentle abrasion of the electrode surface with fine meshed sandpaper helps remove the proteins and other proteinaceous materials from the electrode surface. Electrodes may also be cleaned with distilled water (~100 mL) for at least half an hour using a gentle swirling motion. The electrodes are then dried by patting them with soft paper. A combination of these procedures is also used to resuscitate unstable electrodes. If all fails, it is time to purchase new electrodes. For the storage of electrodes, the manufacturer recommends cleaning up the gunk (see above) and storing the electrodes dry under dark (if the electrodes are composed of Ag/AgCl).

For sterilizing the electrodes, usually 70–80 per cent ethyl alcohol is used. Briefly (one–two seconds) dip the electrodes in ethyl alcohol solution and shake loose the droplets dangling at the tip of the electrodes, since excess alcohol propagated to the culture fluid may be detrimental for the function of cultured cells. Another way to prevent the propagation of excess alcohol is by dipping the sterilized electrodes in a small volume (~2–3 mL) of culture fluid sterilized using a 0.22 µm filter attached to a small syringe. This method is highly recommended, since it not only dilutes excess alcohol, but also stabilizes the electrodes in the same culture fluid prior to moving them into the culture plates containing cell (mono)layers grown in Transwells. At this stage, background level of potential difference (PD) and TEER (or Rt) can be estimated using a blank Transwell bathed with culture fluid. The background PD and TEER levels should be measured before and after the measurements of cell layers grown in Transwells, since background PD tends to drift over time.

One parameter of interest, active ion transport index (alias, Ieq or equivalent short-circuitcurrent), can be estimated from the observed PD and TEER (or Rt). Ohm's law dictates that a voltage (PD, mV) measured across a resistor (TEER or Rt, $k\Omega cm^2$) is associated with a current flowing across the resistor. The current, in this case active ion transport across a biological barrier, is the ratio between the PD and TEER, or Ieq=PD/TEER and the unit is $\mu A/cm^2$. As an example, frog skin generates a PD of approximately 100 mV (lumen negative) and has an Ieq of approximately 50 $\mu A/cm^2$. Thus, the TEER of frog skin is approximately 2 $k\Omega cm^2$. Bullfrog lung generates PD of approximately 10 mV and TEER of approximately 1 $k\Omega cm^2$, yielding Ieq of approximately 10 $\mu A/cm^2$. Some biological barriers do not generate measurable PD, hence the Ieq is close to zero, but they have a finite TEER. One example for the latter case is gallbladder epithelium, where TEER is <100 $\Omega cm^2$.

## Determination of epithelial barrier properties

Description of Ussing chamber set-up is described elsewhere in this book in detail. The methods for mounting tissues and various cells cultured on permeable filters are also contained in other sections. In this chapter, only the pertinent information on how to investigate 'barrier properties' of a given endothelial and/or epithelial barrier using Ussing chamber techniques (and bioelectric measurements) is described. Barrier properties of an epithelial/endothelial barrier include active and passive transport properties, e.g. transtissue or transcell layer $PD$ (measured in mV, lumen or apical side as reference) as a result of active ion absorption and/or secretion across the barrier, the active ion transport rate ($Isc$ or $Ieq$, $\mu A/cm^2$; measured as the short-circuit current), e.g. the current flowing across the barrier when the $PD$ is brought down to zero by generating external current in the voltage clamp unit, and overall barrier resistance (TEER or Rt, measured as $k\Omega cm^2$). The important point here is the unit for TEER (or Rt), which is conventionally defined as the voltage deflection normalized by the current density (e.g. $\mu A/cm^2$) to cause that voltage deflection across the exposed surface area available for the barrier mounted in the Ussing chamber set-up. Details of how to measure/estimate these bioelectric parameters appear elsewhere in this book.

These three parameters usually obey the Ohm's law, $PD=Isc\times Rt$ or $Ieq\times TEER$. It should be pointed out here that Rt can also be estimated by passing a small amount of current (e.g. $dI$ in the unit of $\mu A$) across the barrier (regardless of the state of open- versus short-circuit conditions imposed on the barrier) and measuring the resultant voltage deflection of $dV$ (in the unit of mV). Most modern voltage/current clamp instruments have a built-in function for performing this kind of TEER assessment. If the surface area of the barrier exposed in the Ussing chamber fluids is $S$ ($cm^2$), $Rt=dV/(dI/S)=(dV/dI)\times S$ by Ohm's law. This procedure is very useful when the $PD$ of the barrier is near zero (e.g. in the case of gallbladder epithelium, renal proximal tubular epithelium, or other cell lines that do not generate much $PD$ to begin with) or rendered to become zero (e.g. by inhibiting the sodium pump that sets up the $PD$). One cautionary note for this application is that one should generate an $I$–$V$ curve of the barrier to ensure that the $dI$ is within the linear relation with $dV$, since epithelial barrier shows non-linear $I$–$V$ relations at very high $dV$ (Helman, 1979; Helman and Thompson, 1982; Stoddard and Reuss, 1998). If one already knows the magnitude of $Ieq$ or $Isc$ of the barrier, $dI$ should be chosen as approximately $0.5\times Ieq$ or $0.5\times Isc$ to be in the safe linear zone of the $I$–$V$ relation.

Epithelial/endothelial permeability to ions in the absorptive (i.e. apical-to-basolateral or mucosal-to-serosal or luminal-to-abluminal) or secretory (basolateral-to-apical or serosal-to-mucosal or abluminal-to-luminal) directions is an important constituent of barrier properties. The apparent permeability index is defined as ratio between unidirectional flux and concentration gradient for the specific solute of interest (e.g. Na ion). Asymmetry in ion permeability (or unidirectional fluxes measured at an equal but opposite gradient for sodium radionuclide across the barrier) can be analyzed using Ussing's flux ratio equation to ascertain the active or passive nature of such asymmetry (Lim and Ussing, 1982; Sten-Knudsen and Ussing, 1981; Ussing 1965, 1968, 1971, 1978, 1980a,b, 1988; 1994; Ussing and Zerahn, 1999). In order to perform the analysis under open-circuit conditions, $PD$ across the barrier should be measured during the flux measurement. If $PD$ has changed over the course for permeability coefficient

($P_{app}$) (or flux) measurements, one can use the average behavior of the $PD$ during the time course. Ussing's flux ratio equation states that:

$$\frac{J^{ab}}{J^{ba}} = \left(\frac{C^a}{C^b}\right) \times \exp\left[\frac{-z \times PD \times F}{R \times T}\right], \tag{3.1}$$

where $J$s are radiolabeled fluxes; ab and ba denote apical-to-basolateral direction and opposite direction, respectively; $C$s are concentrations of the charged (radiolabeled) solute of interest; a and b are apical and basolateral fluids, respectively; $z$ is the valence of the charged solute; $PD$ is the transtissue voltage ($=V_s-V_m=V_b-V_a$, e.g. mucosal or luminal or apical side as reference); F is Faraday constant; R is the gas constant; and $T$ is the absolute temperature. Substituting $J^{ab}=(P_{app}$ measured in ab direction)$\times$ $C^a-C^b)$, and $J^{ba}=(P_{app}$ measured in the ba direction)$\times(C^b-C^a)$, one can get:

$$\frac{P^{ab}_{app}}{P^{ba}_{app}} = -\left(\frac{C^a}{C^b}\right) \times \exp\left[\frac{-z \times PD \times F}{R \times T}\right], \tag{3.2}$$

since the magnitude of $C^a-C^b$ and that of $C^b-C^a$ are the same when an equal concentration of the radiolabeled solute is used in the respective upstream (e.g. donor) fluid for the assessment of these fluxes.

A special case for the Ussing's flux ratio analysis is when $PD$ is forced to zero with an equal concentration of the solute in both bathing fluids (e.g. short-circuit conditions), where the right hand side of the equation becomes unity. In other words, if there is only passive diffusional mechanism for translocating the charged solute across the short-circuited barrier, the $P_{app}$ (or unidirectional fluxes) do not exhibit asymmetrical behavior. Thus, asymmetric $P_{app}$ or fluxes under the circumstances are supporting evidence for the presence of active mechanism(s) for transbarrier transport of such a charged solute. One cautionary note to these discussions is the assurance that the barrier is uniformly short-circuited. In some complex epithelial barriers (e.g. intestine), a parallel arrangement of various cell types (e.g. goblet cells, microvilli on the surface epithelial cells, and so forth) makes the uniform short-circuiting difficult. For this reason, the use of Ussing's flux ratio analyses is more appropriate in dissecting information on the involvement of active solute transport across the barrier under open-circuited condition.

## Estimation of tight junctional resistance

As paracellular transport of solutes (including water and ions) is modulated by the magnitude of how tight the paracellular seal is, estimation of tight junctional resistance is a useful tool to understand the paracellular transport of solutes. A classic approach to estimating the paracellular resistance is based on equivalent pore theory, assuming that the paracellular routes allow restricted diffusion of hydrophilic solutes. In this approach, physical pores may not be present in the paracellular region, but rather reflect the 'assumed behavior of such imaginary water-filled cylindrical pores as a route for diffusion of hydrophilic solutes'. In other words, there is no a priori reason for the presence or absence of such porous structure at the paracellular pathways. The equivalent pore radius is just a measure of how restrictive the pathway is towards diffusion of hydrophilic solutes. Detailed analysis schemes can be found elsewhere (Adson *et al.*, 1994; Berg *et al.*, 1989; Crandall and Kim, 1981; Kim and Crandall, 1982a,b, 1983; Kim *et al.*, 1979, 1985; Matsukawa *et al.*, 1997; McLaughlin *et al.*, 1993), and, thus, this subject will not be addressed.

Electrical approaches to estimating the paracellular (e.g. tight junctional) resistance have been mostly successful in tight epithelial barriers, where manipulation of cellular resistance to ion movement renders a sizable change in overall Rt. Assuming that experimental maneuvers (e.g. increasing or decreasing cellular conductance/resistance to ion movement) do not affect tight junctional resistance (and electromotive force for the ion to move in or out of the cell), one can accurately estimate tight junctional resistance. The relation between *PD* on y-axis and Rt on x-axis during the perturbation (e.g. application of *Staphylococcus aureus* toxin to apical cell surfaces) results in a linear relation, where the y-intercept is Ec (cellular electromotive force) and x-intercept is tight junctional resistance (Lewis *et al.*, 1978; Wills *et al.*, 1979a,b) with a relation:

$$\frac{PD}{Ec} + \frac{Rt}{Rtj} = 1. \tag{3.3}$$

If non-linear relation is indicated, the above assumptions for constant Rtj (and/ or Ec) during the maneuver have been violated and estimation of Rtj (or Ec) becomes unreliable. Please note that these maneuvers are done under open-circuit conditions. Under short-circuit conditions, similar approaches can also be utilized (Siegel and Civan, 1976; Yonath and Civan, 1971). Maneuvers to alter (apical, in most cases) cellular (with little effect on junctional) resistance alone include:

1   use of ion channel inhibitors (e.g. amiloride for sodium channels),
2   use of second messenger molecules (e.g. cyclic AMP) to increase cellular ion conductance, or
3   use of cell membrane permeabilizing agent (e.g. toxins, gramicidin D, nystatin, or amphotericin).

It should be noted that pore-forming agents such as amphotericin must use apical solution that simulate the intracellular milieu in order to minimize dilution of intracellular ionic composition. Typically, an apical fluid comprised of high potassium concentration with low calcium, chloride, and sodium concentrations are required to mimic cell interior. In order to prevent cell swelling due to KCl influx into cells, chloride in apical solution should be replaced with a large anion species (e.g. gluconate or isethionate) when permeabilizing the cell membranes.

## Characterization of cation/anion selectivity of tight junctions

Information on charge selectivity of tight junctional routes can be useful to determine/ predict the transport rates of charged solutes through paracellular pathways. Modified equivalent pore theory has been reported to be able to predict reasonably the decreased or enhanced passage of charged solutes by passive diffusion via charge-selective pathways in epithelium (Adson *et al.*, 1994). In this section, bioelectrical approaches to determine the nature of charge selectivity of the paracellular routes are described. For a given barrier, in the absence of active transport of the charged solute (including ions), the only route for passage is the tight junctional pathways. For this purpose, sodium pump activity (e.g. the source of the secondary and tertiary active transport) is inhibited using millimolar range of ouabain for approximately 0.5–1 hour. Under this condition, charged molecules only can traverse the barrier via tight junctional routes. First, a dilution potential is measured across the barrier by imposing (e.g. 100 mM) NaCl gradient across the barrier, by bathing the apical side with a Kreb's phosphate-buffered Ringer's solution (KPBR) while bathing the basolateral side with a modified KPBR containing 100 mM less NaCl than that in the unmodified KPBR. (This process can be done by simply replacing 100 mM NaCl with 200 mM mannitol or sucrose, leaving other things the same). There is a gradient for NaCl to diffuse through the paracellular junctional routes in the apical-to-basolateral direction. The magnitude and sign of the *PD* measured across the barrier will be dependent on the nature of the charge selectivity of the tight junctional routes. If Na and Cl are not segregated by the tight junctional routes, the observed *PD* is solely dependent on the magnitude of the NaCl gradient and the mobility of Na and Cl. The observed dilution *PD* can be described as follows (Andreoli *et al.*, 1986, pages 119–123 and 151–165):

$$PD = V_b - V_a = \frac{-(R \times T)}{(z \times F)} \times \ln\left(\frac{\{P_K \times [K_b] + P_{Na} \times [Na_b] + P_{Cl} \times [Cl_a]\}}{\{P_K \times [K_a] + P_{Na} \times [Na_a] + P_{Cl} \times [Cl_b]\}}\right)$$

$$= \frac{-(R \times T)}{z \times F} \times \left(\frac{\{u_{Na} - u_{Cl}\}}{\{u_{Na} + u_{Cl}\}} \times \ln\left(\frac{C_b}{C_a}\right)\right), \tag{3.4}$$

where $P$ = permeability of ions; $[Na]$ = sodium concentration; $u_{Na}$ = mobility of Na; $u_{Cl}$ = mobility of Cl; $C_b$ = NaCl concentration in serosal (= basolateral = albuminal) fluid; and $C_a$ = that in opposite fluid. As can be seen from the dilution potential measurements, a ratio between one cation and counterion can be estimated when both bathing fluids are simplified to keep other ion concentrations the same (e.g. K and Ca). In the latter case, $P_{Na}/P_{Cl}$ can be estimated from the observed dilution *PD* arising from NaCl gradient.

We next estimate a bi-ionic *PD* across ouabain-treated barrier (e.g. active ion transport is inhibited). Bi-ionic *PD* is measured with an equal concentration of two ionic solutions, e.g. NaCl in one fluid and KCl in the other. (Or one can use NaCl in one fluid, while the other fluid contains Na-isothionate). In the former case, where NaCl and KCl are used in the apical and basolateral fluid, respectively, the observed bi-ionic *PD* can be described as,

$$PD = V_s - V_m = V_b - V_a = \left(\frac{R \times T}{(z \times F)} \times \ln\left(\frac{[u_{Na} + u_{Cl}]}{[u_K + u_{Cl}]}\right)\right). \tag{3.5}$$

The bi-ionic *PD* for the cl vs. isothionate can also be similarly derived. Using these observed dilution and bi-ionic *PD*, one can deduce the charge selectivity and permeability sequences for cations and anions. For example, the following is taken from the author's unpublished observations using bullfrog alveolar epithelium as a biological barrier in the determination of dilution and bi-ionic potentials.

From the dilution potential measurements, $P_{Na}/P_{Cl}$ of 0.53 was estimated using the relation that the tight junctional routes of bullfrog alveolar epithelium are anion-selective by two to one over cations (Crandall *et al.*, 1986). Bi-ionic potential measurements yielded the anionic sequence of SCN:NO$_3$:Br:I:Cl:ClO$_4$:isothionate: gluconate = 1.12:1.07:1.05:1.05:1:0.89:0.29:0.25, while cationic sequence yielded K:Rb:Cs:Na:Li = 1.51:1.46:1.42:1:0.82, suggesting that K, Rb, Cs all diffuse approximately 40–50 per cent faster than Na (or Li).

## Estimation of pump current

Under short-circuit conditions, when apical cell membranes are permeabilized (e.g. using alpha-toxin) (Bhakdi *et al.*, 1993; Boyle and Lieberman, 1999; Chang *et al.*, 1995; Iizuka *et al.*, 1994; Ostedgaard *et al.*, 1992; Reddy and Quinton, 1994; Russo *et al.*, 1997; Schultz, 1990; Tabcharani *et al.*, 1994), the resistance to ion movement across apical cell membranes is lost and the observed short-circuit current is taken as a measure of sodium pump current (Ito *et al.*, 2000; Shi and Canadia, 1995; Vasilets *et al.*, 1990). One cautionary note to this scheme is that tight junctional resistance should be well maintained for this to work. A dose response study of the permeabilizing agent is necessary to ensure the correct assessment of both tight junctional resistance and the pump current, where no overt effect of the permeabilizing agent should be apparent on tight junctional resistance as a prerequisite.

With the permeabilizing agent acting to decrease the resistance of apical cell membranes, short-circuit current of the tissue (or cell (mono)layer) will rise to a certain level and plateau at a certain level. In order to prevent leakage of cellular components (e.g. ATP, ions, and other small solutes), one should use an apical bathing solution that mimics the intracellular milieu as much as possible (see above) when permeabilizing. A precipitous fall in short-circuit current after the peak indicates a decrease of cellular components essential to keep the pump running. It should be remembered that the peak short-circuit current observed is the full capacity of the sodium pump, which is the product of the number of pumps and individual pump capacity. Thus, pump density has to be measured separately to delineate the mechanisms of the observed changes in pump current. The number of sodium pumps expressed on the epithelial basolateral surfaces can be estimated by measuring the specific binding isotherms using radio-labeled ouabain, followed by correction for the non-specific binding of ouabain. If the specific binding of ouabain does not change with experimental maneuvers (e.g. stimulation of active ion transport with epidermal growth factor (EGF) or injury to epithelial transport by hypoxia/hyperemia), one can deduce that the individual pump capacity is increased or decreased by the specific experimental condition.

## SUMMARY AND CONCLUSION

In this chapter, basic techniques pertaining to the characterization of epithelial barrier properties using bioelectrical approaches were introduced. More advanced techniques utilizing ion-sensitive indicators, microelectrodes, patch clamp techniques, current fluctuation analysis for ion channel studies have been left out. Interested readers should consult with the pertinent books and review articles contained in the reference section.

## ACKNOWLEDGMENT

This work was supported in part by research grants (HL38658 and HL64365) from the National Institutes of Health and a grant-in-aid (9950442N) from the American Heart Association – National Center.

## REFERENCES

Adson, A., Raub, T. J., Burton, P. S., Barsuhn, C. L., Hilgers, A. R., Audus, K. L. *et al.* (1994) Quantitative approaches to delineate paracellular diffusion in cultured epithelial cell monolayers. *J. Pharm. Sci.*, **83**, 1529–1536.

Andreoli, T. E., Hoffman, J. F., Fanestil, D. D. and Schultz, S. G. (1986) *Physiology of Membrane Disorder*, NY: Plenum Press, New York.

Berg, M. M., Kim, K. J., Lubman, R. L. and Crandall, E. D. (1989) Hydrophilic solute transport across rat alveolar epithelium. *J. Appl. Physiol.*, **66**, 2320–2327.

Bhakdi, S., Weller, U., Walev, I., Martin, E., Jonas, D. and Palmer, M. (1993) A guide to the use of pore-forming toxins for controlled permeabilization of cell membranes. *Med. Microbiol. Immunol. (Berl).*, **182**, 167–175.

Boyle, R. T. and Lieberman, M. (1999) Permeabilization by streptolysin-o reveals a role for calcium-dependent protein kinase c isoforms alpha and beta in the response of cultured cardiomyocytes to hyposmotic challenge. *Cell. Biol. Int.*, **23**, 685–693.

Chang, C. Y., Niblack, B., Walker, B. and Bayley, H. (1995) A photogenerated pore-forming protein. *Chem. Biol.*, **2**, 391–400.

Crandall, E. D. and Kim, K. J. (1981) Transport of water and solutes across bullfrog alveolar epithelium. *J. Appl. Physiol.*, **50**, 1263–1271.

Crandall, E. D., Kim, K. J. and Goodman, B. E. (1986) Active transport and permeability properties of the alveolar epithelium in the lung. In H. Kazemi, A. L. Hyman and P. J. Kadowitz (eds), *Acute Lung Injury: Pathogenesis of Adult Respiratory Distress Syndrome*, Littleton, MA: PSG Publishing Co., Inc.

Fleischer, S. and Fleischer, B. (1990) *Biomembranes, Part V*, New York, NY: Academic Press, Inc.

Helman, S. I. (1979) Electrochemical potentials in frog skin: inferences for electrical and mechanistic models. *Fed. Proc.*, **38**, 2743–2750.

Helman, S. I. and Thompson, S. M. (1982) Interpretation and use of electrical equivalent circuits in studies of epithelial tissues. *Am. J. Physiol.*, **243**, F519–F531.

Iizuka, K., Ikebe, M., Somlyo, A. V. and Somlyo, A. P. (1994) Introduction of high molecular weight (IgG) proteins into receptor coupled, permeabilized smooth muscle. *Cell. Calcium*, **16**, 431–445.

Ito, Y., Mizuno, Y., Aoyama, M., Kume, H. and Yamaki, K. (2000) CFTR-Mediated anion conductance regulates $Na^+$–$K^+$-pump activity in Calu-3 human airway cells. *Biochem. Biophys. Res. Commun.*, **274**, 230–235.

Kim, K. J. and Crandall, E. D. (1982a) Effects of exposure to acid on alveolar epithelial water and solute transport. *J. Appl. Physiol.*, **52**, 902–909.

Kim, K. J. and Crandall, E. D. (1982b) Effects of lung inflation on alveolar epithelial solute and water transport properties. *J. Appl. Physiol.*, **52**, 1498–1505.

Kim, K. J. and Crandall, E. D. (1983) Heteropore populations of bullfrog alveolar epithelium. *J. Appl. Physiol.*, **54**, 140–146.

Kim, K. J., Critz, A. M. and Crandall, E. D. (1979) Transport of water and solutes across sheep visceral pleura. *Am. Rev. Respir. Dis.*, **120**, 883–892.

Kim, K. J., LeBon, T. R., Shinbane, J. S. and Crandall, E. D. (1985) Asymmetric [$^{14}$C]albumin transport across bullfrog alveolar epithelium. *J. Appl. Physiol.*, **59**, 1290–1297.

Lewis, S. A., Wills, N. K. and Eaton, D. C. (1978) Basolateral membrane potential of a tight epithelium: ionic diffusion and electrogenic pumps. *J. Membr. Biol.*, **41**, 117–148.

Lim, J. J. and Ussing, H. H. (1982) Analysis of presteady-state $Na^+$ fluxes across the rabbit corneal endothelium. *J. Membr. Biol.*, **65**, 197–204.

Liu, Y., Nusrat, A., Schnell, F. J., Reaves, T. A., Walsh, S., Pochet, M. *et al.* (2000) Human junction adhesion molecule regulates tight junction resealing in epithelia. *J. Cell. Sci.*, **113**, 2363–2374.

McLaughlin, G. E., Kim, K. J., Berg, M. M., Agoris, P., Lubman, R. L. and Crandall, E. D. (1993) Measurement of solute fluxes in isolated rat lungs. *Respir. Physiol.*, **91**, 321–334.

Matsukawa, Y., Lee, V. H., Crandall, E. D. and Kim, K. J. (1997) Size-dependent dextran transport across rat alveolar epithelial cell monolayers. *J. Pharm. Sci.*, **86**, 305–309.

Mitic, L. L., Van Itallie, C. M. and Anderson, J. M. (2000) Molecular physiology and pathophysiology of tight junctions I. Tight junction structure and function: lessons from mutant animals and proteins. *Am. J. Physiol.*, **279**, G250–G254.

Ostedgaard, L. S., Shasby, D. M. and Welsh, M. J. (1992) Staphylococcus aureus alpha-toxin permeabilizes the basolateral membrane of a $Cl^-$secreting epithelium. *Am. J. Physiol.*, **263**, L104–L112.

Reddy, M. M. and Quinton, P. M. (1994) Rapid regulation of electrolyte absorption in sweat duct. *J. Membr. Biol.*, **140**, 57–67.

Russo, M. J., Bayley, H. and Toner, M. (1997) Reversible permeabilization of plasma membranes with an engineered switchable pore. *Nat. Biotechnol.*, **15**, 278–282.

Schultz, S. G. and Frizzell, R. A. (1976) Ionic permeability of epithelial tissues. *Biochim. Biophys. Acta*, **443**, 181–189.

Schultz, S. G. (1979) Application of equivalent electrical circuit models to study of sodium transport across epithelial tissues. *Fed. Proc.*, **38**, 2024–2029.

Schultz, S. G. (1980) *Basic Principles of Membrane Transport*. Cambridge, UK: Cambridge University Press.

Schulz, I. (1990) Permeabilizing cells: some methods and applications for the study of intracellular processes. *Methods Enzymol.*, **192**, 280–300.

Schultz, S. G., Andreoli, T. E., Brown, A. M., Fambrough, D. M., Hoffman, J. F. and Welsh, M. J. (eds.) (1996) *Molecular Biology of Membrane Transport Disorders*. New York, NY: Plenum Press.

Schultz, S. G. (1998) A century of (epithelial) transport physiology: from vitalism to molecular cloning. *Am. J. Physiol.*, **274**, C13–C23.

Shi, X. P. and Candia, O. A. (1995) Active sodium and chloride transport across the isolated rabbit conjunctiva. *Curr. Eye. Res.*, **14**, 927–935.

Siegel, B. and Civan, M. M. (1976) Aldosterone and insulin effects on driving force of $Na^+$ pump in toad bladder. *Am. J. Physiol.*, **230**, 1603–1608.

Sten-Knudsen, O. and Ussing, H. H. (1981) The flux ratio equation under nonstationary conditions. *J. Membr. Biol.*, **63**, 233–242.

Stoddard, J. S. and Reuss, L. (1988) Voltage- and time dependence of apical membrane conductance during current clamp in Necturus gallbladder epithelium. *J. Membr. Biol.*, **103**, 191–204.

Tabcharani, J. A., Boucher, A., Eng, J. W. and Hanrahan, J. W. (1994) Regulation of an inwardly rectifying K channel in the T84 epithelial cell line by calcium, nucleotides and kinases. *J. Membr. Biol.*, **142**, 255–266.

Tsukita, S. and Furuse, M. (2000) Pores in the wall: claudins constitute tight junction strands containing aqueous pores. *J. Cell. Biol.*, **149**, 13–16.

Ussing, H. H. (1965) Transport of electrolytes and water across epithelia. *Harvey. Lect.*, **59**, 1–30.

Ussing, H. H. (1969) The interpretation of tracer fluxes in terms of membrane structure. *Q. Rev. Biophys.*, **1**, 365–376.

Ussing, H. H. (1971) A discussion on active transport of salts and water in living tissues. Introductory remarks. *Philos. Trans. R. Soc. Lond. B. Biol. Sci.*, **262**, 85–90.

Ussing, H. H. (1978) Physiology of transport regulation. *J. Membr. Biol.*, **40**, 5–14.

Ussing, H. H. (1980a) Epithelial transport of water and electrolytes. *Contrib. Nephrol.*, **21**, 15–20.

Ussing, H. H. (1980b) Life with tracers. *Annu. Rev. Physiol.*, **42**, 1–16.

Ussing, H. H. (1988) The development of the concept of active transport. *Prog. Clin. Biol. Res.*, 3–16.

Ussing, H. H. (1994) Does active transport exist? *J. Membr. Biol.*, **137**, 91–98.

Ussing, H. H. and Zerahn, K. (1999) Active transport of sodium as the source of electric current in the short-circuited isolated frog skin. (Reprinted from *Acta. Physiol. Scand.*, **23**, 110–127, 1951). *J. Am. Soc. Nephrol.*, **10**, 2056–2065.

Vasilets, L. A., Schmalzing, G., Madefessel, K., Haase, W. and Schwarz, W. (1990) Activation of protein kinase C by phorbol ester induces downregulation of the $Na^+/K^+$-ATPase in oocytes of Xenopus laevis. *J. Membr. Biol.*, **118**, 131–142.

Wills, N. K., Eaton, D. C., Lewis, S. A. and Ifshin, M. S. (1979a) Current-voltage relationship of the basolateral membrane of a tight epithelium. *Biochim. Biophys. Acta.*, **555**, 519–523.

Wills, N. K., Lewis, S. A. and Eaton, D. C. (1979b) Active and passive properties of rabbit descending colon: a microelectrode and nystatin study. *J. Membr. Biol.*, **45**, 81–108.

Wills, N. K., Reuss, L. and Lewis, S. A. (eds.) (1996) *Epithelial Transport: A Guide to Methods and Experimental Analysis*. New York, NY: Chapman & Hall.

Yonath, J. and Civan, M. M. (1971) Determination of the driving force of the $Na^+$ pump in toad bladder by means of vasopressin. *J. Membr. Biol.*, **5**, 366–385.

# Chapter 4

# Characterization of transport over epithelial barriers

*Josef J. Tukker*

## INTRODUCTION

A compound administered for systemic pharmacological effect, released in the gastrointestinal (GI) tract from the dosage or delivery form, will have to pass a series of barriers before it reaches the draining blood vessels and the systemic circulation (Figure 4.1).

For most compounds, in this series of barriers the luminal membrane of the enterocyte is the most predominant resistance in uptake. The transport of a compound over a cellular barrier is dependent on the physicochemical characteristics of both the compound and the cellular barrier. For the majority of therapeutically active compounds their properties are such that they are well absorbed after oral intake, or, if another route of administration is chosen like pulmonal, uptake into the systemic circulation is not hampered. For these compounds the physicochemical parameters are favorable for absorption.

In our daily nutrition, however, the majority of compounds do not follow the simple rule of passive diffusion: the molecular size is too large, or the charge and/ or lipophilicity of the molecules are unfavorable. Among these food components we find carbohydrates (after hydrolysis to glucose), proteins (after hydrolysis to small peptides or amino acids), vitamins, ions like $Ca^{++}$ and phosphate. Still, these compounds are also – sometimes even extremely – well-absorbed from the GI

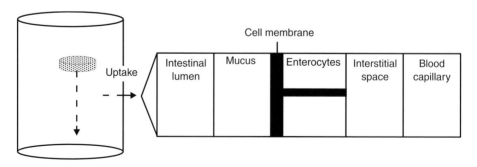

*Figure 4.1* After a drug is released from its dosage form in the gastrointestinal (GI) tract, it has to pass a series of barriers before it reaches the outflowing blood capillaries.

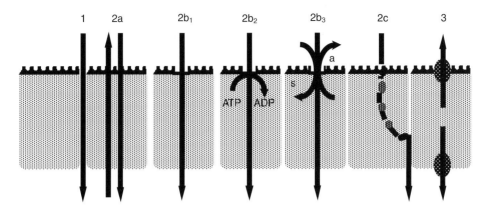

*Figure 4.2* Transport mechanisms in epithelial barriers, like the gastrointestinal (GI) tract. The various numbers denote the different mechanisms: I paracellular transport; 2a transcellular passive diffusion; 2b carrier-mediated transport: $b_1$ facilitated transport, $b_2$ primary active transport, $b_3$ secondary active transport with sym- or antiport (s or a, resp.); 2c (receptor-mediated) endocytosis and pinocytosis; 3 efflux. respectively.

tract. Apparently, not only simple physicochemical parameters rule the distribution of a given compound between the luminal content and the lining enterocytes, but some other processes play an additional role in the transport of compounds over these cellular barriers.

This chapter will focus on the various processes by which compounds can be translocated over cellular barriers. It will focus on the cellular wall of the GI tract, because this wall is a good example for the other barriers between the body and the outside environment: transport routes present in other epithelial barriers are all present in the intestines (though it should be remembered that not all transport routes present in the GI tract are present in other epithelial tissues).

Generally speaking, there are four different ways a compound can be translocated over a cellular barrier, like in the GI-tract. One should keep in mind that the GI tract, like many other epithelia, is covered with a single layer of absorptive cells (in the GI-tract these are called enterocytes); in this layer the cells do not close the layer completely, but leave in between every two cells a small gap, the 'tight junctions'. This space can be different in other epthelia but is typically approximately 1 nm in the GI wall (Macheras *et al.*, 1995).

The different routes of transport in this barrier are depicted in Figure 4.2. From left to right the routes are: the paracellular uptake and transcellular diffusion and carrier-mediated uptake, (receptor-mediated) endocytosis, and carrier-mediated efflux. It should be emphasized that the overall 'absorption' of a given compound over an epithelial barrier is the sum of all subprocesses playing a part in the whole mechanism of uptake.

In this chapter these various routes and mechanisms of uptake will be described and discussed.

## PASSIVE DIFFUSION

A molecule can traverse the epithelial layer by two routes: through the enterocyte (transcellular) and in between the cells (paracellular). Which route will be taken by the compound depends on the physicochemical properties of the permeant, and the properties of the cellular barrier. In principle, the paracellular barrier can be used by all solutes, but is limited by the diameter or size of the compound and the radius of the pore: this radius is estimated between 0.5 and 5 nm (Macheras *et al.*, 1995). This estimate indicates that compounds with a diameter larger than this radius are not likely to follow this route. Indeed, it has been suggested that small molecules (molecular weight ≤ 300 D) like caffeine and atenolol (molecular weight 194 and 266 respectively) can be absorbed via the paracellular route. Using a very water-soluble test compound like polyethylene glycol (PEG) it was shown that the small molecules were significantly absorbed, and a limit of barrier absorption could be set at around 300 D (Artursson *et al.*, 1993). These compounds take this route since they are too hydrophilic to distribute into the lipophilic membranes, which is imperative for the transcellular passive route. A limited group of compounds not withstanding, this route is believed not to play a major role in drug absorption. It has, however, been the subject of many investigations during the past decades where improvement of paracellular transport has been studied, especially for the uptake of small peptides and proteins (de Boer, 1994). An important consequence of this sieve function of the tight junction is the use of paracellular markers like mannitol and fluorescein in *in vitro* transport experiments. In other chapters these compounds will be described and shown to be useful markers for tissue integrity.

But compounds having the proper lipophilic characteristics can traverse over the cellular membranes. These limits for good characteristics are given by the 'rule of 5' (Lipinski *et al.*, 1997) and are approximated by a limit of 15 H-bonds, a molecular weight of approximately 500 D and a log P of approximately 5. Compounds with more hydrophilic properties (H-bonds) or higher lipophilicity (log P) are not likely to be absorbed at a high extent (this rule explicitly excludes substrates for carrier-mediated transport, which will be discussed later).

The process of passive diffusion includes partition of the permeant between the contents of the GI lumen and the lipid bilayer of the enterocyte's wall. A second distribution process takes place at the inner side of the membrane where the permeant again is subject to partition between the bilayer and the cellular contents. The permeant is distributed throughout the enterocyte quickly, presumably by convection, and reaches the basal (serosal) membrane where the same process starts before the compound reaches the capillary blood supply and drainage occurs in the lamina propria.

Overall, the passive absorption process is predominantly governed by the concentration gradients of the permeant existing over the two luminal and basal membranes and might in most cases be well described by Fick's first law of diffusion:

$$J = \frac{D \times A \times K \times C}{h}, \tag{4.1a}$$

where $J$ is the rate of appearance (in mol/hour) of the permeant at the site of absorption; $D$ is the effective diffusion coefficient of the permeant; $A$ is the area available for passive diffusion; $K$ is the apparent distribution coefficient; $C$ is the concentration at luminal side; and $h$ is the thickness of the membrane. The concentration $C$ which is taken as the driving force for permeation is normally taken as the bulk concentration of the permeant. In fact, this should be the concentration at the tissue/solution interface, and this concentration can be much lower than the bulk concentration owing to the stagnant water layer. This layer is very much influenced by the presence of mucus and/or the application of stirring in the model applied. Especially for rather lipophilic compounds, the influence of the stagnant water layer, and thus the difference between bulk and surface concentration, can be large.

Essentially, the rate of appearance $J$ is not dependent on the luminal (or surface) concentration, but on the concentration gradient, or the difference between luminal and serosal concentrations. For most permeants, however, sink conditions are provided since the concentration in the bloodstream is virtually zero ('sink' compared with the luminal and intracellular concentration) because of a rapid drainage from the absorption site. Also, in most permeation models *in vitro* the concentration of the permeant at the acceptor site will be negligible compared with the donor site.

The 'constants' in equation (4.1a) are applied in most equations to *in vitro* permeability testing combined and replaced by the intrinsic membrane permeability constant $P_m$ (unit cm/s); this constant is typical for the given permeant in the given membrane. This constant results in equation:

$$J = P_m \times C. \tag{4.1b}$$

It must be realized that most drugs are weak electrolytes in aqueous solution existing in at least two species, the unionized and the ionized form(s). The ratio between the two species is dependent on both the dissociation constant of the permeant ($pK_a$) and on the ambient pH. The unionized species will show a (much) higher distribution coefficient $K$ and as a consequence permeate at a faster rate than the ionized species. Thus, the intestinal absorption of a drug that is a weak electrolyte will be enhanced as the pH at the site of absorption favors the formation of the unionized fraction. This phenomenon is called the *pH-partition theory* and results, in theory, in faster uptake of salicylic acid from the stomach and a weak base like ketoconazol from the intestines.

In addition to the lipophilicity of the permeant, one more parameter is extremely important and influential in the rate of absorption: the area (A in equation 4.1a). The small intestine looks macroscopically like a long tube. With the naked eye one can see folds (the folds of Kerckring) in the inner wall, but when a magnifying glass is used even further anatomical differences can be observed: the folds are again covered with *villi*, fingershape structures covered with a single layer of cells, the enterocytes. Under a microscope even the presence of numerous microvilli can be shown, covering the enterocytes at the luminal site. These anatomical features cause a huge enlargement of the area for absorption. It has been estimated that the anatomical area available for absorption in the small intestine is increased approximately 600-fold (Wilson, 1962). This enlarged area for absorption in the intestines causes that the absorption of a weak acid, which should be faster in the stomach compared with the intestine, is in fact still faster in the small intestine. It should be emphasized, however, that the estimated anatomical area for absorption does not lead to an effective 600-fold increase in absorption, since this estimation also takes the crypts into account, and hardly any compound will diffuse that deep into this area before getting absorbed.

## CARRIER-MEDIATED TRANSPORT

Many compounds which are part of the normal daily food intake are not able to be absorbed at all, or only to a very low extent, if absorption would be by a passive mechanism and only driven by the concentration gradient and physicochemical characteristics of the compound. Some compounds are present in much too low concentration in the food, and/or are much too hydrophilic to pass the intestinal barrier. Nevertheless, these essential compounds are translocated over the intestinal barrier very effectively by carrier-mediated processes.

Several intestinal transporters are present in the membrane, most of them in the luminal part of the cell membrane and only a few, until now, proved to be present in the serosal membrane. The serosal membrane is supposed to be more leaky or permeable compared with the efficient lipophilic barrier of the luminal section. These carriers are transmembrane proteins with several membrane spanning domains. For example, the peptide carrier (in the intestine called PepT1) is a large protein consisting of 710 amino acids, with 12 membrane spanning domains.

How carriers really work is not yet clear in all cases. Several mechanisms have been hypothesized, such as a vacuum cleaner-like mechanism, or a flip-flop mechanism, but no clear mechanism until now has been postulated or proven for all carriers.

Carrier-mediated transport can be divided roughly into two groups: active transport and facilitated transport. In both types of transport, a membrane-bound carrier is involved. Active transport is essentially characterized by its energy-demanding process, where the energy is delivered by a secondary concomitant transfer of another species downhill (in most cases $Na^+$ or $H^+$) where the gradient in the enterocyte is maintained, for $Na^+$ for instance by the $Na^+/K^+$-ATPase. A $H^+$-gradient is also present in the intestine *in vivo*, since in the enterocyte $H^+$ is exchanged for $Na^+$ by the $Na^+$-$H^+$ exchanger, present in the brush-border membrane, leading to entry of $Na^+$ into and the exit of $H^+$ out of the cell. This results in a microclimate close to the brush-border of the cell, where the pH is 5.5–6 in contrast with the inner pH of 7.2.

Indeed, such an energy supplying $H^+$-gradient is imperative for the continuously active process of the peptide transporter, as will be shown in later sections. Absence of the supporting gradient results in complete abolishment of the active translocation.

Facilitated transport is comparable to active transport in the sense that a membrane-bound transporting protein is also involved. In this transport mechanism, however, no primary or secondary energy source is necessary, since the substrates are only transported downhill from higher to lower concentration comparable to passive diffusion (down their chemical gradient). Whether facilitated transport is

Table 4.1  Summary of the most abundantly identified transporters in the brush-border membrane of the enterocyte (Tsuji and Tamai, 1996)

| Substrate | Symported or antiported species |
|---|---|
| Amino acids | $Na^+$ (s) |
| Oligopeptides | $H^+$ (s) |
| D-glucose | $Na^+$ (s) |
| D-fructose | – |
| Lactic acid | $Na^+$ (s) |
| Bile acids | $Na^+$ (s) |
| Short chain fatty acids | $Na^+$ (s) |
| Phosphate | $Na^+$ (s) |
| Monocarboxylic acids | $HCO_3^-$ (a) |
| Nicotinic acid | $Na^+$ (s) |
| Folic acid | $OH^-$ (a) |
| $Na^+$ | $H^+$ (a) |

playing any role in translocating drugs is yet uncertain. In the intestine many compounds originating from food or internal sources are translocated by carriers. These carriers are mostly present in the small intestine, where most of the food uptake occurs. Carriers have been identified for amino acids, oligopeptides (two–three amino acids), bile acids, glucose, and many other compounds. An enumeration of the most abundant intestinal brush-border epithelial transporters are given in Table 4.1, where both the main natural substrate is indicated and the secondary transported species (mostly symport, in the same direction, e.g. into the cell).

If a compound is transported by an active transport system, then this compound can be translocated against its concentration gradient – 'uphill' – across the cell membrane. This is a useful and necessary characteristic for compounds like food constituents that have only a limited intestinal length available for absorption (a so-called 'absorption window'). Most food components are rather hydrophilic and do not obey the 'rule of five', and are thus unfit for good passive absorption. For example, bile acids (not food components but endogenously excreted into the intestines in the bile) do their solubilizing work in the upper small intestine, but once the fats have been hydrolyzed, the bile acids have to be reabsorbed. This reabsorbtion occurs before the colon is reached in the terminal ileum, in a minor part of the intestine with a total length of approximately 20 cm and an estimated residence time of approximately ten minutes. Active translocation can thus be very efficient for typical substrates.

In passive transport, the rate of transport is governed by the law described in equation 4.1a, and is directly proportional to the concentration of the permeant at the site of the highest concentration (in the intestine, mostly at the luminal side). For active transport, this linear proportionality is only true at (very) low concentrations; at higher concentrations the carrier becomes saturated and transport levels off to a maximum. Further increase of concentration at the luminal site has no further influence on the uptake if active transport only is involved (like in the case for bile acids, as they are so hydrophilic that passive permeation is impossible).

At a certain concentration (different for both the different carriers and their substrates), a maximum in transport rate will be reached. The relationship between the carrier-mediated flux and the concentration is described in equation 4.2:

$$J = \frac{J_{max} \times G}{K_m + C_1},$$

(4.2)

where $J$ is the flux over a membrane (mole/time); $J_{max}$ is the flux at infinite concentration; $K_m$ is the Michaelis constant (in M); and $C_1$ the concentration at receptor site (in M). For this bulk concentration, again, the same restrictions are valid as for the concentration ruling the passive diffusion. This formula is referred to as the Michaelis-Menten equation. The Michaelis constant $K_m$ equals the concentration of the substrate when the transport rate is half of the maximum value, i.e. when $J = J_{max}$. In Figure 4.3 the estimated $K_m$ is indicated with an arrow.

Most of the substrates for active transport, however, have a modest molecular weight and a lipophilicity making passive permeation not completely impossible ('rule of 5'). As a consequence, since the substrates have a tendency to permeate passively and are a substrate for active transport, the resulting transport results in the sum of both mechanisms, passive plus active transport. As a result, if somebody is performing an experiment as described in several other chapters, where the compound is translocated from one side of a membrane to the other and the flux of the compound is followed as a function of concentration at the donor site, one can see a dependency, as depicted in Figure 4.3. The overall transport, consisting essentially of a passive and an active component, follows equation 4.3:

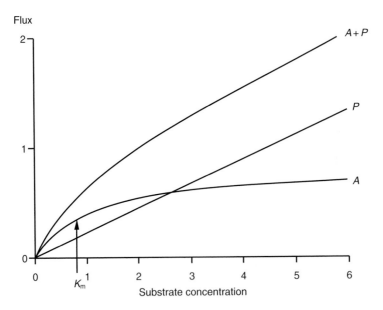

Figure 4.3 In this figure the relation between the flux (appearance of the permeant in the acceptor compartment) and the concentration in the donor phase is depicted for: P, a passive process; A, a carrier-mediated process; and A+P, the combination of the two processes. Most drugs show the latter result. The arrow indicates the approximate value of $K_m$.

*Table 4.2* Characteristics of passive versus carrier-mediated active transport

| Passive | Carrier-mediated/active |
| --- | --- |
| Not-mediated | Carrier-mediated |
| Downhill | Uphill (possible) |
| Not energy-demanding | Energy dependent, e.g. |
| | – $Na^+/K^+$-ATPase |
| | – symport second substrate ($Na^+$ or $H^+$) |
| No saturation | Saturation kinetics |
| No inhibition | Inhibition between substrates |
| Not stereospecific | Stereospecific |
| Not polarized | Polarized |

$$J = \frac{J_{max} \times C_1}{K_m + C_1 + P_m \times C_1}. \tag{4.3}$$

The overall flux at all concentrations is the sum of the flux resulting from the passive permeation based on equation 4.1b and from the active translocation process, described by equation 4.2. This can be seen clearly in Figure 4.3, which depicts the flux increasing with concentration for a separate carrier-mediated ($A$) and passive process ($P$), and in the case both processes play a role in the translocation of a permeant ($A+P$). The active part can be experimentally missed if the $K_m$ has a very low value or only high concentrations of the compound are used.

There are several different characteristics between passive and carrier-mediated, active transport (See Table 4.2). If these differences are taken properly into account in an experimental set-up, it is possible to differentiate between the two transport mechanisms.

A large series of therapeutically active compounds is translocated by intestinal carriers. For most compounds transported by carriers a structure-transport relationship can be established. In Table 4.3 several intestinal carriers are given and characterized by their main natural substrates, and a choice of their artificial substrates is indicated. It should be emphasized that these compounds are orally administered, and if their passive component only were taken into account, the bioavailaibility would have been much lower than with the contribution of active transport.

## CARRIER-MEDIATED EFFLUX

There is clear evidence that a variety of substrates can be transported out of some cells by specialized transport systems belonging to the multidrug resistance family. Several transporters have been identified, such as *p*-glycoprotein (*p*-gp), an efflux transporter which is abundant in barrier tissues like the blood–brain barrier, kidney, liver and intestines (Thiebaut *et al.*, 1987). It was first discovered in cancer cells, where *p*-gp contributes extensively to the resistance of these cells to chemotherapeutic compounds by transporting them out of these cells.

Table 4.3 Examples of intestinal transporters with some of
their commonly used therapeutically active substrates

| Natural substrates | Drug substrates |
| --- | --- |
| Amino acids | Gabapentin |
|  | L-α-methyldopa |
|  | L-dopa |
| Di- and tri-peptides | (several) Cephalosporins |
|  | (several) Aminopenicillins |
|  | (several) ACE-inhibitors |
|  | Bestatin |
| Phosphate | Foscarnet |
|  | Fosfomycin |
| Folic acid | Methotrexate |
| Monocarboxylic acid | Salicylic acid |
|  | Valproic acid |
|  | (several) Statins |

But due to its presence in normal, non-cancerous barrier tissue this transporter plays an important role in the limited oral bioavailability of a large variety of drugs. In the small intestine $p$-gp is present in the apical membrane of the enterocytes, pumping back digoxin, erythromycin, several $\beta$-blockers, antibiotics, cyclosporin B and cytostatics. In later chapters it can be seen that the counter-transport (transport back to the apical side) of $p$-gp can be significant for some compounds.

The transport by the efflux-carriers like $p$-gp can be described by the same equation as for the absorptive carriers (equation 4.2), but now in the opposite direction. This transport was determined for several compounds in various tissues as well as in cultured cells (Makhey et al., 1998). Like other active transport mechanisms, the efflux mechanisms are also saturable in transporting their substrates, and the transport of one substrate can be inhibited by other substrates. Moreover, efflux or carrier-mediated counter-transport, can be shown when transport is studied in both directions over a tissue. If polarization in transport becomes visible, then the presence of carrier-mediated transport can be hypothesized, but has to be proven with proper experiments like saturation and/or inhibition of transport.

## PINOCYTOSIS (PARTICULATE UPTAKE)

In the pinocytotic process, the species that is taken up into the enterocyte or cell as a solid particle or as a large molecule is much larger than the 'rule of five' gives as a limit for passive uptake. The process is comparable to phagocytosis, and the particle is taken up into the cell into vacuoles which are transported across the cell. This process is presumably of little importance for small molecules but might contribute significantly to the uptake process of macro-molecules.

## CARRIER-MEDIATED ENDOCYTOSIS

Comparable to the process of pinocytosis where large molecules and particulates are engulfed by the plasma membrane, some specific compounds follow a comparable mechanism in uptake. For example, cyanocobalamin combines in the stomach with a protein, intrinsic factor, and then interacts with a receptor on the wall of the enterocyte, resulting in vacuole formation and the subsequent uptake of the combination. There is strong evidence that this mechanism is limited for only a few compounds, e.g. cyanocobalamin and iron.

## CONCLUSION

A variety of processes may play a role in the uptake of a compound over an epithelial barrier. As stated before, in the overall uptake process of one particular compound, several mechanisms can be a part. A compound can be transported from the luminal to the serosal side by an active carrier-mediated process, but almost always this goes together with a passive diffusional process. But a specific compound can be a substrate for more than one carrier as well, making the mathematical treatment of the transport data quite complicated. If the same compound is also a substrate for an efflux transporter, the overall process will consist of all three processes, two in one direction, one in the opposite, leading to an even more complicated kinetic mechanism.

## REFERENCES

Artursson, P., Ungell, A.-L. and Löfroth, J.-E. (1993) Selective paracellular permeability in two models of intestinal absorption: cultured monolayers of human intestinal epithelial cells and rat intestinal segments. *Pharm. Res.*, **10**, 1123–1129.

de Boer, A. G. (1994) *Drug Enhancement: Absorption Concepts, Limitations and Trends*. London: Harwood Press.

Lipinski, C. A., Lombardo, F., Dominy, B. W. and Feeney, P. J. (1997) Experimental and computational approaches to estimate solubility and permeability in drug discovery and development settings. *Adv. Drug Del. Rev.*, **23**, 3–25.

Macheras, P., Reppas, C. and Dressman, J. B. (1995) *Biopharmaceutics of Orally Administered Drugs*. London: Ellis Horwood.

Makhey, V. D., Guo, A., Norris, D. A., Hu, P., Yan, J. and Sinko P. J. (1998) Characterization of the regional intestinal kinetics of drug efflux in rat and human intestine and in Caco-2 cells. *Pharm. Res.*, **15**, 1160–1167.

Thiebaut, F., Tsuruo, T., Hamada, H., Gottesman M. M., Pastan, I. and Willingham, M. C. (1987) Cellular localization of the multidrug-resistance gene product P-glycoprotein in normal human tissue. *Proc. Natl. Acad. Sci. USA*, **84**, 7735–7738.

Tsuji, A. and Tamai, I. (1996) Carrier-mediated intestinal transport of drugs. *Pharm. Res.*, **13**, 963–977.

Wilson, T. H. (1962) *Intestinal Absorption*. Philadelphia: Saunders.

# Chapter 5

# Studying cellular binding and uptake of bioadhesive lectins

*Michael Wirth, Carsten Kneuer, Claus-Michael Lehr and Franz Gabor*

## INTRODUCTION

A major challenge confronting pharmaceutical technologists today is the successful delivery of drugs, especially in view of therapeutically active peptides, proteins and other macromolecular agents such as oligonucleotides or gene vectors. Due to hostile environments, conformational and structural changes or proteolytic degradation can inactivate proteinaceous drugs before they reach the site of absorption. Additionally, the large spatial extension and hydrophilicity of biopharmaceuticals restrict their passage of biological barriers. Altogether, these complications result in poor bioavailability of biopharmaceuticals. One promising approach to improve the bioavailability of drugs is to use bioadhesive drug delivery systems (Lehr, 1994).

### Non-specific adhesion – mucoadhesion

About 15 years ago, bioadhesion, a basic interaction well-known from nature, attracted scientific interest for drug delivery (Gurny and Junginger, 1990). As most of the absorptive epithelia are covered by a mucous layer, research initially focused on mucoadhesive polymers such as polyacrylic acid and derivatives of chitosan. Formation of bioadhesive bonds between the viscoelastic mucous layer and the drug delivery system implies a wetting and swelling of the polymeric matrix followed by interpenetration of polymer chains and mucus, which results in the formation of weak chemical bonds between the entangled chains (Chickering and Mathiowitz, 1999). The intimate contact to the underlying mucosal tissue increases the local concentration gradient from the lumen to the absorptive epithelium, resulting in enhanced passive diffusion of the drug incorporated within the delivery system. The therapeutic effect of the drug delivery system is expected to be supported by prolongation of the residence time at the site of absorption. Besides these classic characteristics of mucoadhesiva, polymers of the polycarboxylate type possess calcium-chelating potency, which, on the one hand restricts proteolytic degradation of peptide and protein drugs by inhibition of certain metalloproteases. On the other hand, loosening of tight junctions by depletion of extracellular calcium opens the paracellular pathway for the absorption of drugs. Though feasibility of the concept of mucoadhesion in practice led to commercially available formulations, some questions associated with peroral administration still remain to be answered. First, the relatively fast turnover of the mucus is contradictory to prolonged fixation at the absorptive mucosal tissue. Second, it is unclear to date to what extent the shed off mucus and foodstuff causes premature inactivation of mucoadhesiva (Lehr, 1994).

## Specific adhesion – glycotargeting

Whereas the mucus turnover rate is about 1–4 hours (Lehr et al., 1991), the renewal of epithelial cells occurs within 1–3 days. Out of this reason, it seems to be advantageous for drug delivery purposes to focus upon the epithelial cell membrane itself representing the main barrier to drug absorption.

At the beginning of the last century, a red ring surrounding epithelial cells was observed after staining with ruthenium red. This carbohydrate layer encircling each mammalian cell – the so-called glycocalyx – is predominantly built up of oligosaccharide moieties of proteoglycans, glycolipids and glycoproteins anchored in the lipid bilayer of the cell membrane. Similar to bacterial adhesion in the gut provided by fimbrial proteins (Gabor et al., 1997a), adhesion of bioadhesin-drug conjugates or bioadhesin-modified colloidal drug delivery systems directly to the membrane of absorptive enterocytes promises all the advantages of classic adhesion, but is expected to be more or less independent from mucus turnover. Additionally, the mucus gel may not present a continuous layer and therefore access to the epithelial cell surface at certain sites may be facilitated (Lehr, 1994). According to the literature, the glycosylation pattern of the glycocalyx exhibits site and species related variations, probably enabling site specific targeting to enterocytes or intestinal M cells for drug and vaccine delivery, respectively (Jepson et al., 1995, 1996; Sharma et al., 1996).

## Lectins – cytoadhesion and cytoinvasion

Bioadhesins capable of binding to the glycocalyx of the gut originate either from bacteria such as *Escherichia coli*, *Salmonella typhimurium* and *Yersinia* species or from plants (Singh et al., 1999). According to Peumans and Van Damme, plant lectins are proteins possessing at least one non-catalytic domain, which bind reversibly to a specific mono- or oligosaccharide (Peumans and Van Damme, 1995). Usually the term 'lectin' is associated with the highly toxic *Ricinus communis* agglutinin, but there is also a high number of plant lectins generally regarded as safe. Many plant lectins, but by far not all, are consumed as a part of the western diet, which might indicate their safety as cytoadhesive agents.

Among the dietary lectins, tomato lectin (TL) was the first shown to be a candidate for lectin-mediated bioadhesion. TL was found to be not only bound to rat's small intestine, Caco-2 monolayers as well as porcine enterocytes, but also to be transported across the intestinal epithelium (Kilpatrick et al., 1985; Lehr et al., 1992; Naisbett and Woodley, 1994a,b). After oral administration to humans, peanut agglutinin (PNA) was also detected in peripheral venous blood confirming transcytosis of intact lectin (Wang et al., 1998). Besides carbohydrate specific binding of wheat germ agglutinin (WGA) and potato lectin to Caco-2 cells, human enterocytes and other intestinal cell lines also uptake into Caco-2 cells was assessed

(Gabor *et al.*, 1997b, 1998; Wirth *et al.*, 1998b). There is also evidence that the blood–brain barrier might be surmounted by proteins after conjugation of lectins, especially WGA (Broadwell *et al.*, 1988). Since uptake of *Ulex europaeus* isoagglutinin-I (UEA-I) by murine M cells was demonstrated, this lectin seems to be an appealing candidate for vaccination (Clark *et al.*, 1995). Besides adhesion to epithelial linings of the gastrointestine, some lectins were also found to interact with mucosal tissues of the upper respiratory tract, the oral cavity and the eye, introducing alternative routes for administration (Clark *et al.*, 2000).

To put the concept of lectin-mediated drug delivery into practice, two approaches are pursued to date: the prodrug-concept yielding soluble lectin-drug conjugates, and surface modified carrier systems. In an earlier publication some bacterial and plant lectins were used as hapten-carriers for oral immunization demonstrating high immunogenicity in mice (De Aizpurua, 1988). An acid-labile conjugate of doxorubicin and WGA was proposed for colon cancer therapy, as it exhibited a higher binding rate to cancer cells and a efficacy after uptake into the target cells (Wirth *et al.*, 1998a).

Much more work has been done on lectin-modified particulate drug delivery systems, as this approach combines targeted cytoadhesion/cytoinvasion with the advantages of particulate drug delivery. The higher drug payload might facilitate reaching therapeutical levels, which is supported by protection of the drug against hostile luminal environment (Ponchel and Irache, 1998). Fundamental knowledge in this area was gained by experiments with TL-coated nanoparticles. Despite considerable binding of TL to *N*-acetyl-D-glucosamine containing residues of pig gastric mucin, the uptake rate of TL-coated nanospheres was found to be 20 times higher than that of naked nanospheres, providing for improved absorption of drug to be incorporated in the future (Irache *et al.*, 1994a,b; Florence *et al.*, 1995; Hussain *et al.*, 1997).

A lot of work on lectins as bioadhesive agents has also concentrated on *Phaseolus vulgaris* agglutinin (PHA) from red kidney beans, especially in view of microparticulate formulations (Lehr and Pusztai, 1995). *In vivo* studies show significant retardation of PHA-coupled 2 µm microspheres in the small intestine of rats compared with controls coated with lectins of different sugar specificity. At high concentrations, however, PHA induces morphological and metabolic changes in enterocytes. Although these toxic effects are fully reversible and probably unnoticeable at concentrations required for bioadhesive formulations, it is unlikely that PHA will ever be used in drug delivery formulations (Bardocz *et al.*, 1995). Upon incubation of Caco-2 monolayers with fluorescein-loaded WGA-coated microspheres, higher cellular uptake of the model drug occurs, as compared with the sole solution probably due to intimate contact to the artificial tissue (Ertl *et al.*, 2000).

According to these observations, some lectins can mediate uptake of conjugated drugs or conjugated colloidal carriers into the cell. But at present, much effort will be continually necessary to elucidate the basic mechanisms of lectin-mediated cytoadhesion and cytoinvasion prior to practical utilization of lectins as 'shoehorns' for drug-delivery to improve poor bioavailability of drugs. This chapter is intended to give a short overview of the methods established in our laboratories that have allowed a deeper insight into the basic interactions between lectins and cells.

## SPECIFIC APPLICATIONS

Peroral administration of drugs is the most comfortable way for the patient, offering a lot of advantages compared with injection: peroral administration is painless and simple for the patient, and it is cheap as there is no need for trained personnel. As one of the main tasks of the alimentary channel is to absorb nutrients, drugs are processed like foodstuff resulting in their degradation by enzymes and acid as well as dilution and dispersion within the gut. Additionally, the residence time of drugs in the small intestine is rather short, being 2–3 hours before clearance into the large bowel.

The small intestine is the main target for absorption of drugs regardless of problems associated with the poor bioavailability of drugs. Consequently, we tried to select an appropriate model for the human intestinal epithelium to illustrate the lectin–tissue interaction. Although perfusion models are nearest to *in vivo* conditions, there is less possibility to gain information about events at the cellular level. In our opinion, the Caco-2 model is well-suited since the lectin–cell interaction can be studied using single cells as well as artificial tissue. Additionally, the Caco-2 model offers some advantages such as human origin, uniformity of the cells, polarization and tightness of the artificial tissue, and comparability of results (for a comprehensive review see Chapter 10). There are, however, limitations inherent to the Caco-2 model restraining transfer of results to *in vivo* conditions. In order to overcome the lack of mucus production, which is supposed to greatly influence the lectin-concept, great efforts are made to establish mucus-producing Caco-2 monolayers by co-culture (Artursson and Borchardt, 1997; Walter *et al.*, 1996). Especially in view of potential benefits of the lectin-concept for vaccination, inclusion of follicle-associated epithelial structures would be desirable (Lavelle, 2000).

All the methods presented rely on fluorescence techniques, since rather cheap equipment is required and no far-reaching safety regulations have to be observed, compared with assays using radiolabeled compounds. The detection limits of flow cytometry in case of assaying single cells, or microplate fluorescence readers in case of assaying monolayers, were found to be sufficient at this level of investigation. Additionally, the rapid progress in confocal laser scanning microscopy (CLSM) allows a view of the cellular distribution of fluorescein-labeled lectins (F-lectins) omitting any complicated and probably denaturating sample preparation techniques. Experiments with single cells were used to estimate the binding rate of different lectins, but it should be kept in mind that the overall surface was higher compared to monolayers. Flow cytometry is a quite simple technique and permits elucidation of the interaction between the lectin and a certain cell, as well as viability of this cell according to its position in the histogram. Due to rapid analysis of up to 200 cells per second and measurement at the single cell level, results with high statistical significance are acquired. However, a drawback of this technique is the rather difficult calibration with a given lectin as cell- or particle-bound fluorescence intensity is determined exclusively. Consequently, the results are only valid in comparison with the fluorescence intensity of other cell-associated fluorescent-labeled compounds, but they are well-suited to follow the cell-binding and intracellular fate of a certain lectins.

In contrast to single cells, Caco-2 monolayers represent a long step towards *in vivo* conditions. Due to polarization and differentiation of the tightly packed cells, artificial tissues exhibit major advantages for studying cellular binding and uptake of lectins. As Caco-2 cells form confluent monolayers in wells of microplates, on filters and cover glass slides a wide variety of method is offered for investigations. Whereas monolayers on impermeable plastic surfaces are appropriate for studying lectin-binding and cellular uptake including the influence of inhibitors, filter-grown monolayers are used to simulate absorption. The amount of cell-associated-labeled lectin is easily detected by fluorescence microplate readers using calibration curves for quantification. CLSM of cover glass slide-grown monolayers offers a fascinating look into both the cells and the intracellular distribution of the labeled lectin, the latter of which can be confirmed by co-localization experiments.

The tools and methods described below seem to be appropriate for studying the interaction between single cell or monolayers originating from different sites in the body or cell lines and F-lectins. Moreover, the interactions not only with bio-adhesins but also with any other fluorescent-labeled compound can be elucidated.

## MATERIALS

*Lectins*:

- Wheat Germ Agglutinin (WGA) – fluorescein-labeled, VECTOR Laboratories (Burlingame, USA); biotinylated, VECTOR Laboratories (Burlingame, USA)
- Ulex Europaeus Isoagglutinin-I (UEA-I) – fluorescein-labeled, VECTOR Laboratories (Burlingame, USA)
- *Solanum tuberosum* lectin (STL) – fluorescein-labeled, VECTOR Laboratories (Burlingame, USA)
- Peanut lectin (PNA) – fluorescein-labeled, VECTOR Laboratories (Burlingame, USA)
- Lens culinaris agglutinin (LCA) – fluorescein-labeled, VECTOR Laboratories (Burlingame, USA)
- Dolichos biflorus agglutinin (DBA) – fluorescein-labeled, VECTOR Laboratories (Burlingame, USA).

*Other chemicals*:

- α-lactalbumin – fluorescein-labeled, MOLECULAR PROBES (Leiden, NL)
- dextran – fluorescein-labeled, average molecular weight 21,200, SIGMA (St. Louis, MO, USA)
- tunicamycin – from *Streptomyces* sp., SIGMA (St. Louis, MO, USA)
- N,N',N''-triacetylchitotriose – FLUKA (Buchs, CH)
- avidin – fluorescein-labeled, SIGMA (St. Louis, MO, USA)
- monensin – sodium salt, SIGMA (St. Louis, MO, USA)
- buffer substances – MERCK (Darmstadt, G).

*Cell culture*:

- Caco-2 cells – passage number 25–40, American Type Culture Collection (Rockville, ML, USA)
- RPMI 1640 *cell culture medium* – SIGMA (St. Louis, MO, USA)

- L-glutamine – MERCK (Darmstadt, G)
- gentamicin sulfate – CALBIOCHEM (La Jolla, CA, USA)
- fetal bovine serum – sterile filtered, cell culture tested; SIGMA (St. Louis, MO, USA)
- trypsin-ethylenediaminetetraacetic acid (EDTA) solution – 0.25 per cent, sterile filtered, cell culture tested; SIGMA (St. Louis, MO, USA).

*Cell culture materials*:

- cell culture flasks – $75 \, cm^2$, canted neck, standard cap, sterile; CORNING COSTAR Corporation (Cambridge, UK)
- microplates – 96-well, tissue-culture-treated (TC), sterile; GREINER (Kremsmünster, A)
- pipettes – 1, 2, 5 and 10 ml, sterile; GREINER (Kremsmünster, A)
- vials – 30 ml, screw cap, sterile; GREINER (Kremsmünster, A).

*Reagents related to immunofluorescence*:

- mounting medium – Molecular Probes, Inc., 4849 Pitchford Avenve, Eugene, OR 97402, USA, www.probes.com; Calbiochem–Novabiochem Corp., 10394 Pacific Center Court, San Diego, CA 92121, USA, www.calbiochem.com; Polysciences, Inc., 400 Valley Road, Warrington, PA 18976, www.polysciences.com
- probes for staining of cellular structures/antibodies – Becton Dickinson GmbH, Tullastr. 8–12, 69126 Heidelberg, Germany, www.bdbiosciences.com; Biozol Diagnostika Vertrieb GmbH, Obere Hauptstr., 10b, 85386 Eching, Germany, www.biozol.com; Calbiochem–Novabiochem Corp., 10394 Pacific Center Court, San Diego, CA 92121, USA, www.calbiochem.com; Chemicon International, Inc., 28835 Single Oak Drive, Temecula, CA 92590, USA, www.chemicon.com; DAKO Diagnostika GmbH, Am Stadtrand 52, 22047 Hamburg, www.dako.com, www.dakousa.com; Molecular Probes, Inc., 4849 Pitchford Av., Eugene, OR 97402, USA, www.probes.com; Serotec Ltd., 22 Bankside, Oxford, OX5 1JE, GB, www.serotec.com; Sigma-Aldrich, 3050 Spruce Street, St. Louis, MI 63103 USA, www.sigma-aldrich.com; Upstate Biotechnology, 199 Saranac Av., Lake Placid, NY 12946, USA, www.upstatebiotech.com; Zymed Laboratories, 458 Carlton Ct., So. San Francisco, CA 94080, USA, www.zymed.com.

*Analysis equipment*:

- microplate washer – Columbus Washer, SLT (Grödig/Salzburg, A)
- microplate fluorescence reader – Spectrafluor Fluorometer, TECAN (Grödig/Salzburg, A) equipped with BIOLISE® software
- flow cytometer – EPICS XL-MCL with System II software, COULTER (Miami, USA)
- confocal laser scanning microscope – BIORAD MRC 1024 Laser Scanning Confocal Imaging System (Hemel Hempstead, UK) equipped with an argon ion laser (American Laser Corp., Salt Lake City, USA) and a Zeiss axiovert 100 microscope (Carl Zeiss, Oberkochen, G). The software used for imaging was Laser Sharp MRC-1024 version 3.1 (Bio-Rad, Deisenhofen, G).

## METHODS

## General methods

### *Cultivation of Caco-2 cells*

Caco-2 cells are cultivated in $75\,cm^2$ cell culture flasks. The cells are grown in RPMI 1640 cell culture medium containing ten per cent fetal calf serum (FCS), 4 mM L-glutamine and 150 μg/ml gentamicin in a humidified five per cent CO2/ 95 per cent air atmosphere at 37 °C until they reach 80 per cent confluency. Sub-cultivation is done by trypsination using a 0.25 per cent trypsin/EDTA solution at a splitting rate of 1:7–1:14.

For single cell experiments cells are counted immediately after subcultivation and the concentration of the cell suspension is adjusted to $6 \times 10^6$ cells/ml.

For experiments using monolayers, cells are seeded on TC-treated 96-well microplates at a density of $1.7 \times 10^4$ cells/well. The cells are fed every other day with culture medium and used 12–14 days after seeding.

### *Washing the cells*

Single cell experiments require washing of the cells to remove any unbound lectin only at the end of incubation. For this purpose, the volume of the cell suspension is adjusted to 200 μl and the cells are spun down (five minutes, 4°C, 1000 rpm). After discarding 150 μl of the supernatant the cells are resuspended after addition of 150 μl phosphate buffered saline (PBS). This procedure is repeated twice. The cell suspension is now ready for analysis by flow cytometry.

Using monolayers, the layer is also washed prior to the lectin incubation. Each well containing the cell layer is washed three times with 100 μl PBS/well per wash cycle, either automatically using a microplate washer or manually. At the end of the incubation period, monolayers are washed again to remove any unbound lectin. The washing procedure is the same as described above.

### *Fluorimetry*

In experiments with monolayers, the interaction of fluorescein- or rhodamine-labeled lectins with the cell layer is analyzed using a fluorescence microplate reader equipped with Biolise™ software for calculation. The relative cell-bound fluorescence intensity is determined by the 'bottom' measurement of the washed monolayers using 50 μl PBS as supernatant during analysis.

Both the autofluorescence of the plate and the monolayer serve as negative controls and are substracted from all the binding data quoted.

### *Flow cytometry*

The washed cell suspension is diluted with 1 ml Cell Pack prior to analysis. Cell bound fluorescence intensity of the single cell population is determined using a forward versus side scatter gate for exclusion of debris and cell aggregates. Fluorescence emission is measured at 525 nm (10 nm bandwidth) after excitation at 488 nm. The mean channel number of the logarithmic fluorescence intensities of individual peaks is used for further calculations. Amplification of the fluorescence signals is adjusted to put the autofluorescence signal of unlabeled cells into the first decade of the four-decade log range. For each measurement 5000 single cells are accumulated.

### CLSM – immunofluorescence

Fluorescence microscopy in general allows the detection of fluorescent molecules within biological samples. If suitable equipment, such as confocal scanning microscopy or computational deconvolution, is available, the relative localization of the fluorophore can be assessed. However, these techniques require a number of preparative steps, including fixation of the sample, fluorescent staining of the non-autofluorescent molecules of interest, counter-staining and embedding.

#### Growth support

Cell cultures to be examined need to be grown on supports that are compatible with cell growth and differentiation, as well as fluorescence-microscopical evaluation. Although optical transparency is not necessary for fluorescence microscopy, as excitation and emission pathways are on one side of the sample, it will be advantageous for easy orientation within the sample in transmission mode. Furthermore, it will be possible to monitor the confluency and integrity of the cell layers during the experiment by conventional light microscopy. Also, the support should exhibit low autofluorescence within the spectral range to be used, since high background signals will be either difficult to substract from the probe or make fluorescence microscopy completely impossible.

Suitable growth supports include:

1.  Round coverslips from glass, which are typically 1 cm in diameter and fit into 24-well plates, where they can be coated with desired matrix proteins. The thickness of the cover slips should be below the working distance of the objectives to be used.
2.  Chamber slides representing glass slides, where removable reservoirs for containment of suitable volumes of culture medium are mounted. When seeding chamber slides, it should be taken into account that the distribution of the cells within one chamber might be very irregular due to the small size and unusual shape of the reservoirs. This problem can be avoided if sufficient volumes of culture medium are used (Nalgene, Falcon, Greiner).
3.  Transwell filter inserts. Cells can be cultured on coated or uncoated transwell filter inserts. Immediately prior to microscopical examination, the filter can be cut out using a scalpel and sandwiched between a glass slide and a cover slip with the cells oriented towards the cover slip. Transwells can be obtained in transparent and non-transparent qualities, but the former are certainly recommended. Due to the existence of pores within the filters, one has generally to expect some influence on the optical properties of the samples. In many cases,

the investigator will observe positive fluorescence signals from the pores (Corning Costar).

4. Alternatively to growing cells on the support, where they are stained and examined, a suspension of cells can be spun onto a glass slide. This procedure is called cytospin and requires a centrifuge equipped with a special rotor. Most suppliers will offer an upgrade for their centrifuges in the larger benchtop-class.

### Fixation and permeabilization of cells

There are two principles of fixation compatible with immunofluorescence staining: aldehyde fixation and solvent fixation. Fixation with aldehydes leads to the cross-linking of proteins, allowing a good preservation of the cellular morphology. However, only by solvent fixation permeabilization of the lipid membranes are achieved, which is a prerequisite for efficient diffusion of antibodies to intracellular targets. Therefore, aldehyde fixation has to be followed by treatment with a suitable detergent, usually 0.1 per cent Triton X-100. Alternatively, saponin can be added to all the incubation steps that require permeability. Solvent fixation is usually carried out with cold methanol, acetone or a mixture thereof. Denaturation and fixation of the sample, as well as cell permeabilization, is achieved in one step by application of solvent fixation. Although this may be practical in many cases, this procedure may lead to structural changes such as shrinking and in some cases even the loss of antibody–antigen reactivity. All fixation protocols can reduce access of relatively large antibody molecules to the epitope. A typical example is the detection of bromodeoxyuridine that has been incorporated into genomic DNA during S phase. This may be overcomed by controlled partial digestion of the sample with proteinase K. Generally, the best-suited fixation protocol should be experimentally determined for each cell type and antibody combination.

### (Indirect) Immunofluorescent staining

To detect antigens by immunofluorescence microscopy, a two-step procedure using a primary antibody that binds the epitope, and a secondary antibody that recognizes the primary antibody and carries the fluorophore for detection, is used in most cases. This procedure, called 'indirect' immunofluorescence staining, results in higher sensitivity and does not require a fluorescent-labeled primary antibody. These antibodies are often not available, except for surface antigens that are routinely measured by flow cytometry. Alternatively to the secondary antibody, fluorescence-tagged avidin can be used for detection if the primary antibody is biotinylated. The existence of immunoglobulins (Igs) from different species (mouse, rat, rabbit), of different isotypes (IgG, IgM) and allotypes (IgG1, IgG2), makes it possible to combine different primary antibodies and to detect them selectively with type-specific secondary antibodies. This 'multicolor' immunofluorescence (double- or triple-labeling) allows analysis of two or three molecules in one sample. After careful analysis of such images conclusions about the (co-)localization of these molecules may be drawn (see Figure 5.1).

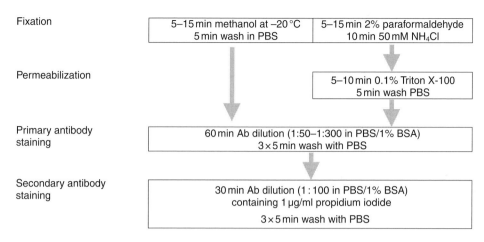

Fixation

| 5–15 min methanol at –20 °C  5 min wash in PBS | 5–15 min 2% paraformaldehyde  10 min 50 mM NH₄Cl |

Permeabilization

5–10 min 0.1% Triton X-100  5 min wash PBS

Primary antibody staining

60 min Ab dilution (1:50–1:300 in PBS/1% BSA)  3 × 5 min wash with PBS

Secondary antibody staining

30 min Ab dilution (1 : 100 in PBS/1% BSA)  containing 1 µg/ml propidium iodide  3 × 5 min wash with PBS

*Figure 5.1* Typical staining protocol for indirect immunofluorescence with counter-staining of the nucleus.

### Counter-staining

In order to provide landmarks for easy orientation within the object, it is very popular to counter-stain major cellular structures. Although this could be achieved with an additional antibody, the use of specific fluorescent dyes is easier and in most cases cheaper. A very broad overview of such dyes for a number of organelles can be found in the catalog of Molecular Probes (www.probes.com). Most frequently, one will counter-stain the nucleus with DAPI (4,6-diamido-2-phenylindolole), resulting in a blue fluorescence, or with propidium iodide (PI), which gives a red signal. If necessary, green-fluorescent DNA dyes, such as YOYO-1, or membrane permeable dyes (e.g. LDS 751) can also be used.

### Embedding

Finally, most samples will need to be embedded and covered with a cover slip in order to allow convenient microscopy with high resolution. However, some equipment, namely inverse fluorescence microscopes equipped with objectives that have a long working distance, allow the inspection of the sample from 'underneath' through the support on which the cells have been grown and stained. In all other cases, it will be necessary to surround the sample with a suitable medium and cover the sample with a cover slip. The embedding medium suitable for fluorescence microscopy will contain an antifade that minimizes photobleaching of the fluorophore. Furthermore, the embedding medium should penetrate well into the sample, mix with the fluid remaining on the sample and should not cause crystallization of salts that might be present. Otherwise, inconsistencies in the diffraction index might result, which negatively affect the optical quality. It is, therefore, advisable to use commercial embedding media, that are either solidifying (easier to handle), or remain fluid.

## CLSM – co-localization analysis

The high spatial resolution of CLSM allows, at least in principle, conclusions to be drawn about the co-localization of two differently labeled antigens within the pixel volume. Each pixel of the image is assigned a value for each signal, usually between 0 and 255. The distribution of pixel values can be visualized in a 2D-dot plot as shown in Figure 5.2. Above certain threshold values for both signals, one can assume that there is a significant amount of both antigens present within the volume that corresponds to the pixel. The population of pixels that satisfies this requirement can now be re-colored, usually yellow or white, to indicate the area of co-localization (Figure 5.2c).

In order to determine reliably the degree of co-localization, very careful attention must be paid to sample preparation and image acquisition. It must be understood that if sufficient care is not taken during sample preparation and acquisition, results may be generated that are misleading. Most problems result from fluorescence bleed-through. This is always a problem when the fluorescence signals are very different in intensity or when there is an overlap in the emission spectra of the fluorescent labels. Bleed-through can be minimized by careful selection of staining and acquisition protocols, but most softwares will allow some additional correction. Figure 5.3 illustrates some common problems that occur when the labels used produce signals that are very different in intensity. Although it is possible to correct bleed-through by post-acquisition processing (Figure 5.3D), the resulting 2D-intensity plot will hardly have the same quality as that obtained after careful optimization of the staining protocol.

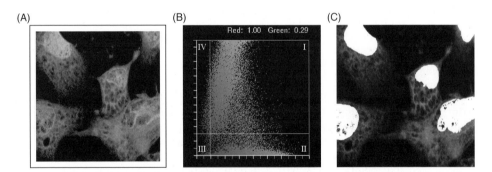

*Figure 5.2* Co-localization analysis. Cos-1 cells treated with an inductor of the transcription factor NFκB were stained for NFκB using a primary anti-NFκB antibody and a secondary fluorescein isothiocyanate (FITC)-labeled anti-mouse immunoglobulin (IgG1) F(Ab)$_2$ fragment. The nucleus was counter-stained with propidium iodide (PI) (red). (A) Confocal cut through the stained cells on the nuclear level. (B) 2D dot-plot of the pixel intensities of Figure A. Each pixel of the image is assigned an intensity value for the green and the red signal. After definition of threshold values, the whole pixel population can be divided into four sub-populations: (I) positive for red and green, (II) positive for green, negative for red, (III) positive for red, negative for green; and (IV) negative for both. (C) Co-localization analysis. (*See Color plate 1*)

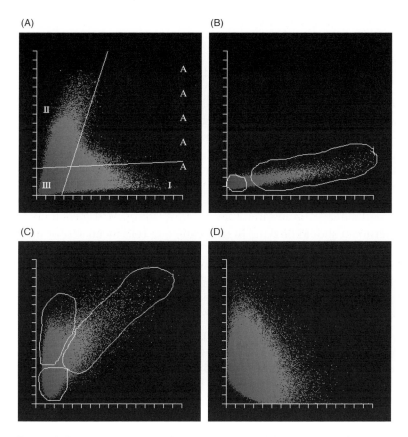

*Figure 5.3* Typical problems in co-localization analysis. (A) A cell stained for the nucleus (red) and caveolae (green) that do not co-localize was analyzed. Population (I) represents green pixels, (II) represents red pixels and (III) represents negative pixels. There are very few positive pixles for the red and green signal. However, the relatively intense red signal caused bleed-through into the green detector channel, resulting in higher values for the green signal. (B–D) A sample with intense green fluorescence, but weak red fluorescence is examined. At low signal amplification, only green and negative pixels are recorded (B). If the signal from the red channel is amplified red pixels also appear, but all the pixels with high green values also have high red values. These values can lead to interpretation as co-localization. However, after careful examination, two sub-populations can be distinguished (C). By compensation of bleed-through from the green signal into the red channel, these two sub-populations are separated. The positive pixels for green fluorescence no longer appear red (D). (*See Color plate 2*)

## Specific methods

### Cytoadhesion assays

#### Saturation assay at 4 °C

To elucidate the interaction between Caco-2 cells and lectins, lectins with different carbohydrate specificities, such as lectins from *Triticum vulgare* (wheat germ),

*Solanum tuberosum* (potatoes), *Arachis hypogea* (peanuts), *Ulex europaeus* (furze seeds), *Lens culinaris* (lentil seeds), and *Dolichus biflorus* (horse gram seeds) are used.

All the incubations and washing steps must be carried out at 4 °C in order to reduce the metabolism of the cells to a minimum. These conditions provide for exclusive binding to the cell membrane and inhibition of active transport processes.

*Single cell experiments*:

- Prepare a dilution series of F-lectins in PBS (320, 160, 80, 40, 20, 10, 5, 2.5 µg/ml) and a Caco-2 cell suspension subcultivated immediately prior to the experiment.
- Mix 50 µl of the cell suspension containing $3 \times 10^5$ cells and 50 µl of the lectin solution.
- Incubate for one hour at 4 °C.
- Wash the cells to remove any unbound lectin as described above (see Washing the cells).
- Each experiment should be done in triplicate. Use cells incubated with PBS as a blank.
- Determine cell-bound fluorescence intensity using flow cytometry (see Flow cytometry).

*Monolayer experiments*:

- Prepare a dilution series of F-lectins in PBS (160, 80, 40, 20, 10, 5, 2.5 µg/ml).
- Wash the confluent cell monolayers as described above (see Washing the cells).
- Add 50 µl lectin-solution to each well and determine the fluorescence intensity using a microplate reader.
- Incubate the monolayers for one hour at 4 °C.
- Suck off the supernatant and wash the monolayers to remove any unbound lectin (see Washing the cells).
- Measure the cell-bound fluorescence intensity as described above (Fluorimetry).
- Each experiment should be done in triplicate. Use cell monolayers incubated with PBS as a control.

Determination of the cell-binding rate of plant lectins with different carbohydrate specificities gives an estimate of the lectin-binding capacity, as well as an indication of the glycosylation pattern of the glycocalyx of the cells of interest. This indication is important for the use of certain lectins as regiospecific bioadhesive tools in pharmaceutical devices. Moreover, the proportionate increase of the cell-bound lectin, with the increasing amounts of lectin points to the specificity of the lectin–cell interaction.

Upon the incubation of Caco-2 single cells with six fluorescein-labeled plant lectins exhibiting different carbohydrate specificities, the mean cell-bound fluorescence intensity increases independent from the type of lectin used and follows the order WGA >> UEA > STL > LCA = PNA > DBA (see Figure 5.4). Using Caco-2 monolayers representing artificial tissue instead of single cells, the lectin-binding capacity order was the same except for LCA and STL, as seen in Figure 5.5. Due to the high Caco-2 binding rate of WGA, it is used in the following experiments to illustrate the lectin–cell interaction in detail.

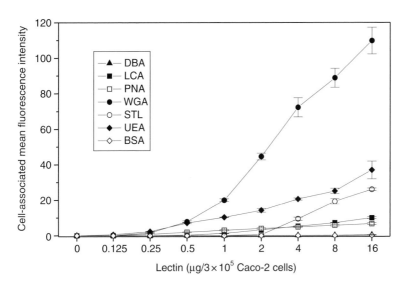

*Figure 5.4* Saturation analysis of lectin-binding sites on Caco-2 single cells with fluorescein-labeled *Dolichos biflorus* agglutinin (DBA), *Lens culinaris* agglutinin (LCA), peanut agglutinin (PNA), wheat germ agglutinin (WGA), *Solanum tuberosum* lectin (STL) and *Ulex europaeus* isoagglutinin I (UEA) in comparison with bovine serum albumin (BSA) by flow cytometry related to an apparent fluorescein/protein ratio = 1. Reproduced with permission (Gabor *et al.*, 1998).

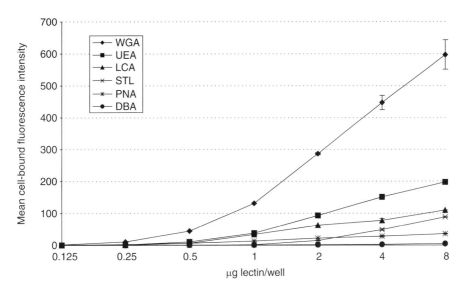

*Figure 5.5* Saturation analysis of lectin-binding sites on Caco-2 monolayers with fluorescein-labeled *Dolichos biflorus* agglutinin (DBA), *Lens culinaris* agglutinin (LCA), peanut agglutinin (PNA), wheat germ agglutinin (WGA), *Solanum tuberosum* lectin (STL) and *Ulex europaeus* isoagglutinin-I (UEA) related to an apparent fluorescein/protein ratio = 1.

## Binding specificity

To assess the specificity of the lectin–Caco-2 interaction, three different experimental set-ups are possible. By comparing the Caco-2 binding capacity of a certain lectin with that of lactalbumin or dextran, the contribution of protein–protein- and non-specific interactions to cell-adhesion of the lectin can be estimated. The carbohydrate specificity of the lectin–Caco-2 interaction can be confirmed by a competitive assay in presence of the complementary carbohydrate. Alternatively, it can be determined by a lectin-binding assay using Caco-2 cells pretreated with tunicamycin, which inhibits *N*-glycosylation, resulting in cells with damaged glycocalyx.

*Caco-2 binding of lectins in comparison with lactalbumin or dextran:*

- Prepare a dilution series of F-lectin, fluorescein-labeled lactalbumin (F-LA) and fluorescein-labeled dextran (F-D) in PBS (80, 40, 20, 10, 5 µg/ml).
- Wash the confluent cell monolayers as described above (see Washing the cells).
- Add 50 µl solution of either F-lectin, F-LA or F-D at five different concentrations to each well for incubation with the monolayers. Each experiment should be done in triplicate. Use cell monolayers incubated with PBS as a control.
- Determine the fluorescence intensity of each well using a microplate reader.
- Incubate the monolayers for one hour at 4 °C.
- Suck off the supernatant and wash the monolayers (see Washing the cells).
- Measure the cell-bound fluorescence intensity as described above (see Fluorimetry).

As indicated by Figure 5.6, the mean cell-bound amount of lactalbumin and dextran at highest concentration represents only four per cent (lactalbumin) or 0.9 per cent (dextran) of the WGA-binding rate. Consequently, lectin binding to the cell surface is not only due to general non-specific proteinaceous interactions, but also due to non-specific carbohydrate-mediated interactions.

Moreover, the percentage of binding is dependent on the amount of lectin added (Figure 5.7). The highest binding rate is achieved at a medium concentration of 20 µg lectin/ml representing about 20 per cent of the total amount added. At higher concentrations, the binding rate decreases due to lack of excessive binding sites. At lower concentrations, the reduced binding rate arises from less frequent contacts between the lectin and the cell surface due to the experimental set-up and the low lectin concentration. In the case of lactalbumin and dextran the binding rate is below one per cent.

*Caco-2 binding of lectins to tunicamycin-pretreated cells:*

Tunicamycin inhibits the cellular *N*-acetylglucosaminyl transferase, which results in production of glycoproteins that are missing some or all of their *N*-linked oligosaccharide side chains. Accordingly, upon the involvement of *N*-linked carbohydrates in the lectin–cell interaction, the lectin-binding to tunicamycin-pretreated Caco-2 cells is reduced, whereas the viability of the cells is maintained.

- Prepare a suspension of tunicamycin-pretreated Caco-2 cells. Add 100 µl tunicamycin-solution (1 mg/ml ethanol (EtOH) 70 per cent) to a cell culture flask

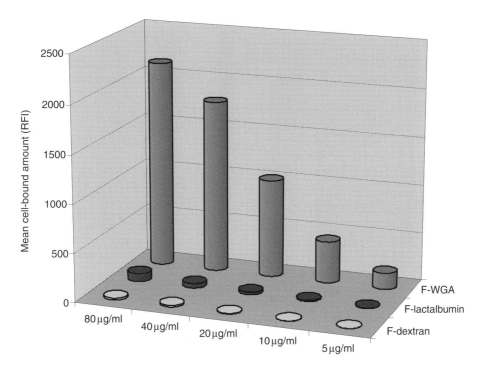

*Figure 5.6* Binding rate of fluorescein-labeled wheat germ agglutinin (F-WGA, molecular weight 37 kDa) in comparison with fluorescein-labeled lactalbumin (F-LA/molecular weight 14.8 kDa) and fluorescein-labeled dextran (F-D/average molecular weight 21.2 kDa) to Caco-2 single cells using flow cytometry for detection.

containing the confluent monolayer and 10 ml cell culture medium. Incubate for 40 hours prior to the subcultivation and counting of cells (see Cultivation of Caco-2 cells). Use cells of the same passage incubated with 10 ml cell culture medium containing 100 µl EtOH 70 per cent as a control.

- Prepare a dilution series of F-lectin in PBS (10, 5, 2.5 µg/ml).
- Mix 50 µl cell suspension ($3 \times 10^5$ cells) and 50 µl F-lectin solution.
- Incubate for one hour at 4 °C.
- Wash the cells to remove any unbound lectin (see Washing the cells).
- Determine the cell-bound fluorescence intensity using flow cytometry (see Flow cytometry).
- Each experiment should be done in triplicate. Use cells incubated with PBS as a blank and cells pretreated with EtOH (70 per cent) as a positive control.

According to the results shown in Figure 5.8, WGA-binding was inhibited to about one third upon the incubation of Caco-2 cells that were pretreated with tunica-mycin in comparison with non-treated cells. This result suggests that the binding of WGA primarily occurs via *N*-glycosylated oligosaccharide moieties present on the cell surface of Caco-2 cells.

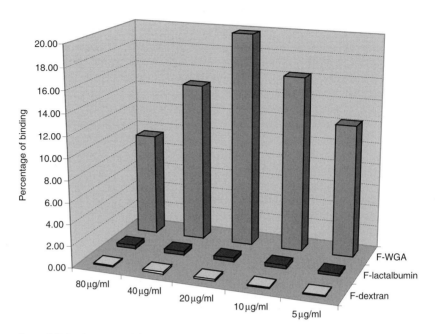

*Figure 5.7* Binding rate of fluorescein-labeled wheat germ agglutinin (F-WGA, molecular weight 37 kDa) in comparison with fluorescein-labeled lactalbumin (F-LA/molecular weight 14.8 kDa) and fluorescein-labeled dextran (F-D/average molecular weight 21.2 kDa) to Caco-2 monolayers using microplate fluorescence reading for detection.

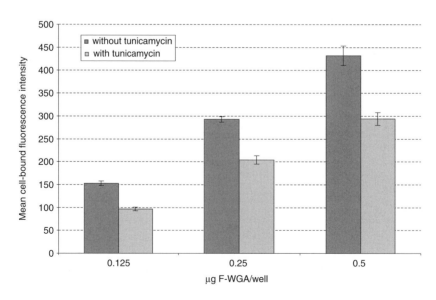

*Figure 5.8* Mean cell-associated relative fluorescence intensities after incubation of tunicamycin-pretreated Caco-2 cells with fluorescein-labeled wheat germ agglutinin (F-WGA) compared with the control omitting tunicamycin-pretreatment.

*Competitive assay with complementary carbohydrate:*

Another method to confirm the carbohydrate specificity of the lectin–cell inter-action is a competitive assay. In the competitive assay the oligosaccharides forming the glycocalyx of Caco-2 cells and the free complementary carbohydrate are allowed to compete for lectin binding. In case of specificity of the lectin–cell inter-action, Caco-2 binding of the lectin is inhibited the more as the concentration of the carbohydrate is increased.

*Single cell experiments*:

- Prepare a dilution series of the lectin-specific carbohydrate (e.g. N,N′,N″-triacetylchitotriose; 5–4000 µg/ml PBS, serial dilution) and a Caco-2 cell sus-pension subcultivated immediately prior to the experiment.
- Mix 50 µl of the cell suspension ($3 \times 10^5$ cells) and 100 µl carbohydrate solution or PBS (positive control) and incubate for five minutes at 4 °C.
- Add 50 µl of a solution containing the F-lectin (e.g. F-WGA; 5 µg/ml PBS).
- Incubate for one hour at 4 °C.
- Wash the cells to remove unbound lectin and soluble lectin-carbohydrate complexes (see Washing the cells).
- Determine cell-bound fluorescence intensity by flow cytometry (see Flow cytometry).
- Each experiment should be done in triplicate. Use cells incubated with PBS as a blank.

*Monolayer experiments*:

- Wash the confluent Caco-2 monolayers (see Washing the cells).
- Add to each well 50 µl of the complementary carbohydrate-solution (e.g. N,N′,N″-triacetylchitotriose in PBS) at four different concentrations (2, 1, 0.2, 0.1 mg/ml) and 50 µl PBS for the positive control, respectively.
- Incubate for five minutes at 4 °C followed by the addition of 50 µl F-lectin-solution (e.g. F-WGA; 10 µg/ml PBS).
- Incubate the monolayers for one hour at 4 °C.
- Suck off the supernatant and wash the monolayers as described above (see Washing the cells).
- Measure the cell bound fluorescence intensity (see Fluorimetry).
- Each experiment should be done in triplicate. Use cell monolayers incubated with PBS as a blank.

As indicated by Figures 5.9 and 5.10, the mean cell-bound amount of lectin is inversely proportional to the increasing amounts of the corresponding com-plementary carbohydrate. Though direct comparison of the results from both experiments (single cells, monolayers) is not reasonable due to different concen-trations and experimental set-ups, the results of both studies confirm the carbohydrate-mediated binding and thus specificity of the interaction between the lectin and the cell surface. According to these experiments, contribution of the non-specific interactions amounts to only one per cent.

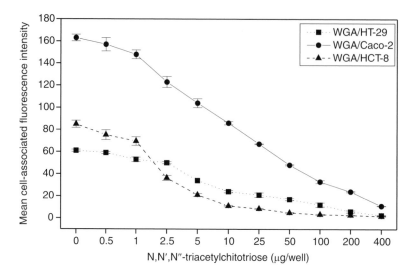

*Figure 5.9* Competitive inhibition of wheat germ agglutinin (WGA)-binding sites on Caco-2, HT-29 and HCT-8 single cells by the addition of increasing amounts of the complementary carbohydrate N,N′,N″-triacetylchitotriose.

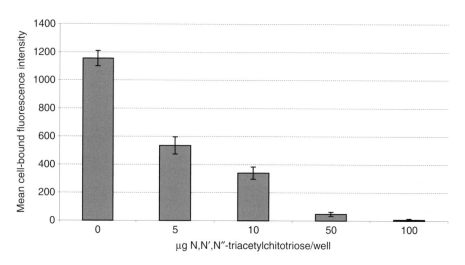

*Figure 5.10* Mean cell-bound relative fluorescence intensity of fluorescein-labeled wheat germ agglutinin after incubation of Caco-2 monolayers with increasing amounts of the complementary carbohydrate N,N′,N″-triacetylchitotriose.

## Cytoinvasion assays

As the lectin is bound to the glycocalyx of Caco-2 cells, membrane-binding can be followed by uptake of the lectin into the cell. This chapter deals with some aspects concerning discrimination between cell-binding and uptake. In this respect, the incubation temperature is highly important. At 4 °C the fluidity of the cell membrane and the metabolism of the cells is reduced, resulting in the repression of energy dependent transport processes. For this reason the amount of cell-associated lectin refers mainly to membrane-bound lectin. In contrast, upon incubation at 37 °C the cell is metabolically active and energy-consuming transport processes take place. Thus, the amount of cell-associated lectin refers to binding and uptake.

### Binding versus uptake by use of an avidin-biotin assay

This assay relies on the use of biotinylated lectin for cytoassociation in conjunction with fluorescein-labeled avidin (F-avidin) for detection of the lectin. As F-avidin neither interacts with Caco-2 cells nor permeates the Caco-2 membrane, the amount of membrane-bound lectin can be distinguished from internalized lectin. Thus, the fluorescence intensity determined refers to cell-bound lectin exclusively, provided that excessive biotinylated lectin is removed and incubation with F-avidin is carried out at 4 °C.

- Prepare a solution of biotinylated lectin in PBS (e.g. biotin-WGA; 40 µg/ml) and a solution of F-avidin (80 µg/ml PBS).
- Mix 50 µl of a Caco-2 cell suspension ($3\times10^5$ cells, subcultivated immediately prior to the experiment) and 50 µl of the biotinylated lectin solution.
- Incubate for 2.5, 5, 10, 20, 40, 60, 90, 120 and 180 minutes at either 4 °C or 37 °C.
- Wash the cells to remove any unbound lectin (see Washing the cells).
- Adjust the volume of the cell suspension to 50 µl and incubate the cells with 50 µl F-avidin solution for 20 minutes at 4 °C.
- Repeat the washing step (see Washing the cells) to remove free F-avidin.
- Determine cell-bound fluorescence intensity by flow cytometry (see Flow cytometry).
- Each experiment should be done in triplicate. Use cells incubated with PBS as a blank.

Provided that binding to the cell surface and no intracellular uptake of the lectin occurs upon incubation at 4 °C, Figure 5.11 shows the ratio between binding to the cell surface and uptake at 37 °C by course of time using the avidin-biotin system. Since the lectin, once bound to the membrane, is not released again upon incubation at 37 °C and only surface-bound lectin is determined by avidin-detection, the decrease in cell-associated fluorescence intensity is due to internalization of initially membrane-bound lectin. Accordingly, uptake of membrane-bound lectin starts about five minutes after contact between the lectin and the cell. More than 83 per cent of initially surface-bound lectin are taken up into the Caco-2 cells within three hours, whereas most of the lectin is internalized during the first 60 minutes.

### Binding versus uptake: influence of incubation-time and -temperature

As mentioned above, the interaction between lectins and cells is strongly influenced by the incubation temperature. Therefore, the experiments below are intended to

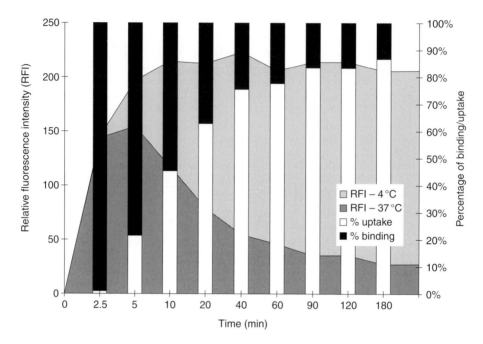

*Figure 5.11*  Mean cell-bound relative fluorescence intensity (RFI) of the biotin-wheat germ agglutinin (WGA) – fluorescein–avidin – complex after incubation with Caco-2 single cells at either 4 °C or 37 °C. Provided that the amount of membrane-bound lectin is comparable at both temperature levels, the difference in cell-bound relative fluorescence intensity between 4 °C and 37 °C corresponds to the amount of lectin taken up into the Caco-2 cells.

elucidate the influence of the incubation temperature with respect to the incubation period. To avoid different starting situations all the cells are loaded with the lectin for five minutes at 4 °C followed by incubation of the cells for different periods at either 4 °C or 37 °C.

*Single cell experiments*:

- Prepare a solution of F-lectin (e.g. F-WGA; 10 μg/ml PBS) and a Caco-2 cell suspension subcultivated immediately prior to the experiment.
- Mix 50 μl of the cell suspension ($3 \times 10^5$ cells) and 50 μl of the labeled lectin-solution and incubate for five minutes at 4 °C.
- Wash the cells to remove any unbound lectin (see Washing the cells).
- Adjust the volume of the cell suspension to 50 μl followed by further incubation at either 4 °C or 37 °C.
- Determine the cell-associated fluorescence intensity after 0, 15, 30, 60, 120, 180 and 240 minutes by flow cytometry (see Flow cytometry).
- Each experiment comprising a certain incubation-time interval and temperature level should be done in triplicate. Use cells incubated with PBS as a blank.

*Monolayer experiments*:

- Wash the confluent Caco-2 cell monolayers (see Washing the cells).
- Add 50 µl of F-lectin solution in the PBS (e.g. F-WGA; 10 µg/ml) to each well and incubate the cell layer for five minutes at 4 °C.
- Suck off the supernatant and wash the monolayers to remove any unbound lectin (see Washing the cells).
- Adjust the supernatant of the cell layer to 50 µl and incubate the cells for 0, 10, 20, 40, 60, 120, 180 and 240 minutes at either 4 °C or 37 °C.
- Suck off the supernatant and add 50 µl PBS.
- Measure the cell-associated fluorescence intensity (see Fluorimetry).
- Each experiment comprising a certain incubation-time and temperature level should be done in triplicate. Use cells incubated with PBS as a blank.

As obvious from the results of the two experiments (Figures 5.12 and 5.13), the cell-associated fluorescence intensity is not altered markedly upon incubation at 4 °C indicating that initially cell-bound lectin is not released again during incubation. At 37 °C the cell-associated fluorescence intensity decreases with increasing incubation time. Since dissociation of cell-bound lectin could not be detected, this decrease is subject to internalization and/or accumulation of the lectin within acidic compartments of the cell. It is well-known that the quantum yield of the fluorescein label is pH-dependent. Compared with pH 7.4 the fluorescence intensity of a fluorescein-solution decreases to about ten per cent at lysosomal pH. So the results indicate uptake and intracellular accumulation within acidic compartments of Caco-2 cells.

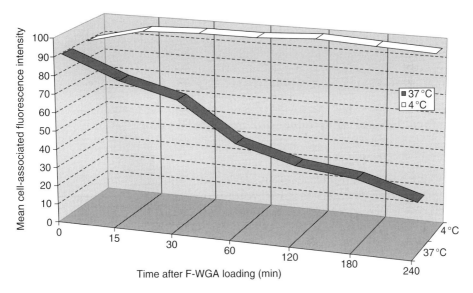

*Figure 5.12* Mean cell-associated relative fluorescence intensity of Caco-2 single cells loaded with equal amounts of fluorescein-labeled wheat germ agglutinin (F-WGA) by course of time after incubation at either 4 °C or 37 °C as determined by flow cytometry.

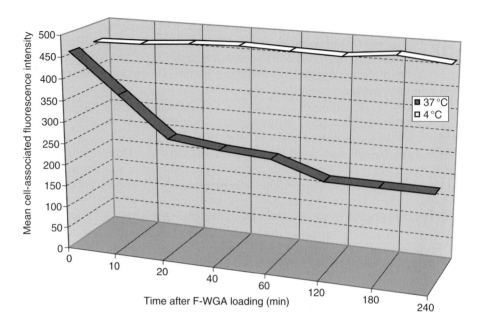

*Figure 5.13* Mean cell-associated relative fluorescence intensity of Caco-2 monolayers loaded with equal amounts of fluorescein-labeled wheat germ agglutinin (F-WGA) by course of time after incubation at either 4 °C or 37 °C as determined by microplate fluorescence reading.

Assaying single cells, the uptake of WGA is rather constant within four hours and results in a decrease in fluorescence emission of about 70 per cent. Using monolayers, an obvious decrease in fluorescence emission is observed within the first 20 minutes. Thereafter, the emission signal is reduced rather slightly, yielding a total decrease in fluorescence intensity of about 50 per cent.

## Confocal imaging of cytoadhesion and cytoinvasion

### Confocal imaging of F-lectin

In addition to flow cytometric analysis of cytoadhesion and cytoinvasion, fluorescence microscopy can be used to visualize these processes. To confirm binding and uptake of F-lectins, single cell-suspensions of Caco-2 cells as well as monolayers were incubated with the lectin under investigation and examined by CLSM.

*Single cell experiments*:

- Prepare a solution of F-lectin (e.g. F-WGA; 15 μg/ml PBS) and a Caco-2 cell suspension ($2 \times 10^6$ cells/ml) subcultivated immediately prior to the experiment.
- Mix 150 μl of the cell suspension and 150 μl of the lectin solution and incubate for one hour at 4 °C and 37 °C, respectively.

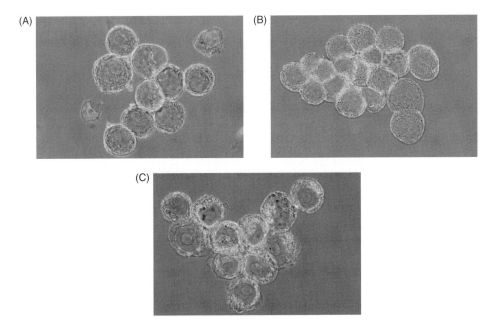

*Figure 5.14* (A) Confocal images of paraformaldehyde-fixed Caco-2 cells stained with fluorescein-labeled wheat germ agglutinin for one hour at 4 °C, (B) for 30 minutes at 37 °C, and (C) for two hours at 37 °C. The cell diameter refers to about 20 mm. Reproduced with permission (Wirth *et al.*, 1998). (*See Color plate 3*)

- After washing the cell suspension (see Washing the cells, but use 250 μl instead of 150 μl), adjust its volume to 50 μl and add 50 μl paraformaldehyde solution (four per cent) to fix the cells.
- Mount the cell suspension for CLSM (see CLSM – immunofluorescence).

As can be seen from the confocal images in Figure 5.14 WGA is bound exclusively to the surface of the cells upon incubation at 4 °C. When the focus is set to the middle of the cell, a fluorescent ring encircling the cell is observed. After incubation at 37 °C the F-WGA is located predominantly within the cytosol. The dot-like distribution pattern points to vesicular accumulation within the cells.

### Confocal imaging of F-lectin and counter-staining

The advantage of microscopy over flow cytometry is that it provides visual information about the distribution of the bound-ligand on the target cell. Phenomena that can be observed directly are preferentially adhesion to certain sites on the target cell, as it may be caused by clustering of the receptor. In polarized cell cultures, potential receptors may also be distributed unevenly between the apical and the basolateral surface. The resulting polarization of ligand adhesion can also be imaged directly.

(A)                                              (B)

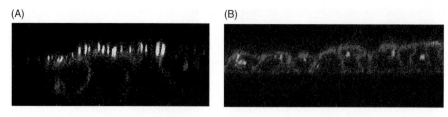

*Figure 5.15* Discrimination between surface binding and internalization. Caco-2 cells are incubated with green fluorescent fluorescein-labeled wheat germ agglutinin (10 μg/ml) for five minutes at 4 °C (A) or five minutes at 4 °C followed by washing and further incubation at 37 °C for 30 minutes (B). Cytoplasmic actin filaments including microvilli are stained with actin-reactive TRITC-phalloidine. After fixation with methanol at −20 °C, the cells are incubated with 0.2 ng/ml of the dye (incubation volume 200 μl) for 30 minutes and washed three times in phosphate buffered saline. Images are obtained by performing a vertical line scan on a confocal scanning microscope. The lectin is observed to be in close contact with the red stained microvilli, but it is not taken up into the cell body at 4 °C (A). At 37 °C, internalization of the lectin occurs (B). (*See Color plate 4*)

In order to distinguish between sole cytoadhesion and internalization (cytoinvasion) of a ligand, it will be necessary to somehow counter-stain the cell body or the cell membrane. Dyes for these purposes are commercially available. One cytoplasmic stain is, for instance, calcein-AM, a substance that freely diffuses through membranes to become hydrolyzed by cytoplasmic esterases of viable cells, resulting in entrapment of the hydrophilic fluorescent-product calcein. When cells forming cilia or microvilli such as differentiated Caco-2 cells are used, stains reacting with cytoskeletal proteins are often better suited to resolve these very fine structural protrusions.

This principle was used to distinguish between surface-bound and internalized lectin (Figure 5.15). The relative localization of a green fluorescent bioadhesin (WGA) on Caco-2 cells is determined after incubation at 4 °C and incubation for different periods at 37 °C. The cell body is counter-stained with TRITC-phalloidin that binds to the actin cytoskeleton. Using this method, it is possible to show that binding to the cell surface only occurs after a short incubation at 4 °C, while at 37 °C rapid internalization into the cell takes place.

### Intracellular localization of lectin

#### Intracellular localization by CLSM

Confocal microscopy can also be applied to test a certain hypothesis about the pathway involved in the intracellular transport of the object under investigation. Such testing requires the attachment of an appropriate fluorescent label onto the object, or alternatively, the possibility of detecting the object by means of immunofluorescence. In addition, a suitable marker for the cellular compartment or pathway of interest has to be identified.

To test the hypothesis about the endocytic uptake of WGA and its routing to the lysosomes, the red fluorescent lectin is incubated with the cells. As flow cytometry suggests uptake of the lectin into acidic compartments of the cell, it is stained for the mature form of the protease Cathepsin D, a selective marker of the lysosome. This is achieved by incubation of methanol-fixed cells with 1:100 dilution of a primary anti-Cathepsin D monoclonal antibody (BD Transduction, Lexington, USA), followed by detection with a fluorescein isothiocyanate (FITC)-conjugated anti-mouse F(Ab)₂ fragment (DAKO, Hamburg, Germany). The analysis of a typical section reveals 27 per cent red-green and 88 per cent green-red co-localization. This means that by the time of fixation, 27 per cent of WGA are taken up by lysosomes and 88 per cent of the lysosome volume is filled with the lectin (see Figure 5.16).

*Lectin–cell association: influence of monensin treatment*

The quantum yield of fluorescein is pH-dependent as it is reduced in acidic milieu. Supposing that cellular uptake of a labeled lectin is followed by accumulation within acidic compartments of the cell such as lysosomes, the quantum yield of cell-associated F-lectin decreases.

Monensin acts as a carboxylic ionophore catalyzing the exchange of protons for potassium ions. Upon treatment of cells with monensin, the pH-gradient between acidic intracellular compartments and the cytosol is compensated, abolishing the quench of the fluorescein label. Thus, the quantum yield of F-lectin associated with the cells should increase after addition of monensin in case of accumulation within acidic compartments of the cells.

(A)    (B)

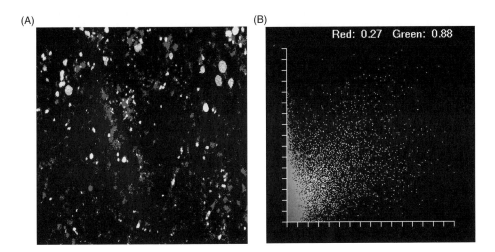

Red: 0.27   Green: 0.88

*Figure 5.16* Co-localization analysis to assess processing of wheat germ agglutinin (WGA) via the endo-/lysosomal pathway. (A) Confocal image of Caco-2 cells incubated with red fluorescent Rh-WGA (10 μg/ml) for five minutes at 4 °C followed by washing and further incubation at 37 °C for 30 minutes. The lysosomes are counterstained with an anti-Cathepsin D antibody and a fluorescein isothiocyanate (FITC)-anti mouse F(Ab)₂ fragment. (B) 2D dot-plot of pixel intensities in image (A). Co-localization analysis shows that 88 per cent of the volume representing green fluorescent lysosomes are filled with red fluorescent lectin, while only 27 per cent of the lectin are found within Cathepsin B positive lysosomes. (*See Color plate 5*)

*Single cell experiments*:

- Prepare a solution of F-lectin (e.g. F-WGA; 10 μg/ml PBS) and a solution of monensin in ethanol (2.42 mM).
- Mix 50 μl of a Caco-2 cell suspension ($3 \times 10^5$cells, subcultivated immediately prior to the experiment) and 50 μl of the labeled lectin solution.
- Incubate for five minutes at 4 °C.
- After washing the cells (see Washing the cells) adjust the volume of the cell suspension to 50 μl.
- Incubate the cell suspension for 0, 15, 30, 60, 120, 180 and 240 minutes at either 4 °C or 37 °C.
- Prepare the cell suspension for flow cytometry and determine the cell-associated fluorescence intensity (see Flow cytometry).
- Add 10 μl monensin solution to the cell-suspension and incubate for three minutes at room temperature.
- Repeat the measurement of the cell-associated fluorescence intensity (see Flow cytometry) to determine the total non-quenched fluorescence intensity.
- Each experiment comprising a certain incubation time interval and temperature level should be done in triplicate. Use cells incubated with PBS as a blank.

*Monolayer experiments*:

- Prepare a 20 μM solution of monensin by dilution of an ethanolic monensin solution (2.42 mM) with PBS.
- Wash the confluent Caco-2 cell monolayers (see Washing the cells).
- Add 50 μl of F-lectin solution in PBS (e.g. F-WGA; 10 μg/ml) to each well.
- Incubate the cell layer for five minutes at 4 °C.
- Wash the monolayers to remove unbound lectin (see Washing the cells) and adjust the supernatant of the cell layer to 50 μl.
- Incubate the monolayers for 0, 10, 20, 40, 60, 120, 180 and 240 minutes at either 4 °C or 37 °C.
- Suck off the supernatant and add 50 μl PBS.
- Determine the cell-associated fluorescence intensity of each well (see Fluorimetry).
- Suck off the PBS, add 50 μl of the 20 mM monensin solution to each well and incubate for three minutes at room temperature.
- Repeat measurement of the cell-associated fluorescence intensity to determine the total non-quenched fluorescence intensity.
- Each experiment concerning a certain incubation time interval and temperature level should be done in triplicate. Use cells incubated with PBS as a blank.

After incubating the cells at 4 °C, no significant change in fluorescence emission is detected after addition of monensin (see Figures 5.17 and 5.18 for single cell and monolayer experiments, respectively). At 37 °C the mean cell-associated fluorescence intensity decreases with increasing incubation time as known from the experiment above. Addition of monensin at the end of the incubation period causes a noticeable increase in fluorescence intensity yielding signals in the

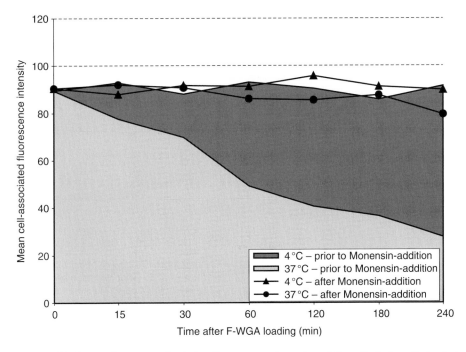

*Figure 5.17* Mean cell-associated relative fluorescence intensity of Caco-2 single cells loaded with equal amounts of fluorescein-labeled wheat germ agglutinin (F-WGA) by course of time after incubation at either 4 °C or 37 °C followed by flow cytometric analysis in presence or absence of monensin.

range of the 4 °C-experiment with the single cells and slightly higher than the 4 °C-values with the Caco-2 monolayers. Keeping in mind that the acidic fluorescence-quenching accounts for up to 90 per cent of the fluorescence emission at pH 7.4, the results indicate that about 65–72 per cent of the total cell-associated WGA are located within acidic compartments of the cell at the end of the incubation period.

## NOTES

- After splitting, the cells should be used for the experiments within one hour to avoid loss of viability.
- To avoid detrimental effects on the cells, they should be centrifuged at 1000 rpm at the maximum.
- Caco-2 monolayers should be cultivated for 12–14 days after seeding prior to the experiments, as differentiation will not start until confluency has been reached. Cultivation for longer periods of time results in susceptibility to detachment of the cells from the supporting matrix.

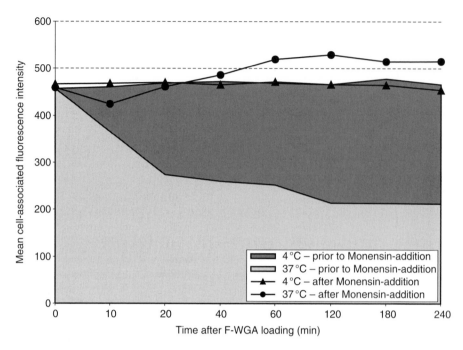

*Figure 5.18* Mean cell-associated relative fluorescence intensity of Caco-2 monolayers loaded with equal amounts of fluorescein-labeled wheat germ agglutinin (F-WGA) by course of time after incubation at either 4 °C or 37 °C followed by fluorescence microplate reading in presence or absence of monensin.

- During washing procedures of monolayers, care should be taken to avoid damage of the cell layer by pipet tips or the nozzles of the washer. Inclining and placing the monolayer-containing device on a dark pad facilitates manual washing. In case of doubt, the monolayer should be microscopically checked.
- When washing single cells pay attention to suck off the supernatant but not the cells.
- Only lectin preparations of the highest purity should be used. When inconsistent results are obtained purity can be simply checked by SDS-PAGE.
- Performing experiments in the presence of additives (ouabain, monensin) requires careful predetermination of the concentration range. In case of underdosing, no effect is observed after addition of the lectin, whereas at too high concentrations viability of the cells might be reduced.

## CONCLUSION

Though there are many questions that still need to be answered, the lectin concept for drug-delivery seems to be more than an interesting idea. Due to selective biorecognition of certain glycocalyx-associated oligosaccharides, lectins provide for selective targeting to different regional compartments of the gut or even cell populations such as M cells. The methods presented above enable elucidation of cytoadhesion, cytoinvasion and intracellular localization of lectins at the cellular level. Use of single cells as well as artificial tissues of different origin leads to a better understanding of the mechanisms involved. Cytoadhesion and cytoinvasion represent a prerequisite for transcellular absorption, leading to the successful delivery of even poorly available drugs. In contrast to animal studies, cell culture models open the possibility of investigating binding and uptake at the cellular level. Basic knowledge of the molecular mechanisms of the interaction between drugs and cells leads to new strategies in the development of drug-delivery systems that are capable of overcoming biological barriers.

Nevertheless, it would be necessary to follow a more systematic approach to glycotargeting by the establishment of a 'glycosylation map' of the gut as postulated by Pusztai *et al.* (1999). Besides biorecognition, lectins may also improve delivery of drugs into the systemic circulation by inducing receptor-mediated endo- or transcytosis. Thus, lectins conjugated to drugs or immobilized at the surface of particular drug delivery systems may enhance the efficacy of mucosally delivered drugs by delaying the transit time and enhancing the uptake into absorptive enterocytes.

As a main requirement for peroral administration some lectins such as WGA or TL are stable against proteolytic degradation and acidic conditions in the gut (Gabor, 1997b; Nachbar *et al.*, 1981; Naisbett and Woodley, 1995). Prior to use of lectins as pharmaceutical excipients, some items concerning safety have to be investigated in future. On the one hand, some lectins are only reputed to be non-toxic as part of the regular diet of humans. Although the amount of lectins necessary for drug-delivery is supposed to be in the nanogram-range, safety was not, so far, been confirmed *in vivo*. On the other hand, antigenicity of the lectins might represent an obstacle towards practical utility in pharmaceutical formulations. To date our knowledge about immunological tolerance relies on a few animal experiments and occurrences in common foodstuff, but is insufficient to guarantee immunological tolerance.

At present, hololectins are predominantly investigated for pharmaceutical purposes, which represent proteins composed of several carbohydrate-binding domains. Consequently, they might agglutinate blood cells due to their multivalency, which probably limits practical application. Out to this reason it might be advantageous to investigate the biorecognitive properties of merolectins such as hevein from the rubber tree or mannose-binding proteins from orchids. They represent small proteins consisting of a single carbohydrate binding domain exclusively omitting hemagglutination (Van Damme *et al.*, 1998).

The lectin-concept is still in its infancy, but hurdles limiting feasibility of the lectin-concept today probably can be surmounted by genetic engineering and biotechnology. Expression of recombinant proteins representing a binding domain, which targets an absorption window and induces specific endo- or transcytosis, is a science fiction at the present but seems to be feasible in the future. Nevertheless, the concept of lectin-mediated drug delivery needs a sophisticated understanding of basic lectin–cell interactions at the cellular level in order to be pursued in future.

## REFERENCES

Artursson, P. and Borchardt, R. T. (1997) Intestinal drug absorption and metabolism in cell cultures: Caco-2 and beyond. *Pharm. Res.*, **14**, 1655–1658.

Bardocz, S., Grant, G., Ewen, S. W. B., Duguid, T. J., Brown, D. S., Englyst, K. *et al.* (1995) Reversible effect of phytohaemagglutinin on the growth and metabolism of rat gastrointestinal tract. *Gut*, **37**, 353–360.

Broadwell, R. D., Balin, B. J. and Saleman, M. (1988) Transcytotic pathway for blood-borne protein through the blood-brain barrier. *Proc. Natl. Acad. Sci. USA*, **85**, 632–636.

Chickering III, D. E. and Mathiowitz, E. (1999) Definitions, mechansims and theories of bioadhesion. In E. Mathiowitz, D. E. Chickering III and C. M. Lehr (eds) *Bioadhesive Drug Delivery Systems – Fundamentals, Novel Approaches and Development*, Marcel Dekker, New York, pp. 1–11.

Clark, M. A., Jepson, M. A., Simmons, N. L. and Hirst, B. H. (1995) Selective binding and transcytosis of Ulex europaeus I lectin by mouse Peyer's patch M cells in vivo. *Cell Tissue Res.*, **282**, 455–461.

Clark, M. A., Hirst, B. H. and Jepson, M. A. (2000) Lectin-mediated mucosal delivery of drugs and microparticles. *Adv. Drug Del. Rev.*, **43**, 207–223.

De Aizpurua, H. D. and Russell-Jones, G. L. (1988) Identification of classes of proteins that provide an immune response upon oral feeding. *J. Exp. Med.*, **167**, 440–451.

Ertl, B., Heigl, F., Wirth, M. and Gabor, F. (2000) Lectin-mediated bioadhesion: preparation, stability and Caco-2 binding of wheat germ agglutinin-functionalized poly(d, l-lactic-co-glycolic acid)-microspheres. *J. Drug Target.*, **8**, 173–184.

Florence, A. T., Hillery, A. M., Hussain, N. and Jani, P. U. (1995) Factors affecting the oral uptake and translocation of polystyrene nanoparticles: histological and analytical evidence. *J. Drug Target.*, **3**, 65–70.

Gabor, F., Bernkop-Schnürch, A. and Hamilton, G. (1997a) Bioadhesion to the intestine by means of E. coli K99-fimbriae: gastrointestinal stability and specificity of adherence. *Eur. J. Pharm. Sci.*, **5**, 233–242.

Gabor, F., Wirth, M., Jurkovich, B., Haberl, I., Theyer, G., Walcher, G. *et al.* (1997b) Lectin-mediated bioadhesion: proteolytic stability and binding-characteristics of wheat germ agglutinin and Solanum tuberosum lectin on Caco-2, HT-29 and human colonocytes. *J. Contr. Rel.*, **49**, 27–37.

Gabor, F., Stangl, M. and Wirth, M. (1998) Lectin-mediated bioadhesion: binding characteristics of plant lectins on the enterocyte-like cell lines Caco-2, HT-29 and HCT-8. *J. Contr. Rel.*, **55**, 131–142.

Gurny, R. and Junginger, H. E. (eds) (1990) *Bioadhesion – Possibilities and Future Trends*, Paperback APV, Stuttgart: Wissenschaftliche Verlagsgesellschaft.

Hussain, N., Jani, P. U. and Florence, A. T. (1997) Enhanced oral uptake of tomato lectin-coated nanoparticles in the rat. *Pharm. Res.*, **14**, 613–618.

Irache, J. M., Durrer, C., Duchêne, D. and Ponchel, G. (1994a) *In vitro* study of lectin-latex conjugates for specific bioadhesion. *J. Contr. Rel.*, **31**, 181–188.

Irache, J. M., Durrer, C., Duchêne, D. and Ponchel, G. (1994b) Preparation and characterization of lectin-latex conjugates for specific bioadhesion. *Biomaterials*, **15**, 899–904.

Jepson, M. A., Mason, C. M., Simmons, N. L. and Hirst, B. H. (1995) Enterocytes in the follicle-associated epithelia of rabbit small intestine display distinctive lectin-binding properties. *Histochemistry*, **103**, 131–134.

Jepson, M. A., Clark, M. A., Foster, N., Mason, C. M., Bennett, M. K., Simmons, N. L. *et al.* (1996) Targeting to intestinal M cells. *J. Anat.*, **189**, 507–516.

Kilpatrick, D. C., Pusztai, A., Grant, G., Graham, C. and Ewen, S. W. B. (1985) Tomato lectin resists digestion in the mammalian alimentary canal and binds to intestinal villi without deleterious effects. *FEBS Lett.*, **185**, 299–305.

Lavelle, E. (2000) Targeted mucosal delivery of drugs and vaccines. *Exp. Opin. Ther. Patents*, **10**, 179–190.

Lehr, C. M., Poelma, F. G. J., Junginger, H. E. and Tukker, J. J. (1991) An estimate of turnover time of intestinal mucus gel layer in the rat in situ loop. *Int. J. Pharm.*, **70**, 235–240.

Lehr, C. M., Bouwstra, J. A., Kok, W., Noach, A. B. J., deBoer, A. G. and Junginger H. E. (1992) Bioadhesion by means of specific binding to tomato lectin, *Pharm. Res.*, **9**, 547–553.

Lehr, C. M. (1994) Bioadhesion technologies for the delivery of peptide and protein drugs to the gastrointestinal tract. *Crit. Rev. Ther. Drug Carrier Syst.*, **11**, 119–160.

Lehr, C. M. and Pusztai, A. (1995) The potential of bioadhesive lectins for the delivery of peptide and protein drugs to the gastrointestinal tract. In A. Pusztai and S. Bardocz (eds) *Lectins – Biomedical Perspectives.*, Taylor & Francis, London, pp. 117–140.

Nachbar, M. S., Oppenheim, J. D. and Thomas, J. O. (1981) Lectins in the U.S. diet: isolation and characterization of a lectin from the tomato (Lycopersicum esculentum). *J. Biol. Chem.*, **5**, 2056–2061.

Naisbett, B. and Woodley, J. (1994a) The potential use of tomato lectin for oral delivery: 1. Lectin binding to rat small intestine *in vitro. Int. J. Pharm.*, **107**, 223–230.

Naisbett, B. and Woodley, J. (1994b) The potential use of tomato lectin for oral delivery: 2. Mechanism and uptake in vitro. *Int. J. Pharm.*, **110**, 127–136.

Naisbett, B. and Woodley, J. (1995) The potential use of tomato lectin for oral drug delivery: 3. Bioadhesion in vivo. *Int. J. Pharm.*, **114**, 227–236.

Peumans, W. J. and Van Damme, E. J. M. (1995) Lectins as plant defense proteins. *Plant Physiol.*, **109**, 347–352.

Ponchel, G. and Irache, J.-M. (1998) Specific and non-specific bioadhesive particulate systems for oral delivery to the gastrointestinal tract. *Adv. Drug Del. Rev.*, **34**, 191–219.

Pusztai, A., Bardocz, S. and Ewen, S. W. B. (1999) Plant lectins for oral drug delivery to different parts of the gastrointestinal tract. In E. Mathiowitz, D. E. Chickering III and C. M. Lehr (eds), *Bioadhesive Drug Delivery Systems – Fundamentals, Novel Approaches and Development*, Marcel Dekker, New York, pp. 387–407.

Sharma, R., Van Damme, E. J. M., Peumans, W. J., Sarsfield, J. and Schumacher, U. (1996) Lectin binding reveals divergent carbohydrate expression in human and mouse Peyer's patches. *Histochem. Cell Biol.*, **105**, 459–465.

Singh, R. S., Tiwary, A. K. and Kennedy, J. F. (1999) Lectins: sources, activities and applications. *Crit. Rev. Biotech.*, **19**, 145–178.

Van Damme, E. J. M., Peumans, W. J., Pusztai, A. and Bardocz, S. (1998) *Handbook of Plant Lectins: Properties and Biomedical Applications*, Chichester: Wiley & Sons, pp. 3–9.

Walter, E., Janich, S., Roessler, B. J., Hilfinger, J. M. and Amidon, G. L. (1996) HT-29-MTX/Caco-2 cocultures as an in vitro model for the intestinal epithelium: in vitro–in vivo correlation with permeability data from rats and humans. *J. Pharm. Sci.*, **10**, 1070–1076.

Wang, Q., Yu, L. G., Campbell, B. J., Milton, J. D. and Rhodes, J. M. (1998) Identification of intact peanut agglutinin in peripheral venous blood. *Lancet*, **352**, 1831–1832.

Wirth, M., Fuchs, A., Wolf, M., Ertl, B. and Gabor, F. (1998a) Lectin-mediated drug targeting: characteristics and antiproliferative activity of wheat germ agglutinin conjugated doxorubicin on Caco-2 cells. *Pharm. Res.*, **7**, 1031–1037.

Wirth, M., Hamilton, G. and Gabor, F. (1998b) Lectin-mediated drug targeting: quantification of binding and internalization of wheat germ agglutinin and Solanum tuberosum lectin using Caco-2 and HT-29 cells. *J. Drug Targeting*, **6**, 95–104.

Chapter 6

# High-throughput epithelial cell culture systems for screening drug intestinal permeability

*Laurie Withington*

## INTRODUCTION

Within the pharmaceutical industry, the assessment of a compound's intestinal permeability remains one of the key determinants in the drug development process. Bioavailability measurements are determined by measuring the drug in the blood, hours following oral administration to an animal. How well the drug is absorbed across the intestinal epithelia is difficult to determine with these *in vivo* measurements. With the advancement of new and improved cell culture systems, intestinal cell lines are available to study the transport of therapeutics across the epithelia, although these cell culture systems can also be a very time-consuming, labor-intensive process. Due to the logarithmic increase in pharmaceutical leads from the drug discovery stages of the pipeline, there is, in turn, a serious demand to increase throughput of *in vitro* cell-based assays for evaluating intestinal permeability. A significant number of pharmaceutical companies have altered their drug development testing strategy, such that they are now conducting primary screens of compounds for various ADME (absorption, distribution, metabolism, excretion) parameters to decrease the amount of compounds advancing into traditional, more time-consuming secondary assays. The goal is to rank order drug-permeability into groups, so that only the more 'drug-like' compounds will continue in the drug development pipeline, minimizing the bottleneck of excessive number of compounds in the drug development phase.

To increase throughput of these cell-based assays, modifications can be applied to traditional cell culture protocols and traditional assay platforms. This chapter focuses on the types of modifications that can be made for creating high-throughput screens to test for intestinal permeability.

## *IN VITRO* CELL CULTURE MODELS

Within the pharmaceutical industry, traditional *in vitro* assays for assessing intestinal permeability involve the use of either primary or transformed intestinal cell types that are seeded onto microporous membrane insert systems. The membranes used for drug permeability contain pore sizes ranging in diameter from 0.4 μm to 1 μm, to allow drug compounds to pass through them. Basically, cells are cultured on the membrane inserts, and drugs diluted in a transport buffer are added to the upper chamber (apical compartment) of the system. Transport buffer is added to the lower chamber (basal compartment) of the insert system. The movement of drug from the apical compartment, through the cell monolayer and membrane insert, to the basal compartment is intended to mimic the *in vivo* movement of drug from the intestinal lumen into the bloodstream.

Various types of membrane inserts have been used over the years to culture cells for drug transport studies. Low-throughput insert systems such as Boyden chambers and Snapwell™ diffusion systems were traditionally used to conduct *in vitro* assays. As the technology developed, commercially available insert systems (BD Biosciences, Corning) were introduced as individual insert chambers in a variety of pore sizes and densities (6-well, 12-well or 24-well). More recently, with the introduction of the 24- and 96-multiwell format (BD Biosciences), all 24 or 96 inserts are part of a single unit suitable for both manual and robotic compound screening.

Various cell types that attempt to mimic the intestine have been used in the inserts. Primary cells have been isolated from various non-human species. Obviously, one restriction of these cells is that they are non-human, and another is that they tend to lose intestinal-like characteristics after limited passages in culture (Pageot *et al.*, 2000). Various spontaneously transformed human intestinal cell lines have been identified for potential use in models. Two cell lines from a human colon adenocarcinoma have been identified, HT-29 and Caco-2, that are capable of enterocytic differentiation with prolonged time in culture (Blais *et al.*, 1997). A subclone of Caco-2 cells, TC7, and a colonic crypt cell line, T84, have also been evaluated (Delie and Rubas, 1997). Within the literature, the majority of drug permeability studies have been conducted with the Caco-2 cell line. The more commonly used Caco-2 cell line has been shown to differentiate into a polarized, epithelial cell monolayer containing brush-border membranes and intercellular tight junctions after approximately 21 days in culture.

Generally within pharmaceutical companies, drug transport studies are conducted to determine basic characteristics of unknown lead compounds, for example: to what extent is the compound permeable through a monolayer; is the compound passively or actively transported; and, is the compound's permeability transporter-dependent. Numerous studies show that the permeability of passively absorbed compounds through Caco-2 monolayers correlates with the percentage of oral absorption seen in humans (Delie and Rubas, 1997). Movement of molecules via specific transporters in Caco-2 monolayers, such as glucose, peptides, amino acids, nucleosides, bile acids, protons, monocarboxylic acids, is shown to exemplify what occurs in the intestine (Delie and Rubas, 1997). The presence and the activity of specific efflux transporters are shown to be functional in Caco-2 monolayers. The role of efflux proteins is meant to prevent molecules from entering the bloodstream, thus maintaining them in the intestinal lumen. Efflux transporters associated with the intestine as well as Caco-2 cells include multidrug-resistance (MDR1) protein-1 (Gutmann *et al.*, 1999), MDR1-associated protein-1 and -2 (Makhey *et al.*, 1998), organic anion transport protein-1 and -3, (Makhey *et al.*, 1998), and organic cation transport protein-1 (Bleasby *et al.*, 2000). One of the major downfalls of the 21-day Caco-2 culture is that mold or bacterial contaminations are common. Unfortunately, once bacterial contamination occurs in culture, it can spread easily to other cultures in the incubator. In addition, a three-week culture period is not feasible for running numerous drug-transport studies, thus an alternative model may prove worthwhile.

A non-intestinal cell line, the Madin–Darby canine kidney (MDCK) cell line, has also been recently evaluated as a model for drug-transport studies (Irvine *et al.*, 1999). The use of MDCK cells are highlighted by the fact that monolayers with mature intercellular junctions are obtained after five days in culture. Some negative factors for using these cells are that they are non-human, and they are of a non-intestinal origin. In addition, renal cells are known to contain different transporters than intestinal cells, creating many more discrepancies. In brief, the short culture period is a definite advantage, although a human intestinal *in vitro* model would be superior.

For the above reasons, the BD BioCoat™ three-day Caco-2 Assay System has become commercially available, and its mechanisms are discussed throughout this chapter (BD, Bedford, MA).

## HIGH-THROUGHPUT CACO-2 CULTURES

### Differentiation agonists

Seeing the benefit for creating an alternative high-throughput cell culture model, methods for generating an accelerated Caco-2 culture were pursued. Briefly, the *in vivo* process of intestinal differentiation is thought to proceed as follows: as absorptive cells pass from the neck of the crypt in the intestine to the surface of the crypt, well-developed brush-borders containing microvilli are formed, as well as mature intercellular tight junctions. As differentiation proceeds from the intestinal crypt to the surface compartment, cells show considerable vacuolation, increased urokinase expression, decreased adhesion to the basement membrane, and apoptosis and/or cell shedding occurs (Hall *et al.*, 1994). The intracellular mechanisms in epithelial cells that induce differentiation are not completely understood, although numerous factors have been shown to induce such processes *in vitro*. Differentiation of intestinal epithelial cells is shown to be regulated by agents such as growth factors (Hardin *et al.*, 1993), sodium butyrate (Souleimani and Asselin, 1993), neurohumoral peptides (Gomez *et al.*, 1995), luminal nutrients (Turowski *et al.*, 1994) and extracellular matrices (Basson *et al.*, 1996; Darimont *et al.*, 1998). The fact that phenotypic changes occur as cells migrate from the neck of the crypt to the surface compartment suggests that contents in the intestinal lumen may influence epithelial cell differentiation.

Short-chain fatty acids, such as butyrate, are bacterial products of carbohydrate fermentation and are present in millimolar concentrations in the lumen (Bond and Levitt, 1976). Whether butyrate actually plays a role in intestinal differentiation *in vivo* is unknown, but it has been shown to differentiate Caco-2 cells *in vitro*. In response to butyrate exposure, Caco-2 monolayers have been shown to induce urokinase activity, cell cycle arrest, and display increased transepithelial electrical resistance (TEER) values (Mariadason *et al.*, 2000). At similar dose and time intervals, sodium butyrate also reduces c-myc mRNA levels in Caco-2 cells (Souleimani and

Asselin, 1993). It is possible that post-transcriptional modification of gene expression could be one of the major targets of butyrate in Caco-2 cells, which in turn induces specific protein modifications.

Extracellular matrix proteins have also been shown to induce differentiation in Caco-2 cells. Caco-2 cells cultured on either collagen-I, -IV, or laminin-coated plasticware significantly increases activity levels of alkaline phosphatase (AKP), dipeptidyl peptidase (DPP), lactase, sucrase-isomaltase, as well as cell spreading (Basson *et al.*, 1996). These brush-border enzymes are common markers used to determine Caco-2 cell differentiation, and a monolayer is considered differentiated when these enzymes are active.

The combination of two differentiation factors, sodium butyrate and collagen-I, are used to create the BD BioCoat™ three-day Caco-2 Assay System, which speeds the differentiation of the Caco-2 culture from 21 days to three days (BD Biosciences, Bedford, MA). This system contains a proprietary enterocyte differentiation media containing sodium butyrate, and a multiwell insert plate coated with collagen-I by a proprietary process to form fibrils. This combination of fibrillar collagen-coated inserts with the proprietary media formulation, provides a patented process to differentiate Caco-2 cells in three days versus the conventional 21-day culture.

## Protocols

### Stock cultures of Caco-2 cells

One of the most critical factors for maintaining Caco-2 cells for routine transport studies are standardized protocols. It is vital that one consistently passages Caco-2 cells the same way so that discrepancies in transport assays cannot be attributed to how the stock cultures were handled. Cells are generally grown in cell culture flasks ranging in size from 25 to 175 cm$^2$. Cells are subcultured by splitting when they are 80–90 per cent confluent. This level of confluency in Caco-2 cells is reached when cell numbers are $0.2–0.25 \times 10^6$ cells/cm$^2$ (Delie and Rubas, 1997). This number of cells is equivalent to $35–44 \times 10^6$ cells/175 cm$^2$ flask. The effect of cell density in stock cultures prior to the assay is critical for establishing optimum cell barrier formation. Figure 6.1 compares the appearance of Caco-2 cultures grown at 85,000 cells/cm$^2$ and 250,000 cells/cm$^2$ (Asa and Timmins, 1998). Figure 6.2 compares mannitol permeability using cells that were grown at the varying densities. Cells cultured at 250,000 cells/cm$^2$ produced cell monolayers with significantly lower mannitol permeability values, and thus with tighter monolayers.

Cells are routinely split at a ratio of 1:10 every seven days for seeding in new flasks. Caco-2 cells are subcultured using 0.05 per cent trypsin/0.53 mM ethylenediaminetetraacetic acid (EDTA) in Hank's balanced salt solution (HBSS) (without Ca$^{2+}$ and Mg$^{2+}$). For cells used in the accelerated BD Caco-2 system, it is important to passage cells using trypsin at this low concentration, as opposed to traditionally used higher concentrations of 0.25–0.5 per cent trypsin. Stock cultures are maintained in Dulbecco's modified Eagle's medium (DMEM) containing glutamine in Falcon™ tissue culture flasks. Fetal bovine serum (FBS) is added at 10–20 per cent, and antibiotics are not supplemented.

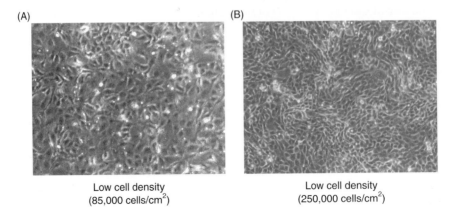

Low cell density
(85,000 cells/cm²)

Low cell density
(250,000 cells/cm²)

*Figure 6.1* Caco-2 cells are cultured in Dulbecco's modified Eagle's medium (DMEM)+20 per cent fetal bovine serum (FBS) for different lengths of time. The cells in (A) are grown to a density of 85,000 cells/cm². The cells in (B) are grown to a density of 250,000 cells/cm². Note that even at lower densities the Caco-2 cells can completely cover the growth surface of the tissue culture vessel (×10).

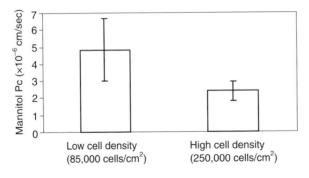

*Figure 6.2* Caco-2 cells are cultured in tissue culture flasks to either low (85,000 cells/cm²) or high (250,000 cells/cm²) density prior to use in the BD BioCoat™ HTS Caco-2 assay system. Cells are seeded and cultured as per manufacturers' instructions. Mannitol permeability ($n$=12)±standard deviation is determined.

### Using the accelerated Caco-2 cell model for transport assays

Following trypsinization of a stock culture of Caco-2 cells, the cells can be used for seeding onto membrane inserts for transport studies. The reagents required for the accelerated BD BioCoat™ Caco-2 Assay System (BD Biosciences, Bedford, MA) are:

- Basal seeding media containing MITO+™ serum extender,
- Entero-STIM™ differentiation media,
- HTS fibrillar collagen 24-multiwell insert systems (1 μm pore size).

The serum-free Basal Seeding Media is DMEM-based and used for growing Caco-2 cells prior to differentiation. MITO+™ Serum Extender is a concentrated, lyophilized, formulation of hormones, growth factors, and other metabolites, required for the maintenance of cells under serum-free conditions. Entero-STIM™ Differentiation Media is a serum-free fully defined media containing butyric acid. The 24-multiwell insert systems have been treated with type 1 rat tail collagen under conditions that allow *in situ* formation of large collagen fibrils.

In the accelerated Caco-2 model, cells are seeded onto the inserts at a higher density than the traditional 21-day model. Briefly, cells are seeded (day zero) at $2–2.5 \times 10^5$ cells/insert ($0.3\,cm^2$) in Basal Seeding Media containing MITO+™ Serum Extender and incubated for 24 hours. Media is then changed (day one) to Entero-STIM™ Differentiation Media containing MITO+™ Serum Extender and incubated for an additional 48 hours. Drug transport assays are performed at day three.

*Helpful Hint*: All cell types that are cultured on membrane inserts should be treated with care upon the addition and removal of media to the insert. When changing media, do not remove all the media from the apical compartment so as to prevent drying of the cells, but leave a small amount on the cells prior to the addition of the new media. One may argue that the remaining spent media in the insert will prevent optimum cell growth, although we have found that as long as it is a small amount, cell growth is not compromised. When adding media to the apical compartment, add the media slowly down the side of the insert to prevent physical disruption of the cell monolayer.

## Morphology

The morphological structure of Caco-2 monolayers in the accelerated protocol was prepared for electron microscopic evaluation (Woods and Asa, 1997). Post-fixation of specimens was in 2 per cent weight/volume $OsO_4$ in 0.2 M sodium cacodylate buffer, pH 7.2. After dehydration through a graded series of ethanol steps, the specimens were embedded in a firm recipe of Spurr's embedding resin. Gold sections were obtained using an LKB ultra-microtome. Sections stained with uranyl acetate and lead citrate were examined in a Hitachi 7100 transmission electron microscope. As seen in Figure 6.3, the ultrastructural components of Caco-2 cells grown in the accelerated three-day environment display characteristics of a differentiated Caco-2 monolayer. A monolayer of cells containing surface specialized microvilli, intercellular tight junctional processes, and desmosomes are readily apparent. Magnification of the intercellular junctions is highlighted in Figure 6.4. Tight junctional complexes form toward the apical portion of the cells closest to the microvilli, while desmosomes are periodic intercellular attachments located in regions between the tight junctions and the basal membrane. It has been stated that a differentiated Caco-2 monolayer must express the tight junctional complexes and desmosomes, as well as reveal typical brush-border microvilli, characteristic of morphological polarization of cells (Delie and Rubas, 1997).

Recently, a study in Switzerland qualified Caco-2 tight junctional protein expression in the accelerated three-day model and the traditional 21-day model (Rothen-Rutishauser *et al.*, 2000). Briefly, cells were fixed and labeled with anti-zonulae

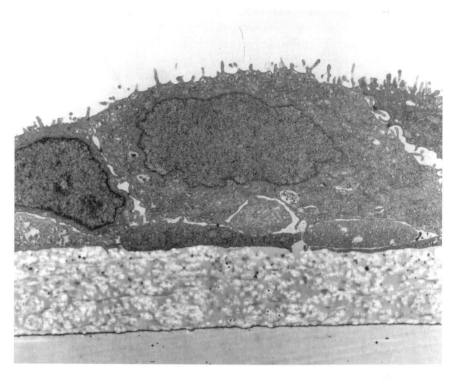

*Figure 6.3* Electron micrograph of a Caco-2 cell monolayer cultured using the BD BioCoat™ HTS Caco-2 assay system. Differentiation characteristics such as microvilli, cellular interdigitations, and tight junction formation are readily apparent (× 14,400).

occludentes (ZO)-1/anti-rat cyanine-5, and confocal data sets were sampled on a Zeiss LSM 410 inverted microscope. A single monolayer of cells was visualized in the accelerated Caco-2 model with a complete network of tight junctional complexes between cells. The authors noted that Caco-2 cells cultured for 21 days sometimes produced multilayers on standard membranes (Rothen-Rutishauser *et al.*, 2000). The multilayers were distinguished from monolayers by the expression of tight junctional proteins in the lower layers as well as at the apical surface.

## Biochemistry

The evolution of traditional 21-day Caco-2 cultures undergoing spontaneous differentiation, grown either on plastic or on a permeable filter, is characterized by the event sequence of proliferation, confluency and differentiation. After seeding cells at $4 \times 10^4$ cells/cm$^2$, proliferation starts after a lag time of 48 hours and confluency is reached after five days, although proliferation may continue up to nine days (Delie and Rubas, 1997). The differentiation process spontaneously follows the proliferation and confluency phases, and is generally identified after a 21-day

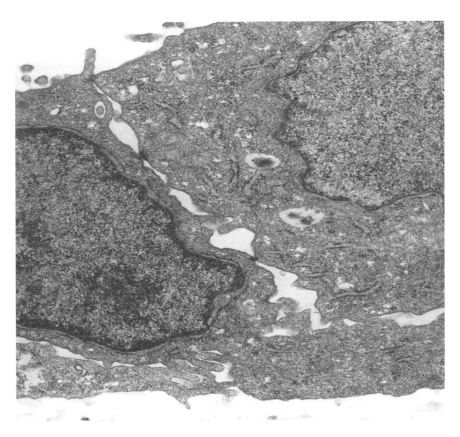

*Figure 6.4* Electron micrograph of a Caco-2 cell monolayer cultured using the BD BioCoat™ HTS Caco-2 assay system. This micrograph isolates the intercellular junction formation between two adjacent cells (×41,400).

culture. A differentiated monolayer is in part characterized by upregulation of brush-border enzymes on the surface of microvilli. Such brush-border peptidase activities include AKP, DPP-IV, and aminopeptidase-N (APN). The activity of these enzymes is low during the proliferative phase, and increased once confluency is reached. In Caco-2 cells, the maximum enzyme activity is about half of what is observed with small intestinal cells for AKP and ten per cent of that seen for APN (Pinto *et al.*, 1983). Despite their colonic origin, Caco-2 cells appear to express enzymes typically found in small intestinal cells.

In the accelerated three-day Caco-2 model, cells are seeded at a significantly higher density of $6-7.5 \times 10^5$ cells/cm$^2$ as compared with the traditional 21-day model. This higher seeding density is necessary for the cells to reach confluency after one day in culture. Differentiation of the three-day model is not spontaneous, but is initiated on the second day of culture by the addition of butyric acid and by the specialized fibrillar collagen matrix. BD Biosciences in collaboration with Setsunama University has shown that DPP and APN are upregulated in the accelerated model to the same extent as the 21-day Caco-2 culture (Taki *et al.*, 1999). In addition, AKP is significantly upregulated compared with undifferentiated Caco-2 controls (Asa and Timmins, 1998).

## DRUG TRANSPORT STUDIES USING HIGH-THROUGHPUT CACO-2 CELL CULTURES

It has been shown that a differentiated Caco-2 monolayer suitable for drug transport studies can be produced in three days. The most important determinate of an intestinal cell culture model is the use and ability to classify a set of representative compounds that are generally used as standards by pharmaceutical companies for the classification of drugs. Such classifications include low permeability, high permeability, carrier-mediated permeability and efflux activity.

### Monolayer integrity

In order to classify drug permeability, one must ensure that the cell culture model used contains functional intercellular junctions that have the ability to regulate paracellular permeability. Charged and highly water soluble compounds, including peptides, are typically unable to penetrate cell membranes, preferring the paracellular route. The limited space inhibits transport through this pathway, and the presence of functional tight junctions further limits passive diffusion of larger molecules. Typically, hydrophilic, low molecular weight compounds such as mannitol and PEG are tested to ensure that a significantly low amount permeates through the paracellular junctional pathway. If the amount transported through this route is at a low enough level, then the integrity of that monolayer is considered intact. The acceptance level for mannitol permeability may differ from laboratory to laboratory; thus there is no standardized mannitol permeability coefficient. Reported mannitol permeability values obtained through traditional 21-day cultured Caco-2 monolayers are seen to range from $0.18$–$9.70 \times 10^{-6}$ cm/sec (Deli and Rubas, 1997). Reported mannitol values obtained using the BD BioCoat™ cultured Caco-2 cell monolayers have been shown to range from $0.7 \times 10^{-6}$ cm/sec (Taki *et al.*, 1999), to $1.0 \times 10^{-6}$ cm/sec (Withington and Asa, 1999), to $4.0 \times 10^{-6}$ cm/sec (Chong *et al.*, 1997).

Electrophysiology studies are used to evaluate ion transport across mucosal barriers. In Caco-2 monolayers, electrical conductivity is almost limited to the paracellular ion flux; thus the TEER across the monolayer is a good indicator for the development of tight junctions and is commonly used to describe monolayer integrity. The drawback of TEER measurements as a standardized measurement for monolayer integrity is that many different factors can alter TEER measurements. TEER increases with increasing Caco-2 cell passage number (Deli and Rubas, 1997). Transport studies are generally conducted with cells cultured at passages ranging from 20 to 80, creating the potential for extensive variability among the reported TEER measurements. TEER also decreases with increasing temperature (Deli and Rubas, 1997). Thus, TEER measurements reported using monolayers ranging in temperature from 25 to 37 °C also create extensive variability. TEER measurements can also vary based on the plastic support that is used (Deli and Rubas, 1997). Insert diameter, pore size and pore number can all lead to cell monolayer TEER variability. Taking this into account, it seems difficult to compare TEER values reported by different authors, yet it may be valuable to provide evidence of an eventual toxicity provoked by a studied compound. Caco-2 monolayers cultured in the accelerated three-day system generally display TEER measurements ranging from 200 to 1000 $\Omega cm^2$ at the time of transport study (unpublished data). Another study comparing TEER values of Caco-2 monolayers using the accelerated three-day method from day one to day three indicates values of 100 ($\Omega cm^2$) at day one and 400 ($\Omega cm^2$) at day 3, the final day of culture (Taki *et al.*, 1999). These data lend further evidence to the claim that tight junctional development occurs in the accelerated three-day Caco-2 culture to create monolayer integrity.

## Passive permeability

Almost all drugs that are given orally are absorbed across the intestinal mucosa by passive diffusion. A drug that has high passive, transcellular permeability in a cell culture model tends to be a better candidate for further drug development. The Food and Drug Administration (FDA) has listed a set of drug standards of known permeability and solubility in a Biopharmaceutics Classification System (BCS) that can be used to validate new assay systems. Table 6.1 illustrates the comparison of permeability values between Caco-2 cells cultured for 21 days and those cultured in the accelerated three-day culture using drugs listed in the BCS (Withington and Asa, 1999). In the three-day system compounds of high-expected permeability had high permeability coefficient values ($24.6-60.4 \times 10^{-6}$ cm/sec), while compounds of low-expected permeability had low permeability values ($0.5-4.9 \times 10^{-6}$ cm/sec).

*Table 6.1* Drug permeability comparison between BioCoat™ three-day and traditional 21-day Caco-2 assay

| Drug | BioCoat three-day Pc ($\times 10^{-6}$ cm/sec) | Traditional 21-day Pc ($\times 10^{-6}$ cm/sec) |
|---|---|---|
| Propranolol | 34.8 ± 13.9 | 43.0 ± 3.8 |
| Metoprolol | 35.0 ± 11.5 | 18.1 ± 2 |
| Ketoprofen | 30 ± 1 | 28.2 ± 2 |
| Carbamazepine | 35 ± 6 | 22.8 ± 4 |
| Caffeine | 41.8 ± 2 | 35.5 ± 2 |
| Theophylline | 24.6 ± 2 | 20.9 ± 1 |
| Antipyrine | 60.4 ± 4 | 69.4 ± 3 |
| Naproxen | 42.5 ± 2 | 47 ± 2 |
| Atenolol | 4.5 ± 2.3 | 0.1 ± 0.07 |
| Hydrochlorothiazide | 2.2 ± 0.01 | 6.6 ± 1.9 |
| Furosemide | 4.9 ± 2 | 4.8 ± 1 |
| Ranitidine | 0.5 ± 0.1 | 1.3 ± 1 |
| Mannitol | 1.0 ± 0.1 | 3.0 ± 1.0 |
| Verapamil (A > B) | 51.9 ± 6 | 69.4 ± 0.1 |
| Verapamil (B > A) | 89.7 ± 2 | 81.3 ± 2 |
| Lucifer yellow | 0.8 ± 0.1 | 0.9 ± 0.1 |

In order to rank compound permeability in a set of unknown drugs, it is important to be able to differentiate the different levels of permeability in a model assay system.

## Efflux activity

Many studies on intestinal absorption of drugs suggest that compound bioavailability may be limited by the resistance of permeability across the intestinal barrier. The MDR1 gene product $p$-gp was originally found to cause drug resistance in cancer chemotherapy, as well as in Caco-2 cells, upon treatments with anticancer drugs (Hunter *et al.*, 1993). At low drug concentrations, the $p$-gp transporter may secrete drugs out of the epithelium back into the intestinal lumen. At high drug concentrations, the secretion may be saturable, leading to an apparent increase in absorption. Although other efflux transporters are functionally expressed in the intestine, such as organic cation transporter, organic anion transporter, and multiple resistance protein (MRP), most studies have focused on $p$-gp activity.

The level of $p$-gp activity in a cell culture model is commonly determined by initially adding drug compound to the apical compartment of the insert such that the drug is in contact with the cells, while buffer alone is added to the basal compartment. In another insert, the same drug is added to the basal compartment, while buffer alone is added to the apical compartment. The drug flux is compared between the two inserts, and if the monolayer actively expresses $p$-gp, there will be higher drug flux from the basal compartment into the apical compartment, than the apical-to-basal direction. If the drug is passively absorbed, there will be equivalent drug flux in both directions. One study compared the flux of two $p$-gp specific substrates, propranolol and quinidine, in this manner (Sweetland and Polzer, 1998). In the accelerated three-day Caco-2 culture, compound flux was four–five times greater in the basal-to-apical direction. In the traditional 21-day Caco-2 culture, compound flux was six–ten times greater in the basal-to-apical direction. Although both models effectively express $p$-gp, the 21-day culture showed higher expression than the three-day culture. Recently, it has been identified that if the Caco-2 culture is extended to five days in the BD BioCoat™ Caco-2 system, the monolayers show increased $p$-gp activity (See Variations in assay preparation).

## Transporter activity

Intestinal cells are equipped with an array of transport proteins to facilitate uptake of otherwise poorly absorbed compounds. Fortunately, the Caco-2 cell expresses many of the important carrier proteins associated with drug transport and delivery (Delie and Rubas, 1997). Benzoic acid is a common substrate used to test for the presence of the monocarboxylic acid transporter. In both the three-day culture and the traditional 21-day culture, the transport of this substrate was found to be more than double in the apical-to-basal direction than in the basal-to-apical direction, indicating carrier-mediated transport (Sweetland and Polzer, 1998). Phenylalanine, a substrate for the amino acid transporter, was also greater in the apical-to-basal direction in both culture systems, indicating transporter activity (Sweetland and Polzer, 1998). In addition, permeability of a substrate for the dipeptide carrier, glycyl-sarcosine, was found to be similar between the two assay systems (Taki *et al.*, 1999).

## Variations in assay preparation

Variations of the accelerated three-day Caco-2 culture protocol have been employed to determine if there is an effect on drug permeability values. One such variation was to extend the culture time of the three-day model to five days. At the end of three days, cells were refed with Entero-STIM™ Differentiation Media containing MITO+™ Serum and incubated for two additional days. The extension of the culture period to five days was found to increase the TEER measurements (Taki *et al.*, 1999), decrease mannitol permeability (Sweetland and Polzer, 1998), increase *p*-gp activity (Taki *et al.*, 1999; Withington and Asa, 1999), increase dipeptide carrier activity (Taki *et al.*, 1999), increase amino acid transporter activity (Sweetland and Polzer, 1998), and increase monocarboxylic acid transporter activity (Sweetland and Polzer, 1998).

## SPECIALIZED ENABLING PLATFORMS TO INCREASE THROUGHPUT

In recent years, the use of cell-based assays for drug-discovery screening has increased tremendously, although some cell-based assays remain difficult to automate, owing either to special cell handling conditions or to general format incompatibility with robots. This is particularly true of membrane-based cellular assays where separate housings (or inserts) support individual membranes. These inserts are difficult to manipulate, making automation complicated and expensive. To resolve this problem, BD Biosciences has developed automation-compatible HTS insert systems that enable pharmaceutical companies to increase throughput of intestinal drug absorption testing. BD has developed both 24-multiwell insert systems, as well as 96-multiwell insert systems, where all 24 and 96 inserts are part of a single unit.

## HTS 24-multiwell insert system

At the time BD Biosciences designed the 24-multiwell insert system in 1997, most Caco-2 users were using individual inserts for placement in either 6-well, 12-well or 24-well polystyrene plates. In designing this insert system for automation and throughput, the system was optimized for many of the following characteristics:

- ease of set-up or use
- ease of robotic manipulation
- fluid handler access to the upper and lower chambers for sampling
- prevent cross-contamination between chambers
- consistent performance.

The BD Falcon™ HTS 24-multiwell insert system consists of a multiwell insert plate housing, a feeder tray, and a non-directional lid composed of polyethylene terephthalate (PET) (Figure 6.5). PET material was chosen for its excellent cell compatibility and optical characteristics. The insert plate is designed for use with the standard BD Falcon 24-well™ plate. This system has many features designed to facilitate automation. Twenty-four inserts were integrated into a one-piece design

<u>LID:</u>
Material: PET (Polyethylene Terepthalate)
Length: A = 129.57 mm (5.101 inches)
Width:  B = 86.82 mm (3.418 inches)
Height:  C = 8.20 mm (0.323 inches)

<u>INSERT PLATE HOUSING:</u>
Material: PET (Polyethylene Terepthalate)
Length: D = 127.61 mm (5.024 inches)
Width:  E = 85.01 mm (3.347 inches)
Height:  F = 18.14 mm (0.714 inches)

  *Insert Well*:
  Top Interior Diameter: 12.50 mm (0.492 inches)
  Bot. Interior (membrane) Diameter: 6.50 mm
                (0.250 inches)
  Bot. Exterior Diameter: 10.00 mm (0.394 inches)
  Total Well Depth: 18.14 mm (0.714 inches)
  Well to Well Distance: 19.30 mm (0.760 inches)
  Sampling Port Length: 9.50 mm (0.37 inches)
  Sampling Port Width: 4.00 mm (0.16 inches)

<u>FEEDER TRAY:</u>
Material: PS (Polystyrene)
Length:  G = 127.86 mm (5.034 inches)
Width:   H = 85.47 mm (3.365 inches)
Height:  I = 19.94 mm (0.785 inches)

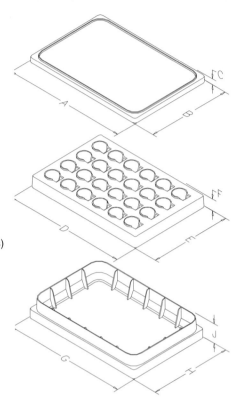

*Figure 6.5* Composed of three components, the 24-multiwell insert system has been designed to be compatible with common laboratory automation. Basic dimensions are shown above.

to facilitate manipulation. Flanges on each component have at least a 6 mm vertical gripping surface for easy robotic handling. The insert plate flange can be gripped with the lid on and has a readable surface for labeling or identification. Furthermore, the inner edge of this flange is beveled to aid alignment during assembly. The lid is rectangular, free of interior features, smooth on top for vacuum delidders, and has minimal system overhang. The assembled system height is 2 mm higher than a Falcon 24-well™ plate and lid. Numerous cellular based assays have been validated using this system (Henderson and Asa, 1998).

## Cell culture

Optimized cell culture is one of the most important determinants of obtaining consistent data with cell-based assays, while at the same time the most labor-intensive. For this reason, BD Biosciences designed the feeder tray, which eliminates some of this burden. When culturing the cells, instead of placing the insert housings in 24-well plates, the inserts are placed in a tray containing one compartment. The user simply adds the appropriate volume to the feeder tray, and thus does not have to add media to 24 individual compartments. Feeder tray ribs, which were originally designed as side wall supports, additionally were observed to reduce liquid sloshing during system movement. The insert housings contain access ports, which allow the addition and removal of liquids from the basal compartments, to be done by both robotic and manual manipulation.

### Transport assay

After the cells have been in culture for appropriate time, the insert housings are transferred to a 24-well polystyrene plate. When measuring the compound's permeability, it is important to separate the apical compartment from the basal compartment by placing the insert system into a 24-well plate. Both during the assay, and after completion of the assay, samples in the basal compartments are taken robotically or manually through the access ports.

## HTS 96-multiwell insert system

Due to the high demand for more high-throughput platforms than the 24-Multiwell Insert System, BD Biosciences developed a 96-multiwell insert system (Figure 6.6). Similar to the 24-well system, the 96-multiwell insert system is automation-compatible and can be used with robotic systems that are currently being used with the 24-multiwell insert system. Both the 24- and the 96-multiwell insert systems are currently being run on robots from such companies as TECAN, Beckman Coulter and Brandel. The 96-multiwell insert platform provides many advantages. The smaller wells enable five-fold less sample to be used, which in itself is a significant cost-reduction to pharmaceutical companies, since the developed drug candidates are costly. The system is not only amenable to robotic manipulation, but also to manual use with multipipettors. The user can run significantly more samples per assay, which is the main reason that companies switch from the 24- to the 96-multiwell insert system. Aside from all the advantages of the system, the main concern of users is if the rank order of drug permeability can still be achieved with the 96-well insert system.

### Cell culture

Like the 24-multiwell insert system, the 96-multiwell system also has a feeder tray to ease the burden of culturing the cells. The appropriate amount of media is added to the feeder tray, and the insert housing fits securely on top of the feeder tray. The insert housing also contains access ports that allow access to the basal compartment for robotic or manual addition of media.

### Transport assay

After the cells have been in culture for appropriate time, the 96-multiwell insert platform is transferred to a 96-well plate designed specifically for the insert housing. One of the main modifications to the plate was to create square wells as opposed to round wells. This modification allows enough space for the insert compartment as

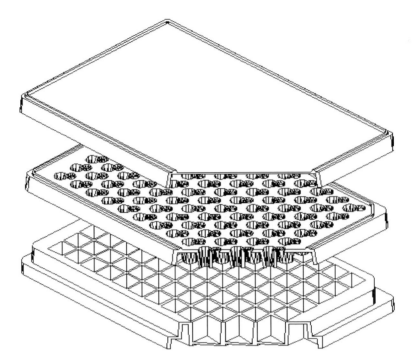

*Figure 6.6* Composed of four components, the 96-multiwell insert system includes the non-directional lid, the insert plate with access ports, the custom receiver tray, and the custom feeder tray. The system has been designed to be compatible with common laboratory automation.

well as accessibility through the access port. The square well was further modified to prevent wicking of liquid up the sides of the well. An automation-compatible modification was to slant the bottom of each well. The slanted bottom enables a robot or user to extract samples from the basal compartment either during the assay or at the end of the assay. The slant allows the entire sample to be removed, and for those collecting samples at various time points, this feature is crucial.

Permeability values for a subset of compounds listed in the BCS were compared using the BD BioCoat™ Caco-2 Assay System in both the 24-multiwell and the 96-multiwell insert systems (Table 6.2) (Withington *et al.*, 1999). About four–five-fold less cells were used to seed the inserts in the 96-multiwell inserts compared with the 24-multiwell inserts. In addition, about six-fold less media were used in the apical compartment, and four-fold less media in the basal compartment. Nonetheless, the values obtained using the different plates were comparable, and rank ordering of the permeability of the compounds was achieved. Permeability values of four different drugs were also compared between the two sized inserts using the traditional 21-day Caco-2 culture (Figure 6.7) (Withington and Asa, 2000). Again, the rank ordering of the compound's permeability remains intact with both systems. For drug absorption screening purposes, the rank order of compound's permeability is the most crucial element that must prevail in a cell culture model.

*Table 6.2* Comparison of drug permeability obtained in 96-multiwell and 24-multiwell insert systems

| Drug | 24-multiwell assay system | | 96-multiwell assay system |
|------|---------------------------|---|---------------------------|
| | BioCoat three-day Pc ($10^{-6}$ cm/sec) | 21-day Pc ($10^{-6}$ cm/sec) | BioCoat three-day Pc ($10^{-6}$ cm/sec) |
| Propranolol | 52±0.9 | 43.0±3.8 | 31.5±6 |
| Theophylline | 24.6±2 | 20.9±1 | 30±4 |
| Antipyrine | 60.4±4 | 69.4±3 | 33±1 |
| Naproxen | 42.5±2 | 47±2 | 29±2 |
| Caffeine | 41.8±2 | 35.5±2 | 54±4 |
| Furosemide | 4.9±2 | 4.8±1 | 2±0.5 |
| Ranitidine | 0.5±0.1 | 1.3±1 | 0.6±1 |
| Mannitol | 1.0+0.1 | 3.0±1.0 | 1.2±1 |
| Lucifer yellow | 0.8±0.1 | 0.9±0.1 | 0.63+0.2 |

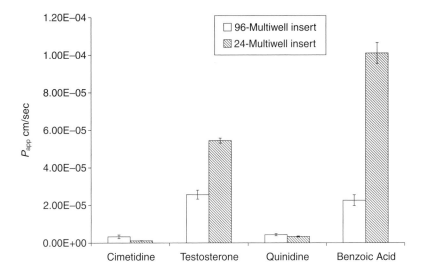

*Figure 6.7* Drug permeability (A>B) was compared between cells cultured in the 96-multiwell insert system and the 24-multiwell insert system. Caco-2 cells were cultured for 21 days (n=8), and mannitol and lucifer yellow flux were <0.5 per cent per hour and <3 per cent per hour, respectively.

## CONCLUDING REMARKS

One of the major bottlenecks in the pharmaceutical industry today is created from the logarithmic increase of compounds emerging from primary screens of new drug candidates. This in turn produces an increased number of compounds merging into the traditional secondary assays that are required for further drug development. Intestinal drug-absorption screening is a fundamental component of the drug-development pipeline. The ability to create high-throughput assays within drug-absorption screening will lessen pharmaceutical bottlenecks.

The model discussed within this chapter utilizes differentiation agonists to create an accelerated Caco-2 culture, as well as insert platforms in 24 and 96-multiwell formats to increase assay throughput. The main purpose of these increased throughput assay systems is to narrow down the number of drug candidates that proceed into more time-consuming secondary assays. Drug developers basically now have the ability to screen out the more 'drug-like' compounds faster and more efficiently. The idea of high-throughput drug-absorption assays is not meant to be the sole assay for drug-absorption determination, but to be a screen to narrow down the number of compounds that do need further drug-development testing.

## REFERENCES

Asa and Timmins (1998) *Poster Presentation at AAPS*, Washington, DC.

Basson, M., Turowski, G. and Emenaker, N. (1996) *Exp. Cell Res.*, **225**, 301–305.

Blais, A., Aymard, P. and Lacour, B. (1997) *Eur. J. Physiol.*, **434**, 300–305.

Bleasby, K., Chauhan, S. and Brown, C. (2000) *Br. J. Pharmacol.*, **129**, 619–625.

Bond, J. and Levitt, M. (1976) *J. Clin. Invest.*, **57**, 1158–1164.

Chong, S., Dando, S. and Morrison, R., (1997) *Pharm. Res.*, **14**, 1835–1837.

Darimont, C., Gradoux, N., Cumin, F., Baum, H. and Pover, A. (1998) *Exp. Cell Res.*, **244**, 441–447.

Delie, F. and Rubas, W. (1997) *Crit. Rev. Ther. Drug Carrier Syst.*, **14**, 221–286.

Gomez, G., Zhang, T., Rajarman, S., Thakore, K., Yanaihara, N., Townsend, C., *et al.* (1995) *Am. J. Physiol.*, **268**, G71–G81.

Gutmann, H., Fricker, G., Torok, M., Michael, S., Beglinger, C. and Drewe, J. (1999) *Pharm. Res.*, **16**, 402–407.

Hall, P., Coates, P., Ansari, B. and Hopwood, D. (1994) *J. Cell Sci.*, **107**, 3569–3577.

Hardin, J. A., Buret, A., Meddings, J. B. and Gall, D. G. (1993) *Am. J. Physiol.*, **264**, G312–G318.

Henderson and Asa (1998) *Poster Presentation at Lab Automation*, San Diego, CA.

Hunter, J., Hirst, B. and Simmons, N. (1993) *Pharm. Res.*, **10**, 743–749.

Irvine, J., Takahashi, L., Lockhart, K., Cheong, J. and Tolan, J. (1999) *J. Pharm. Sci.*, **88**, 28–33.

Makhey, V., Guo, A., Norris, D., Hu, P., Yan, J. and Sinko, P. (1998) *Pharm. Res.*, **8**, 1160–1167.

Mariadason, J., Rickard, K., Barkla, D., Augenlicht, L. and Gibson, P. (2000) *J. Cell Physiol.*, **183**, 347–354.

Pageot, L., Perreault, N., Basora, N., Francoeur, C., Magny, P. and Beaulieu, J. (2000) *Microsc. Res. Tech.*, **49**, 394–406.

Pinto, M., Robine-Leon, S., Appay, M., Kedinger, M., Triadou, N., Dussaulx, E. *et al.* (1983) *Biol. Bell*, **47**, 323.

Rothen-Rutishauser, B., Braun, A., Gunthert, M. and Wunderliallenspach, H. (2000) *Pharm. Res.*, 4899.

Souleimani, A. and Asselin, C. (1993) *FEBS*, **326**, 45–50.

Sweetland, R. and Polzer, R. (1998) *Poster Presentation at AAPS*, San Francisco, CA.

Taki, Y., Konishi, K., Yamashita, S., Yata, N., Sezaki, H. and Asa, D. (1999) *Poster Presentation at Oral Drug Delivery Conference*, Kobe, Japan.

Turowski, G., Rashid, Z., Hong, F., Madri, J. and Basson, M. (1994) *Cancer Res.*, **54**, 5974–5980.

Withington, L. and Asa, D. (1999) *Poster Presentaion at AAPS*, New Orleans, LA.

Withington, L. and Asa, D. (2000) *Poster Presentation at Millenial World Congress*, San Francisco, CA.

Withington, L., Asa, D. and Henderson, D. (1999) *Poster Presentation at Society for Biomolecular Screening*, Edinburgh, Scotland.

Woods, B. and Asa, D. (1997) *Poster Presentation at the American Association of Pharmaceutical Scientists*, Boston, MA.

Chapter 7

# Good cell culture practice: good laboratory practice in the cell culture laboratory for the standardization and quality assurance of *in vitro* studies

*Gerhard Gstraunthaler and Thomas Hartung*

## BACKGROUND AND RATIONALE

Cultured human and animal cells are increasingly used as the basis for simplified, direct biological systems in basic science that have the potential to be more controllable and more reproducible than *in vivo* systems, e.g. laboratory animals. However, if a biological experimental system is simplified to fundamental levels then it is paramount that the essential components of such a reduced system are closely defined and reproducible.

The maintenance of high standards is fundamental to all good scientific practice and is essential to securing the reproducibility, credibility, acceptance and proper application of any results produced (Cooper-Hannan *et al.*, 1999). Therefore, minimal requirements for quality standards in cell and tissue culture have to be defined (Froud and Luker, 1994; Gstraunthaler, 2000).

## INTRODUCTION

In recent years, cell and tissue cultures have gained considerable importance in basic research, and their use had rapidly expanded in applied biotechnology. *In vitro* systems are finding increasing application as production systems for various kinds of materials, including monoclonal antibodies, vaccines, hormones, drugs, and nutrients, for use in research, diagnosis and therapy. In addition, there is an increasing use of genetically modified human and animal cells, and of cells derived from genetically modified animals. New therapies are also becoming more widely used, which are based on cell and gene therapy and tissue engineering, where *in vitro* methodologies play a vital role.

Further significant developments are certain to result from the use of *in vitro* systems for high-throughput screening, from the human genome project, to the emerging fields of genomics and proteomics, and from the use of biomarkers of disease, susceptibility, exposure and effect.

The *in vitro* systems themselves are also becoming more varied and more sophisticated, with the development of epithelial cultures on permeable supports, co-cultures of various cell types, sandwich cultures, perfusion cultures, hanging-drop methods, air–liquid interface maintenance, long-term maintenance, controlled cell differentiation, and tissue reconstruction approaches.

Regardless of the ultimate goal of the tissue culture work, the type of cell(s) being cultured, or the level of sophistication of the culture methods used, the job of the culturist is always the same: to maintain viable, differentiated, functional cells outside of their normal, *in vivo* environment. To accomplish this, an artificial (*in vitro*) environment must be created to replace the function of the missing tissue, organ, and whole organism. Thus, the cultured cells have to be kept clean, supplied with nutrients, free of waste, and kept in a milieu that will foster their survival, growth, and proliferation. Aseptic techniques and sterile reagents prevent unintended infections. Cultured cells are fed at regular intervals to maintain both nutrition supply and waste removal, and they are housed in an environment that will maintain the correct temperature, moisture, and gas levels.

Furthermore, every cell exhibits a typical behavior in culture. This behavior will be specific for the cell or organ system that is being studied, and can be considerably different when the culture conditions are varied. Thus, the exact requirements for each cell or organ system are quite different and specific.

Thus, cell culture techniques represent versatile but often ambiguous models of living organisms. The uncertainty resulting from many factors which can cause inherent artifacts demands the highest level of standardization, definition and control for the achievement of meaningful results. As far as *in vitro* systems are used to model the *in vivo* situation, all efforts should be undertaken to approximate in cell behavior relevant for the *in vivo* situation. Therefore, guidelines for *good cell culture practice* (GCCP) have been elaborated defining minimum quality standards for the work with *in vitro* systems based on cells, tissues and organs obtained from humans and animals. This work was initiated at the 3rd World Congress on Alternatives and Animal Use in the Life Sciences, Bologna, Italy, 1999 (Gstraunthaler and Hartung, 1999; Hartung and Gstraunthaler, 2000; Hartung *et al.*, 2000, 2001), and pursued as a task force installed by the ECVAM, *European Centre for the Validation of Alternative Methods*. The issues and topics that have been stressed by these guidelines are as follows: (a) origin and nomenclature of cultured cells in use; (b) basal characterization of cell culture systems and maintenance of differentiated functions; (c) culture media, culture conditions, and handling of cell culture systems; (d) storage of cells, cell culture collections, and cell line banking; (e) education and training of cell culture personnel; (f) safety; (g) patenting; and (h) ethical issues. Here, a summary of the current status of elaboration is given.

## STANDARDIZATION

The standardization of *in vitro* systems first depends on control of the starting material. The starting material for an *in vitro* system essentially comprise the cultured cells, the culture medium, and the culture substratum. These three components interact and the total system variance is undoubtedly a result of this interaction. Nevertheless, the potential for variation can first be considered for each separate component.

## The cells

The cells used in test systems may be freshly isolated from animal tissue (primary cultures) or, at the other extreme, the cells used may comprise a laboratory adapted cell strain/cell line that has been serially propagated and maintained in continuous culture for long periods of time (continuous cell lines). Freshly isolated primary cultures will rapidly de-differentiate in culture and have a limited lifespan. The immortalization of cells, either spontaneously or induced – resulting in the ability to proliferate *in vitro* (almost) indefinitely, as shown by continuous cell lines – is associated with genetic and phenotypic cell transformations of the type commonly associated with tumor cells (Freshney, 1994; Shay *et al.*, 1991). Continuous cell lines are poorly differentiated and lose many of the phenotypic characteristics of the ancestor cell type *in vivo*. Even continuous cell lines have a variable capacity for serial propagation and many cell lines senesce and may progressively lose the ability to multiply. The potential for cell-associated variability is different for primary cultures compared with continuous cell lines.

### Primary cultures

Primary cultures crudely isolated from animal tissue represent a heterogeneous population of different cell types. Each isolate will be unique and impossible to exactly reproduce. Specific isolation procedures and selective tissue culture conditions have to be applied in order to obtain the most uniform cell population (Gstraunthaler and Pfaller, 1993). The process of de-differentiation commences the moment the cells are separated from the parent tissue and so a primary cell culture is a dynamic system in a constant state of change. Primary cell cultures commonly require complex nutrient media supplemented with animal serum and other non-defined components, and a specific extracellular matrix. Consequently, primary cell culture systems are extremely difficult to standardize.

### Continuous cell lines

Continuous cell lines are able to multiply for extended periods *in vitro* and can be expanded and cryopreserved as cell banks. Most of the fundamental phenotypic changes that occur closely following original isolation from the parent animal tissue are complete so that a continuous cell line is more homogeneous, more stable and hence more reproducible than a population of primary cells. The outstanding disadvantage of most continuous cell lines is that they retain little phenotypic differentiation and poorly represent the *in vivo* situation. The opportunity for variability of continuous cell lines stems from the very fact they can be cultivated indefinitely. Continuous cell lines have been distributed worldwide in a totally uncontrolled manner with a massive potential for identification errors and the introduction of contaminants. Therefore, it is recommended that authenticated stocks of a continuous cell line are purchased from a recognized national animal cell culture repository such as the European Collection of Cell Cultures (ECACC), the German Collection of Microorganisms and Cell Cultures (DSMZ) or the American Type Culture Collection (ATCC).

Nevertheless, it is possible that even an authenticated continuous cell line with the same nominal identity purchased from different sources may exhibit phenotypic differences that reflect divergence owing to different culture histories. It is even possible that different banks of the same cell line manufactured within the same repository may exhibit minor differences that could be significant in the context of a specific study.

## The culture medium

Cell culture medium is usually comprised of a defined base (e.g. BME, basal medium eagle; or MEM, minimum essential medium) that includes salts, sugars and amino acids mixed with a variety of supplements that depend on the culture requirements of the cell type. Animal serum, available from different species and from various developmental states of an animal (fetal, newborn, adult) provides the necessary hormones, growth factors and mitogens, attachment factors, proteins, trace elements and lipids that most cells need to survive in an external environment. However, along with the known elements of serum are the undefinable components that add to the variability and inconsistencies of serum from batch to batch. Therefore, the medium supplements comprise the most significant, potential source of variability. The approach of serum-free cell culture is the ultimate goal of standardized, well-defined culture conditions (Barnes *et al.*, 1987; Taub, 1990).

In most cases, animal serum is manufactured as a pool of donations taken from a large number of animals. Such a pooling strategy results in a measure of homogeneity between different batches of animal serum produced by the same manufacturer. However, there are likely to be qualitative differences between serum collected in different geographical regions.

## CELLS

## Any report on cell culture experiments should include a basic description of the cultured cells (Schaeffer, 1984, 1990; Fedoroff, 1967)

*   *Nomenclature* of cell type or cell line in use (code, e.g. ATCC No.).
*   *Origin* and *mode of culture initiation* (species, organ, tissue, lineage, mode of transformation, sublines/hybrid cells; in case of humans: donor, disease, biopsy, tumor) (Anderson *et al.*, 1998).
*   *Source*, e.g. cell bank (ATCC, ECACC, DMSZ, Riken Gene Bank, etc.), laboratory of origin, original publication/patent (Hay, 1988; Hay *et al.*, 1996).
*   Basic *morphological description* of cultured cells, including stability of the phenotype.
*   *Differentiation state* of cells must be controlled (e.g. by morphology, histochemistry, enzyme/gene expression, growth rates, viability, sensitivity to toxins, scope to stimulate cell functions, surface markers, adherence to matrices). The appropriate measure of differentiation should be independent of the cellular

function under study; it should be assessed at least at the beginning and at the end of each series of experiments. Wherever available and feasible, organotypic culture systems should be employed (e.g. sandwich cultures, air–liquid interface cultures, perfusion cultures, cultures on microporous supports).

Measures undertaken for cell line *identification* and *authenticity*, e.g. karyotyping, DNA analysis, fingerprinting, testing for cross-contamination (Hay, 2000; Markovic and Markovic, 1998; Nelson-Rees *et al.*, 1981).

- *Risk assessment*: (risk group and biosafety level, e.g. genetic modification (Wiebel *et al.*, 1997), special care for human and primate cultures, exclusion of viral infections such as hepatitis B virus (HBV), hepatitis C virus (HCV) and human immunodeficiency virus (HIV) parasites (Döhmer *et al.*, 1991; Johannsen *et al.*, 1988).

## CULTURE METHODS

## Culture methods should be precisely defined in standard operating procedures (Freshney, 1994; McAteer and Davis, 1994)

### Choice of culture media (Cartwright and Shah, 1994)

To grow and propagate cells *in vitro* the extracellular milieu must mimic the *in vivo* situation with respect to temperature, oxygen supply, $CO_2$ concentration, ambient pH, osmolarity. Furthermore, the extracellular milieu must supply nutrients and remove metabolic end products. Thus, the main functions of cell culture media are to maintain ambient pH and osmolarity essential for cell viability and to provide the nutrients and energy needed for cell growth and proliferation. Temperature, oxygen and $CO_2$ contents of the cultures must be controlled via the incubator atmosphere.

The efficiency of a mammalian cell culture system depends on the interaction of many different factors. The nutritional and mitogenic support that a culture medium provides for the cells is a crucial element, but this in turn is influenced by the type of culture system in which the cells are propagated.

Cultures may be operated in a simple batch mode, or as fed batches, or as a perfused reactor system. Furthermore, cells may be grown attached to surfaces (monolayer cultures) or grown in suspension.

Composition of *culture media* for routine cultures (maintenance media) and/or experimental cultures and supplements/additives (e.g. serum, growth factors, hormones, antibiotics) should be defined, including name of the supplier; medium volumes used and feeding cycles should be defined; changes of batches of material should be controlled with regard to their influence on the principal end points of the study (Darling and Morgan, 1994; Gstraunthaler *et al.*, 1999).

*Culture vessels* (flasks, Petri dishes, bottles, roller cultures, etc.) should be defined, with the name of the manufacturer/supplier (Doyle *et al.*, 1998).

*Culture substratum* (coating material, e.g. collagen, fibronectin or laminin, coating procedure) should be defined, with the name of the manufacturer/supplier.

*Subcultivation* intervals (cell density, confluent/subconfluent cultures, split ratio, initial passage number, number of passages in culture) should be defined.

## Specialized culture techniques

A number of novel cell culture techniques, e.g. co-cultures, sandwich cultures, air–liquid interface cultures, perfusion cultures and cultures on microporous supports, offer new opportunities to approximate cellular responses similar to the *in vivo* situation. With regard to barrier functions, especially epithelial cultures on microporous membrane supports have to be considered (Gstraunthaler and Pfaller, 1993; Handler *et al.*, 1989; Pitt *et al.*, 1987). In recent years, cell culture devices with membrane supports have provided researchers with greater flexibility to culture and investigate epithelial monolayer cultures. The use of microporous culture inserts offers a more physiologically relevant environment in which to culture and investigate epithelial cells. This environment is evidenced by a higher degree of differentiation of epithelial cells because nutrients, hormones, and other factors readily gain access to the basal surface of the epithelium. Furthermore, the cultured epithelium forms a barrier, separating the two physiological fluid compartments, the apical (mucosal) from the basolateral (serosal) compartment, which enables manipulation of selecting the composition of culture medium in either compartment, or collection of media samples independently.

These features of the filter insert culture techniques are ideal for studies of vesicular trafficking and transcytosis, transepithelial solute and fluid transport, the sidedness of cell response to agents and toxicants, and the maintenance of integrity and barrier function of the cultured epithelia.

### Types of microporous culture inserts

Today, a variety of membrane materials are available in different thickness, pore sizes, and pore densities. Since these physical attributes affect permeability, durability, transparency for light microscopic investigation and protein binding properties, selection of the most appropriate membrane becomes a critical factor in designing the optimal culture system for a specific application. Inserts are either designed for free-standing, so that they can be placed in large culture vessels, and thus have short feet to hold them off the floor of the outer well, or they hang suspended from the lip of the outer well. The latter design is well-suited for co-cultures, since the outer (basolateral) cell culture will not be disturbed or damaged by the insert feet. Thus, a variety of parameters have to be taken into consideration, and have to be standardized when culturing epithelial cells on microporous filter inserts.

## MATERIALS

## Materials employed must be of the highest quality

The quality of equipment, donor animals/animal facilities, cells and cell culture materials should be such as to guarantee reproducible and reliable results. The impact of variation in these materials should be controlled and documented. Equipment and instruments should be maintained and calibrated properly (e.g. control of temperature and $CO_2$ levels of incubators). All materials employed should be stored under appropriate conditions to protect them from damage, infestations or contaminations.

Donor animals from an outside source should be kept in quarantine for an appropriate period. Similarly, measures should be taken when a cell line is introduced into the laboratory to assure that no infection/contamination of cell lines already present can occur (certificate from supplier, test for most common contaminations, e.g. mycoplasms, growth in antibiotic-free medium for a specified period).

Tests for frequent contaminations, e.g. mycoplasms, must be performed on a regular basis. Results must be discarded in the event of any evidence of contamination of materials.

All wastes must be treated properly, thereby minimizing the threat to humans (e.g. toxicity, mutagenicity, teratogenicity), as well as to other cells and animals under study.

## DOCUMENTATION AND REPORTING

Data must be analyzed and documented properly. Data should be subjected to adequate statistical analysis without the subselection or neglecting of data sets. A distinction should be made between variation among replicates within a single experiment and variation among replicate-independent experiments.

A complete documentation of raw data, analysis procedures, lists of materials and equipment employed, as well as all procedures including derivations from the protocol should be maintained and stored for at least ten years by the primary investigator.

## EDUCATION AND TRAINING

Since cell culture is a sophisticated technique, proper education and training of cell culture personnel is a prerequisite for the state-of-the-art culturing of cells, and for maintaining cultures at the highest reachable quality level, respectively. In addition, specific safety precautions have to be defined. Depending on the cells (primary cultures, transformed and/or transfected cells or cell lines, etc.) and depending on the techniques applied, a risk assessment has to be conducted, e.g. the evaluation of the risk group and the biosafety level.

In daily work in the tissue culture laboratory, one has to distinguish between routine protocols and specialized culture techniques. Therefore, education and training of laboratory personnel will vary.

## ACKNOWLEDGMENT

The authors are indebted to Drs Michael Balls, Sandra Coecke, David Lewis and Olivier Blanck, members of the ECVAM Task Force on GCCP, for their valuable contributions which formed the basis of this paper.

# REFERENCES

Anderson, R., O'Hare, M., Balls, M., Brady, M., Brahams, D., Burt, A. *et al.* (1998) The availability of human tissue for biomedical research. ECVAM Workshop Report 32, *ATLA*, **26**, 763–777.

Barnes, D., McKeehan, W. L. and Sato, G. H. (1987) Cellular endocrinology: integrated physiology in vitro. *In Vitro Cell. Dev. Biol.*, **23**, 659–662.

Cartwright, T. and Shah, G. P. (1994) Culture media. In J. M. Davis (ed), *Basic Cell Culture, A Practical Approach*, pp. 57–91, Oxford: Oxford University Press.

Cooper-Hannan, R., Harbell, J. W., Coecke, S., Balls, M., Bowe, G., Cervinka, M. *et al.* (1999) The principles of Good Laboratory Practice: application to *in vitro* toxicology studies. ECVAM Workshop Report 37, *ATLA*, **27**, 539–577.

Darling, D. C. and Morgan, S. J. (1994) *Animal Cells: Culture and Media, Essential Data*. Chichester-New York: John Wiley & Sons.

Döhmer, J., Erfle, V., Hunsmann, G., Johannsen, R., Miltenburger, H. G., Rüger, R. and Schlumberger, H. D. (1991) Gefährdungspotential durch Retroviren beim Umgang mit tierischen Zellkutturen. *BIO forum*, **11**, 428–436.

Doyle, A. and Griffiths, J. B. (1997) *Mammalian Cell Culture: Essential techniques*. Chichester-New York: John Wiley & Sons, Inc.

Doyle, A., Griffiths, J. B. and Newell, D. G. (1998) *Cell & Tissue Culture: Laboratory Procedures*. Chichester-New York: John Wiley & Sons, Ltd.

Fedoroff, S. (1967) Proposed usage of animal tissue culture terms. *Exp. Cell Res.*, **46**, 642–648.

Freshney, R. I. (1994) *Culture of Animal Cells. A Manual of Basic Techniques*, 3rd edition, New York-Chichester: Wiley & Sons, Inc.

Froud, S. J. and Luker, J. (1994) Good laboratory practice in the cell culture laboratory. In J. M. Davis (ed), *Basic Cell Culture, A Practical Approach*, pp. 273–286, Oxford: Oxford University Press.

Gstraunthaler, G. (2000) Standardisierung in der Zellkultur – wo fangen wir an? In H. Schöffl, H. Spielmann, F. P. Gruber, H. Appl, F. Harrer, W. Pfaller and H. A. Tritthart (eds), *Forschung ohne Tierversuche 2000*, pp. 40–49 Wien-New York: Springer.

Gstraunthaler, G. and Hartung, T. (1999) Bologna Declaration toward Good Cell Culture Practice. 3rd World Congress on Alternatives and Animal Use in the Life Sciences, Bologna, Italy. *ATLA*, **27**, 206.

Gstraunthaler, G. and Pfaller, W. (1993) The Use of Cultured Renal Epithelial Cells in *in Vitro* Assessment of Xenobiotic-Induced Nephrotoxicity. In M. W. Anders, W. Dekant, D. Henschler, H. Oberleithner and S. Silbernagl (eds), *Renal Disposition and Nephrotoxicity of Xenobiotics*, pp. 27–61, New York: Academic.

Gstraunthaler, G., Seppi, T. and Pfaller, W. (1999) Impact of culture conditions, culture media volumes and glucose content on metabolic properties of renal epithelial cell cultures. Are renal cells in tissue culture hypoxic? *Cell. Physiol. Biochem.*, **9**, 150–172.

Handler, J. S., Green, N. and Steele, R. E. (1989) Cultures as Epithelial Models: Porous-Bottom Culture Dishes for Studying Transport and Differentiation. *Methods Enzymol.*, **171**, 736–744.

Hartung, T. and Gstraunthaler, G. (2000) The standardisation of cell culture procedures. In M. Balls, A.-M. van Zeller and M. E. Halder (eds), *Progress in the Reduction, Refinement and Replacement of Animal Experimentation*, pp. 1655–1658, Amsterdam: Elsevier Science B. V.

Hartung, T., Gstraunthaler, G. and Balls, M. (2000) Bologna Statement on Good Cell Culture Practice (GCCP). *ALTEX*, **17**, 38–39.

Hartung, T., Gstraunthaler, G., Coecke, S., Lewis, D., Blanck, O. and Balls, M. (2001) Good Cell Culture Practice (GCCP) – eine Initiative zur Standardisierung und Qualitätssicherung von *in vitro* Arbeiten. Die Etablierung einer ECVAM Task Force on GCCP. *ALTEX*, **18**, 75–78.

Hay, R. J. (1988) The seed stock concept and quality control for cell lines. *Anal. Biochem.*, **171**, 225–237.

Hay, R. J. (2000) Cell line authentication and the seed stock concept. In H. Schöffl, H. Spielmann, F. P. Gruber, H. Appl, F. Harrer, W. Pfaller and H. A. Tritthart (eds), *Forschung ohne Tierversuche 2000*, pp. 275–281, Wien-New York: Springer.

Hay, R. J., Reid, Y. A., McClintock, P. R., Chen, T. R. and Macy, M. L. (1996) Cell line banks and their role in cancer research. *J. Cell. Biochem. Suppl.*, **24**, 107–130.

Johannsen, R., Albert, W., Hunsmann, F., Krämer, P., Noe, W., Schirrmacher, V., Schlumberger, H. D. and Streissle, G. (1988) Chancen und Risiken durch Säugerzellkulturen, *Forum Mikrobiologie*, **11**, 359–367.

Markovic, O. and Markovic, N. (1998) Cell cross-contamination in cell cultures: the silent and neglected danger. *In Vitro Cell. Dev. Biol.*, **34**, 1–8.

McAteer, J. A. and Davis, J. (1994) Basic cell culture technique and maintenance of cell lines. In J. M. Davis (ed.), *Basic Cell Culture, A Practical Approach*, pp. 93–148, Oxford: Oxford University Press.

Nelson-Rees, W. A., Daniels, D. W. and Flandermeyer, R. R. (1981) Cross-contamination of cells in culture. *Science*, **212**, 446–452.

Pitt, A. M., Gabriels, J. E., Badmington, F., McDowell, J., Gonzales, L. and Waugh, M. E. (1987) Cell Culture on a Microscopically Transparent Microporous Membrane. *BioTechniques*, **5**, 162–171.

Schaeffer, W. I. (1984) Usage of vertebrate, invertebrate and plant cell, tissue and organ culture terminology. *In Vitro*, **20**, 19–24.

Schaeffer, W. I. (1990) Terminology associated with cell, tissue and organ culture, molecular biology and molecular genetics. *In Vitro Cell. Dev. Biol.*, **26**, 97–101.

Shay, J. W., Wright, W. E. and Werbin, H. (1991) Defining the molecular mechanisms of human cell immortalization. *Biochim. Biophys. Acta*, **1072**, 1–7.

Taub, M. (1990) The use of defined media in cell and tissue culture. *Toxicol. in Vitro*, **4**, 213–225.

Wiebel, F. J., Andersson, T. B., Casciano, D. A., Dickins, M., Fischer, V., Glatt, H. *et al.* (1997) Genetically engineered cell lines: characterization and application in toxicity testing. ECVAM Workshop Report 26, *ATLA*, **25**, 625–639.

# Chapter 8

# Regulatory acceptance of *in vitro* test systems as an alternative to safety testing in animals

*Horst Spielmann and Manfred Liebsch*

## INTERNATIONAL HARMONIZATION OF ANIMAL TEST IN REGULATORY TOXICOLOGY

National Centre for Documentation and Evaluation of Alternative Methods to Animal Experiments (ZEBET) was established in 1989 at the Federal Health Office (BGA) in Berlin as the National German Centre for Documentation and Evaluation of Alternatives to Animal Testing. ZEBET's mission is to reduce animal testing for regulatory purposes. The only concept available in 1989 to reduce testing in animals was the Three Rs principles of Russell and Burch (Russell and Burch, 1959)

In the process of producing chemicals it is the goal of regulatory toxicology to ensure the occupational safety of workers, to ensure the safety of food and beverages, to protect patients against possible hazards represented by drugs and medical devices, and to protect humans and the environment against possible hazards posed by residues of chemicals, e.g. pesticides. The standard approach in regulatory toxicology to assess the toxicity of chemicals is the determination of toxic properties in standardized animal tests, as described in the *OECD Guidelines for Testing of Chemicals* (OECD, 1983). This information is then used by regulators to classify each chemical according to internationally harmonized guidelines in the first step, e.g. as harmful, toxic, irritant. The second step is to label each chemical according to European Union (EU) risk (R) phrases, e.g. 'R-41: risk of serious damage to the eye'. The consequences of classification and labeling are the restricted use of the tested chemical in finished products (depending on exposure), and safety and labeling recommendations.

The international harmonization of toxicity tests by the OECD in 1982 was the first, and so far, the most effective step in reducing duplication of testing in animals for regulatory purposes, since a toxicity test conducted according to the OECD guidelines will be accepted by regulatory agencies in all OECD member states. These member states are the world's major industrial nations. A similar approach has been used for the safety and efficacy testing of drugs by the International Conference on Harmonization (ICH), which represents the three major economic regions, namely Europe, Japan and the USA. Since 1990, the ICH has accepted harmonized guidelines for efficacy and safety testing of drugs and medicines, including animal tests. Again, the harmonization of test guidelines has led to significant reduction of testing in animals, since regulatory agencies around the world now accept the results of a test conducted according to ICH guidelines.

## EVOLUTION OF THE PRINCIPLES OF SCIENTIFIC VALIDATION I: FIRST AMDEN WORKSHOP ON VALIDATION

Regulators will only accept alternatives to animal tests in toxicology if the new tests will allow them to classify and label chemicals in the same way as the results of current animal tests allow them to do. The OECD has, therefore, indicated that *in vitro* toxicity tests can be accepted for regulatory purposes only after a successful experimental validation study. This procedure is essential to prove that the new *in vitro* toxicity tests will provide the same level of protection as the animal tests are currently providing.

To approach this problem scientifically, in 1990 European and American scientists interested in the validation of toxicity tests met in Amden, Switzerland, to agree on a definition of experimental validation and to define the essential steps in this process. In the workshop report of the 1st Amden validation workshop, validation was defined as the process by which reproducibility and relevance of a toxicity testing procedure are established for a particular purpose (Balls *et al.*, 1990), regardless of whether the method is an *in vitro* or *in vivo* test. In addition, at this workshop the essential steps of the experimental validation process were defined in the following manner:

1   test development in a single laboratory;
2   experimental validation under blind conditions in several laboratories in a ring trial;
3   independent assessment of the results of the validation trial; and
4   regulatory acceptance.

Steps 2 and 3 were identified as the core part of a formal validation study conducted for regulatory purposes. The report of the 1st Amden workshop on validation encouraged scientists to start formal validation studies. Since the Draize eye test has been the most widely criticized toxicity tests, several international validation studies on alternatives to the test were initiated:

1   BGA/Bundesministerium für Bildung und Forschung (BMBF) study: national validation study in Germany 1988–1995 (Spielmann *et al.*, 1996).
2   Inter-Regulatory Agency Group (IRAG) study: retrospective international study, organized by US regulatory agencies 1991–1994 (Spielmann *et al.*, 1997).
3   European Commission (EC)/Home Office (HO) study: international validation study organized by the UK, sponsored by EU 1992–1995 (Balls *et al.*, 1995a).
4   Japanese study: national validation study 1991–1995 (Ohno *et al.*, 1994).
5   European Cosmetics, Toiletry and Perfumery Association (COLIPA) study: international validation study 1994–1997 (Brantom *et al.*, 1997).

The management team of the EC-Home Office (EC/HO) validation study, in which nine alternatives to the Draize eye test were tested under blind conditions with 60 carefully selected test chemicals in 36 laboratories, concluded at the end of the study in 1995 that none of the *in vitro* alternatives were able to completely replace the Draize eye test, and that the validation process had to be improved (Balls *et al.*, 1995a).

# EVOLUTION OF THE PRINCIPLES OF SCIENTIFIC VALIDATION II: SECOND AMDEN WORKSHOP ON VALIDATION

Despite the joint efforts of many scientists around the world, the first attempt at validation failed and the leading scientists involved met for a 2nd validation workshop in Amden in 1994 to learn from the unsuccessful attempts and to improve the validation procedure. Taking this experience into account, the participants in the 2nd Amden validation workshop recommended the inclusion of new elements into the validation process (Balls *et al.*, 1995b), which had not been sufficiently identified in the 1st Amden validation workshop. The three essential elements recommended were the definition of a biostatistically based *prediction model* (PM), the inclusion of a *prevalidation stage* between test development and formal validation under blind conditions, and a well-defined *management structure*.

A PM should allow the prediction of *in vivo* end points in animals or humans from the end points determined in the *in vitro* test. The PM must be defined mathematically in the standard operation procedure of the test that will undergo experimental validation under blind conditions with coded chemicals (Balls *et al.*, 1995b). In order to assess the limitations of a new test before it can be evaluated in a validation study, the test should be standardized in a *prevalidation study* with a few test chemicals in a few laboratories (Curren *et al.*, 1995). This will ensure that the *in vitro* test method, including the PM, is robust and the formal validation study with coded chemicals is likely to be successful. Finally, the goal of a validation study has to be defined clearly, and the *management structure* has to ensure that within the study the scientists who are responsible for essential tasks can conduct their duties independently of the sponsors and the managers of the study, e.g. biostatistical analysis, and the selection, coding and shipment of the test chemicals.

The improved concept of experimental validation for regulatory purposes defined in the 2nd validation workshop in Amden was accepted by the EU validation center, ECVAM, in 1995, and in 1996 it was accepted by US regulatory agencies (NIEHS, 1997) and also by the OECD (1996). After this agreement at the international level, scientists have tried to follow the ECVAM/US/OECD principles for validation in new validation trials. The improved validation concept was immediately introduced into ongoing validation studies, such as the ECVAM/COLIPA validation study on *in vitro* phototoxicity tests.

# SUCCESSFUL VALIDATION AND REGULATORY ACCEPTANCE OF *IN VITRO* TOXICITY TESTS

## Validation of the 3T3 NRU *in vitro* phototoxicity test

Phototoxicity is an acute reaction, which can be induced by a single treatment with a chemical and ultra violet (UV) or visible radiation. Since no standard guideline for the testing of photoirritation potential, either *in vivo* or *in vitro*, had been accepted for regulatory purposes at the international level by the OECD, the EC and the COLIPA established a joint program in 1991 dedicated to developing and validating *in vitro* photoirritation tests. In the first phase of the study, which was funded by Direction General Environment (DG XI) of the EC and coordinated by ZEBET, *in vitro* phototoxicity tests established in laboratories of the cosmetic industry were evaluated, and a new assay, the 3T3 neutral red uptake (NRU) PT test, which is a photocytotoxicity test using the mouse fibroblast cell line 3T3 and NRU as the end point for cytotoxicity was developed.

In the prevalidation study, which was conducted with 20 test chemicals (11 phototoxic and nine non-phototoxic ones), quite unexpectedly, the 3T3 NRU PT *in vitro* phototoxicity test was the only *in vitro* test in which all of the test chemicals were correctly identified as phototoxic or non-phototoxic (Spielmann *et al.*, 1994a). Independently of this prevalidation exercise, a laboratory in Japan subsequently obtained the same correct results in the 3T3 NRU PT, when testing the same set of 20 test chemicals.

In the second phase of the study, which was funded by ECVAM and coordinated by ZEBET, the 3T3 NRU PT test was validated with 30 carefully selected test chemicals in 11 laboratories in a blind trial. A special ECVAM workshop was held to independently select a representative set of test chemicals covering all major classes of phototoxins, selected according to results from standardized photopatch testing in humans (Spielmann *et al.*, 1994b). The results obtained in this *in vitro* test under blind conditions were reproducible, and the correlation between *in vitro* and *in vivo* data was almost perfect (Spielmann *et al.*, 1998a). Therefore, the ECVAM Scientific Advisory Committee (ESAC) concluded that the 3T3 NRU PT is a scientifically validated test which is ready to be considered for regulatory acceptance (Balls and Corcelle, 1998). However, the EU expert committee on the safety of cosmetics, the Scientific Committee on Cosmetology and Non-Food-Products (SCCNFP), criticized the fact that there was an insufficient number of UV-filter chemicals (widely used as sunblockers) tested in the formal validation study. In the blind trial on UV-filter chemicals, which was again funded by ECVAM and coordinated by ZEBET, the phototoxic potential of all of the 20 test chemicals (ten UV-filter chemicals, which were non-phototoxic, and ten phototoxic test chemicals) was predicted correctly in the 3T3 NRU PT *in vitro* phototoxicity test (Spielmann *et al.*, 1998b).

Therefore, in 1998 the EU, having accepted the 3T3 NRU PT test as the first experimentally validated *in vitro* toxicity test for regulatory purposes, officially applied to the OECD for worldwide acceptance of this *in vitro* toxicity test. Early in 2000 the EC officially accepted and published the 3T3 NRU PT phototoxicity test in *Annex V* of *Directive 67/548 EEC on the Classification, Packaging and Labelling of Dangerous Substances* (European Commission, 2000a). Thus, this *in vitro* test is the first formally validated *in vitro* toxicity test that has been accepted into Annex V, and it is the only phototoxicity test that is accepted for regulatory purposes in Europe. Meanwhile experts of the OECD are recommending the acceptance of this *in vitro* toxicity test for regulatory purposes.

## Validation of two *in vitro* skin corrosivity tests

Two *in vitro* tests for skin corrosivity testing, applying a human skin model EPISKIN™ and excised rat skin, were successfully validated in an ECVAM validation study from 1996 to 1998 (Fentem *et al.*, 1998). The ESAC concluded in 1998 that the results obtained with the EPISKIN™ test involving the use of a reconstructed human skin model and the rat skin transcutaneous electrical resistance (TEER) test in the international ECVAM validation study on *in vitro* tests for skin corrosivity were reproducible, both within and among laboratories that performed the test (ESAC, 1998). The tests were able to distinguish between corrosive and non-corrosive chemicals for all of the chemical types studied. ESAC, therefore, agrees with the conclusions from the formal validation study that the EPISKIN™ test and the TEER test are scientifically validated and can be used as replacements for the animal test for distinguishing between corrosive and non-corrosive test chemicals, and that the tests are ready to be considered for regulatory acceptance. As a result, the two *in vitro* corrosivity tests were accepted by the EC for regulatory purposes in the year 2000 (European Commission, 2000b).

## Validation of the EpiDerm™ human skin model for corrosivity testing

Since the EPISKIN™ human skin model was not commercially available after it had been experimentally validated, a second human skin model, the EpiDerm™, was validated in an ECVAM study from 1998 to 2000. This short study, which was coordinated by ZEBET and conducted in three laboratories with chemicals from the previous validation study, proved that the EPISKIN™ human skin model met the acceptance criteria of the TER and EPISKIN™ *in vitro* corrosivity tests. Therefore, ESAC concluded in 2000 that the EpiDerm™ human skin model can be used for distinguishing between corrosive and non-corrosive chemicals within the context of the EU test guideline for skin corrosion (ESAC, 2000a).

## ECVAM validation study of three *in vitro* embryotoxicity tests

In an ECVAM validation study, three *in vitro* embryotoxicity tests were validated under blind conditions from 1997–2000. In the EU, there is a strong demand for validated *in vitro* tests in developmental toxicity testing using mammalian embryos, as well as primary cultures of embryonic cells and permanent cell lines. The most important result of the present validation study was that, for the first time three *in vitro* embryotoxicity tests have been established that are backed by validated test protocols which will be available through ECVAM as INVITTOX protocols. These are: (1) the whole embryo culture (WEC) test using cultures of whole rat embryos; (2) the micro mass (MM) test employing primary cultures of dissociated limb bud cells of rat embryos; and (3) the embryonic stem cell test (EST), which uses two established mouse embryonic cell lines and which does necessitase sacrificing pregnant animals.

In the ECVAM validation study 20 test chemicals were tested that were backed by high quality *in vivo* data in humans and animals. Each *in vitro* test was experimentally validated and evaluated under blind conditions in four laboratories. All of the *in vitro* embryotoxicity tests met three essential criteria of validated alternative toxicity tests. First, standard operation procedures (SOPs) were established, which are now available to the public. Second, sound biostatistical prediction models (PMs) have been established and validated (Genschow *et al.*, 2000). The PMs for all of the three tests provide an accuracy of close to 80 per cent and, more importantly, 100 per cent predictivity for strong embryotoxic chemicals. Thus, they can routinely be used to identify strongly embryotoxic chemicals, e.g. when screening new substances. Third, the three *in vitro* tests were experimentally validated in a blind ring trial according to the validation scheme recommended by the EU, the OECD and the US National Institute of Environmental Health NIEHS (Balls *et al.*, 1990; NIEHS, 1997; OECD, 1983, 1996).

Thus, this study clearly demonstrates that the ECVAM strategy for prevalidation and validation of *in vitro* toxicity tests is sound. This conclusion, based on the final report of the study, was accepted by ECVAM in June 2000 (Spielmann *et al.*, 2001).

## Validation of the local lymph node assay (LLNA) for sensitizing properties

The LLNA for the evaluation of sensitizing properties, which was developed and validated in laboratories of the chemical industry in the UK, was accepted for regulatory purposes in 1999 by the federal regulatory authorities of the USA under the chairmanship of the US Validation Centre, Inter-agency Co-ordinating Committee on the Validation of Alternative Methods (ICCVAM), at the NIEHS (NIEHS, 1999). In the year 2000, ESAC concluded from reviewing this report that the LLNA is a scientifically validated test which can be used to assess the skin sensitization potential of chemicals. Therefore, ESAC recommended that the LLNA should be the preferred method for sensitization testing since it used fewer animals and causes less distress than the conventional guinea-pig methods (ESAC, 2000b). However, in some instances and for scientific reasons ESAC accepted the use of the conventional methods.

## Ban of the ascites method for the production of monoclonal antibodies

Taking into account the studies conducted in several EU member states, ESAC has also recommended banning the production of monoclonal antibodies using the *in vivo* ascites mouse technique. Several companies have developed bioreactors which culture mouse hybridoma cells *in vitro* and thus produces monoclonal antibodies. A few EU member states are strictly enforcing the ban of the ascites mouse method, which causes pain, suffering and death to the mice, e.g. Germany, The Netherlands, Sweden and the UK, while other EU member states have not yet implemented the ban. Taking into account the progress in Europe, meanwhile, the US NIH (2000) has stated that any research institute that continues to approve the routine use of ascites in producing monoclonal antibodies will no longer be eligible to receive a US government research grant. The US NIH is also recommending the use of bioreactors rather than ascites mice for the production of monoclonal antibodies (ESAC, 2000b).

## Regulatory acceptance of four alternatives to the Draize eye irritation test

Several validation studies of *in vitro* alternatives to the Draize eye test have been conducted in Europe during the past decade. As a result, four *in vitro* alternatives have been accepted for regulatory purposes to identify severely eye irritating materials according to EU Directive 86/906/EEC for the classification and labeling of hazardous chemicals: the Hens egg test at the chorion allantoic membrane (HET-CAM) assay on the chicken egg; the bovine cornea opacity and permeability (BCOP) test on the isolated bovine cornea from slaughterhouse; and, two *in vitro* tests on isolated chicken and rabbit eyes from animals that have been sacrificed for other purposes. Chemicals which provide a negative reaction in any of the four *in vitro* tests still have to be tested in the Draize eye test in one–three rabbits in order to confirm the absence of eye irritation potential. In several EU member states, e.g. France and Germany, the HET-CAM test is accepted by the national authorities for the safety testing of cosmetics.

## CONCLUSION AND RECOMMENDATIONS

The successful validation and regulatory acceptance of several *in vitro* toxicity tests in the EU proves that the validation procedure recommended by ECVAM and the OECD (Balls *et al.*, 1990; NIEHS, 1997; OECD, 1983, 1996;) is the most appropriate for the validation of *in vitro* toxicity tests. However, taking into account both time-frame and costs, e.g. of the ECVAM/COLIPA validation study of *in vitro* embryotoxicity tests, the formal validation procedure must be improved in order to reduce both costs and duration of the studies. To illustrate the problem, this particular study required funding of more than one million ECU ($\cong$ US $) and the ECVAM validation study of three *in vitro* embryotoxicity was funded with a budget of 1.6 million ECU. The two examples illustrate that validation studies are very expensive and time-consuming, since it appears to take, on average, ten years from test development to regulatory acceptance.

Until today under the chairmanship of Professor Michael Balls and ECVAM, the EU has taken the lead in the experimental validation of *in vitro* test methods for regulatory purposes. In Japan several validation studies have been conducted on alternatives to the Draize eye test and on acute local irritation testing on the skin. In contrast, the ICCVAM has focused its activity on reviewing the results of validation studies that were funded by other institutions, since money of the National Toxicology Program has so far not been set aside for the validation of *in vitro* methods, but the situation may improve in the very near future.

Taking into account the lessons learned during the past decade, progress in the acceptance of *in vitro* toxicity tests at an international level will be achieved only if Europe, Japan and the USA share the burden of funding validation studies in a coordinated manner. The OECD might provide an appropriate forum for this important international activity, in particular since the OECD has meanwhile accepted the *in vitro* toxicity tests that are used for regulatory purposes in Europe after successful validation.

## ACKNOWLEDGMENT

The progress in the validation and acceptance of *in vitro* methods for toxicity testing could only be achieved with the continuous support of the staff of ZEBET, and the high priority that our work has been given both by ECVAM and COLIPA. I am particularly indebted to my colleagues in the management team of the ECVAM/COLIPA validation project on *in vitro* phototoxicity tests, namely Michael Balls (ECVAM, Ispra, Italy), Wolfgang Pape (Beiersdorf, Hamburg, Germany), Odile de Silva (L'Oréal, Paris, France) and Jack Dupuis (COLIPA, Brussels, Belgium), who passed away quite unexpectedly one month ago and to whom we want to dedicate the present manuscript.

## REFERENCES

Balls, M. and Corcelle, G. (1998) Statement on the scientific validity of the 3T3 NRU PT test (an in vitro test for phototoxic potential). *ATLA*, **26**, 7–8.

Balls, M., Blaauboer, B., Brusik, D., Frazier, J., Lamp, D., Pemberton, M. *et al.* (1990) Report and recommendations of the CAAT/ERGATT workshop on the validation of toxicity test procedures. *ATLA*, **18**, 313–337.

Balls, M., Botham, P. A., Bruner, L. H. and Spielmann, H. (1995a) The EC/HO international validation study on alternatives to the Draize eye irritation test. *Toxicol. In Vitro*, **9**, 871–929.

Balls, M., Blaauboer, B. J., Fentem, J., Bruner, L., Combes, R. D., Ekwal, B. *et al.* (1995b) Practical aspects of the validation of toxicity test procedures. The report and recommendations of ECVAM Workshop 5. *ATLA*, **23**, 129–147.

Brantom, P. G., Bruner, L. H., Chamberlain, M., de Silva, O., Dupuis J., Earl, L. K. *et al.* (1997) A summary report of the COLIPA international validation study on alternatives to the Draize rabbit eye irritation test. *Toxicol. In Vitro*, **11**, 141–179.

Curren, R. D., Southee, J. A., Spielmann, H., Liebsch, M., Fentem, J. and Balls, M. (1995) The role of prevalidation in the development, validation and acceptance of alternative methods. *ATLA*, **23**, 211–217.

ESAC – ECVAM Scientific Advisory Committee (1998) Statement on the scientific validity of the rat skin transcutaneous resistance (TER) test and the EPISKIN™ test (in vitro tests for skin corrosivity). *ATLA*, **26**, 275–280.

ESAC – ECVAM Scientific Advisory Committee (2000a) Statement on the application of the EpiDerm™ human skin model for skin corrosivity testing. *ATLA*, **28**, 365–367.

ESAC – ECVAM Scientific Advisory Committee (2000b) Statement on the validity of the local lymph node assay for skin sensitisation testing. *ATLA*, **28**, 365–367.

European Commission (2000a) Test guideline B-41 'phototoxicity – *in vitro* 3T3 NRU phototoxicity test' of Annex V of the EU Directive 86/906/EEC for classification and labelling of hazardous chemicals, O.J. of the European Communities, June 8, 2000, L136, 98–107.

European Commission (2000b) Test guideline B-40 'skin corrosivity – *in vitro* method' of Annex V of the EU Directive 86/906/EEC for classification and labelling of hazardous chemicals, O.J. of the European Communities, June 8, 2000, L136, 85–97.

Fentem, J. H., Archer, G. E. B., Balls, M., Botham, P. A., Curren, R. D., Earl, L.K. *et al.* (1998) The ECVAM international validation study on in vitro test for skin corrosivity. 2. Results and evaluation by the Managment Team. *Toxicol. In Vitro*, **12**, 483–524.

Genschow, E., Scholz, G., Brown, N., Piersma, A., Brady, M. and Clemann, N. (2000) Development of prediction models for three *in vitro* embryotoxicity tests in an ECVAM validation study. *In Vitro Mol. Toxicol.*, **13**, 51–65.

National Institute of Environmental Health (NIEHS) (1997) Validation and regulatory acceptance of toxicological test methods: a report of the ad hoc interagency co-ordinating committee on the validation of alternative methods. NIEHS NIH Publication No. 97-3981, Research Triangle Park, USA.

National Institute of Environmental Health (NIEHS) (1999) The murine local lymph node assay. The results of an independent peer review evaluation co-ordinated by the interagency Co-ordinating Committee on the Validation of Alternative Methods (ICCVAM) and the National Toxicology Program Center for the Evaluation of Alternative Toxicological methods (NICEATM). NIEHS NIH Publ. No. 99-4494, Research Triangle Park, USA.

OECD (1983) *OECD Guideline for Testing of Chemicals*. OECD Publications Office, Paris, France.

OECD (1996) Final report of the OECD workshop on harmonization of validation and acceptance criteria for alternative toxicological tests methods. OECD Publication Office, Paris, France.

Ohno, Y., Kaneko, T., Kobayashi, T., Inoue, T., Kuroiwa, Y., Yoshida, T. *et al.* (1994) First-phase validation of the in vitro eye irritation test for cosmetic ingredients. *In Vitro Toxicol.*, **7**, 89–94.

Russell, W. M. S. and Burch, R. L. (1959) *The Principles of Humane Experimental Technique*. London, Methuen.

Spielmann, H., Balls, M., Brand, M., Döring, B., Holzhütter, H. G., Kalweit, S. *et al.* (1994a). EC/COLIPA project on in vitro phototoxicity testing: first results obtained with the Balb/c 3T3 cell phototoxicity assay. *Toxicol. In Vitro*, **8**, 793–796.

Spielmann, H., Lovell, W. W., Hölzle, E., Johnson, B. E., Maurer, T., Miranda, M. *et al.* (1994b) In vitro phototoxicity testing. The report and recommendations of ECVAM Workshop 2. *ATLA*, **22**, 314–348.

Spielmann, H., Liebsch, M., Kalweit, S., Moldenhauer, F., Wirnsberger, T., Holzhütter H.-G. *et al.* (1996) Results of a validation study in Germany on two *in vitro* alternatives to the Draize eye irritation test, the HET-CAM test and the 3T3 NRU cytotoxicity test. *ATLA*, **24**, 741–858.

Spielmann, H., Liebsch, M., Moldenhauer, F., Holzhütter, H.-G. Bagley, D. M., Lipman, J. M. *et al.* (1997) IRAG working group 2 report: CAM-based assays. *Food Chem. Toxicol.*, **35**, 39–66.

Spielmann, H., Balls, M., Dupuis, J., Pape, W. J. W., Pechovitch, G., de Silva, O. (1998a) The international EU/COLIPA in vitro phototoxicity validation study: results of Phase II (blind trial); part 1: the 3T3 NRU phototoxicity test. *Toxicol. In Vitro*, **12**, 305–327.

Spielmann, H., Balls, M., Dupuis, J., Pape, W. J. W., de Silva, O., Holzhütter, H. G. *et al.* (1998b) A study on UV filter chemicals from Annex VII of Euroepan Union Directive 76/768/EEC, in the *in vitro* 3T3 NRU phototoxicity test. *ATLA*, **26**, 679–708.

Spielmann, H., Genschow, E., Scholz, G., Brown, N.A., Piersma, A., Brady, G. *et al.* (2001) Preliminary results of the ECUAM validation study on three in vitro embryotoxicity tests. *ATLA* **29**, 301–303

US NIH (2000) NIH Guide to grants and contracts. Published on 3 February 2000 on the internet: http.//grants.gov/grants/guide/notice.-files/NOT-OD-00-019.html

# Regulatory acceptance of *in vitro* permeability studies in the context of the biopharmaceutics classification system

*Donna A. Volpe, Lawrence X. Yu and Helga Möller*

## INTRODUCTION

The biopharmaceutics classification system (BCS) is a drug development tool that allows estimation of the relative contributions of the three major factors that govern the rate and extent of oral absorption of a drug from solid oral dosage forms, namely dissolution, solubility and permeability. The permeability of drugs can be ascertained by *in vitro* methods with epithelial cell cultures. However, there are a number of cell lines used for evaluating intestinal absorption, and the protocols for conducting the permeability experiments differ among laboratories. Therefore, regulatory acceptance of *in vitro* permeability data in the context of the BCS is based on method suitability where the goal is to classify a drug as having high permeability (HP) or low permeability (LP). A cell culture method is considered to be suitable when it has established a rank-order relationship between experimental permeability values of model drugs and the extent of intestinal absorption in humans. After demonstrating method suitability and maintaining the same protocol, the method can be used to classify a drug's permeability when used in conjunction with standard compounds. The method must also take into consideration a drug's stability and solubility, whether the drug is passively or actively absorbed, and if the cell line expresses efflux pumps. The BCS relies on method suitability and internal standards for *in vitro* permeability assays, as opposed to a standardized protocol, thus allowing for variability in cell lines and methods among laboratories while encouraging advancements in the existing technology.

## THE BCS

Dissolution, solubility and permeability are the three major factors that influence the rate and extent of oral absorption of a drug that is stable in the gastrointestinal tract. The BCS is a drug development tool that allows estimation of the relative contributions of these three factors that affect drug absorption from immediate release (IR) solid oral dosage forms (Amidon *et al.*, 1995). In terms of bioequivalence, if two products of the same drug substance have the same concentration-time profile at the intestinal membrane surface, then they will have the same rate and extent of oral absorption. Additionally, if two drug products have the same *in vivo* dissolution profile, under all luminal conditions, they will also have the same rate and extent of oral absorption (Amidon *et al.*, 1995).

## Food and Drug Administration (FDA)s BCS guidance

The FDA (2000) has applied the above principles in a Guidance for Industry (BCS Guidance) to the regulatory application of bioavailability and bioequivalence and to recommend methods for classifying drugs and IR drug products. The BCS Guidance seeks to:

- improve efficiency of drug development and review process by proposing a method to identify expendable bioequivalence trials;
- recommend a class of IR solid oral dosage forms for which bioequivalence may be determined by *in vitro* dissolution tests;
- propose methods for the classification of drug product dissolution and drug substance solubility and permeability.

### Class boundaries and membership

The BCS Guidance sets boundaries for drugs in terms of solubility, permeability and dissolution. A drug substance is considered to be highly soluble when the highest dose strength is soluble in greater than or equal to 250 mL water over a pH range of 1–7.5 at 37 °C. A drug substance is regarded as highly permeable when the extent of intestinal absorption in humans is greater than 90 per cent of an administered dose based on mass balance or in comparison with an intra-venous reference dose. A drug product is regarded as rapidly dissolving when greater than or equal to 85 per cent of the labeled amount of drug substance dissolves within 30 minutes using the United States Pharmacopeia (USP) appar-atus-I or -II in a volume of less than or equal to 900 mL buffer solution (Hussain *et al.*, 1999). High solubility (HS) and HP ensure that the extent of absorption of two IR products containing the same drug will be equivalent, provided the products dissolve with sufficient rapidity. Based upon a drug substance's solubility and permeability, it is assigned a BCS class membership (Amidon *et al.*, 1995):

- Class-I – high permeability, high solubility HP/HS
- Class-II – high permeability, low solubility HP/LS
- Class-III – low permeability, high solubility LP/HS
- Class-IV – low permeability, high solubility LP/LS.

### Candidates for biowaiver

The BCS Guidance serves to modify the regulatory methods to establish bioequiva-lence in the United States by setting drug product dissolution standards to reduce the requirement for clinical bioequivalence trials (Löbenberg and Amidon, 2000).

In the current BCS Guidance, the following criteria are required for justifying a request (biowaiver) for clinical bioequivalence/bioavailability studies.

- drug substance defined as a class-I drug (i.e. HP/HS);
- IR drug product is rapidly dissolving and meets dissolution criteria comparison;
- drug substance not a narrow therapeutic index drug;
- excipients used in dosage form previously used in an FDA-approved IR solid dosage form.

## European perspective in context of BCS

The BCS concept is included in the European Agency for the Evaluation of Medicinal Products (EMEA) draft guidance for the investigation of bioavailability and bioequivalence (EMEA, 2000). This note focuses on the comparison of therapeutic performances of two medicinal products containing the same active substance by considering the possibility of utilizing *in vitro* instead of clinical studies with pharmacokinetic end points. Among others, validated *in vitro* dissolution and permeability studies are accepted as an alternative to clinical bioavailability studies to assess bioequivalence. Although this current EMEA draft guidance addresses BCS concepts, certain aspects (e.g. permeability determination, class boundaries) remain to be specified. According to the draft guideline recommendations, justification of *in vivo* bioequivalence study waivers should be based upon concepts underlying BCS, e.g. high solubility and permeability of the active pharmaceutical ingredient and rapid dissolution of the drug product (H. Möller, personal communication).

## *IN VITRO* PERMEABILITY EXPERIMENTS ACCORDING TO THE BCS GUIDANCE

In the framework of the BCS Guidance, the goal of permeability evaluations is to classify a drug substance as either HP or LP. According to the BCS Guidance, permeability can be assessed in human or animal studies, or with *in vitro* permeation experiments across an epithelial cell monolayer.

## Cell culture methods

### Epithelial cell lines

Although a number of epithelial cell lines have been used for *in vitro* intestinal permeability assays, the Caco-2 colon adenocarcinoma line has become the most utilized model since its described use by Hidalgo *et al.* (1989). However, other epithelial cell lines can be used in developing permeability model systems, including clones of Caco-2 (e.g. TC7) (Grés *et al.*, 1998), the small intestinal HT29 line and its variants (e.g. HT29-18-C$_1$, HT29-MTX) (Barthe *et al.*, 1999; Collett *et al.*, 1996; Wils *et al.*, 1994), Caco-2/HT29 co-cultures (Hilgendorf *et al.*, 2000), other colon cell lines (e.g. T84, HCT-8) (Barthe *et al.*, 1999; Collington *et al.*, 1992), transformed cell lines (e.g. 2/4/A1) (Crespi *et al.*, 1996, 2000; Tavelin *et al.*, 1999), and the Madin–Darby kidney cell line (MDCK) (Irvine *et al.*, 1999; Ranaldi *et al.*, 1992).

## *Aspects of methods*

An *in vitro* model of oral absorption should reflect, morphologically and functionally, the *in vivo* biological barrier of the intestinal mucosa (Borchardt, 1994; Hillgren *et al.*, 1995, Wilson, 1990). The predictive goal of the permeability model within a laboratory influences the choice of assay protocol. The protocol depends upon the level of *in vitro–in vivo* correlation the model is intended to provide: categorization of a drug substance as HP or LP, qualitative or quantitative approximation of the extent of intestinal absorption, rank-order permeability of a series of drugs, or determination of transport mechanism(s).

A review of the literature yields a framework as to how *in vitro* permeability assays are performed to assess and predict oral drug absorption. The model is comprised of three phases: cell culture, transport experiment and data analysis. The differences and similarities as to how these assays are performed affect the outcome of the transport experiments (Artursson *et al.*, 1996b; Audus *et al.*, 1990; Bailey *et al.*, 1996; Borchardt, 1994; Delie and Rubas, 1997; Meunier *et al.*, 1995). The differences in culture and transport techniques have resulted in different measurements of *in vitro* permeability ($P_{app}$) for the same drug (Delie and Rubas, 1997). Culture and transport conditions have great effect on the morphology and absorptive characteristics of the cell monolayer (Briske-Anderson *et al.*, 1997; Hillgren *et al.*, 1995; Hosoya *et al.*, 1996; Wilson, 1990). The differences in *in vitro–in vivo* correlation goals and methodologies suggest a basic approach to developing a model that can be applied to any circumstance and epithelial cell type.

## *Standardization*

Since there is variability in methodology between laboratories, a standard protocol would not be suitable for all cases. The variety of culture and transport parameters (e.g. growth conditions, filters, monolayer age, transport buffers, etc.) used by researchers reveal that an optimal set of conditions have yet to be determined for this model. The parameters in the three phases of the permeability model need to be carefully optimized and controlled to best imitate the *in vivo* intestinal barrier (Audus *et al.*, 1990; Bailey *et al.*, 1996; Borchardt, 1994).

## Establishment of a cell culture model

The BCS Guidance allows the use of cell culture methods to classify a drug substance as HP or LP. The variety of protocols in the literature does not provide absolute values to categorize HP and LP drugs. To preclude the inclusion of a standard protocol, the BCS Guidance recommends the development of an *in vitro* permeability model within a laboratory based on method suitability.

*Table 9.1* Method suitability for *in vitro* drug permeability methods according to the BCS Guidance

| Methods development | Establish a rank-order relationship between experimental permeability values and the extent of drug |
|---|---|
| | Model drugs ($n=20$) should represent a range of low ($<50\%$), moderate (50–89%), and high ($\geq 90\%$) |
| Use of method | There is no need to retest all model compounds to classify a drug substance when maintaining the same protocol |
| | Standard compounds are used when evaluating the test drug |
| | Reference – intra-laboratory variation, presence of efflux/active transporters; |
| | Internal – classify as HP or LP; |
| | Molecular marker – monolayer integrity. |
| | The permeability values of internal standards should not differ greatly between different tests, including those conducted to demonstrate suitability of the method. |
| Classification criteria | A test drug substance is highly permeable when its permeability is greater than or equal to that of the HP internal standard. |

## Method suitability

Method suitability allows for the use of any epithelial cell line and protocol within a laboratory. Employing the developed protocol, the researcher establishes a rank-order relationship between experimental $P_{app}$ values and the extent of intestinal absorption in humans ($\%f_a$) with model drugs. This relationship between $P_{app}$ and $\%f_a$ should clearly differentiate between HP and LP drug substances. For *in vitro* cell methods, it is recommended that 20 model drugs be used to establish method suitability. These model drugs should represent a range of *in vivo* human intestinal absorption, i.e. drugs with low ($f_a<50$ per cent), moderate ($f_a=50$–89 per cent), and high ($f_a\geq 90$ per cent) permeability. Due to the variability in experimental methods, the BCS Guidance advises the use of a sufficient number of cell monolayers (e.g., six) for each drug. After demonstrating method suitability, there is no need to re-evaluate the model drugs for future studies to classify any test drug as HP or LP provided the same assay protocol is used for each study (Table 9.1).

## Standard compounds

The routine use of standard compounds in a laboratory generates an acceptance criteria which can be monitored on a regular basis (Lee *et al.*, 1997; Hidalgo *et al.*, 1998). The standards can be divided into three categories:

- Reference standard – screen intra-laboratory variation, evaluate presence of efflux/active transporters;
- Internal standard – classify test drug as HP or LP;
- Molecular marker – appraise monolayer integrity.

The reference compounds are known LP, moderate permeability and HP drugs, along with substrates for active transporters and efflux mechanisms.

Internal standards are used in the study to determine permeability class membership and to ensure reproducibility. The internal standards should be compatible with the test drug, i.e. no significant physical, chemical or permeation interactions. An ideal internal standard is transported by passive diffusion, has no metabolism in cells, lacks cellular toxicity, has a known *in vivo* human absorption, and is not a substrate of any efflux mechanism.

The flux of a low permeability paracellular marker molecule (e.g. mannitol, PEG-4000, dextran, inulin, lucifer yellow) is used in each study to provide evidence for monolayer integrity. Such markers demonstrate the tight junctions of a cell monolayer, thus providing a physical barrier for drug transport (Borchardt, 1994; Hidalgo *et al.*, 1989).

The permeability of standard compounds (reference, internal, marker) should not differ considerably between distinct studies, including those that demonstrated method suitability. Ideally, the internal standard and molecular marker are added to cell monolayers along with the test drug during a transport experiment. Alternatively, the internal standard and molecular marker are evaluated in other monolayers at the same time.

### Assigning class membership

For a given model with demonstrated method suitability, selection of a highly permeable internal standard with permeability in close proximity to the high–low class boundary may facilitate classification of a test drug substance. A test drug is considered to be highly permeable when its *in vitro* $P_{app}$ value is greater than or equal to that of the HP internal standard (Table 9.1).

## Other considerations

Cell culture models with proven method suitability and standard compounds are acceptable processes to classify a drug substance for the intention of a waiver of bioequivalence or bioavailability clinical trials according to the BCS Guidance. However, there are other issues to take into account when using cell culture assays to classify a drug's permeability.

### Drug stability

The principle of the BCS, i.e. dissolution, permeability and solubility governs that the rate and extent of absorption applies to drugs that are stable in the gastrointestinal tract (Amidon *et al.*, 1995). In the BCS Guidance, *in vivo* or *in situ* human or animal permeability studies require documentation that the drug loss in the gastrointestinal tract or tissue is due to intestinal membrane permeation rather than a degradation process. It may be useful to determine the stability of a drug under the cell culture conditions utilized during the transport study (e.g. temperature, time, buffer, pH). This determination increases confidence that the resultant $P_{app}$ values are not skewed from drug loss due to instability.

### Drug solubility

The use of cell culture techniques can be limiting for drugs poorly soluble in an aqueous solution. The need for cell viability and integrity during the transport experiment curtails the use of organic solvents which can damage the cell monolayer. In addition, organic solvents may modify the physical barrier of the monolayers, thus altering the permeation of drugs through the cells (Pauletti *et al.*, 1998). Organic co-solvents (e.g. polyethylene glycol, dimethyl sulfoxide, ethanol) can be used at low concentrations when there is demonstration that the co-solvent does not alter the permeability of several model drugs.

### Efflux pumps

Low *in vivo* permeability of some drug substances may be the result of efflux mechanisms, such as *p*-glycoprotein (*p*-gp) (Hidalgo and Li, 1996; Hosoya *et al.*, 1996). When efflux transporters are absent or have low expression in cell culture models, there exists a potential for misclassification of a drug subject to efflux. Therefore, expression of known transporters in a selected permeability model should be characterized. Functional expression of efflux mechanism is verified with bidirectional transport studies with recognized substrates (e.g. cyclosporin A, vinblastine, rhodamine 123) at non-saturating concentrations. Efflux is revealed by a higher rate of substrate's permeability in the secretive (basolateral-to-apical) direction as compared with the absorptive (apical-to-basolateral) direction. In addition, other efflux transporters in the multidrug resistance protein (MRP) family may be involved in the secretion of drugs from the gastrointestinal tract (Gutman *et al.*, 1999; Suzuki and Sugiyama, 2000).

### Passive versus active transport

*In vitro* cell culture models are acceptable methods to classify a drug's permeability in the BCS Guidance, provided that the test drug is passively absorbed. Passively absorbed drugs have better correlations between $f_a$ and $P_{app}$ in cell permeability models (Artursson and Borchardt, 1997; Artursson *et al.*, 1996b; Lennernäs *et al.*, 1996; Palm *et al.*, 1999). In the BCS Guidance, passive transport of a drug is shown by satisfying one of the following conditions:

- linear relationship of the dose and measured bioavailability of the drugs in humans;
- lack of dependence on initial concentration of *in vivo* or *in situ* permeability in an animal model; or
- lack of dependence on initial concentration and directional transport in an *in vitro* model.

If there is a need to demonstrate passive transport of a drug substance, permeability can be measured at several concentrations (e.g. 0.01×, 0.1×, 1× the highest dose strength dissolved in 250 mL aqueous media) along with bidirectional transport of the test drug (i.e. basolateral-to-apical versus apical-to-basolateral).

The level of expression of carrier and efflux proteins can differ on intestinal cell lines as a function of time in culture and culture conditions (Artursson *et al.*, 1996a; Hidalgo and Li, 1996; Hosoya *et al.*, 1996; Meunier *et al.*, 1995). Reference compounds can be used to determine the presence or absence of active drug transporters (e.g. amino acids, di/tripeptides, monocarboxylic acids, nucleosides) in the cell permeability model.

## CONCLUSION

The BCS Guidance allows for the use of cell culture methods to classify a drug substance as HP or LP. The guidance does not endorse a standard method but recommends the development of an *in vitro* permeability model based on method suitability. The variety of parameters in the cell culture, transport and analysis phases of the model used by researchers confirm that the definitive model has not been established. The differences in laboratory methodologies suggest that a basic outline be utilized for regulatory purposes that can be applied to a variety of cell types and protocols. The BCS Guidance utilizes method suitability to classify drugs with the use of reference standards. The advantages of such a system are that it accounts for inter-laboratory variability, allows for improvements in technology (e.g. automation, miniaturization, acceleration), and is applicable to a variety of epithelial cell types. It is important for a laboratory to be consistent in its assay methods and establish validation criteria (Bailey *et al.*, 1996; Hidalgo *et al.*, 1998; Hillgren *et al.*, 1995; Lee *et al.*, 1997; Sulzbacher *et al.*, 1998). An assay with established method suitability, standard compounds and criteria for classifying drugs improves the reliability for such assays for regulatory utility (Table 9.1).

## REFERENCES

Amidon, G. L., Lennernäs, H., Shah, V. P. and Crison, J. R. (1995) A theoretical basis for a biopharmaceutic drug classification: the correlation of *in vitro* drug product dissolution and *in vivo* bioavailability. *Pharm. Res.*, **12**, 413–420.

Artursson, P., Karlsson, J., Ocklind, G. and Schipper, N. (1996a) Studying transport processes in absorptive epithelia. In A. Shaw (ed.), *Cell Models of Epithelial Tissues – A Practical Approach*, pp. 111–133, Oxford: IRL.

Artursson, P., Palm, K. and Luthman, K. (1996b) Caco-2 monolayers in experimental and theoretical prediction of drug transport. *Adv. Drug Deliv. Rev.*, **22**, 67–84.

Artursson, P. and Borchardt, R. T. (1997) Intestinal drug absorption and metabolism in cell cultures: Caco-2 and beyond. *Pharm. Res.*, **14**, 1655–1658.

Audus, K. L., Bartel, R. L., Hidalgo, I. J. and Borchardt, R. T. (1990) The use of cultured epithelial and endothelial cells for drug transport and metabolism studies. *Pharm. Res.*, **7**, 435–451.

Bailey, C. A., Bryla, P. and Malick, A. W. (1996) The use of the intestinal epithelial cell culture model, Caco-2, in pharmaceutical development. *Adv. Drug Deliv. Rev.*, **22**, 85–103.

Barthe, L., Woodley, J. and Houin, G. (1999) Gastrointestinal absorption of drugs: methods and studies. *Fundam. Clin. Pharm.*, **13**, 154–168.

Borchardt, R. T. (1994) Rational delivery strategies for the design of peptides with enhanced oral delivery. *Drug Dev. Ind. Pharm.*, **20**, 469–483.

Briske-Anderson, M. J., Finely, J. W. and Newman, S. M. (1997) The influence of culture time and passage number on the morphological and physiological development of Caco-2 cells. *Proc. Soc. Exp. Biol. Med.*, **214**, 248–257.

Center for Drug Evaluation and Research, Food and Drug Administration. Guidance for Industry. (2000) 'Waiver of *in vivo* Bioavailability and Bioequivalence Studies for Immediate Release Solid Oral Dosage Forms Based on a Biopharmaceutics Classification System'. August 2000.

Collett, A., Sims, E., Walker, D., He, Y. L., Ayrton, J., Rowland, M. *et al.* (1996) Comparison of HT29–18-C$_1$ and Caco-2 cell lines as models for studying intestinal paracellular drug absorption. *Pharm. Res.*, **13**, 216–221.

Collington, G. K., Hunter, J., Allen, C. N., Simmons, N. L. and Hirst, B. H. (1992) Polarized efflux of 2′,7′-bis(2-carboxyethyl)-5(6)-carboxyfluorescein from cultured epithelial cell monolayers. *Biochem. Pharmacol.*, **44**, 417–424.

Crespi, C. L., Penman, B. W. and Hu, M. (1996) Development of Caco-2 cells expressing high levels of cDNA-derived cytochrome P4503A4. *Pharm. Res.*, **13**, 1635–1641.

Crespi, C. L., Fox, L., Stocker, P., Hu, M. and Steimel, D. T. (2000) Analysis of drug transport and metabolism in cell monolayer systems that have been modified by cytochrome P4503A4 cDNA-expression. *Eur. J. Pharm. Sci.*, **12**, 63–68.

Delie, F. and Rubas, W. (1997) A human colonic cell line sharing similarities with enterocytes as a model to examine oral absorption: advantages and limitations of the Caco-2 model. *Crit. Rev. Ther. Drug Carrier Syst.*, **14**, 221–286.

European Agency for the Evaluation of Medicinal Products, Committee for Proprietary Medicinal Products. (2000) 'Note for Guidance on the Investigation of Bioavailability and Bioequivalence'. CPMP/EWP/QWP/1401/98.

Grés, M. C., Julian, B., Bourrié, M., Meunier, V., Rogues, C., Berger, M. *et al.* (1998) Correlation between oral drug absorption in humans, and apparent drug permeability in TC-7 cells, a human epithelial intestinal cell line: comparison with the parental Caco-2 cell line. *Pharm. Res.*, **15**, 726–733.

Gutman, H., Fricker, G., Török, M., Michael, S., Beglinger, C. and Drewe, J. (1999) Evidence for different ABC-transporters in Caco-2 cells modulating drug uptake. *Pharm. Res.*, **16**, 402–407.

Hidalgo, I. J., Raub, T. J. and Borchardt, R. T. (1989) Characterization of the human colon carcinoma cell line (Caco-2) as a model system for intestinal epithelial permeability. *Gastroenterology*, **96**, 736–749.

Hidalgo, I. J. and Li, J. (1996) Carrier-mediated transport and efflux mechanisms in Caco-2 cells. *Adv. Drug Deliv. Rev.*, **22**, 53–66.

Hidalgo, I. J., Windisch, V., Hu, H., Furukawa, D. and Kardos, P. (1998) Importance of establishing acceptance criteria for Caco-2 monolayers used in permeability studies. *PharmSci*™, **1**, 1197.

Hilgendorf, C., Spahn-Langguth, H., Regårdh, C. G., Lipka, E., Amidon, G. L. and Langguth, P. (2000) Caco-2 versus Caco-2/HT29-MTX co-cultured cell lines: permeabilities via diffusion, inside- and outside-directed carrier-mediated transport. *J. Pharm. Sci.*, **89**, 63–75.

Hillgren, K. M., Kato, A. and Borchardt, R. T. (1995) *In vitro* systems for studying intestinal drug absorption. *Med. Res. Rev.*, **15**, 83–109.

Hosoya, H. I., Kim, K. J. and Lee, V. H. L. (1996) Age dependent expression of P-glycoprotein gp170 in Caco-2 cell monolayers. *Pharm. Res.*, **13**, 885–890.

Hussain, A. J., Lesko, L. J., Lo, K. Y., Shah, V. P., Volpe, D. and Williams, R. L. (1999) The biopharmaceutics classification system: highlights of FDA's draft guidance. *Dissolution Technol.*, **6**, 5–9.

Irvine, J. D., Takahashi, L., Lockhart, K., Cheong, J., Tolan, J. W., Selick, H. E. *et al.* (1999) MDCK (Madin–Darby canine kidney) cells: a tool for membrane permeability screening. *J. Pharm. Sci.*, **88**, 28–33.

Lee, C. P., deVrueh, R. L. A. and Smith, P. L. (1997) Selection of development candidates based on in vitro permeability measurements. *Adv. Drug Deliv. Rev.*, **23**, 47–62.

Lennernäs, H., Palm, K., Fagerholm, U. and Artursson, P. (1996) Comparison between active and passive drug transport in human intestinal epithelial (Caco-2) cells *in vitro* and human jejunum *in vivo. Int. J. Pharm.*, **127**, 103–107.

Lögenberg, R. and Amidon, G. L. (2000) Modern bioavailability, bioequivalence and bio-pharmaceutics classification system. New scientific approaches to international regulatory standards. *Eur. J. Pharm. Biopharm.*, **50**, 3–12.

Meunier, V., Bourrié, M., Berger, Y. and Fabre, G. (1995) The human intestinal epithelial cell line Caco-2; pharmacological and pharmacokinetic applications. *Cell. Biol. Toxicol.*, **11**, 187–194.

Palm, K., Luthman, K., Ros, J., Gråsjo, J. and Artursson, P. (1999) Effect of molecular charge on intestinal drug transport: pH dependent transport of cationic drugs. *J. Pharmacol. Exp. Ther.*, **291**, 435–443.

Pauletti, G. M., Audus, K. L. and Hidalgo, I. J. (1998) Effect of co-solvents on the physical barrier properties of Caco-2 cell monolayers. *PharmSci*™, **1**, 1182.

Ranaldi, G., Islam, K. and Sambuy, Y. (1992) Epithelial cells in culture as a model for the intestinal transport of antimicrobial agents. *J. Antimicrob. Agents Chemother.*, **36**, 1374–1381.

Sulzbacher, A., Jarosch, A., Schuler R., Acerbi, D., Ventura, P., Puccini, R. *et al.* (1998) Validation of a Caco-2 cell monolayer culture for drug transport studies. *Int. J. Clin. Pharmacol. Ther.*, **36**, 86–89.

Suzuki, H. and Sugiyama, Y. (2000) Role of metabolic enzymes and efflux transporters in the absorption of drugs from the small intestine. *Eur. J. Pharm. Sci.*, **12**, 3–12.

Tavelin, S., Milovic, V., Ocklind, G., Olsson, S. and Artursson, P. (1999) A conditionally immortalized epithelial cell line for studies of intestinal drug transport. *J. Pharm. Exp. Ther.*, **290**, 1212–1221.

Wils, P., Warnery, A., Phung-Ba, V. and Scherman, D. (1994) Differentiated epithelial cell lines as *in vitro* models for predicting the intestinal absorption of drugs. *Cell Biol. Toxicol.*, **10**, 393–397.

Wilson, G. (1990) Cell culture techniques for the study of drug transport. *Eur. J. Drug Metabol. Pharmacokinet.*, **15**, 159–163.

# Part 2

# Models of specific epithelial and endothelial barriers relevant to drug delivery

# Chapter 10

# Caco-2 cell monolayers as a model for studies of drug transport across human intestinal epithelium

*Weiqing Chen, Fuxing Tang, Kazutoshi Horie and Ronald T. Borchardt*

## INTRODUCTION

The gastrointestinal (GI) tract has always been of interest for drug delivery owing to the convenience of oral administration. However, the successful development of oral drugs has been challenged by a variety of barriers in the GI tract. One of the barriers that restrict the bioavailability of orally administered drugs is the epithelium of the intestinal mucosa that lines the surface of the GI tract. The intestinal mucosa is both a physical and a biochemical barrier, separating the external environment from the internal milieu of the body (Pauletti *et al.*, 1997).

The physical barrier arises from cell membranes and the intercellular junctions between the cells (e.g. tight junctions). Permeation of drugs across the intestinal epithelium is restricted to paracellular and transcellular pathways depending on their physicochemical properties (e.g. size, charge, lipophilicity, and conformation). Most large molecules that are hydrophilic are prevented from passing across the cell membranes unless some specific membrane proteins are involved to serve as channels, carriers, or transporters. Only lypophilic molecules may directly pass across the lipid bilayer of the cell membranes by passive diffusion. In addition to the physical barrier, the intestinal epithelium also possesses various metabolic enzymes (e.g. intestinal peptidases, cytochrome P450) and polarized efflux systems (e.g. *p*-glycoprotein, *P*-gp) which act as biochemical barriers further limiting drug absorption in the intestine. Consequently, many drug candidates are restricted from oral dosing in clinical development owing to this biological barrier.

Pharmaceutical scientists have dedicated significant amount of time and effort to developing either delivery systems (e.g. formulation strategies) that could facilitate intestinal mucosal permeation of drug candidates or to designing structural features that confer drug candidates with good absorption characteristics. Executing these strategies requires access to technologies that can be used for assessing of the intestinal permeation of new chemical entities and for basic research aimed at mechanistic understanding of the barrier properties.

Many intestinal mucosal models have been developed for *in vitro* studies during the last several decades. These include utilizing excised intestinal segments (Fisher and Parsons, 1949; Parsons, 1968), everted intestinal sacs (Wilson and Wiseman, 1954), everted intestinal rings (Porter *et al.*, 1985), isolated mucosal sheets (Csáky, 1984), isolated epithelial enterocytes (Moyer, 1983), cell membrane vesicles (Hopfer *et al.*, 1973), and cultured cells (Dharmsathaphorn and Madara, 1990; Hidalgo *et al.*, 1989; Kreusel *et al.*, 1991; Wikman *et al.*, 1993). All of these models mimic the characteristics of the intestinal mucosa to a varying extent and have provided different approaches for studies of intestinal drug absorption.

The isolated intestinal segments and the everted intestinal sacs have been used to study drug transport across the intestinal mucosa by monitoring the presence of test compounds on both the luminal and serosal sides when the tissues were perfused or incubated with drug solutions (Wilson and Wiseman, 1954). In some cases, the intestine was cut open as mucosal sheets with either musculature stripped or non-stripped and clamped between the two compartments of a diffusion chamber so that drug transport could be studied in both directions (Csáky, 1984). Although these freshly excised intestinal segments or mucosal sheets offer the tissue structures and components very similar to the *in vivo* biological barrier, there are many limitations and disadvantages regarding using these models for transport studies. The maintenance of tissue viability and barrier integrity over time during experiments is the major concern of using isolated intestinal tissues (Levine *et al.*, 1970; Parsons and Paterson, 1965). A large quantity of test compounds is usually required. In addition, the complexity of the tissue components involved makes it difficult to interpret experimental results conclusively.

Utilizing membrane vesicles prepared from brush-border or basolateral (BL) cell membranes of fresh intestinal enterocytes can simplify the assay systems for absorption studies. However, membrane vesicles are only useful for determination of membrane uptake, not for transepithelial transport processes.

To better assess the absorption of compounds across intestinal epithelium, the ideal models would be isolated intestinal barriers, such as the cell culture systems, with defined monolayers of normal polarized human enterocytes. Assays utilizing cultured cells require relatively small amounts of a compound and can be performed in large numbers. The experimental conditions, such as pH, temperature, drug concentrations, and presence of the inhibitors of various enzyme and transporters can be precisely controlled. Such defined systems are more suitable for automation in large scale, which not only allows high-throughput screening of lead compounds in drug discovery and development but also permits mechanistic studies of drug transport and metabolism. However, attempts to grow monolayers of isolated normal human intestinal enterocytes have failed owing to low viability and difficulties in cell attachment, monolayer formation, and differentiation (Goodwin *et al.*, 1993; Kimmich, 1990; Moyer, 1983). Therefore, the interest in the applications of human colonic tumor cell lines for drug transport studies has increased during the last several decades. The most commonly used cell lines that have been reported in the literature are Caco-2 cells, HT-29 cells, and T84 cells. Among them, the Caco-2 cell line is the best-characterized and most widely-used culture system in both academia and industry (Artursson *et al.*, 1996; Audus *et al.*, 1990; Bailey *et al.*, 1996; Burton *et al.*, 1997; Gan and Thakker, 1997; Hillgren *et al.*, 1995).

Caco-2 cells were originally derived from human colon adenocarcinoma. When grown under standard culture conditions the cells can spontaneously undergo enterocytic differentiation after reaching confluency, which gives the Caco-2 cells many properties resembling the small intestinal epithelium (Hidalgo *et al.*, 1989; Quaroni and Hochman, 1996). For example, the Caco-2 cells become columnar in shape and form junctional complexes with well-developed brush-border facing the medium (Hidalgo *et al.*, 1989). Several typical small intestinal brush-border enzymes such as sucrase-isomaltase, dipeptidylpeptidase-IV, and alkaline phosphatase are present in differentiated Caco-2 cells (Hidago *et al.*, 1989; Zweibaum *et al.*, 1984). The cells also express oligopeptide transporter (Saito and Inui, 1993; Thwaites *et al.*, 1994), *p*-gp (Hunter *et al.*, 1993), and many other protein carriers or transporters for nucleosides, amino acids, glucose, and vitamin B12 (Blais *et al.*, 1987; Dantzig and Bergin, 1990; Dix *et al.*, 1990; Harris *et al.*, 1992; Hu and Borchardt, 1992; Ward and Tse, 1999). In addition, the cells have been shown to express many enzymes that are involved in phase-I and phase-II reactions, including phenol sulfotransferase (Baranczyk-Kuzma *et al.*, 1991), UPD-glucuronyl transferase (Peter and Reolofs, 1992), and P450 enzymes (Boulenc *et al.*, 1992; Peter and Reolofs, 1992). The structural and biochemical characteristics of Caco-2 cells, as well as their ease of handling, have made Caco-2 cell line the best intestinal mucosal model for studies of drug absorption and drug metabolism. Many of our insights into the intestinal barrier properties and understanding of the chemical and physiological constraints of oral drug absorption have been derived from the experimental results using Caco-2 cell model (see Table 10.1 for summary of the application of the Caco-2 cell culture system).

In this chapter, a detailed methodology of growing Caco-2 cells on polycarbonate membrane is described. The protocols of using Caco-2 cells for conducting uptake and transport studies of test compounds will also be addressed, along with the theoretical considerations of data analysis.

*Table 10.1* Applications of Caco-2 cell model in drug absorption studies

| Applications | References |
|---|---|
| Assess potential toxic effects of drug candidates or formulation components on the intestinal mucosa | Tang *et al.*, 1993 |
| Characterize the optimal physicochemical properties (e.g. hydrogen bonding potential, conformation) of a drug for passive diffusion via the paracellular or transcellular pathways across the intestinal mucosa | Burton *et al.*, 1996 |
| Determine the structure–transport relationships for carrier-mediated pathways (e.g. peptide transporter) of drug transport | Hidalgo and Li, 1996 |
| Determine how the components of a formulation (e.g. adjuvants) might influence the intestinal mucosa transport of drug candidates | Nerurkar *et al.*, 1996 |
| Investigate potential drug–drug interactions during intestinal mucosal transport | Wacher *et al.*, 1996 |
| Determine the structure–transport relationships for the apically polarized efflux systems (e.g. *p*-glycoprotein) in the intestinal mucosa | Burton *et al.*, 1997 |
| Elucidate the pathways of drug transport (e.g. paracellular versus transcellular; passive versus carried-mediated) across the intestinal mucosa | Knipp *et al.*, 1997 |
| Elucidate potential pathways of drug metabolism in the intestinal mucosa. | Schmiedlin-Ren *et al.*, 1997 |

## CELL CULTURE

### Protocol 1: starting and maintaining Caco-2 cells in culture

Newly started Caco-2 cells from frozen conditions (cryogenic vials) should be grown in the culture medium containing 20 per cent serum in flasks for two or three passages until the cells are growing normally. The cells are then cultured in the medium containing ten per cent serum. When the cell density in flasks reaches 80 per cent confluency the cells need to be split into new flask(s) for further growing or they can be seeded onto Transwells® for transport experiments.

*Materials*

- frozen Caco-2 cells (ATCC #HTB-37)
- complete Dulbecco's modified Eagle's medium (DMEM)-10 (37 °C, see REAGENT AND SOLUTIONS)
- complete DMEM-20 (37 °C, see REAGENT AND SOLUTIONS)
- ethylenediaminetetraacetic acid (EDTA)/Phosphate buffered saline (PBS) solution (37 °C, see REAGENT AND SOLUTIONS)
- trypsin solution (37 °C, JRH Biosciences, No. 59228-78P)
- 50 ml plastic centrifuge tubes (Fisher)
- 2, 5, 10, and 25 ml sterile disposable pipettes (Fisher)
- Pipet-aid® (Drummond Scientific)
- 25 and 150 cm² T-flasks (Corning)
- reciprocal shaking water-bath (Precision Scientific, Model 25)
- vacuum pump (Precision Scientific, Model DD 20)
- centrifuge (e.g. Centra® CL2, International Equipment Company)
- cell culture incubator (Forma Scientific)
- laminar flow hood equipped with ultraviolet (UV) lighting (Bellco Glass, Inc.)
- microscope (TMS, Nicon).

*Start seeding with frozen Caco-2 cells in 25 cm² T-flask*

1   Take a vial of the frozen cells (1 ml of $10^6$ cells/ml) from a liquid nitrogen tank and thaw it as quickly as possible in a water-bath (37 °C). Do not keep the cells at this temperature for more than five minutes.
2   Transfer the cell suspension to a 50 ml centrifuge tube containing 40 ml of complete DMEM-20 and mix the cell suspension gently.
3   Centrifuge the cell suspension at 150×g for five minutes to get cell pellets and then gently resuspend the cells in 40 ml complete DMEM-20.
4   Repeat centrifugation and resuspend the cells in 10 ml of complete DMEM-20.
5   Place the entire cell suspension in a 25 cm² T-flask and incubate the flask in a culture incubator (37 °C, 95 per cent relative humidity, and 5 per cent $CO_2$).

*Transfer and grow Caco-2 cells in 150 cm² T-flask*

6   Feed the cells with 10 ml of complete DMEM-20 when cells become attached on the flask (~ day four) and every three days thereafter. Split the cells when 80 per cent confluency is reached (about 12–15 days).

7   To split the cells, replace the medium with 10 ml of EDTA/PBS solution (37 °C) for five minutes. Remove EDTA/PBS solution. Treat the cells with 3 ml of trypsin solution (37 °C, warm trypsin solution in a water-bath for less than 30 minutes to prevent self-digestion) for one minute and then remove the solution.

8   Keep the trypsin-treated flask (with no medium) in the incubator until the cells detach from the flask (about 10–15 minutes).

9   Gently mix the detached cells with 12 ml of complete DMEM-20. Place all of the cell suspension in a new 150 cm$^2$ flask containing 12 ml of complete DMEM-20 and keep it in an incubator for further growing.

10  Feed the cells with complete DMEM-20 the next day and thereafter every other day. Split the cells when approximately 80 per cent confluency is reached (steps 7–9, except that 1/2 or 1/3 of the cell suspension is placed into each new flask). Use complete DMEM-20 to feed the cells until they grow at a normal rate (about five–seven days to reach 80 per cent confluency after each splitting).

*Maintain Caco-2 cells in 150 cm$^2$ flask*

Once the newly started Caco-2 cells reach the normal growth rate, the cells can be maintained in 25 ml of complete DMEM-10 instead of complete DMEM-20.

*Split the cells*

11  Treat the cells with EDTA/PBS and trypsin steps 7 and 8.

12  Add 12 ml of complete DMEM-10 to the flask and mix the cells gently. Place 1 ml of the cell suspension in a new 150 cm$^2$ flask. Add 25 ml of complete DMEM-10. Label the current passage number and keep it in the incubator.

13  Feed the flask the day after splitting and every other day thereafter until the cells have reached >80 per cent confluency.

Continue splitting the flask after reaching the confluency until it is time to start new frozen cells (~ passages 70–80). Caco-2 cell monolayers with high passage numbers exhibit lower expression of efflux and other active transporters. It may be necessary to perform routine bidirectional transport experiments using certain substrates (e.g. [$^3$H] digoxin, a *p*-gp substrate) when reaching higher passage numbers prior to restarting new cells.

## Protocol 2: growing Caco-2 cell monolayers on Transwell® filters

### Seeding and feeding the Caco-2 cells

To perform transport studies with Caco-2 cell monolayers, cells must be seeded onto Transwells®. These Transwells® are equipped with a sterile polycarbonate filter that divides the well into apical (AP) and BL compartments (Figure 10.1). Cells must be grown on the membrane support to full differentiation (days 21–28 after seeding).

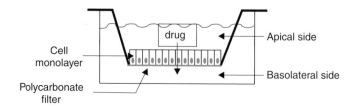

*Figure 10.1* The diagram of Transwell®.

*Materials*

- ethanol/collagen solution (see Reagents and Solutions)
- incomplete and complete DMEM-10 (37 °C)
- Caco-2 cells in 150 cm² flask, >80 per cent confluent, growing in complete DMEM-10 (see Protocol 1)
- EDTA/PBS solution (37 °C)
- trypsin solution (37 °C)
- 6-well clusters with Transwell® (24 mm diameter, 3 μm pore size; #3414, Costar)
- Hemacytometer (e.g. Reichert Bright-Line, Fisher).

*Coat the Transwells®*

1   Add 275 μl of ethanol/collagen solution to the center of each Transwell®.
2   Distribute the solution over the filter by tilting the Transwell® trays.
3   Keep the Transwell® trays uncovered in the laminar flow hood and allow Transwells® to dry overnight under UV light.

*Prepare the Transwells® for seeding*

4   Add 2.6 ml of incomplete media to the bottom (basolateral side) and 1.5 ml to the top (AP side) of each Transwell®. Incubate the Transwells® in an incubator for 2 hours.
5   Remove the incomplete media from both the AP and BL sides with a nine-inch sterile disposable Pasteur pipette.
6   Add 2.6 ml of prewarmed complete DMEM-10 to the bottom of each well and return the Transwells® to the incubator.

*Seed the Caco-2 cells onto Transwells®*

7   When the flask has reached >80 per cent confluency, detach the cells from the flask as described (Protocol 1: starting and maintaining Caco-2 cell in culture, steps 7 and 8).
8   Add 12 ml of prewarmed complete DMEM-10 to the flask. Mix the cells gently by drawing the cell suspension in and out of a 10 ml pipette (mixing is critical to assure separation of cells from cell clusters).
9   Transfer the cell suspension to a sterile 50 ml centrifuge tube containing 24 ml of complete DMEM-10 (1:3 dilution). Place 2.5–3 ml of the diluted cell

suspension into a new $150\,cm^2$ flask containing 25 ml of complete DMEM-10 if the Caco-2 cells are to be maintained in culture on a regular basis.

10  Count the density of the diluted Caco-2 cell suspension using a Reichert Bright-Line Hemacytometer®.

11  Further dilute the Caco-2 cell suspension with complete DMEM-10 to achieve a density of $2.5 \times 10^5$ cells/ml (about $10^5$ cells/cm$^2$).

12  Seed 1.5 ml of the diluted cell suspension onto the top of each Transwell® prepared in step 6. Distribute the cells evenly onto the Transwells® by swirling the Transwells® frequently while seeding the cells.

### Feed the Caco-2 cells in Transwells®

Feed the top and the bottom of each Transwell® the day after seeding and then every other day through day seven. After day seven, feed the tops of the Transwells® every day and the bottoms every other day until days 21–28.

13  Remove the culture media from the bottoms of all Transwells® using a 9″ sterile glass Pasteur pipette connected to a vacuum pump. If feeding of the tops is needed, carefully remove the culture media from the tops of the Transwells®.

14  Add 2.6 ml of prewarmed complete DMEM-10 to the bottom of each well and 1.5 ml to the top of each well with a 25 ml pipette. Repeat for all seeded trays and return the trays to the incubator.

## Protocol 3: testing Caco-2 monolayer integrity

### Mannitol flux experiment

It is important to validate the integrity of Caco-2 monolayers prior to experimentation by testing the permeation of an impermeable substance (e.g. radiolabeled mannitol) across the Caco-2 monolayers. The use of an epithelial volt ohm meter is another means of testing the monolayer integrity, which measures the potential across the cell monolayer as well as the resistance encountered in the system.

### Materials

- D-[$^{14}$C] mannitol (51 mCi/mmole; NEN Life Sciences)
- prewarmed Hank's balanced salt solution (HBSS) (see Reagent and Solutions)
- Caco-2 cell monolayers grown on Transwells® for 21–28 days (see Protocol 2)
- scintillation cocktail (e.g. Research Products International)
- 6 ml scintillation vials and caps (Fisher)
- 6-well cell culture clusters (Costar).

### Prepare sample vials

1  Arrange and label scintillation vials as triplicate for certain time points (e.g. 0, 30 minutes). Samples are taken every 30 minutes from the basolateral side and every 60 minutes from the AP side, up to 180 minutes.

2   Label one vial for stock solution of mannitol and another vial for HBSS buffer
solution only as blank.

*Prepare the stock and blank solutions*

3   Take the source, D-[$^{14}$C] mannitol, from the freezer to the designated bench
for radioactive materials and allow it to warm to room temperature.
4   Meanwhile, label a 6-well cluster with numbers 1–3 and place the tray under
UV light in the aseptic hood to decontaminate.
5   Place 6 ml of HBSS (37 °C) buffer solution in the vials labeled as 'stock' and
'blank'.
6   Add 10 µl of D-[$^{14}$C] mannitol to the 'stock' vial. Cap and mix it well.
7   Place the 'blank' and 'stock' in a water-bath (37 °C).

*Prepare the Transwells*®

8   Select three Transwells® of Caco-2 monolayers and remove the growth media
from both sides.
9   Transfer the Transwells® to the labeled 6-well cluster.
10   Wash the tops and bottoms with warm HBSS. Repeat three times.
11   Place 2.6 ml of HBSS buffer solution in the receiver compartment (the bottom
in this case). Quickly move to the next step.

*Perform the experiment*

12   Slowly deliver 1.5 ml of mannitol 'stock' solution to the donor compartments
by placing the pipette tip against the Transwell® wall to avoid disrupting the
monolayer. Note the time as time-zero.
13   Cover the tray and place it in the shaking water-bath (55 rpm). *Do not spill the
solution out of the donor compartment when placing the tray.*
14   Pipette 100 µl of mannitol 'stock' solution into the scintillation vials designated
as time-zero for each donor compartment. Pipette 100 µl of 'blank' into the
vials designated as time-zero for each receiver compartment. Add 3 ml of
scintillation cocktail to these vials, cap, and mix.
15   At 30 minutes, remove the 'blank' solution and cluster from the shaking water-
bath. Take 100 µl of sample from the receiver compartment of each well into
the appropriate vials prepared and replenish each of the receiver compartments with 100 µl of blank. Place the blank solution and cluster back to the
water-bath after sampling.
16   At 60 minutes, pipette 100 µl from the donor compartment of each well into
appropriate vials, being careful not to disrupt the monolayer. Then take the
samples from receiver compartments as step 15. Add 3 ml of cocktail to each of
these vials, cap and mix.
17   Repeat steps 15 and 16 every 30 minutes and every 60 minutes, respectively,
to collect samples into the appropriate vials, up to 180 minutes.
18   After finishing sampling, mix all vials well. Then place the vials in the scintillation counter to quantify radioactivity (dpm).

*Data analysis (per cent of transported mannitol and its apparent permeability coefficient ($P_{app}$))*

19  The background radioactivity in each sample must be first corrected by subtracting the radioactivity at time-zero from that at the various time points ($dpm_{t(C)} = dpm_t - dpm_0$, $t = 0, 30, 60, 90, 120, 180$ minutes).

20  Calculations should be performed for each of the triplicates at certain time points and the results are then used to determine the average.

21  Calculate the cumulative amount of D-[$^{14}$C] mannitol ($Q_{Bt}$, in dpm) which appears in the receiving (basolateral) compartment at each time point. A series of amount (100 µl) of D-[$^{14}$C] mannitol removed from the receiving compartment during sampling needs to be added back when calculating the total radioactivity in the receiving compartment at each time point:

$$Q_{B0} = dpm_{B0(C)} \times \left(\frac{V_R}{V_S}\right) \tag{10.1}$$

$$Q_{B30} = dpm_{B30(C)} \times \left(\frac{V_R}{V_S}\right) + dpm_{B0(C)} \tag{10.2}$$

$$Q_{B180} = dpm_{B180(C)} \times \left(\frac{V_R}{V_S}\right) + dpm_{B0(C)} + \cdots + dpm_{B180(C)} \tag{10.3}$$

($V_R$ = the volume of fluid in the receiving compartment; $V_S$ = sample volume).

22  Calculate the total amount of radioactivity in the donor compartment at time-zero ($Q_{A0}$, in dpm):

$$Q_{A0} = dpm_{A0(C)} \times \left(\frac{V_D}{V_{S'}}\right), \tag{10.4}$$

($V_D$ = the volume of donor compartment; $V_{S'}$ = sample volume from donor compartment).

23  Calculate the per cent of D-[$^{14}$C] mannitol transported at each time point:

$$\text{per cent of mannitol transported at the time } t = \left(\frac{Q_{Bt}}{Q_{A0}}\right) \times 100. \tag{10.5}$$

24  Calculate the apparent $P_{app}$, which is $\Delta Q/\Delta t$, with respect to the surface area of the polycarbonate filters of the Transwells® ($A$, in cm² for Transwells of 24 mm diameter, $A = 4.71$ cm²) and the initial concentration of D-[$^{14}$C] mannitol in the donor compartment ($C_0$, in dpm/ml).

$$C_0 = \frac{dpm_{A0(C)}}{V_{S'}} \tag{10.6}$$

$$P_{app} = \left| \frac{\left(\frac{\Delta Q}{\Delta t}\right)}{(60 \times A \times C_0)} = \frac{m}{(60 \times A \times C_0)} \right| \tag{10.7}$$

(here '60' is the second/minute conversion factor).

The $P_{app}$ value should be in the range of $10^{-7}$ cm/second for this paracellular marker, and an average per cent transport of 0.5–1.5 per cent in three hours is indicative of a confluent monolayer with the appropriate tight junction integrity.

## Protocol 4: measurement of the $P_{app}$ of a compound across Caco-2 cell monolayers

This protocol describes a procedure for measuring the permeability of a test compound across Caco-2 cell monolayers from the AP (donor) to BL (receiver) side using Transwells®, which is similar to the procedure described above for mannitol flux. However, this protocol can be adapted for mechanistic studies of drug transport by manipulating the conditions of the transport systems utilized. For example, one can include a known substrate of a certain transporter with a test compound in the transport system to see whether they compete for the same transport processes (Leibach and Ganapathy, 1996). One can also change the pH or a certain ion concentration in the transport assay buffer to modify the proton or the ion gradient that may be related to the activity of the transporters interested (Ganapathy and Leibach, 1983). In addition, a transport experiment can be performed in the BL-to-AP direction in which the BL side is employed as the drug donor compartment and the AP side as the receiving compartment to test if a compound is a substrate for an efflux transporter (e.g. *p*-gp) (Walle and Walle, 1998). The integrity of the cell monolayers must be validated prior to experimentation (see Protocol 3).

*Materials*

- test compound: e.g. 1.0 mM L-carnosine (Aldrich) dissolved in Earle's balanced salt solution (EBSS) (see REAGENT AND SOLUTIONS)
- prewarmed HBSS
- Caco-2 cell monolayers grown on Transwells® for 21–28 days and validated for monolayer integrity (see Protocol 2 and 3)
- high-performance liquid chromatography (HPLC) system (e.g. Shimadzu LC-6A)
- reverse-phase HPLC column (e.g. Vydac $C_{18}$, 4.6×250 mm)
- HPLC sample vials and caps (Sun International Trading, Ltd. #200, 264 and #500, 206).

*Prepare standards*

Serially dilute a known concentration of a test compound with HBSS to prepare 200 µl each of five standard solutions in HPLC sample vials at a concentration range of 0–1/10th the donor concentration. Add 100 µl of 0.05 N HCl into the vials to quench any possible enzymatic reaction.

*Prepare sample vials*

Label the sample vials in triplicates according to the side of the compartments and the time points of sampling. For example, A01, A02, and A03 stand for sample one, two, and three at time zero from AP side, respectively. B201, B202, and B203 stand for sample one, two, and three at 20 minutes from BL side, respectively.

Add 180 µl of HBSS and 100 µl of 0.05 N HCl into the vials designated for AP samples and 100 µl of 0.05 N HCl into the vials designated for BL samples. Keep the vials in order in collecting tray(s).

*Conduct the transport experiment*

(All compound solutions and buffer added to the assay system are prewarmed to 37 °C unless specified.)

1 Wash the cell monolayers with HBSS (both the BL and the AP side) twice. Transfer the Transwells® into a 6-well cluster.
2 Incubate the cells with HBSS (2.6 ml on the BL side and 1.5 ml on the AP side) for 30 minutes in a water-bath. Then remove the HBSS.
3 Add 2.6 ml of HBSS to the BL side and 1.5 ml of the test compound solution to the AP side. Incubate the Transwells® in a shaking water-bath.
4 Take aliquots (20 µl) from the AP side at designated time points and aliquots (200 µl) from the BL side at the same time points. Replenish the BL side after each sampling with the same volume of HBSS.
5 Analyze standards and the samples from both donor and receiver sides by HPLC.

*Data analysis*

6 Plot the standard curve (concentration versus peak areas) based on the standard concentrations. Perform linear regression.
7 Use the parameters of the standard curve to calculate the concentrations of samples from both the donor side and the BL side at various time points (the peak areas of the apical samples need to be multiplied by ten).
8 Calculate the cumulative amount ($Q_t$, in mmole; $t$=sampling time in minutes) of the compound transported from the AP to the BL side at each time point. The amounts that are taken away from the receiver side at each sampling time need to be added back as performed as in Protocol 3 (mannitol flux experiment).
9 Plot the cumulative amount versus time and use the linear portion of the plot to determine the appearance rate ($\Delta Q/\Delta t$, in mmole/second) of the compound on the BL (receiver) side.
10 Calculate the apparent $P_{app}$ coefficient (in cm/second) according to the following equation:

$$P_{app} = \frac{(\Delta Q/\Delta t)}{(60 \times A \times C_0)}, \qquad (10.8)$$

($A$ is the surface area of the cell monolayers in cm²; $C_0$ is the initial concentration of the compound on the donor side in mmole/cm³).

## Protocol 5: uptake studies

Caco-2 cells may be seeded in 24-well tissue-culture plates at a density of $1.7 \times 10^5$ cells per well. The cells will be ready for uptake studies after 21–28 days. The uptake studies can be used to assess the binding affinity of a test compound for a certain transporter. In this case, intestinal oligopeptide transporter is used as an example. The uptake system includes a test compound and L-[$^3$H] carnosine, a known substrate for the oligopeptide transporter. Inhibition of the test compound on carnosine binding to the transporter indicates the affinity of this compound for the same transporter. Another carnosine uptake system containing GlyPro, which is also a known substrate for the oligopeptide transporter, is included in the experiment to serve as a control.

*Materials*

- L-carnosine (0.1 mM L-carnosine containing 1 μCi/ml L-[$^3$H] carnosine)
- 10 mM GlyPro
- 10 mM Test compound
- Caco-2 cells grown on 24-well plate for 21–28 days
- EBSS buffer solution (37 °C)
- liquid scintillation cocktail and counter (e.g. LS6000IC, Beckman)
- 6 ml scintillation vials and caps (Fisher)
- 0.3 M NaOH.

*Set-up uptake systems*

1   Prepare the uptake stock solutions for the following groups in EBSS. The pH of each group needs to be checked and adjusted if necessary:
    Group 1: 0.1 mM carnosine only
    Group 2: 0.1 mM carnosine + 10 mM GlyPro
    Group 3: 0.1 mM carnosine + 10 mM test compound.
2   Label the scintillation vials for each group in triplicates:
    T11, T12, T13: total radioactive carnosine added to Group 1
    T21, T22, T23: total radioactive carnosine added to Group 2
    T31, T32, T33: total radioactive carnosine added to Group 3
    B1, B2, B3: blanks
    11, 12, 13: carnosine uptake of Group 1
    21, 22, 23: carnosine uptake of Group 2
    31, 32, 33: carnosine uptake of Group 3

*Uptake experiments*

3   Wash the 24-well plate containing Caco-2 cells with prewarmed HBSS three times and label the wells according to the groups designated above.
4   Incubate with 1.0 ml of HBSS for 20 minutes in a shaking water-bath and then remove the HBSS from the wells.
5   Add 150 μl of uptake stock solutions into the wells (triplicates for each group). Mix the wells by swirling and then put the plate back to the water-bath.
6   Take 20 μl of the stock solution of each group into scintillation vials labeled as $T_{ij}$ (total radioactivity added in dpm; i=group number; j=replicate number) and 100 μl HBSS into $B_j$. Add 5 ml scintillation cocktail and count. The values of total radioactivity added to each group need to be corrected for blanks ($T_{ij(C)} = T_{ij} - Blank_{average}$).

7   After 20 minutes of incubation, add 1.5 ml of cold HBSS into each well to stop uptake. Remove the solution and wash the wells with cold HBSS three more times.
8   Add 200 μl of 0.3 M NaOH to each well. Scrape the cells off.
9   Remove the cell extracts into the scintillation vials. Wash the residue materials in the wells with another 100 μl of 0.3 M NaOH and collect into the vials.
10  Add 5 ml of scintillation cocktail and count for radioactivity of uptake ($U_{ij}$, in dpm).

*Data analysis*

11  Determine the radioactivity (dpm) per picomole of carnosine in each group:

$$\text{Radioactivity per picomole of carnosine} = T_{ij(C)}/(0.1 \times V_S),\qquad(10.9)$$
[$V_S$ = sample volume; $T_{ij(C)}$ = corrected total dpm of added;
0.1 = carnosine concentration (picomole/μl) used for uptake].

12  Calculate the amount of [$^3$H] carnosine taken up in each group ($Q_{ij}$, in picomole):

$$Q_{ij} = \frac{U_{ij(C)}}{\left[\dfrac{T_{ij(C)}}{(0.1 \times V_S)}\right]},\qquad(10.10)$$

($U_{ij(C)}$ = corrected dpm of a sample).

One may take the protein concentration of each well into consideration and calculate carnosine uptake in terms of picomoles of carnosine per milligram protein. However, the variations in the amount of protein among each well are usually very small if the Caco-2 cells are seeded under the same conditions and preparations.
13  Calculate the average of carnosine uptake in Group1 ($Q_1$, picomole).
14  Calculate the percentage of carnosine uptake in the groups containing test compound or inhibitor versus the average of carnosine uptake of the control (Group 1):

$$\text{carnosine uptake (\%)} = \frac{Q_{ij}}{Q_1} \times 100.\qquad(10.11)$$

Calculations should be performed for each individual triplicate and the results are then used to determine the average of each group.

If we consider carnosine uptake of control group as 100 per cent, reductions of carnosine uptake by test compounds indicate affinities of the test compounds for the same transporters as carnosine.

## QUALITY CONTROL

Successful use of Caco-2 model relies heavily on the quality of the cells used for drug transport studies. There are many factors affecting the quality of Caco-2 monolayers including culture media, stratum, coating materials, seeding density, cell passage numbers, and ages of the seeded cell monolayers. Several issues need to be taken for consideration. First, the growth medium should be kept in as much consistency as possible, especially the batch of fetal bovine serum (FBS) used. Whenever a new batch of FBS is used, growth as well as biochemical characteristics of the cells (e.g. enzyme activity, protein expression levels of specific transporters and *p*-gp) may need to be checked. Second, it is important to keep splitting Caco-2 cells into new flask when 80 per cent confluency is reached during culture maintenance to prevent the cells from spontaneous differentiation. Third, when seeding cells to Transwells® the support membrane (culture stratum) of the Transwells® and seeding density can influence cell growth, cell differentiation, and formation of cell monolayers (Braun *et al.*, 2000; Rothen-Rutishauser *et al.*, 2000). In general, polycarbonate filters with a seeding density of $10^5$ cells/cm$^2$ are recommended. Fourth, it has been reported that cell passage number is related to the expression levels of various transporters (Anderle *et al.*, 1998). The optimum passage range recommended for experimental purposes is around 28–65 (Rothen-Rutishauser *et al.*, 2000). Furthermore, membrane integrity should always be checked before using the monolayers for transport studies. Finally, it is advised to perform control experiments using a known substrate (e.g. L-carnosine for oligopeptide transporters, digoxin for *p*-gp) either periodically for the ongoing Caco-2 cells or parallel to the experiments for new compounds.

## TROUBLESHOOTING

When measured $P_{app}$ are not in good agreement with known data, refer to the troubleshooting guide (Table 10.2) to find possible causes and solutions for the problems.

*Table 10.2* Troubleshooting guide for transport studies using Caco-2 cell monolayers

| Problem | Possible cause | Solution |
|---|---|---|
| Very high $P_{app}$ | Leaky cell monolayers | Validate the integrity of the cell monolayers (see Support Protocol 3) |
| | High expression level of transporter protein for facilitated transport process | Protein expression level may vary with changes in passage number, source of cells, and culturing conditions (Anderle *et al.*, 1998). Try to use same batch of cells for a set of related experiments |
| Very low $P_{app}$ | Degradation | Check the stability of the test compound by incubating a solution of the test compound (1/100 the donor concentration) on both sides of the cells. Monitor the concentration changes on both sides of the cell monolayers |
| | Poor solubility | Be certain that the compound is completely dissolved. For a compound with poor water solubility, dimethyl sulfoxide (DMSO) (less that 1 per cent v/v) may be used to dissolve the compound |

*Table 10.2* Continued

| | For transporter-mediated transport, the transporter has become saturated | Check the concentration of the test compound. Lower the concentration (e.g. use concentration lower than or equal to the compound's $K_m$ vlaue if this value is known) to avoid saturation |
|---|---|---|
| Poor mass balance | High degradation | Check stability as mentioned above |
| | Significant cellular uptake | Check the amount of the test compound in the cell monolayer. Collect and lyse the cells, then extract the compound with a suitable solvent. Analyze the amount of the compound that was taken up by the cells (Camenisch *et al.*, 1998). |
| | Adsorption of the compound to Transwells® and containers | Check adsorption of the drug solution to the Transwells® by monitoring the concentration before and after a solution is transferred to the Transwells® Try to use containers that do not adsorb the test compound significantly or use higher concentration of the test compound to reduce the problem |
| High standard deviation (e.g. >30 per cent) | Damaged cell monolayers | Avoid disturbing the cell monolayers during sampling. Take caution when maintaining the cell monolayers in culture |
| | Cell monolayers from different batches | Try to use the same batch of cells for related experiments |

## *IN VITRO/IN VIVO* CORRELATION AND LIMITATIONS OF USING Caco-2 MODEL

The major purpose of using Caco-2 cell monolayers for drug absorption studies is to evaluate the permeability of new chemical entities of interest and use the *in vitro* permeability parameters to predict their *in vivo* absorption in humans. Good correlation between the apparent drug permeability in Caco-2 model and human oral absorption has been demonstrated by the early work done by Artursson (1990) on a series of passive diffusion compounds with homologous structural features, as well as those with different physicochemical properties (Artursson and Karlsson, 1991). Similar conclusions have been drawn by many other researchers using similar strategies (Pade and Stavchansky, 1998; Rubas *et al.*, 1993; Stewart *et al.*, 1995). However, poor correlation was found with the permeability of some compounds that are carrier-mediated (Chong *et al.*, 1996; Grès *et al.*, 1998). It is proposed that permeability obtained from Caco-2 cell model may be used to predict human GI absorption for compounds with $P_{app}$ greater than $5 \times 10^{-6}$ cm/second (Chong *et al.*, 1996; Ren and Lien, 2000). Cautions must be taken when $P_{app}$ measured is smaller than $10^{-6}$ cm/second because good drug candidates might be excluded owing to incorrect prediction of *in vivo* absorption. The limitation of using Caco-2 cells has been attributed to the colonic tumor origin of this cell line (Hilgendorf *et al.*, 2000). Caco-2 cells tend to have more tightened intercellular junctions resembling more colonic than small intestine, resulting in poor permeability for hydrophilic compounds passing across the cells through the paracellular pathway (Artursson *et al.*, 1996). Over-expression of *P*-gp and under expression of absorptive transporters, such as oligopeptide transporters, may lead to higher secretion of *P*-gp substrates and lower permeability of carrier-mediated compounds. Furthermore, Caco-2 cell model is composed of the absorptive enterocytes only, whereas the intestinal epithelium consists of several different cell types which may affect drug absorption under *in vivo* situation.

## REAGENT AND SOLUTIONS

*Incomplete DMEM*

- 67.4 g DMEM powder (JRH Biosciences)
- 18.5 g sodium bicarbonate (Sigma)
- 7.1 g HEPES (Sigma).

Place approximately 4.8 L autoclaved water and add all ingredients in 6-L flask while stirring Titrate to pH 7.35 with 10 N and 1.0 N aqueous HCl. Add water to 5.0 L. Transfer medium to filtration tank. Filter medium through a 0.22 μm disposable filter (Millipore) in a laminar flow hood and dispense 450 ml aliquots into individual sterile bottles. Store medium at 4 °C and use within two months.

*Complete DMEM-10 (10 per cent serum complete DMEM)*

- 45 ml heat-inactivated FBS (Atlanta Biologicals)
- 5 ml penicillin/streptomycin solution (supplied at 10,000 U/ml penicillin and 10,000 μg/ml streptomycin; Life Technologies)
- 5 ml 200 mM (100X) L-glutamine solution (Life Technologies)
- 5 ml 10 mM (100X) minimum essential medium (MEM) nonessential amino acids solution (Life Technologies)
- 450 ml incomplete DMEM.

Add the first four components to the bottle containing the incomplete DMEM and swirl gently to mix. Store at 4 °C and use within two weeks.

*Complete DMEM-20 (20 per cent Serum complete DMEM)*

Prepare as for complete DMEM-10, but bring to 20 per cent serum by using 90 ml heat-inactivated FBS instead of 45 ml.

*EDTA/PBS solution*

- 48.5 g Dulbecco's phosphate-buffered saline powder (Sigma)
- 0.5 g EDTA (Sigma).

Place approximately 4.8 L of autoclaved water and the other ingredients in 6-L flask while stirring. Titrate to pH 7.35 with 10 N and 1.0 N aqueous HCl. Add water to 5.0 L. Transfer medium to filtration tank. Filter the solution through a 0.22 μm disposable filter (Millipore) in a laminar flow hood and dispense 450 ml aliquots into individual sterile bottles. Store the solution at 4 °C and use within one year.

*EBSS*

- 8.70 g EBSS (Sigma)
- 1.95 g 2-(*N*-morpholino)ethanesulfonic acid (MES; Sigma)
- 4.50 g D-(+)-glucose (Sigma)
- 2.20 g sodium bicarbonate (Sigma).

Dissolve components in approximately 950 ml of autoclaved water and titrate to pH 6.0 with 10 N and 1.0 N aqueous HCl solutions. Add water to 1.0 L. Store at 4 °C and use within one year.

*Ethanol/collagen solution*

- Type-I rat tail collagen (Collaborative Biomedical/Becton Dickenson)
- 60 per cent (v/v) ethanol (diluted using autoclaved water).

Filter 60 per cent ethanol using a 20 ml disposable syringe (e.g. Becton Dickenson) fitted with a 0.22 μm disposable filter unit (Millipore) into a sterile 50 ml plastic centrifuge tube. Dilute collagen 1:3 with the filtered 60 per cent ethanol and mix thoroughly (collagen is 0.9 mg/ml final). Store at 4 °C and use within one week.

*HBSS*

- 5 L autoclaved water
- 49 g HBSS powder (Sigma)
- 1.85 g sodium bicarbonate (Sigma)
- 17.5 g D-(+)-glucose (Sigma)
- 14.3 g HEPES (Sigma).

Combine ingredients and most of the water; stir well to dissolve. Titrate solution to pH 7.35 with 10 N and 0.1 N aqueous HCl and add water to 5.0 L. Filter using a 0.22 μm disposable filter unit (Millipore) and dispense 450 ml aliquots into sterile bottles. Store the solution at 4 °C and use within six months.

## CONCLUSION

Caco-2 cell line is very useful as an *in vitro* model for studying intestinal drug absorption and metabolism owing to its unique biological characteristics and ease of handling. However, currently it is still not a perfect model. Alterations of phenotypes of the cells exist from time to time upon passaging and subculturing, which result in large variation among laboratories. Therefore, refinements are needed to better predict human intestinal permeability.

## REFERENCES

Anderle, P., Niederer, E., Rubas, W., Hilgendorf, C., Spahn-Langguth, H., Wunderli-Allenspach, H. *et al.* (1998) *p*-glycoprotein (*p*-gp) mediated efflux in Caco-2 cell monolayers: the influence of culturing conditions and drug exposure on *p*-gp expression levels. *J. Pharm. Sci.*, **87**, 757–762.

Artursson, P. (1990) Epithelial transport of drugs in cell culture. I: A model for studying the passive diffusion of drugs over intestinal absorptive (Caco-2) cells. *J. Pharm. Sci.*, **79**, 476–482.

Artursson, P. and Karlsson, J. (1991) Correlation between oral drug absorption in humans and apparent drug permeability coefficients in human intestinal epithelial (Caco-2) cells. *Biochem. Biophys. Res. Commun.*, **175**, 880–885.

Artursson, P., Palm, K. and Luthman, K. (1996) Caco-2 monolayers in experimental and theoretical predictions of drug transport. *Adv. Drug Deliv. Rev.*, **22**, 67–84.

Audus, K. L., Bartel, R. L., Higalgo, I. J. and Borchardt, R. T. (1990) The use of cultured epithelial and endothelial cells for drug transport and metabolism studies. *Pharm. Res.*, **7**, 435–451.

Bailey, C. A., Bryla, P. and Malick, A. W. (1996) The use of the intestinal epithelial cell culture model Caco-2, in pharmaceutical development. *Adv. Drug Del. Rev.*, **22**, 85–103.

Baranczyk-Kuzma, A., Garren, J. A., Hidalgo, I. J. and Borchardt, R. T. (1991). Substrate specificity and some properties of phenol sulfotransferase from human intestinal Caco-2 cells. *Life Sci.*, **49**, 1197–1206.

Blais, A., Bissonnette, P. and Berteloot, A. (1987) Common characteristics for Na$^+$-dependent sugar transport in Caco-2 cells and human fetal colon. *J. Membr. Biol.*, **99**, 113–125.

Boulenc, X., Bourrie, M., Fabre, I., Roque, C., Joyeux, H., Berger, Y. *et al.* (1992) Regulation of cytochrome P450IA1 gene expression in a human intestinal cell line, Caco-2. *J. Pharmacol. Exp. Ther.*, **263**, 1471–1478.

Braun, A., Hämmerle, S., Suda, K., Rothen-Rutishauser, B., Günthert, M., Krämer, S. D. *et al.* (2000) Cell cultures as tools in biopharmacy. *Eur. J. Pharm. Sci.*, **11**, S51–S60.

Burton, P. S., Conradi, R. A., Ho, N. F. H., Hilgers, A. R. and Borchardt, R. T. (1996) How structural features influence the permeability of peptides. *J. Pharm. Sci.*, **85**, 1336–1340.

Burton, P. S., Goodwin, J. T., Conradi, R. A., Ho, N. F. H. and Hilgers, A. R. (1997) *In vitro* permeability of peptidomimetics: the role of polarized efflux pathways as additional barriers to absorption. *Adv. Drug Deliv. Rev.*, **23**, 143–156.

Camenisch, G. P., Wang, W., Wang, B. and Borchardt, R. T. (1998) A comparison of the bioconversion rates and the Caco-2 cell permeation characteristics of coumarin-based cyclic prodrugs and methyl ester-based linear prodrugs of RGD peptidomimetics. *Pharm. Res.*, **15**, 1174–1181.

Chong, S., Dando, S. A., Soucek, K. M. and Morrison, R. A. (1996) *In vitro* permeability through caco-2 cells is not quantitatively predictive of in vivo absorption for peptide-like drugs absorbed via the dipeptide transporter system. *Pharm. Res.*, **13**, 120–123.

Csáky, T. Z. (1984) Methods for investigation of intestinal permeability. In T. Z. Csáky (ed.), *Pharmacology of Intestinal Permeation I*, pp. 91–111, Springer-Verlag, Berlin.

Dantzig, A. and Bergin, L. (1990) Uptake of the cephalosporin, cephalexin, by a dipeptide transport carrier in the human intestinal cell line, Caco-2. *Biochim. Biophys. Acta*, **1027**, 211–217.

Dharmsathaphorn, K. and Madara, J. L. (1990) Established intestinal cell lines as model systems for electrolyte transport studies. In S. Fleischer and B. Fleischer (eds), *Methods in Enzymology*, pp. 354–389, Academic Press, Inc., New York.

Dix, C. J., Hassan, I. F., Obray, H. Y., Shah, R. and Wilson, G. (1990) The transport of vitamin B12 through polarized monolayers of Caco-2 cells. *Gastroenterology*, **98**, 1272–1279.

Fisher, R. B. and Parsons, D. S. (1949) A preparation of surviving rat small intestine for the study of absorption. *J. Physiol.*, **110**, 36–46.

Gan, L. S. L. and Thakker, D. R. (1997) Application of the Caco-2 model in the design and development of orally active drugs: elucidation of biochemical and physical barriers posed by the intestinal epithelium. *Adv. Drug Deliv. Rev.*, **23**, 77–98.

Ganapathy, V. and Leibach, F. H. (1983) Role of pH gradient and membrane potential in dipeptide transport in intestinal and renal brush-border membrane vesicles from the rabbit: studies with L-carnosine and glycyl-L-proline. *J. Biol. Chem.*, **258**, 14189–14192.

Goodwin, T. J., Schroeder, W. F., Wolf, D. F. and Moyer, M. P. (1993) Rotating-wall vessel coculture of small intestine as a prelude to tissue modeling: aspects of simulated microgravity. *Proc. Soc. Exp. Biol. Med.*, **202**, 181–192.

Grès, M. C., Julian, B., Bourriè, M., Meunier, V., Roques, C., Berger, M. *et al.* (1998) Correlation between oral drug absorption in humans and apparent drug permeability in TC-7 cells, a human epithelial intestinal cell line: comparison with the parental Caco-2 cell line. *Pharm. Res.*, **15**, 726–733.

Harris, D. S., Slot, J. W., Geuze, H. J. and James, D. E. (1992) Polarized distribution of glucose transporter isoforms in Caco-2 cells. *Proc. Natl. Acad. Sci. USA*, **89**, 7556–7560.

Hidalgo, I. J. and Li, J. (1996) Carrier-mediated transport and efflux mechanisms in caco-2 cells. *Adv. Drug Deliv. Rev.*, **22**, 53–66.

Hidalgo, I. J., Raub, T. J. and Borchardt, R. T. (1989) Characterization of the human colon carcinoma cell line (Caco-2) as a model system for intestinal epithelial permeability. *Gastroenterology*, **96**, 736–749.

Hilgendorf, C., Spahn-Langguth, H., Regardh, C. G., Lipka, E., Amidon, G. L. and Langguth, P. (2000) Caco-2 versus Caco-2/HT29-MTX co-cultured cell lines: permeabilities via diffusion, inside- and outside-directed carrier-mediated transport. *J. Pharm. Sci.*, **89**, 63–75.

Hillgren, K. M., Kato, A. and Borchardt, R. T. (1995) In vitro systems for studying intestinal drug absorption. *Med. Res. Rev.*, **15**, 83–109.

Hopfer, U., Nelson, K., Perrotto, J. and Isselbacher, K. J. (1973) Glucose transport in isolated brush border membrane from rat small intestine. *J. Biol. Chem.*, **248**. 25–32.

Hu, M. and Borchardt, R. T. (1992) Transport of a large neutral amino acid in a human intestinal epithelial cell line (Caco-2): uptake and efflux of phenylalanine. *Biochim. Biophys. Acta*, **1135**, 233–244.

Hunter, J., Jepson, M. A., Tsuruo, T., Simmons, N. L. and Hirst, B. H. (1993) Functional expression of P-glycoprotein in apical membranes of human intestinal Caco-2 cells: Kinetics of vinblastine secretion and interaction with modulators. *J. Biol. Chem.*, **268**, 14991–14997.

Kimmich, G. A. (1990) Isolation of intestinal epithelial cells and evaluation of transport functions. In *Methods in Enzymology*, S. Fleischer and B. Fleischer (eds), pp. 324–340, Academic Press, Inc., San Diego.

Knipp, G. T., Ho, N. F. H., Barsuhn, C. I. and Borchardt, R. T. (1997) Paracellular diffusion in caco-2 monolayers: Effect of pertubants on the transport of hydrophilic compounds that vary in charge and size. *J. Pharm. Sci.*, **86**, 1105–1110.

Kreusel, K. M., Fromm, M., Schulzke, J. D. and Hegel, U. (1991) Cl⁻ secretion in epithelial monolayers of mucus-forming human colon cells (HT-29/B6). *Am. J. Physiol.*, **261**, C574–C582.

Leibach, F. H. and Ganapathy, V. (1996) Peptide transporters in the intestine and the kidney. *Annu. Rev. Nutr.*, **16**, 99–119.

Lennernäs, H., Palm, K., Fagerholm, U. and Artursson, P. (1996) Comparison between active and passive drug transport in human intestinal epithelial (Caco-2) cell *in vitro* and human jejunum *in vivo*. *Int. J. Pharm.*, **127**, 103–107.

Levine, R. R., McNary, W. R., Kornguth, P. J. and LeBlanc, R. (1970) Histological reevaluation of everted gut technique for studying intestinal absorption. *Eur. J. Pharmacol.*, **9**, 211–219.

Moyer, M. P. (1983) Culture of human gastrointestinal epithelial cells. *Proc. Soc. Exp. Biol. Med.*, **174**, 12–15.

Nerurkar, M. M., Burton, P. S. and Borchardt, R. T. (1996) The use of surfactants to enhance the permeability of peptides through Caco-2 cells by inhibition of an apically polarized efflux system. *Pharm. Res.*, **13**, 528–534.

Pade, V. and Stavchansky, S. (1998) Link between drug absorption solubility and permeability measurements in Caco-2 cells. *J. Pharm. Sci.*, **87**, 1604–1607.

Parsons, D. S. (1968) Methods for investigation of intestinal absorption. In C. F. Code (ed.), *Alimentary Canal*, Sect. **6**, pp. 1177–1216, American Physiology Society, Washington, D. C.

Parsons, D. S. and Paterson, C. R. (1965) Fluid and solute transport across rat colonic mucosa. *Q. J. Exp. Physiol.*, **50**, 220–231.

Pauletti, G. M., Gangwar, S., Siahaan, T. J., Aubé, J. and Borchardt, R. T. (1997) Improvement of oral peptide bioavailability: peptidomimetics and prodrug strategies. *Adv. Drug Deliv. Rev.*, **27**, 235–256.

Peters, W. H. M. and Roelofs, H. M. J. (1992) Biochemical characterization of resistance to mitoxantrone and adriamycin in Caco-2 human colon adenocarcinoma cells: a possible role for glutathione S-transferases. *Cancer Res.*, **52**, 1886–1890.

Porter, P. A., Osiecka, I., Borchardt, R. T., Fix, J. A., Frost, L. and Gardner, C. (1985) *In vitro* drug absorption models. II. Salicylate, cefoxitin, $\alpha$-methyldopa and theophylline uptake in cells and rings: correlation with *in vivo* bioavailability. *Pharm. Res.*, **1**, 293–298.

Quaroni, A. and Hochman, J. (1996) Development of intestinal cell culture models for drug transport and metabolism studies. *Adv. Drug Deliv. Rev.*, **22**, 3–52.

Ren S. and Lien E. J. (2000) Caco-2 cell permeability vs human gastrointestinal absorption: QSPR analysis. *Prog. Drug Res.*, **54**, 1–23.

Rothen-Rutishauser, B., Braun, A., Günthert, M. and Wunderli-Allenspach, H. (2000) Formation of multilayers in the caco-2 cell culture model: a confocal laser scanning microscopy study. *Pharm. Res.*, **17**, 460–465.

Rubas, W., Jezyk, N. and Grass, G. M. (1993) Comparison of the permeability characteristics of a human colonic epithelial (Caco-2) cell line to colon of rabbit, monkey and dog intestine and human drug absorption. *Pharm. Res.*, **10**, 113–118.

Saito, H. and Inui, K. (1993) Dipeptide transporters in apical and basolateral membranes of the human intestinal cell line caco-2. *Am. J. Physiol.*, **265**, G289–G294.

Schmiedlin-Ren, P., Thummel, K. E., Fisher, J. M., Paine, M. F., Lown, K. S. and Watkins, P. B. (1997) Expression of enzymatically active CYP3A4 by Caco-2 cells grown on extracellular matrix-coated permeable supports in the presence of $1\alpha$, 25-dihydroxyvitamin D3. *Mol. Pharmacol.*, **51**, 741–754.

Stewart, B. H., Chan, O. H., Lu, R. H., Reyner, E. L., Schmid, H. L., Hamilton, H. W. *et al.* (1995) Comparison of intestinal permeabilities determined in multiple *in vitro* and *in situ* models: relationship to absorption in humans. *Pharm. Res.*, **12**, 693–699.

Tang, A. S., Chikhale, P. J., Shah, P. K. and Borchardt, R. T. (1993) Utilization of a human intestinal epithelial cell culture system (Cacco-2) for evaluating cytoprotective agents. *Pharm. Res.*, **10**, 1620–1626.

Thwaites, D. T., Hirst, B. H. and Simmons, N. L. (1994) Substrate specificity of the di/tripeptide transporter in human intestinal epithelia (Caco-2): identification of substrates that undergo $H^+$-coupled absorption. *Br. J. Pharmacol.*, **113**, 1050–1056.

Walle, U. K. and Walle, T. (1998) Taxol transport by human intestinal epithelia Caco-2 cells. *Drug Metab. Dispos.*, **26**, 343–346.

Wacher, V. J., Salphati, L. and Benet, L. Z. (1996) Active secretion and enterocytic drug metabolism barriers to drug absorption. *Adv. Drug Deliv. Rev.*, **20**, 99–112.

Ward, J. L. and Tse, C. M., (1999) Nucleoside transport in human colonic epithelial cell lines: evidence for two $Na^+$-independent transport systems in T84 and caco-2 cells. *Biochim. Biophys. Acta*, **1419**, 15–22.

Wikman, A., Karlsson, J., Carlstedt, I. and Artursson, P. (1993) A drug absorption model based on the mucus layer producing human intestinal goblet cell line HT29-H. *Pharm. Res.*, **10**, 843–852.

Wilson, T. H. and Wiseman, G. (1954) The use of sacs of everted small intestine for the study of the transference of substances from the mucosal to the serosal surface. *J. Physiol.*, **123**, 116–125.

Zweibaum, A., Hauri, H. P. and Sterchi, B. (1984) Immunohistological evidence, obtained with monoclonal antibodies, of small intestinal brush border hydrolases in human colon cancer and foetal colons. *Int. J. Cancer*, **34**, 591–598.

# Chapter 11

# Transport studies using intestinal tissue *ex vivo*

*Anna-Lena Ungell*

## INTRODUCTION

The oral route is considered to be a complex and difficult route for administration and far from all drugs will enter the systemic circulation via the gastrointestinal (GI) membranes. This preclusion is due to the fact, in physiological aspects, that the GI tract has a function of absorbing food nutrients, electrolytes and water, yet at the same time is a very effective barrier against toxins, bacteria and foreign material (Kararli, 1989). In general, a large number of drugs entering the development phase or rather, during the phase-II–phase-III clinical trials, fail owing to lack of sufficient ADME (absorption, distribution, metabolism, excretion) and biopharmaceutical properties. The high attrition rate has highlighted the importance of screening for good pharmacokinetic properties as early as possible in the discovery process, e.g. during the lead identification phase. Known failures are mainly due to low solubility, poor permeability and extensive first-pass metabolism, both in the GI membranes and in the liver. Low oral availability of a drug can be associated with significant variability both between different treatment occasions and between different patients. Knowledge of the reasons for low bioavailability is, therefore, of great importance both for efficacy and safety in the treatment and for the development of follow-up candidates.

Lately, the importance of transporters in the intestinal membrane, liver and kidneys has also become clearer and high non-metabolic clearance rates can now be evaluated as well as drug–drug interactions during multiple therapies. The importance of transporters in the GI membranes and their influence on drug absorption, e.g. uptake and efflux transporters, however, still remains an open question.

Delivery of drugs via the GI membranes is usually for targeting receptors within the systemic circulation, which implies that the drugs have to have properties for entering the circulation and to be able to reach the target at a certain concentration without formation of toxic metabolites. Lately, an interest in mucosal targeting for local administration to transporters/receptors or inflammatory targets in the GI membrane has also been evident. The drug that is intended for a local mucosal target within the GI tract has other pharmacokinetic/structural properties than those intended for the systemic circulation.

The ideal absorption model/method for studying pharmacokinetic and biopharmaceutical properties of drugs, in general, needs to have all the physiological and biochemical properties of the true barrier. It must also be easy to use. The method also needs to have low variability between experiments and should be unbiased by the experimentalists. As it is not possible to use human tissues as frequently as animal tissues, there is a need to be aware of species differences with respect to carrier-mediated transport and/or regional differences in transport and metabolism. This is extremely important for scaling of *in vitro* data from animals to humans. If the intention is to use animal tissue for the targeting of a drug to the mucosa, there is also a need to evaluate whether the pharmacological response is valid in humans, e.g. target receptor specificity.

As the handling of the tissue is important *ex vivo*, the viability and integrity as well as variability becomes an issue. Ungell (1997) has suggested some factors that can influence the viability of the *ex vivo* preparation, and also, subsequently, the results. Among other factors, starvation or anesthesia of the rats prior to the experiments has been reported to increase the permeability of the intestinal segments (Uhing and Kimura, 1995a,b; Wirén *et al.*, 1999). The extent of variability and lack of success in predicting human fraction absorbed (fabs%) using the different absorption models, with respect to these two factors, have not yet been fully evaluated.

## THE COMPLEXITY OF THE ABSORPTION PROCESS

The chemical structure and property of the drug molecule itself, and the physiology within the environment of the GI tract, affect the permeation of the molecule through the intestinal membrane. Factors such as drug solubility, partition coefficient (Log D, Log P), pKa, molecular weight/volume, aggregates, particle size on the one hand, and pH in the lumen and at the surface of the membrane, absorptive surface area, blood flow, membrane permeability, and enzymes on the other, have been discussed as important for drug absorption (for more factors see Ungell, 1997; Ungell and Abrahamsson, 2001).

The uptake of drugs across the intestinal membrane can occur through several processes: transcellularly across the lipid membrane or paracellularly between the epithelial cells in the tight junctional gap (Ungell, 1997; Ungell and Abrahamsson, 2001). The transcellular route is generally via carrier proteins or by passive diffusion. The molecular properties resulting in transport via carriers are, in most cases, completely different from those favoring simple passive diffusion. Additionally,

*Table 11.1* A list of *ex vivo* methods for studying drug absorption from the gastrointestinal (GI) tract. The methods are arranged according to screening capacity and increased complexity towards the *in vivo* situation. Adapted from Ungell, 1997

| Ex vivo method | Number of experiments/animal | Complexity vs. invivo |
|---|---|---|
| BBMV | High | Low |
| Everted rings | High to intermediate | |
| Everted sacs | Intermediate | |
| Gut-loop | Intermediate | |
| Ussing chamber | Intermediate to low | |
| *In vitro* and *in situ* perfusions | Low | High |

passive diffusion through the lipid bilayer represents a completely different barrier and complexity when compared with the aqueous pathway of the paracellular route. The structural requirements for these two routes are, therefore, different. For guiding the synthesis of new structural analogs more knowledge about these mechanisms results in a higher success rate and is one important rationale for the development of *in silico* models (theoretical) to predict human fabs% (Zamora *et al.*, 2000; Zamora and Ungell, 2001). The basic rules around a structure–absorption relationship have not yet been established, but some helpful guidance can be derived from polar surface area calculations (Palm *et al.*, 1996, 1997) and 'Rule of Lipinski' (Lipinski *et al.*, 1996). There are few studies in the literature showing structural changes correlating to changes in permeability over the intestinal membrane, and most studies have been performed using cell monolayers (e.g. Burton *et al.*, 1992; Hidalgo *et al.*, 1995; Kamm *et al.*, 1999; Lang *et al.*, 1997; Walter *et al.*, 1995; ). It has also become clear that many drugs which have been thought to be transported by passive transcellular diffusion possess properties that involve carrier-mediated transport either as uptake or efflux, e.g. some beta-blockers, angiotensin-connecting enzyme (ACE) inhibitors, beta-lactam antibiotics, (3-hydroxy-3-methyl-glutaryl)-CoA HMG-CoA reductase inhibitors, prodrugs of antiviral agents (Chong *et al.*, 1996; Karlsson *et al.*, 1993; Kim *et al.*, 1994; Sinko and Amidon, 1989; Sinko and Balimane, 1998; Wu *et al.*, 2000). At low concentrations within the intestinal lumen, i.e. for drugs with very high potencies, and for prodrugs, this could imply a need for information of the involvement of active transport processes in the GI absorption. The concentration is related to the effective dose, the solubility of the drug in the intestinal fluids, the volume and composition of the fluid (Hörter and Dressman, 1997; Raoof *et al.*, 1998), and if there is saturation of drug transporter systems, such as efflux proteins carrying the drug from the inside back into the lumen (e.g. *p*-glycoprotein (*p*-gp), multidrug resistance protein (MRP1-6)). The latter has recently been proposed to be important for the overall absorption of drugs in the GI tract (Hunter and Hirst, 1997; Makhey *et al.*, 1998; Saitoh and Aungst, 1995; Wacher *et al.*, 1998). The concentration of the free drug at the uptake site is equally important for transport processes in the absorptive direction, such as oligopeptide transporters, dipeptide transporters, amino acid transporters, monocarboxylic transporters, etc. (Tsuji and Tamai, 1996).

Apart from membrane permeability and solubility, the time the molecule spends in the region of absorption, i.e. transit time or exposure time, is an additionally important factor for intestinal absorption. Generally, transit times in humans are seconds in the esophagus, 0.5–1.5 hours in the stomach, 3–4 hours in the small intestine and 8–72 hours in the colon. In addition, the solubility of the drug in the luminal environment and in the *in vitro* or *ex vivo* system is a critical issue for the correct evaluation of the *in vivo* dosing to man and for understanding the *in vitro* data (Ungell and Abrahamsson, 2001). Complete absorption can be said to occur when the drug has maximum permeability coefficient ($P_{app}$) and maximum solubility at the site of absorption (Pade and Stavchansky, 1998).

Regionally, these different physiological factors will change owing to alterations of both the intestinal milieu and the membranes of the GI tract. (Dressman and Yamada, 1991; Hörter and Dressman, 1997; Raoof *et al.*, 1998; Sinko and Amidon, 1989; Sinko and Hu, 1996; Ungell *et al.*, 1997). If the regional difference in absorption probability of the drug is known (regional permeability and interactions), increased absorption can be achieved by the use of an absorption window, e.g. targeting the drug to a specific region to avoid critical regions of enzymes or low permeability.

## MODELS FOR STUDYING THE INTESTINAL ABSORPTION OF DRUGS

Due to the complexity of the absorption process, there is no single method representing all the barriers of the intestinal membranes. Instead, there is a need for more than one screening method for both the drug discovery and rational drug development. Models available at present, for both theoretical prediction and for the study of drug absorption, are mainly: computational methods, partitioning between water and oil, cell cultures, membrane vesicles, intestinal rings or sacs, excised segments from animals in the Ussing chamber, gut loops, *in vitro* and *in situ* intestinal perfusions, *in vivo* cannulated or fistulated animals, and *in vivo* gavaged animals (see reviews, such as Borchardt *et al.*, 1996; Hillgren *et al.*, 1995; Stewart *et al.*, 1995; Ungell, 1997). Examples of the use of these *in vitro/ex vivo* models in the evaluation of oral drug candidates have been reported (LeCluyse and Sutton, 1997; Lee *et al.*, 1997; Stewart *et al.*, 1997 see Table 11.1 and 11.2).

*Table 11.2* Advantages and disadvantages using different *ex vivo* methods for drug absorption

| Absorption model | *+ or −* |
|---|---|
| Log P, Log D (pH buffer) or other physicochemical predictive model | 'Rule of thumb', no animals needed, only the transcellular pathway described, no carrier mediated process involved |
| BBMV or BLMV | Easy to use, part of the absorption process described, lack of either basolateral or brush-border functions, unspecific binding, both animal and human tissue |
| Intestinal rings | Easy to use, transport not directed, unspecific binding, integrity < 10–20 minutes, both animal and human tissue |
| Everted sacs | Easy to use, directed transport, bad stirring conditions, integrity < 30 minutes, both animal and human tissue |
| Gut loop | Requires animal surgery and anesthetics, directed transport, blood flow intact, disappearance of drug measured, stirring conditions bad in loop |
| Cell lines | Easy to use, good stirring, directed transport, human or animal cell lines available, tightness of the tight junction higher, transporters?, appearance rate |
| Ussing chamber | Easy to use, good stirring, directed transport, animal consuming, integrity limited for two–three hours, appearance rate, both animal and human tissue |
| Intestinal perfusions | Good stirring, good oxygenation, directed transport, animal consuming, disappearance rate, anesthesia, both with and without intact blood flow, complex |

## Ex vivo methods

*Ex vivo* methods can be used when the mechanisms of absorption (paracellular, transcellular or carrier-mediated) and the enzymatic degradation or regional difference in permeability are to be evaluated. The name *ex vivo* refers to that an animal has been sacrificed or to human tissues that have been excised during cancer surgery. In general, these types of methods can offer several advantages over cell monolayers or physicochemical methods. The main advantages are the possibility of pretreatment of the animal *in vivo* before *ex vivo* evaluation of the absorption within an *in vitro* system and the use of human tissue. Some of these methods can be used in high-throughput format (HTS), e.g. brush-border membrane vesicles (BBMV), but in general these methods are used in a more reduced screening mode (Table 11.1). A short description of the best known biological *ex vivo* methods follows below and more detailed information on each of the methods can be found in the references (Borchardt *et al.*, 1996; Hillgren *et al.*, 1995; Kararli, 1989, 1995; Stewart *et al.*, 1997; Ungell, 1997; Ungell and Abrahamsson, 2001).

### Methods describing drug uptake

#### Membrane vesicles and intestinal rings

This group of methods represent drug uptake into the enterocyte. A trait common to both methods is that they are both technically quick and easy to use, even for persons not very skilled in using biological material.

BBMV are usually used in the discovery or development of drugs for the evaluation of enzyme interactions or ion transport coupled transport processes and have been widely used historically in physiological studies. The BBMV method is based on a homogenization of an inverted frozen intestine to give a purified fraction of the apical cell membranes from a chosen part of the GI tract (Alcorn *et al.*, 1991; Kararli, 1989; Kessler *et al.*, 1978; Sinko *et al.*, 1995; Stewart *et al.*, 1997). The method can frequently be used for isolated studies of the brush-border membrane transport without any basolateral membrane influence. BBMV has been used for studies concerning the intestinal peptide carrier system (Yuasa *et al.*, 1993, 1994), the transport of nucleosides analogs (Sinko *et al.*, 1995), and to clarify the mechanism of absorption of fosfomycin (Ishizawa *et al.*, 1992). It has also been used for studies of glucose, amino acids and salicylate uptake (Osiecka *et al.*, 1985). Membrane vesicles have been isolated from numerous animals, including man (Hillgren *et al.*, 1995). The functionality of the preparation (i.e. whether the membrane is closed) is assessed by using substrates with specific carriers, such as glucose, phosphate or amino acids. The orientation of the membrane, i.e. right side or inside out, is assessed by enzyme markers. Recently, BBMV was tested as a screening model for a large number of compounds using 96-well Multiscreeen Filtration plates (Quilianova *et al.*, 1999). After correcting the unspecific binding to the tissue, the permeability values were found to show a good correlation to the *in vivo* human fabs%.

A part of the BBMV method represents lipid membrane extraction and can be used in drug absorption studies for the evaluation of a biological LogD value (Alcorn *et al.*, 1991). Different regions of the GI tract can be used to evaluate the influence of regional differences in lipid composition on the permeability of drugs, as has been suggested (Kim *et al.*, 1994; Thomson *et al.*, 1986; Ungell *et al.*, 1997). Brush-border membranes can also be used for measuring drug stability, and the enzyme distribution has been widely studied (Stewart *et al.*, 1997).

The major disadvantage of these processes is that they represent only a fraction of the complete absorption process, i.e. into the cell or out of the cell via the apical membrane (or basolateral membrane if basolateral membrane vesicles (BLMV) is used). No paracellular process can be studied, nor can processes that need the opposite membrane to function, i.e. the basolateral membrane and its function for absorption, e.g. processes linked to the active transport of $Na^+$ by the basolateral $Na^+/K^+$-ATPase (Kararli, 1989) (Table 11.2). There may be a day-to-day variation in vesicle preparation and a leakage of drugs from the vesicles during washing and filtration, which can affect the drug concentration (Osiecka *et al.*, 1985). The method is best suited for use of radioactively labeled substances since the analysis requires high sensitivity when determining the difference between compound taken up into the vesicles and unspecifically bound (Table 11.2). Despite these drawbacks, it can be used for mechanistic studies of the drug absorption process, for which human intestinal tissues can be used. However, the data describing a direct correlation of different mechanistic results to human *in vivo* absorption values are few (Quilianova *et al.*, 1999).

The second method belonging to this group is the *intestinal rings* or *slices*. This method for studying drug absorption has been used extensively for the kinetic analysis of carrier-mediated transport of glucose, amino acids and peptides (Kararli, 1989; Kim *et al.*, 1994; Leppert and Fix, 1994; Osiecka *et al.*, 1985; Porter *et al.*, 1985). The method is very easy to administer; the intestine of the animal is cut into rings or slices of approximately 30–50 mg (2–5 mm in width) which are put into an incubation medium for a very short period of time (often up to one minute) with agitation and oxygenation. Samples of the incubation medium and rings are analyzed for drug content after the incubation. The intestine is sometimes everted on a glass rod before cutting, and different regions of the intestinal tract can be used.

The main advantage of this method is its ease of preparation. As in the BBMV, this method can also be used for testing many different drugs simultaneously. The intestinal rings have several disadvantages, however. Diffusion into the tissue slices takes place on the side of the tissue (although not entirely through the lipid membrane) as the connective tissue and muscle layers are exposed to the incubation solution. Correction is not always made for the adsorption of a drug on the surface of the tissue, and the slices do not maintain their integrity for more than 20–30 minutes (Levine *et al.*, 1970; Osiecka *et al.*, 1985). The method is also restricted by the limits of the analytical methods. Nevertheless, a good mechanistic correlation to *in vivo* measurements has been achieved with the method in kinetic studies of carrier-mediated mechanisms of peptides (Kim *et al.*, 1994). The method was evaluated for the prediction of *in vivo* absorption potential (Leppert and Fix, 1994) and it was shown that, under appropriate conditions, uptake into everted intestinal rings closely parallels known *in vivo* bioavailability. The intestinal ring method has also recently been experimentally improved for better hydrodynamics and requires lower volumes during the incubation period (Uch and Dressman, 1997; Uch *et al.*, 1999).

## Methods with well-defined transport direction

The cell monolayer systems and *ex vivo* models, such as the Ussing chamber technique, gut loops and intestinal perfusions belong to the group of absorption methods with well-defined transport direction. The cell monolayer techniques are described elsewhere (see Chapter 10). Although cell monolayers offer a rapid tool to determine intestinal permeability (i.e. HTS), the optimization of pharmacokinetic properties in the later phases of drug discovery and the relevance of the absorption mechanism(s) or the involvement of carriers require further understanding. To facilitate this further understanding, more time- and animal-consuming methods, e.g. excised tissues, gut loops and intestinal perfusions, are best used. These methods are closer to the more complex *in vivo* situation and are also more sensitive to changes in the handling of the tissue and surrounding milieu (Polentarutti *et al.*, 1999; Wirén *et al.*, 1999).

### Excised intestinal segments

The *everted sac* (everted intestine) method is based on the preparation of a 2–3 cm long tube of the gut which is tied off at the ends after evertion on a glass rod (Kararli, 1989). The serosa becomes the inside of the sac and the mucosa faces the outer buffer solution. As a modification of this procedure, the serosal layer and muscular layers can also be stripped off before evertion on the glass rod (Hillgren *et al.*, 1995). The presence or absence of the serosal layer may give different transport rates of compounds, e.g. salicylic acid (Hillgren *et al.*, 1995).

An oxygenated buffer solution is injected into the sac, which is put into a flask containing the drug compound. Samples of fluid are taken from the buffer solution in the flask. The sac is weighed before and after the experiments to compensate for fluid movement. In one modification of the method, one end of the tissue is cannulated with a polyethylene tubing (Kararli, 1989), which makes it easier to withdraw samples from the serosal side of the intestine.

An advantage of this method is that it is rapid and many drugs can be tested simultaneously, especially low permeability drugs, owing to the low volume of the serosal compartment. There is good performance as regards stirring conditions on the mucosal side, although the oxygenation of the tissue is poor as a result of the unstirred and unoxygenated serosa inside the uncannulated sac. Another advantage of this method is that it needs no specialized equipment, in contrast to the Ussing chamber and cell culture models.

Disadvantages are mainly the viability issue and the diffusion through the lamina propria. Histological studies have shown that structural changes start as early as five minutes after the start of incubation and a total disruption of the epithelial tissue can be seen after one hour (Levine *et al.*, 1970) (Table 11.2). As for intestinal rings, there is no correction for the binding of drug substance onto the surface of the mucosa (when uncannulated sacs are used). In addition to the measurement of drug uptake, this technique can also report the measurement of the  uptake of nano particles coated with tomato lectins (Correno-Gomez *et al.*, 1999).

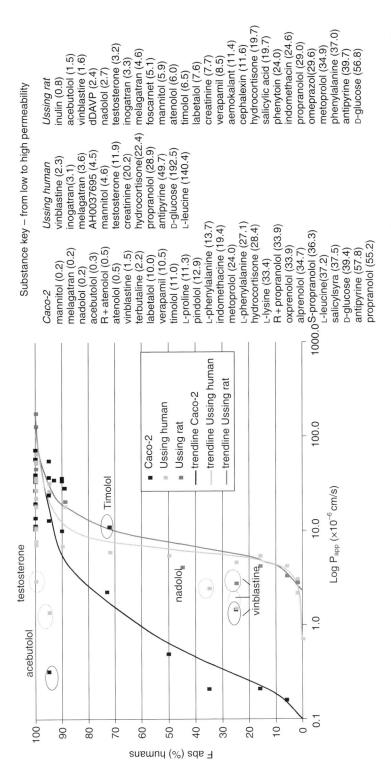

Substance key – from low to high permeability

*Caco-2*
mannitol (0.2)
melagatran (0.2)
nadolol (0.2)
acebutolol (0.3)
R + atenolol (0.5)
atenolol (0.5)
vinblastine (1.5)
terbutaline (2.2)
labetalol (10.0)
verapamil (10.5)
timolol (11.0)
L-proline (11.3)
pindolol (12.9)
L-phenylalanine (13.7)
indomethacine (19.4)
metoprolol (24.0)
L-phenylalanine (27.1)
hydrocortisone (28.4)
L-lysine (33.4)
R + propranolol (33.9)
oxprenolol (33.9)
alprenolol (34.7)
S-propranolol (36.3)
L-leucine(37.2)
salicylsyra (37.5)
D-glucose (39.4)
antipyrine (57.8)
propranolol (55.2)

*Ussing human*
vinblastine (2.3)
inogatran(3.1)
melagatran (3.6)
AH0037695 (4.5)
mannitol (4.6)
testosterone (11.9)
creatinine (20.2)
hydrocortisone(22.4)
propranolol (28.9)
antipyrine (49.7)
D-glucose (192.5)
L-leucine (140.4)

*Ussing rat*
inulin (0.8)
acebutolol (1.5)
vinblastire (1.6)
dDAVP (2.4)
nadolol (2.7)
testosterone (3.2)
inogatran (3.3)
melagatran (4.6)
foscarnet (5.1)
mannitol (5.9)
atenolol (6.0)
timolol (6.5)
labetalol (7.6)
creatinine (7.7)
verapamil (8.5)
aemokalant (11.4)
cephalexin (11.6)
hydrocortisone (19.7)
salicylic acid (19.7)
phenytoin (24.0)
indomethacin (24.6)
propranolol (29.0)
omeprazol(29.6)
metoprolol (34.9)
phenylalanine (37.0)
antipyrine (39.7)
D-glucose (56.8)

*Figure 11.1* Correlation of permeability data from rat and human intestinal segments in the Ussing chamber and Caco-2 to fraction absorbed in humans. Data compiled from references Ungell et al., 1998; Sjöström et al., 2000. The different compounds tested in each of the models can be found in the list to the right. Permeability coefficients ($P_{app}$) are expressed as mean values from two–five experiments, and the different colored circles indicate outlayers which are compounds transported actively that were measured at different concentrations and compounds which are unstable during transport.

*Ussing chamber*

The Ussing chamber technique is an old technique for studying transport across an epithelium and was developed by Ussing and Zerhan in 1951. It has been used extensively in physiological studies concerning the pharmacology and physiology of ion and water fluxes across the intestinal wall. It has also recently been used for drug absorption studies using excised intestinal tissues from different animals – rabbits, dogs, rats or monkeys (Artursson *et al.*, 1993; Jezyk *et al.*, 1992; Palm *et al.*, 1996; Polentarutti *et al.*, 1999; Rubas *et al.*, 1993; Ungell *et al.*, 1997). The Ussing chamber technique has also recently been used for human biopsies (Bijlsma *et al.*, 1995; Sjöström *et al.*, 2000; Söderholm *et al.*, 1998, 1999). The method is generally based on the excision of intestinal segments from the animals. These segments are cut open into planar sheets which may be stripped of the serosa and the muscle layers and mounted between two-diffusion half-cells (Grass and Sweetana, 1988) containing stirred and oxygenated buffer solutions. The $P_{app}$ ($P_{app}=dQ/dt \times 1/A \times C_0$) of the compounds are calculated from the measurement of the rate of transport, $dQ/dt$, of the molecules from one side of the segment to the other (either mucosa to serosa or serosa to mucosa), divided by the exposed area of the segment (A) and the donor concentration of the drug ($C_0$). Sink conditions are criteria for good data analysis and are obtained by maintenance of a concentration in the receiver chamber of below 10 per cent of the concentration in the donor compartment during the course of the experiment.

Stirring the solutions on both sides of the membrane is very important, especially for lipophilic drugs (Karlsson and Artursson, 1991). Similar permeability dependency on the stirring conditions has not yet been performed using Ussing chambers. However, stirring can be achieved by a gas-lift system, as originally proposed by Ussing and Zerhan (1951), by a more refined gas-lift system, as shown by Grass and Sweetana (1988), or by stirring with rotors (Polentarutti *et al.*, 1999; Ungell and Abrahamsson, 2001). The Ussing chamber technique has also been used for identifying metabolites formed during transport (Ungell *et al.*, 1992; Sjöström *et al.*, 2000) and for the evaluation of carrier-mediated transport of prodrugs (Schwaan *et al.*, 1995; Tukker and Schoenmakers, 1998) and for regional differences in *p*-gp-mediated efflux (Saitoh and Aungst, 1995), although not as extensively as the Caco-2.

The integrity and viability of the tissue must be verified simultaneously when using the Ussing chamber technique, because the it on integrety will strongly impair the transport of the drug molecules (Table 11.2). The viability of the tissues is verified with the measurement of potential difference (PD), short-circuit current and calculation of the transepithelial electrical resistance by Ohm's law (Bijlsma *et al.*, 1995; Polentarutti *et al.*, 1999; Ungell *et al.*, 1992; Sutton *et al.*, 1992; Söderholm *et al.*, 1998;). The values that set the limits of viability should be in the range of what have been measured *in vivo*, e.g. the range for rat jejunum and ileum should be between –5 and –6 mV (Podesta and Mettrick, 1977) and for human jejunum, ileum and colon the ranges should be –3, –6 and –17 to –35 mV, respectively (Davis *et al.*, 1982). Extracellular marker molecules such as mannitol, inulin, Na-fluorescein and polyethylene glycol (PEG) 4000 have been used to verify a tight epithelium (Pantzar *et al.*, 1994) and for testing effects of enhancers and increased fluid absorption (Borchardt *et al.*, 1996; Karlsson *et al.*, 1994; 1999). It has also become very popular to verify a viable and intact epithelium using biochemical markers such as lactate dehydrogenase (LDH) release (Oberle *et al.*, 1995) and morphology evaluations (Polentarutti *et al.*, 1999; Söderholm *et al.*, 1998). The more viable the segment, the better the interpretation of the results. Extremes in permeability values can be discarded from the data set giving better and more reliable results and a better overall understanding of drug transport (Polentarutti *et al.*, 1999).

This method has several advantages for predicting *in vivo* drug absorption in humans. First, there is a good correlation with the $P_{app}$ of human jejunum *in vivo* (Lennernäs *et al.*, 1997) for both passively transported low and high permeability compounds (Figure 11.1). Second, the technique can be used for different regions of the GI tract, evaluating the regional absorption characteristics of drugs (Jezyk *et al.*, 1992; Pantzar *et al.*, 1993; Narawane *et al.*, 1993; Polentarutti *et al.*, 1999; Ungell *et al.*, 1997). Furthermore, mucosa other than the intestine can be used, making it possible to evaluate other administration sites with the same model, e.g. nasal tissue, esophageal tissue and buccal tissue. The method using diffusion cells can also be employed for cultured monolayers using a modified insert for the monolayer membrane, e.g. Snapwell®. The method is very useful for evaluating mechanisms of absorption. It can shed light on the importance of ionic transport processes on the transport of drug molecules due to the physiological presence of a crypt-villus axis and a heterogeneous population of cells (mature and immature cells are present, as well as cells with different functions). The method also has the advantage of being available for human tissues, slices or biopsies from surgically removed tissues (Bijlsma *et al.*, 1995; Sjöström *et al.*, 2000; Söderholm *et al.*, 1998) , and also for mouse intestine, which represents the most challenging developments of this method for future screening of drugs, especially for mechanistic studies and enzymatic evaluation of drugs and prodrugs, as well as the use of knock-out animals.

The major disadvantage of this absorption method is the diffusion pathways for the molecules, which are unphysiological, i.e. the lack of vascular supply forces the molecules to diffuse through the lamina propria and, in the case that unstripped tissues also are used, through the serosal layer (Table 11.2). It was recently proposed that the presence of the serosal and muscular layers might have different impacts on the transport of molecules with different physicochemical characteristics, which are both size and lipophilicity dependent (Breitholtz *et al.*, 1999). The lamina propria, muscle layers and serosal layer can also be different in different animals and regional segments. Some reports have also proposed that there may be difficulties with the unstirred water layers and there is concern regarding the stirring conditions, especially of the solution in the donor compartment. The model is probably not designated as an HTS tool (Table 11.1). However, when correctly used as a mechanistic secondary screening tool, well-suited for evaluation of new candidate drugs (Stewart *et al.*, 1997), data from the Ussing chamber technique are more closely related to the human situation than many of the other biological methods available.

### Gut loop model

The gut loop can be described as a simplified perfusion model. A segment of the intestine is isolated under anesthesia, still attached to blood circulation *in situ* in the live animal (Mirchandani and Chien, 1995; Nakayama *et al.*, 2000). After rinsing, a small volume of physiological buffer containing the drug to be tested is injected into the intestinal loop of approximately 10 cm and the loop is sealed. The loop (or loops) is then restored into the animal's abdominal cavity. After some time and careful washing, the content of the loop is analyzed for drug content and, additionally, blood samples from the anesthetized animal can be taken (Mirchandani and Chien, 1995).

The main disadvantages of this model are the stirring conditions of the injected volume in the intestinal segment and the anesthesia (Mirchandani and Chien, 1995). There are, however, several advantages with the technique, e.g. the method is very simple, specific and no expensive equipments like cell monolayers, Ussing chambers or perfusions are needed and the method can be performed by an experimentalist trained in experiments with *in vivo* animals. An additional advantage is that regional assessment of a drug can be performed in the same animal, excluding the large intervariability between rats otherwise seen. The method is, however, not so frequently used as too many animals are consumed compared with other techniques (Caco-2) (Table 11.1).

### Intestinal perfusions

There are reports in the literature on the *isolated perfused* intestine as a technique for absorption studies as well as *in situ* perfusions (Blanchard *et al.*, 1990; Chiou, 1995; Fagerholm *et al.*, 1996; Krugliak *et al.*, 1994; McCarthy and Sutton, 1998; Oeschsenfahrt and Winne, 1974; Raoof *et al.*, 1998; Sinko *et al.*, 1995; Sinko and Balimane, 1998; Sinko and Hu, 1996; Winne, 1979). A segment of 10–30 cm of the intestine is cannulated on both ends and perfused with a buffer solution at a flow rate of 0.2 ml/minute (Fagerholm *et al.*, 1996). The blood side can also be cannulated through the mesenteric vein and artery (*in vitro* isolated; double perfusion). The difference between *in situ* and *in vitro* is the use of the rat circulation *in vivo* (which is a vascular perfusion in the *in vitro* situation) (Fagerholm *et al.*, 1996; Kim *et al.*, 1993; Stewart *et al.*, 1997; Windmueller *et al.*, 1970). A comparison between *in situ* and *in vitro* perfusion gives the opportunity to evaluate the influence of the hepatic clearance on the absorption of drugs.

Both perfusion methods can use different evaluation systems to test the drug absorption, using the difference between 'in' and 'out' concentrations in the perfusion solutions, and/or disappearance and appearance on both sides of the membrane, and also by analysing the drug concentration on the blood side. The permeability, usually called the $P_{eff}$, is calculated from the following equation, $P_{eff} = [-Q_{in} \times \ln(C_{out}/C_{in})]/2\pi r L$ (parallel tube model), where $Q$ is the flow rate, $C_{in}$ and $C_{out}$ are the inlet and outlet concentrations of each drug, and $2\pi r L$ is the mass transfer surface area within the intestinal segment. Different lengths are used between 10–30 cm, but the best flow characteristics are achieved with 10 cm (Fagerholm *et al.*, 1996). PEG 4000 is used for corrections of fluid flow and to verify the absence of leakage in the model. In addition, mannitol may be used as a permeability marker molecule (Krugliak *et al.*, 1994). Mannitol is more sensitive to changes in the intestinal barrier function, compared with PEG 4000 alone. *In situ* perfusions have also been used for mechanistic studies of the efflux of drugs (Lindahl *et al.*, 1999).

The major advantage of this type of absorption method is the blood supply giving the tissue oxygen and the correct flow characteristics on the serosal side of the membrane, e.g. there is less diffusion through the lamina propria. Second, different parts of the GI tract can be used, as in the Ussing chamber technique. Good stirring, i.e. flow characteristics, of the mucosal/luminal solution has been reported (Fagerholm *et al.*, 1996). A very good correlation with perfusions has been found to the fabs% absorbed and human permeability of different types of drugs (Amidon *et al.*, 1988; Fagerholm *et al.*, 1996). A third, very important advantage of this technique is the presence of enteric nerves and endocrine input, resulting in a more physiological control of viability and transport processes (Stewart *et al.*, 1997).

The disadvantage of this method is its use of anesthesia, which has been reported to affect drug absorption (Uhing and Kimura, 1995a,b; Wirén *et al.*, 1999). PEG 4000 is used to verify the integrity of the barrier, which can lead to misinterpretation of the integrity of the tissue owing to the high molecular weight of the marker (Table 11.2). An additional disadvantage, although less important for mechanistic studies, is that the method is time- and animal-consuming, which makes it less useful for screening purposes for large compound libraries. Some discrepancies between the disappearance rates of drugs and their appearance on the blood side have also been reported, indicating a loss of the drug in the system, either by enzymatic degradation or by adsorption to the plastic catheters. High disappearance rates yield high rates of falsely positive results when predicting the fabs% absorbed.

### *In vitro–in vivo* correlations

Methods primarily used for the purpose of *in vitro–in vivo* correlations are *in situ* perfusions of rat gut, regionally cannulated/fistulated rats and dogs, bioavailability models in different animals, intestinal perfusions in man (Lennernäs, 1997; Tamamatsu *et al.*, 1997), triple-lumen perfusions (Gramatte *et al.*, 1994), and bioavailability studies in man (Buch and Barr, 1998; Godbillon *et al.*, 1985; Ungell and Abrahamsson, 2001). Early in the clinical phase, *in vivo* experiments in animals and humans are needed to facilitate the understanding of a particular drug's mechanism of absorption. These experiments will also provide information to support the pharmaceutical dosage form program and allow for the correlation of the performance of the more simple animal models, e.g. membrane $P_{app}$ assessment, ADME studies, dose and concentration dependency, food interactions, and absorption performance.

All *in vitro* methods used, regardless of what mechanism or part of the absorption process they may represent, must be correlated to the *in vivo* situation and, if possible, also to absorption in humans (Dowty and Dietsch, 1997) (Figure 11.1). This correlation process is not a simple evaluation since different methods represent different parts of the total process and the main barrier will affect the main part of the results. Values for *in vivo* absorption in humans are not easy to obtain from clinical studies, and the available values are often a result of a recalculation of data obtained for other purposes during the clinical trials. The values of fabs%, for drugs in the literature are therefore in many cases uncertain. Published, compiled data on bioavailability, on the other hand, can be found in Benet *et al.* (1996).

Correlations have been made in different laboratories using different models (cells, rings, Ussing, perfusion and Caco-2) (Artursson *et al.*, 1993; Fagerholm *et al.*, 1996; Kim *et al.*, 1993; Lennernäs *et al.*, 1997; Porter *et al.*, 1985; Rubas *et al.*, 1993; Tanaka *et al.*, 1995) and in different laboratories using the same model (Caco-2, intestinal perfusion) (Artursson *et al.*, 1996; McCarthy and Sutton, 1998). Additionally, kinetic data for transport via carrier systems using different models have also employed correlations (Hillgren *et al.*, 1995) as has the evaluation of the absence or presence of mucus (Matthes *et al.*, 1992). Only one record is available showing the interlaboratory comparison of *in vitro/in situ* perfusions of the gut (McCarthy and Sutton, 1998). An evaluation of the predictability of *in vitro* rat permeability to predict *in vivo* rat and human absorption has also been performed (Dowty and Dietsch, 1997), in which the need for information regarding the variability in the prediction from the *ex vivo* models is addressed, as is the use of internal standards to improve their predictability. The largest difference between models of absorption is the expression of transporters (Makhey *et al.*, 1998) which has been found during the comparison of rat perfusion and Caco-2 model, with respect to D-glucose and L-leucine uptake (Lennernäs *et al.*, 1996) or efflux transporters, as reported by Saitoh and Aungst (1995).

Many different *in vitro* and *ex vivo* models have been used for the evaluation of drug enhancement in the GI tract (Borchardt *et al.*, 1996; Surendran *et al.*, 1995; Swenson *et al.*, 1994; van Hoogdalem *et al.*, 1989; Yeh *et al.*, 1994; Yodoya *et al.*, 1994) as well as for the influence of different unphysiological vehicles for sparingly soluble drugs (Hanisch *et al.*, 1998, 1999).

Although a good correlation with human fabs% can be obtained for each of the different *ex vivo* methods, absolute permeability values vary over different ranges. The threshold value for complete absorption, therefore, varies, and for excised jejunal segments of the rat in the Ussing chamber is approximately $10 \times 10^{-6}$ cm/ seconds (Lennernäs *et al.*, 1997); for perfusion of the rat jejunum and for the perfused human jejunum *in vivo* the threshold value is around $50 \times 10^{-6}$ cm/seconds (Fagerholm *et al.*, 1996); and for human tissue in the Ussing chamber the threshold value is approximately $10 \times 10^{-6}$ cm/seconds (Sjöström *et al.*, 2000) (Figure 11.1). These values indicate a small parallel shift for different methods concerning the predictive permeability versus fabs% *in vivo*, which recently has been suggested for the methods of *in situ* rat perfusion, Ussing chamber with rat jejunal segments and the perfusion of the human jejunum (Lennernäs *et al.*, 1997). This parallel shift between different methods and animals is expected since the lipid membrane composition can vary both with species and diet (Thomson *et al.*, 1986; Ungell *et al.*, 1997). This shift can also be attributed to differences in measuring and calculating the permeability coefficients (see below). This is, however, of no concern if the ranking order is the same between the methods used.

The values of the permeability coefficients also indicate experimental windows of different sizes. The Caco-2 cell technique seems to operate roughly between 0.1 to $200 \times 10^{-6}$ cm/seconds (Artursson and Karlsson, 1991), the excised rat and human intestinal segments in the Ussing chamber between 1 to $200 \times 10^{-6}$ cm/seconds (Figure 11.1), and the perfused rat intestine and perfused human jejunum between 10 to $1000 \times 10^{-6}$ cm/seconds. These differences might not only be due to the differences in species and lipid composition in the membranes, as suggested earlier

(Ungell *et al.*, 1997; Thomson *et al.*, 1986), but also to the techniques themselves and the calculations of $P_{app}$. Human intestinal segments studied in the Ussing chambers show a similar range of permeabilities to the rat intestinal segments using the Ussing chamber technique (Sjöström *et al.*, 2000) (Figure 11.1). While the disappearance rates of the compounds are measured in the perfusion systems, the appearance rates are determined in the Ussing chambers and Caco-2 systems. If the compound is metabolized or absorbed into the tissue or other surfaces in the perfusion model, it will give a false high permeability value. There could also be some differences, physically, between a drug entering the bilayer membranes (uptake) versus a drug which is diffusing through the lipid bilayer of the epithelium (Bassolino *et al.*, 1996; Jacobs and White, 1989; Marrink and Berendsen, 1994). In addition, an *in situ* perfusion *in vivo* has optimal 'sink conditions' and plasma pro-teins are present on the serosal/basolateral side, creating a gradient for many fast transported compounds better than would be expected in a static system like Caco-2 monolayers, rings, everted sacs or the Ussing chamber.

Regarding the simpler models for drug absorption, such as BBMV, rings and sacs, the data for comparison with *in vivo* in humans are less substantial. BBMV has been isolated from numerous species, including man (Hillgren *et al.*, 1995), but a direct comparison with fabs% in man and its suitability as a screening model for drug absorption has only been reported by Quilianova *et al.*, (1999). Kaplan (1972) assessed the use of the cannulated everted sac as a screen for drug permeability. In this paper, a series of compounds were studied in everted sacs from rat ileum and then compared with *in vivo* dog absorption. Non-absorbable drugs on one side and absorbable drugs on the other side could be evaluated.

Owing to factors such as differences in the handling of animals and their age and species, food and tissues, laboratories will have different prediction factors for absorption when they use the different methods available (Thomson *et al.*, 1986; Ungell *et al.*, 1997; Ungell, 1997). The ranking order between the different drugs might also be different between laboratories because of the different levels of viability and integrity of the biological systems used (Ungell, 1997). The integrity of the tissue change is time related, which means that there is a limit in time for use of the different systems (Levine *et al.*, 1970; Polentarutti *et al.*, 1999), and there are important effects of anesthesia and food prior to the sacrifice of the animal (Wirén *et al.*, 1999). Integrity may also be related to buffer solutions and pH, oxygenation of the solutions, stirring conditions, preparation of the tissues and other physical handling, and temperature (Ungell, 1997). The surface exposed to the drug is different in different models and for high and low permeability drugs, as suggested by Artursson *et al.* (1996), Oliver *et al.*, (1998), and Strocchi and Levitt (1993). The 'true' exposed surface area is the same as the serosal surface area for cultured monolayers (Artursson *et al.*, 1996), but may vary for excised segments for the Ussing chamber or perfusions, depending on which region of the GI tract is used (Collett *et al.*, 1997; Oliver *et al.*, 1998; Polentarutti *et al.*, 1999). As a result of different handling during the preparation of the tissues, the effective surface area for absorption may also be different, and a time-dependent change in surface area during the course of the experiments has been reported (Polentarutti *et al.*, 1999). This change in absorptive area will affect high and low permeability drugs differ-ently (Artursson *et al.*, 1996; Strocchi and Levitt, 1993). For a full understanding of the differences in results between laboratories and between species, these parameters may be useful as a complement to other valuable information regarding the performance of the experiments and the technique used.

The variability between experiments and laboratories makes it very difficult when comparing values from different laboratories. Each laboratory should, therefore, be careful in standardizing and correlating their own models to human absorption values before using them as predictive tools. Indeed, the guidelines from Food and Drug Administration (FDA) regarding the use of a compound data set for classification of a candidate drug (CD) using the biopharmaceutical classification system (BCS) could be helpful in standardizing *ex vivo* and *in vitro* techniques (Amidon, 1996).

## *Ex vivo* models for studying local targets of drugs in the intestinal mucosa

Recently, more importance is given to the fact that enterocytes may act as antigen-presenting cells to the lymphocytes of the intestinal epithelium and surrounding lymphoid tissue and may be crucial to the maintenance of the pool of peripheral T lymphocytes. Several data link the integrity of the mucosal epithelium on one side and the immune system on the other. Crohn's disease is thought to be associated with increased intestinal permeability (Hollander, 1992; Söderholm *et al.*, 1999), but it is still unclear if this barrier effect is connected to the inflammatory process, per se. Intestinal models used to study the local changes in permeability and the barrier function related to different diseases are mainly *in vivo* models pre-treated with an immunological challenge, *ex vivo* models i.e. Ussing chambers or perfusions, and cell monolayers, either as they are or transfected/cloned with appropriate receptors or transporters (Crissinger *et al.*, 1990; McKay *et al.*, 1997).

The targets of these concepts are located within or close to the intestinal mucosa and no drug has to/should enter the systemic circulation, i.e. the passive permeability should be low or the drug should be extensively metabolized in the liver to reduce systemic side effects Figure 11.2. These targets have been discussed for treatment of inflammatory bowel disease (IBD) (Hamedani *et al.*, 1997). Examples of such drugs are sulfasalazine, olsalazine, and budesonide (Hamedani *et al.*, 1997).

The initial screening of such a concept would, therefore, be mainly focused using *ex vivo* models of animals or cells instead of *in vivo* bioavailability (Bjarnason *et al.*, 1995; McKay *et al.*, 1997; Söderholm *et al.*, 1999). Instead of screening for high permeability, the strategic and experimental difficulties would be to prove a true low intestinal absorption in the *in vitro* models, since many other factors will also give the same result, such as adsorption to plastic, low solubility, etc. (Ungell and Abrahamsson, 2001). In addition, the animal or *ex vivo* model will require a knowledge of where the target is situated and how the drugs would permeate to the target of interest Figure 11.2. The analytical detection limits within the experimental systems become extremely important as do the species and regional differences in metabolic capacity. The drug has to be soluble and chemically/metabolically stable in the hostile environment of the gut lumen fluid, not necessarily absorbed over the GI membranes or effectively metabolized by the enterocyte or the liver and still be potent for its target (Hamedani *et al.*, 1977). Apart from the difficulties with respect to the chemistry and pharmacology around the synthesis of such drugs,

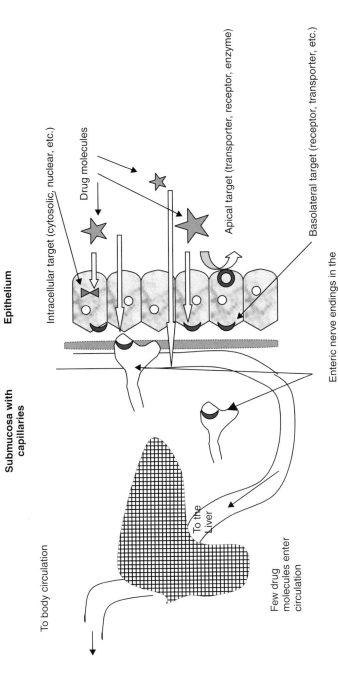

*Figure 11.2* Concept of systemic versus mucosal target delivery. The properties of the drug molecule are determined by the localization of the target. Very small amounts of the drug are thought to reach the systemic circulation. If the target is intracellular, active transport into the cell can be the rationale, and if the target is located in the submucosa a high permeability and low metabolic stability can be two of the requirements.

there is also the need for a very specific view on the development of easy biological *in vitro* techniques for evaluation of pharmacokinetic properties and, additionally, to the use of these models as pharmacological tools and disease models. If the drug is not absorbed or if the bioavailability is very low in the systemic circulation, the traditional *in vivo* measurements of plasma concentrations in rats are unusable as a screening model. More specific models representing both the target membrane and barrier at the same time are needed. Suitable models for mucosal deliverable drugs could also be chosen from the generally used models for drug absorption. BBMV have been isolated from diseased tissues for comparison with the effect of the disease on carrier-mediated transport, e.g. glucose uptake (Maenz and Cheeseman, 1986), and Ussing chambers with excised tissues and Caco-2 cells have been used for evaluation of IBD-related diseases (McKay *et al.*, 1997).

## Biological method supporting QSAR (Quantitative Structure–Activity Relationship) modeling in early discovery

The most frequently used biological model within this respect is the Caco-2 mono-layers (Palm *et al.*, 1996, 1997; Stenberg, 2001), but excised rat intestinal segments in the Ussing chamber (Palm *et al.*, 1996; Zamora and Ungell, 2001) have also been used, as has human jejunal perfusions (Winiwarter *et al.*, 1998). Also, when the carrier-mediated transport is concerned, monolayers of cultured cells seem to be the preferred method (Österberg and Norinder, 2000). However, Caco-2 mono-layers might not represent human fabs% for all drugs now entering the lead dis-covery programs, and the amalgamating of research data to build larger data bases for modeling can be misleading. For instance, the Caco-2 model fails to pre-dict absorption via the dipeptide carrier system (Chong *et al.*, 1996). Different expressions of transporters (Makhey *et al.*, 1998; Stephens *et al.*, 2001) as well as discrepancies regarding the experimental set-up and laboratories can give a huge diversity in the results. Small changes such as pH gradients, stirring conditions, sampling times and concentrations used are only a few factors that have a large impact on the transport rates of drugs.

Recently, Zamora and Ungell (2001) have shown that the information on import-ant structural properties of a set of betablockers were differing when data were obtained from different models of drug absorption, such as Ussing chambers and Caco-2 cells. The Caco-2 cell monolayers showed a more hydrophobic nature similar to LogD, while the excised segments, especially the small intestinal segment, represented a more hydrophilic barrier and were the farthest away from the physico-chemical methods (Zamora and Ungell, 2001).

## REGIONAL ASSESSMENT OF INTESTINAL DRUG ABSORPTION

For developing extended oral drug release dosage forms, knowledge of the regional differences in the absorption pattern becomes very important in evalu-ation and success (Gardner *et al.*, 1997; Kararli, 1995; Narawane *et al.*, 1993;

Pantzar *et al.*, 1993; Thomson *et al.*, 1986; Ungell *et al.*, 1997). In addition, these mechanisms are also species dependent (Kararli, 1995) and must be correlated to the human situation (Gardner *et al.*, 1997; Buch and Barr, 1998). In addition, regional differences in the intestinal permeability seem to be unclear since contradictory results have been achieved from different techniques used by different laboratories (Artursson *et al.*, 1993; Narawane *et al.*, 1993; Rubas *et al.*, 1993; Ungell *et al.*, 1998).

Regional absorption differences can be seen for a compound as regards $P_{app}$, mechanisms of absorption (Makhey *et al.*, 1998; Nakayama *et al.*, 2000; Raoof *et al.*, 1997; Stephens *et al.*, 2001; Ungell *et al.*, 1997) and metabolism in the intestinal lumen, in the brush-border region or within the cell of the epithelium (Sinko and Hu, 1996). The importance of good regional absorption performance (e.g. high and similar absorption throughout the GI tract) of a selected compound may be crucial for the development of extended release formulations and should, therefore, be evaluated early in the screening phase for optimal drug candidate selection. For better prediction of the *in vitro* values of regional permeability of drugs, correlations to regional *in vivo* data in animals or humans must be obtained, although no such record can be found in the literature up to date.

## CONCLUSION

Some advantages and disadvantages using commonly known *ex vivo* absorption models are presented here and their prediction value for fabs% in humans are discussed. Evaluation of the differences between the models and importance of intralaboratory standard markers is suggested, as well as ensuring the viability and integrity of the tissues for good performance.

## REFERENCES

Alcorn, C. J., Simpson, R. J., Leahy, D. and Peters, T. (1991) *In vitro* studies of intestinal drug absorption: determination of partition and distribution coefficients with brush border membrane vesicles. *Biochem. Pharmacol.*, **42**, 2259–2264.

Amidon, G. L. (1996) A Biopharmaceutic Classification System: update May 1996. *In Biopharmaceutics Drug Classification and International Drug Regulation*, pp. 11–30, Capsugel symposium services.

Amidon, G. L., Sinko, P. J. and Fleisher, D. (1988) Estimating human oral fraction dose absorbed: a correlation using rat intestinal membrane permeability for passive and carrier-mediated compounds. *Pharm. Res.*, **5**, 651–654.

Artursson, P. and Karlsson, J. (1991) Correlation between oral drug absorption in humans and apparent drug permeability coefficients in human intestinal epithelial (Caco-2) cells. *Biochem. Biophy. Res. Comm.*, **175**, 880–885.

Artursson, P., Ungell, A.-L. and Löfroth, J.-E. (1993) Selective paracellular permeability in two models of intestinal absorption: cultured monolayers of human intestinal epithelial cells and rat intetsinal segments. *Pharm. Res.*, **10**, 1123–1129.

Artursson, P., Palm, K. and Luthman, K. (1996) Caco-2 monolayers in experimental and theoretical predictions of drug transport. *Adv. Drug Deliv.*, **22**, 67–84.

Bassolino, D., Alper, H. and Stouch, T. R. (1996) Drug-membrane interactions studied by molecular dynamics simulation: size dependence of diffusion. *Drug Des. Disc.*, **13**, 135–141.

Benet, L. Z., Oie, S. and Schwartz, J. B. (1996) Design and optimization of dosage forms regimens: pharmacokinetic data. In Goodman and Gilman's: *Pharm. Basis of Therap.*, 9th edition, pp. 1707–1792.

Bijlsma, P. B., Peeters, R. A., Groot, J. A., Dekker, P. R., Taminiau, J. A. and Van der Meer, R. (1995) Differential *in vivo* and *in vitro* intestinal permeability to lactulose and mannitol in animals and humans: a hypothesis. *Gastroenterology*, **108**, 687–696.

Bjarnason, I., Macpherson, A. and Hollander, D. (1995) Intestinal permeability: an overview. *Gastroenterology*, **108**, 1566–1581.

Blanchard, J., Tang, L. M. and Earle, M. E. (1990) Reevaluation of the absorption of carbenoxolone using an *in situ* rat intestinal technique. *Pharm. Sci.*, **79**, 411–414.

Borchardt, R. T., Smith, P. L. and Wilson, G. (1996) 'Models for assessing drug absorption and metabolism'. *Pharm. Biotech.*, **8**, Plenum Press, New York.

Breithotz, K., Hägg, U., Utter, L., Wiklander, K. and Ungell, A.-L. (1999) The influence of the serosal layer on viability and permeability of rat intestinal segments in the Ussing chamber. *AAPS Pharm. Sci.*, Suppl., **1**, S653.

Buch, A. and Barr, W. H. (1998) Absorption of propranolol in humans following oral, jejunal and ileal administration. *Pharm. Res.*, **15**, 953–957.

Burton, P. S., Conradi, R. A., Hilgers, A. R., Ho, N. F. H. and Maggiora, L. L. (1992) The relationship between peptide structure and transport across epithelial cell monolayers. *J. Control. Release*, **19**, 87–97.

Chiou, W. L. (1995) The validation of the intestinal permeability approach to predict oral fraction of dose absorbed in humans and rats. *Biopharm. Drug Dis.*, **16**, 71–75.

Chong, S., Dando, S. A., Soucek, K. M. and Morrison, R. A. (1996) *In vitro* permeability through Caco-2 cells is not quantitatively predictive of *in vivo* absorption for peptide-like drugs absorbed via the dipeptide transporter system. *Pharm. Res.*, **13**, 120–123.

Collett, A., Walter, D., Sims, E., He, Y.-L., Speers, P., Ayrton, J. *et al.* (1997) Influence of morphometric factors on quantitation of paracellular permeability of intestinal epithelia *in vitro*. **14**, 767–773.

Correno-Gomez, B., Woodley, J. F. and Florence, T. (1999) Studies on the uptake of tomato lectin nanoparticles in everted gut sacs. *Int. J. Pharm.*, **183**, 7–11.

Crissinger, K. D., Kvietys, P. R. and Granger, D. N. (1990) Pathophysiology of gastrointestinal mucosal permeability. *J. Inter. Med.*, **228**(1), 145–154.

Davis, G. R., Santa Ana, C. A., Morawski, S. G. and Fordtran, J. S. (1982) Permeability characteristics of human jejunum, proximal colon and distal colon: results of potential difference measurements and unidirectional fluxes. *Gastroenterology*, **83**, 844–850.

Dowty, M. E. and Dietsch, C. R. (1997) Improved prediction of *in vivo* peroral absorption from *in vitro* intestinal permeability using an internal standard to control for intra- and inter-rat variability. *Pharm. Res.*, **14**, 1792–1797.

Dressman, J. B. and Yamada, K. (1991) Animal models for oral drug absorption. In , Welling and Tse, (eds), *Pharmaceutical Bioequivalence*. Dekker, New York pp. 235–266.

Fagerholm, U., Johansson, M. and Lennernäs, H. (1996) The correlation between rat and human small intestinal permeability to drugs with different physico-chemical properties. *Pharm. Res.*, **13**, 1335.

Gardner, D., Casper, R., Leith, F. and Wilding, I. (1997) Noninvasive methodology for assessing regional drug absorption from the gastrointestinal tract. *Pharm. Techno.*, **21**, 82–89.

Godbillon, J., Evard, D. and Vidon, N. (1985) Investigation of drug absorption from the gastrointestinal tract of man. III Metoprolol in the colon. *Br. J. Clin. Pharmacol.*, **19**, 113S–118S.

Gramatte, T., Desoky, E. E. and Klotz, U. (1994) Site-dependent small intestinal absorption of ranitidin. *Eur. J. Clin. Pharmacol.*, **46**, 253–259.

Grass, G. M. and Sweetana, S. A. (1988) *In vitro* measurement of gastrointestinal tissue permeability using a new diffusion cell. *Pharm. Res.*, **5**, 372–376.

Hamedani, R., Feldman, R. D. and Feagan, B. G. (1997) Review article: drug development in inflammatory bowel disease: budesonide-a model of targeted therapy. *Aliment. Pharmacol. Ther.*, **11**(3), 98–108.

Hanisch, G., von Corswant, C., Breitholtz, K., Bergstrand, S. and Ungell, A.-L. (1998) Can mucosal damage be minimised during permeability measurements of sparingly soluble compounds? Fourth international conf. on drug absorption: towards prediction and enhancement of drug absorption, Edinburgh.

Hanisch, G., Kjerling, M., Pålsson, A., Abrahamsson, B. and Ungell, A.-L. (1999) Effects of vehicles for sparingly soluble compounds on the drug absorption *in vivo*. Elderly people & Medicines, Stockholm conference center, Älvsjö 11th–13th, October 1999.

Hidalgo, I. J., Bhatnagar, P., Lee, C.-P., Miller, J., Cucullino, G. and Smith, P. L. (1995) Structural requirements for interaction with the oligopeptide transporter in Caco-2 cells. *Pharm. Res.*, **12**, 317–319.

Hillgren, K. M., Kato, A. and Borchardt, R. T. (1995) *In vitro* systems for studying intestinal drug absorption. *Med. Res. Rev.*, **15**, 83–109.

Hollander, D. (1992) The intestinal permeability barrier: a hypothesis as to its regulation and involvement in Crohn's disease. *Scand. J. Gastroenterol.*, **27**, 721–726.

Hörter, D. and Dressman, J. B. (1997) Influence of physicochemical properties on dissolution of drugs in the gastrointestinal tract. *Adv. Drug Del. Rev.*, **25**, 3–14.

Hunter, J. and Hirst, B. H. (1997) Intestinal secretion of drugs: the role of P-glycoprotein and related drug efflux systems in limiting oral drug absorption. *Adv. Drug Del. Rev.*, **25**, 129–157.

Ishizawa, T., Sadahiro, S., Hosoi, K., Tamai, I., Terasaki, T. and Tsuji, A. (1992) Mechanisms of intestinal absorption of the antibiotic, fosfomycin, in brush-border membrane vesicles in rabbits and humans. *J. Pharmacobio. Dyn.*, **15**, 481–489.

Jacobs, R. E. and White, S. E. (1989) The nature of the hydrophobic binding of small peptides at the bilayer interfaces: implications for the insertion of transbilayer helices. *Biochemistry*, **28**, 3421–3437.

Jezyk, N., Rubas, W. and Grass, G. M. (1992) Permeability characteristics of various intestinal regions of rabbit, dog and monkey. *Pharm. Res.*, **9**, 1580–1586.

Kamm, W., Raddatz, P., Gante, J. and Kissel, T. (1999) Prodrug approach for $alpha_{IIb}beta_3$-peptidomimetic antagonists to enhance their transport in monolayers of a human intestinal cell line (Caco-2): comparison of *in vitro* and *in vivo* data. *Pharm. Res.*, **16**, 1527–1533.

Kaplan, S. A. (1972) Biopharmaceutical considerations in drug formulation design and evaluation. *Drug Metab. Rev.*, **1**, 15–34.

Kararli, T. T. (1989) Gastrointestinal absorption of drugs. *Crit. Rev. Ther. Drug Carrier Syst.*, **6**, 39–86.

Kararli, T. T. (1995) Comparison of the gastrointestinal anatomy, physiology, and biochemistry of humans and commonly used laboratory animals. *Biopharm. Drug Dispos.*, **16**, 351.

Karlsson, J. and Artursson, P. (1991) A method for the determination of cellular permeability coefficients and aqueous boundary layer thickness in monolayers of intestinal epithelial (Caco-2) cells grown in permeable filter chambers. *Int. J. Pharm.*, **71**, 51–64.

Karlsson, J., Kuo, S. M., Ziemniak, J. and Artursson, P. (1993) Transport of celiprolol across human intestinal epithelial (Caco-2) cells: mediation of secretion by multiple transporters including P-glycoprotein. *Br. J. Pharmacol.*, **110**, 1009–1016.

Karlsson, J., Ungell, A.-L. and Artursson, P. (1994) Effect of an oral rehydration solution on paracellular drug transport in intestinal epithelial cells and tissues: assessment of charge and tissues selectivity. *Pharm. Res.*, **11**, S248.

Karlsson, J., Ungell, A.-L., Gråsjö, J. and Artursson, P. (1999) Paracellular drug transport across intestinal epithelia: influence of charge and induced water flux. *Eur. J. Pharm. Sci.*, **9**, 47–56.

Kessler, M., Acuto, O., Storelli, C., Murer, H., Muller, M. and Semenza, G. (1978) A modified procedure for the rapid preparation of efficiently transporting vesicles from small intestinal brush border membranes. *Biochim. Biophys. Acta.*, **506**, 136–154.

Kim, D. C., Burton, P. S. and Borchardt, R. T. (1993) A correlation between the permeability characteristics of a series of peptides using an *in vitro* cell culture model (Caco-2) and those using an *in situ* perfused rat ileum model of the intestinal mucosa. *Pharm. Res.*, **10**, 1710–1714.

Kim, J. S., Oberle, R. L., Krummel, D. A., Dressman, J. B. and Fleischer, D. (1994) Absorption of ACE inhibitors from small intestine and colon. *J. Pharm. Sci.*, **83**, 1350–1356.

Krugliak, P., Hollander, D., Schlaepfer, C. C., Nguyen, H. and Ma, T. Y. (1994) Mechanisms and sites of mannitol permeability of small and large intestine in the rat. *Dig. Dis. Sci.*, **39**, 796–801.

Lang, V. B., Langguth, P., Ottiger, C., Wunderli-Allenspach, H., Rognan, D., Rothen-Rutishauser, B., *et al.* (1997) Structure-permeation relations of met-enkephalin peptide analogues on absorption and secretion mechanisms in Caco-2 monolayers. *J. Pharm. Sci.*, **86**, 846–853.

LeCluyse, E. L. and Sutton, S. C. (1997) *In vitro* models for selection of development candidates. Permeability studies to define mechanisms of absorption enhancement. *Adv. Drug Del. Rev.*, **23**, 163–183.

Lee, C.-P, de Vrueh, R. L. A. and Smith, P. L. (1997) Selection of development candidates based on *in vitro* permeability measurements. *Adv. Drug Del. Rev.*, **23**, 47–62.

Lennernäs, H. (1997) Human jejunal effective permeability and its correlation with preclinical drug absorption models. *J. Pharm. Pharmcol.*, **49**, 627–638.

Lennernäs, H., Palm, K. and Artursson, P. (1996) Comparison between active and passive drug transport in human intestinal epithelial (Caco-2) cells *in vitro* and human jejunum *in vivo*. *Int. J. Pharm.*, **127**, 103–107.

Lennernäs, H., Nylander, S. and Ungell, A.-L. (1997) Jejunal permeability: a comparison between the ussing chamber technique and the single-pass perfusion in humans. *Pharm. Res.*, **14**(5), 667–671.

Leppert, P. S. and Fix, J. A. (1994) Use of everted intestinal rings for *in vitro* examination of oral absorption potential. *J. Pharm. Sci.*, **83**, 976–981.

Levine, R. R., McNary, W. F., Kornguth, P. J. and LeBlanc, R. (1970) Histological reevaluation of everted gut technique for studying intestinal absorption. *Eur. J. Pharmacol.*, **9**, 211–219.

Lindahl, A., Persson, B., Ungell, A.-L. and Lennernäs, H. (1999) Surface activity and concentration dependent intestinal permeability in the rat. *Pharm. Res.*, **16**, 97–102.

Lipinski, C. A., Lombardo, F., Dominy, B. W. and Feency, P. J. (1996) Experimental and computational approaches to estimate solubility and permeability in drug discovery and development settings. *Adv. Drug Deliv. Rev.*, **23**, 3–25.

Maenz, D. D. and Cheeseman, C. I. (1986) *Biochim. Biophys. Acta.*, **860**, 277.

Makhey, V. D., Guo, A., Norris, D. A., Hu, P., Yan, J. and Sinko, P. J. (1998) Characterization of the regional intestinal kinetics of drug efflux in rat and human intestine and in Caco-2 cells. *Pharm. Res.*, **15**, 1160–1167.

Marrink, S. J. and Berendsen, H. J. C. (1994) Simulation of water transport through a lipid membrane. *J. Phys. Chem.*, **98**, 4155–4168.

Matthes, I., Nimmerfall, F., Vonderscher, J. and Sucker, H. (1992) Mucus models for investigation of intestinal absorption. Part 4: comparison of the *in vitro* mucus model with absorption models *in vivo* and *in situ* to predict intestinal absorption. *Pharmazie.*, **47**, 787–791.

McCarthy, J. M. and Sutton, S. (1998) Validation of a correlation between rat and human intestinal permeability. Abstract 3311 AAPS. *Pharm. Sci.*, Suppl., 1 (No. 1), S452.

McKay, D. M., Philpott, D. J. and Perdue, M. H. (1997) *In vitro* models in inflammatory bowel disease research–a critical review. *Aliment. Pharmacol. Ther.*, **11**(3), 70–80.

Mirchandani, H. L. and Chien, Y. W. (1995) A multi-loop *in situ* technique to study intestinal drug absorption. *STP Pharm. Sci.*, **5**(2), 145–151.

Narawane, M., Podder, S. K., Bundgaard, H. and Lee, V. H. L. (1993) Segmental differences in drug permeability, esterase activity and ketone reductase activity in the albino rabbit intestine. *J. Drug Target.*, **1**, 29–39.

Nakayama, A., Saitoh, H., Oda, M., Takada, M. and Aungst, B. (2000) Region-dependent disappearance of vinblastin in rat small intestine and characterization of its P-glycoprotein-mediated efflux system. *Eur. J. Pharm. Res.*, **11**, 317–324.

Oberle, R. L., Moore, T. J. and Krummel, D. A. P. (1995) Evolution of mucosal damage of surfactants in rat jejunum and colon. *J. Pharmacol. Toxicol. Methods*, **33**, 75–81.

Oeschsenfahrt, H. and Winne, D. (1974) The contribution of solvent drug to the intestinal absorption of the basic drugs amidopyridine and antipyrine from the jejunum of the rat. *Nauynun-Schmiedeberg's Arch. Pharmacol.*, **281**, 175–196.

Oliver, R. E., Jones, A. F. and Rowland, M. (1998) What surface of the intestinal epithelium is effectively available to permeating drugs?. *J. Pharm. Sci.*, **87**, 634–639.

Osiecka, I., Porter, P. A., Borchardt, R. T., Fix, J. A. and Gardner, C. R. (1985) *In vitro* drug absorption models. I. Brush border membrane vesicles, isolated mucosal cells and everted intestinal rings: characterization and salicylate accumulation. *Pharm. Res.*, **2**, 284–293.

Österberg, T. and Norinder, U. (2000) Theoretical calculation and prediction of P-glycoprotein-interacting drugs using MolSurf parametrization and PLS statistics. *Eur. J. Pharm. Sci.*, **10**, 295–303.

Pade, V. and Stavchansky, S. (1998) Link between drug absorption solubility and permeability measurements in Caco-2 cells. *J. Pharm. Sci.*, **87**, 1604–1607.

Palm, K, Luthman, K., Ungell, A.-L., Strandlund, G. and Artursson, P. (1996) Correlation of drug absorption with molecular surface properties. *J. Pharm. Sci.*, **85**, 32.

Palm, K., Stenberg, L. P., Luthman, K. and Artursson, P. (1997) Polar molecular surface properties predict the intestinal absorption of drugs in humans. *Pharm. Res.*, **14**, 568–571.

Pantzar, N., Weström, B. R., Luts, A. and Lundin, S. (1993) Regional small-intestinal permeability *in vitro* to different sized dextrans and proteins in the rat. *Scand. J. Gastroenterol.*, **28**, 205–211.

Pantzar, N., Lundin, S., Wester, L. and Weström, B. R. (1994) Bidirectional small intestinal permeability in the rat to common marker molecules *in vitro*. *Scand. J. Gastroenterol.*, **29**, 703–709.

Podesta, R. B. and Mettrick, D. F. (1977) $HCO_3$ transport in rat jejunum: relationship to NaCl and $H_2O$ transport *in vivo*. *Am. J. Physiol.*, **232**(1), E62–E68.

Polentarutti, B., Peterson, A., Sjöberg, Å., Anderberg, E.-K., Utter, L. and Ungell, A.-L. (1999) Evaluation of viability of excised rat intestinal segments in the ussing chamber: investigation of morphology, electrical parameters and permeability characteristics. *Pharm. Res.*, **16**, 446–454.

Porter, P. A., Osiecka, I., Borchardt, R. T., Fix, J. A., Frost, L. and Gardner, C. R. (1985) *In vitro* drug absorption models. II: Salicylate, cefoxitin, alphamethyl dopa and theophylline uptake in cells and rings: correlation with *in vivo* bioavailability. *Pharm. Res.*, **2**, 293–298.

Quilianova, N., Chen, Y., Richard, A. and Hu, Z. (1999) Drug absorption screening model using rabbit intestinal brush border membrane vesicles (BBMV). Abstract from AAPS Meeting in Washington on Membrane transporters. Washington, April 1999.

Raoof, A. A., Moriarty, D., Brayden, D., Corrigan, O. I., Cumming, I., Butler, J. *et al.* (1997) Comparison of methodologies for evaluating regional intestinal permeability. *Adv. Exp. Med. Biol.*, **423**, 181–189.

Raoof, A. A., Butler, J. and Devane, J. G. (1998) Assessment of regional differences in intestinal fluid movement in the rat using a modified *in situ* single pass perfusion model. *Pharm. Res.*, **15**, 1314–1316.

Rubas, W., Jezyk, N. and Grass, G. M. (1993) Comparison of the permeability characteristics of a human colonic epithelial (Caco-2) cell line to colon of rabbit, monkey and dog intestine and human drug absorption. **10**, 113–118.

Saitoh, H. and Aungst, B. J. (1995) Possible involvement of multiple P-glycoprotein mediated efflux systems in the transport of verapamil and other organic cations across rat intestine. *Pharm. Res.*, **12**, 1304–1310.

Schwaan, P. W., Stehouwer, R. C. and Tukker, J. J. (1995) Molecular mechanism for the relative binding affinity to the intestinal peptide carrier. Comparison of three ACE inhibitors: enalapril, enalaprilat and lisinopril. *Biochem. Biophys. Acta-Biomembranes*, **1236**, 31–38.

Sinko, P. J. and Amidon, G. L. (1989) Characterization of the oral absorption of beta-lactam antibiotics. II: competitive absorption and peptide carrier specificity. *J. Pharm. Sci.*, **78**, 723–727.

Sinko, P. J. and Balimane, P. V. (1998) Carrier-mediated intestinal absorption of valcyclovir, the L-valyl ester prodrug of acyclovir: 1. interactions with peptides, organic anions and organic cations in rats. *Biopharm. Drug Dispos.*, **19**, 209–217.

Sinko, P. J. and Hu, P. (1996) Determining intestinal metabolism and permeability for several compounds in the rats. Implications on regional bioavailability in humans. *Pharm. Res.*, **13**, 108–113.

Sinko, P. J., Hu, P., Waclawski, P. and Patel, N. R. (1995) Oral absorption of anti AIDs nucleoside analogues. 1. Intestinal transport of didandosine in rat and rabbit preparations. *J. Pharm. Sci.*, **84**, 959–965.

Sjöström, M., Lindfors, L. and Ungell, A.-L. (1999). Inhibition of binding of an enzymatically stable thrombin inhibitor to luminal proteases as an additional mechanism of intestinal absorption enhancement. *Pharm. Res.*, **16**, 74–79.

Sjöström, M., Sjöberg, Å., Utter, L., Hyltander, A., Karlsson, J., Stockman, E. *et al.* (2000) Excised human intestinal segments as a mechanistic tool for verifying transport properties of drug candidates. Abstract AAPS 2000. *Pharm. Sci.*, Suppl.

Söderholm, J. D., Hedamn, L., Artursson, P., Franzen, L., Larsson, J., Pantzar, N. *et al.* (1998) Integrity and metabolism of human ileal mucosa *in vitro* in the Ussing chamber. *Acta. Physiol. Scand.*, **162**, 47–56.

Söderholm, J. D., Holmgren Petersson, K., Olaison, G., Franzén, L. E., Weström, B., Magnusson, K.-E. *et al.* (1999) Epithelial permeability to proteins in the noninflamed ileum of Crohn's disease? *Gastroenterology.*, **117**, 65–72.

Stenberg, P. (2001) Computational models for the prediction of intestinal membrane permeability. Doctoral thesis Uppsala University, Faculty of Pharmacy 247, ISBN 91-554-4934-4.

Stephens, R. H., O'Neill, C. A., Warhurst, A., Carlson, G. L., Rowland, M. and Warhurst, G. (2001) Kinetic profiling of P-glycoprotein-mediated drug efflux in rat and human intestinal epithelia. *J. Pharmacol. Exp. Therap.*, **296**, 584–591.

Stewart, B. H., Chan, O. H., Lu, R. H., Reyner, E. L., Schmid, H. L. and Hamilton, H. W. (1995) Comparison of intestinal permeabilities determined in multiple *in vitro* and *in situ* models: relationship to absorption in humans. *Pharm. Res.*, **12**(5), 693–699.

Stewart, B. H., Chan, O. H., Jezyk, N. and Fleischer, D. (1997) Discrimination between drug candidates using models for evaluation of intestinal absorption. *Adv. Drug Del. Rev.*, **23**, 27–45.

Strocchi, A. and Levitt, M. D. (1993) Role of villus surface area in absorption. *Dig. Dis. Sci.*, **38**, 385.

Surendran, N., Ugwu, S. O., Nguyen, L. D., Sterling, E. J., Dorr, R. T. and Blanchard, J. (1995) Absorption enhancement of melanotan-1: comparison of the Caco-2 and rat *in situ* models. *Drug Deliv.*, **2**, 49–55.

Sutton, S. C., Forbes, A. E., Cargyll, R., Hochman, J. H. and Le Cluyse, E. L. (1992) Simultaneous *in vitro* measurement of intestinal tissue permeability and transepithelial electrical resistance (TEER) using Sweetana-Grass diffusion cells. *Pharm. Res.*, **9**, 316–319.

Swenson, E. S., Milisen, W. B. and Curatolo, W. (1994) Intestinal permeability enhancement: efficacy, acute toxicity and reversibility. *Pharm. Res.*, **11**, 1132–1142.

Tamamatsu, N., Welage, L. S., Idkaldek, N. M., Liu, D.-Y., Lee, P. I.-D., Hayashi, Y. *et al.* (1997) Human intestinal permeability of piroxicam, propranolol, phenylalanine, and PEG 400 determined by jejunal perfusion. *Pharm. Res.*, **14**, 1127–1132.

Tanaka, Y., Taki, Y., Sakane, T., Nadai, T., Sezaki, H. and Yamashita, S. (1995) Characterization of drug transport through tight-junctional pathway in Caco-2 monolayer: comparison with isolated rat jejunum and colon. *Pharm. Res.*, **12**, 523–528.

Thomson, A. B. R., Keelan, M., Clandinin, M. T. and Walker, K. (1986) Dietary fat selectively alters transport properties of rat jejunum. *J. Clin. Invest.*, **77**, 279–288.

Tsuji, A. and Tamai, I. (1996) Carrier-mediated intestinal transport of drugs. *Pharm. Res.*, **13**, 963–977.

Tukker, J. J. and Schoenmakers, R. G. (1998) Relevance of the peptide-bond carbonyl for affinity for the small-peptide carrier. Abstract 3291. *AAPS Pharm. Sci.*, Suppl., **1**(1), S-447.

Uch, A. S. and Dressman, J. (1997) Improved methodology for uptake studies in intestinal rings. *Pharm. Res.*, **14**(11), S-29.

Uch, A. S., Hesse, U. and Dressman, J. B. (1999) Use of 1-methyl-pyrrolidone as a solubilising agent for determining the uptake of poorly soluble drugs. *Pharm. Res.*, **16**(6), 968–971.

Uhing, M. R. and Kimura, R. E. (1995a) The effect of surgical bowel manipulation and anaesthesia on intestinal glucose absorption in rats. *J. Clin. Invest.*, **95**, 2790–2798.

Uhing, M. R. and Kimura, R. E. (1995b) Active transport of 3-O-methyl-glucose by the small intestine in chronically catheterised rats. *J. Clin. Invest.*, **95**, 2799–2805.

Ungell, A.-L. (1997) *In vitro* absorption studies and their relevance to absorption from the GI tract. *Drug Dev. Indust. Pharm.*, **23**(9), 879–892.

Ungell, A.-L. and Abrahamsson, B. (2001) Biopharmaceutical support in candidate drug selection. In Pharmaceutical Preformulation and Formulation. A Practical Guide from Candidate Drug Selection to Commercial Dosage Formulation. Chapter 2.3, M. Gibson, (ed) Interpharm Press.

Ungell, A.-L., Andreasson, A., Lundin, K. and Utter, L. (1992) Effects of enzymatic inhibition and increased paracellular shunting on transport of vasopressin analogues in the rat. *J. Pharm. Sci.*, **81**, 640.

Ungell, A.-L., Nylander, S., Bergstrand, S., Sjöberg, Å. and Lennernäs, H. (1998) Membrane transport of drugs in different regions of the intestinal tract of the rat. *J. Pharm. Sci.*, **87**, 360–366.

Ussing, H. H. and Zerahn, K. (1951) Active transport of sodium as the source of electric current in the short-circuited isolated frog skin. *Acta. Physiol. Scand.*, **23**, 110–127.

van Hoogdalem, E. J., de Boer, A. G. and Breimer, D. D. (1989) Intestinal drug absorption enhancement. *Pharm. Ther.*, **44**, 407–443.

Wacher, V. J. Silvermann, J. A., Zhang, Y. and Benet, L. Z. (1998) Role of P-glycoprotein and cytochrome P450 3A in limiting oral absorption of peptides and peptidomimetics. *J. Pharm. Sci.*, **87**, 1322–1330.

Walter, E., Kissel, T., Reers, M., Dickneite, G., Hoffmann, D. and Stuber, W. (1995) Transepithelial transport properties of peptidomimetic thrombin inhibitors in monolayers of a human intestinal cell line (Caco-2) and their correlation to *in vivo* data. *Pharm. Res.*, **12**, 360.

Windmueller, H. G., Spaeth, A. E. and Ganote, C. E. (1970). Vascular perfusion of isolated rat gut norepinephrine and glucocorticoid requirements. *Am. J. Physiol.*, **218**, 197–204.

Winiwarter, S., Bonham, N. M., Ax, F., Hallberg, A., Lennernäs, L. and Karle'n, A. (1998) Correlation of human jejunal permeability (*in vivo*) of drugs with experimentally and theoretically derived parameters. A multivariate data analysis approach. *J. Med. Chem.*, **41**, 4939–4949.

Winne, D. (1979) Rat jejunum perfused *in situ*: effect of perfusion rate and intraluminal radius on absorption rate and effective unstirred layer thickness. *Naunyn-Schmiedeberg's Arch. Pharm.*, **307**, 265–274.

Wirén, M., Söderholm, J. D., Lindgren, J., Olaison, G., Permert, J., Yang, H. *et al.* (1999) Effects of starvation and bowel resection on paracellular permeability in rat small-bowel mucosa *in vitro*. *Scand. J. Gastroenterol.*, **34**, 156–162.

Wu, X., Whitfield, L. R. and Stewart, B. H. (2000) Atorvastin transport in the Caco-2 cell model: contributions of P-glycoprotein and the proton-monocarboxylic acid co-transporter. *Pharm. Res.*, **17**, 209–215.

Yeh, P. Y., Smith, P. L. and Ellens, H. (1994) Effect of medium-chain glycerides on physiological properties of rabbit intestinal epithelium *in vitro*. *Pharm. Res.*, **11**, 1148–1153.

Yodoya, E., Uemura, K., Tenma, T., Fujita, T., Murakami, M., Yamamoto, A., *et al.* (1994) Enhanced permeability of tetragastrin across the rat intestinal membrane and its reduced degradation by acylation with various fatty acids. *J. Pharmacol. Exp. Ther.*, **27**, 1509.

Yuasa, H., Amidon, G. L. and Fleischer, D. (1993) Peptide carrier-mediated transport in intestinal brush border membrane vesicles of rats and rabbits: cephradine uptake and inhibition. *Pharm. Res.*, **10**, 400–404.

Yuasa, H., Fleischer, D. and Amidon, G. L. (1994) Noncompetetive inhibition of cephradine uptake by enalapril in rabbit intestinal brush border membrane vesicles an enalapril specific inhibitory binding site on the peptide carrier. **269**, 1107–1111.

Zamora, I. and Ungell, A.-L. (2001) Comparison between different absorption models using QSAR. *Pharm. Sci.*, (in press).

Zamora, I., Oprea, T. and Ungell, A.-L. (2000) Prediction of oral drug permeability. In proceedings of the 13th European QSAR meeting, Dusseldorf, March 2000.

# Chapter 12

# Models of the alveolar epithelium

*S. Fuchs, M. Gumbleton, Ulrich F. Schäfer and Claus-Michael Lehr*

## INTRODUCTION

### Anatomy of the lung

The lung is a complex heterogeneous tissue adapted for gaseous exchange. As shown in the schematic in Figure 12.1, the lung comprises a lobular structure with each lobe having its own blood supply originating from the bronchial arteries. The lungs are contained within the thoracic cavity and are surrounded by the parietal and visceral pleural membranes. The latter membrane is directly adherent to

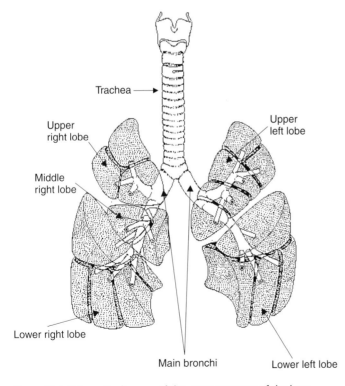

Figure 12.1 Schematic diagram of the gross anatomy of the lung.

the lung surface tissue, while the parietal pleura lines the chest wall. The two pleurae are lubricated by serous fluid and the region between them – intrapleural space – has a subatmospheric pressure. When the muscles of the diaphragm and the intercostal muscles of the ribcage contract there is an increase in intrathoracic volume. Correspondingly, during quiet breathing the intrapleural pressure (at the start of inspiration approximately −2.5 mm Hg relative to atmospheric) decreases to approximately −6 mm Hg and the lungs expand, and the pressure in the airways becomes slightly negative and air flows into the lungs – 'inspiration'.

The respiratory system consists of two functionally and structurally distinct regions known as the upper and lower respiratory tracts. The upper respiratory tract consists of the nasal and paranasal passages, and the pharynx (collectively termed the nasopharyngeal region). The nasopharyngeal region serves to filter, warm and humidify inhaled air, thus protecting the respiratory membranes of the lower tract from damage. The trachea connects the nasopharyngeal region to the lower respiratory tract. Bifurcation of the trachea leads to the formation of the left and right main bronchi which enter the lung parenchymal tissue.

The main bronchi are often considered as the start of the lower respiratory tract, functioning in the conduction of inspired air through to the gas exchange region of the alveoli. Between the trachea and the alveolar sacs there are 23 generations of bifurcation. The first 16 generations form the conducting zone and comprise the main bronchi, small bronchi, brochioles and terminal bronchioles. The remaining seven generations form the transitional and respiratory zones, ultimately comprising the respiratory bronchioles, alveolar ducts, and the alveoli themselves. The aggregate circumference of the 16th generation of air passages (terminal bronchioles) has been determined to be approximately 2000 times the circumference of the trachea. The epithelium changes markedly in structure from the main bronchi through to the alveolar epithelium (Figure 12.2).

The alveoli, or air sacs, are organized as clusters continuous with alveolar ducts. The diameter of an average alveolus has been calculated at 250 μm. Each alveolus is surrounded by many blood capillaries in such an arrangement that constitutes an extensive air–blood interface separated by a thin tissue barrier and which, therefore, allows for the optimal diffusion of gases across the respiratory membrane. The alveolar epithelial-pulmonary capillary barrier comprises mainly the alveolar epithelium and pulmonary capillary endothelium (Figure 12.3). In parts of the barrier, the basal membranes of the epithelial and endothelial cells are directly in contact, while in other parts they are separated by interstitium. The total alveolar epithelial surface area within an average adult human lung has been estimated, using electron microscopy techniques, to be as large as 140 m$^2$ (Gehr *et al.*, 1978). The alveolar epithelial surface is covered with a film of surfactant that lowers the surface tension in the lungs and is essential if the alveolar sacs are to expand during inspiration. The volume of this alveolar film has been calculated at between 7–20 ml per 100 m$^2$ alveolar surface area (Bastacky *et al.*, 1995; Macklin, 1955; Weibel, 1963).

The alveolar epithelium is comprised predominantly of two specialized epithelial cell types; the terminally differentiated squamous alveolar epithelial type-I (ATI) cell which constitutes approximately 93 per cent of the alveolar epithelial surface area (33 per cent of alveolar epithelial cells by number) and the surfactant-

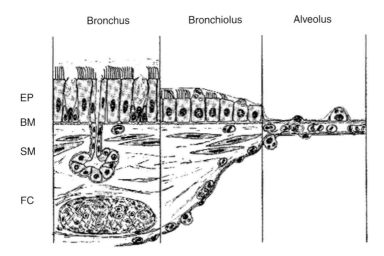

*Figure 12.2* Schematic diagram of the varying mucosal barriers within the lung (EP: epithelial layer; BM: basement membrane; SM: smooth layer; FC: fibrous coat). The wall of the conducting airways consists of three major components: (1) a mucosa composed of epithelial and connective tissue lamina; (2) a smooth muscle sleeve, and; (3) an enveloping connective tissue tube partly with cartilage.

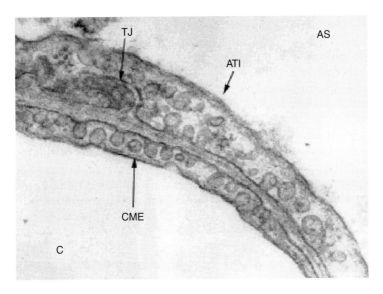

*Figure 12.3* Electron micrograph of the alveolar–capillary barrier in the rat lung, showing 'flask-shaped' caveolae invaginations and free cytosolic vesicles within both capillary microvascular endothelial (CME) cells and alveolar type-I epithelial (ATI) cells. AS = alveolar airspace; C = capillary lumen, and; TJ = tight junction.

producing cuboidal alveolar epithelial type-II (ATII) cell comprising the remaining 7 per cent by surface area and 67 per cent by epithelial cell number (Crapo et al., 1982). It has been calculated that the average human alveolus contains approximately 40 ATI cells (Patton, 1996). Morphometric data (Crapo et al., 1982) indicate that the human ATI cell has an abluminal or interstitial membrane surface area averaging 5,000 μm², with an average cell thickness of 0.35 μm ranging from 2–3 μm in the perinuclear region of the cell to approximately 0.2 μm in the peripheral attenuated regions of the cell. Between species, the dimensions of the ATI cell appear similar (Crapo et al., 1982; Haies et al., 1981). The large surface area of the ATI cell provides for a calculated cell surface density of approximately 20,000 ATI cells per cm² of alveolar epithelium. The extremely thin cytoplasmic attenuated regions extending away from the nuclear body and which contain only very sparse numbers of cellular organelles (the majority being present within the perinuclear region) are a characteristic feature of the ATI cell.

Another characteristic feature of the ATI cells is that they possess a large number of plasmalemmal invaginations or vesicles, termed caveolae (Gumbleton et al., 2000; Newman et al., 1999), whose precise function remains to be elucidated but may be of significance in maintaining the barrier properties of the epithelium and the removal of endogenous and exogenous substrates from the alveolar air space. Some of these morphometric features of the ATI cell exemplify the *favorable* anatomical determinants that have driven interest in the alveolar type-I epithelium as a barrier across which to deliver systemically active proteins and peptides (see Section 1.2).

The cuboidal ATII cell is considerably smaller than the ATI cell (e.g. basal surface area averaging 180 μm² and a uniform cell thickness of approximately 10 μm (Crapo et al., 1982)) and is richly endowed with organelles and microvilli on its apical membrane. These cells are situated in the corners of the alveolus where their physiological functions include, among others, surfactant production and secretion, and to serve as progenitor cells for regeneration of type-I cell epithelium. Current evidence would support a role for the ATII cell serving as the sole *in vivo* progenitor for, and transdifferentiating into, the terminally differentiated ATI cell (reviewed in Uhal, 1997). The pulmonary capillary microvascular endothelial cells are the third major cell type of the air capillary barrier. These cells possess a very similar thin attenuated squamous morphology to the ATI cell, although their cell surface area is reported to be up to three–four times smaller, e.g. luminal surface area of the pulmonary capillary cells averaging 1,350 μm² (Crapo et al., 1982). Like the ATI cell the pulmonary capillary cells possess numerous caveolae that have been biochemically identified as caveolae and are known to contain specific binding sites and receptors for select macromolecules (Schnitzer et al., 1994).

The interstitium lying between the alveolar epithelium and pulmonary capillary endothelium is made up of several different cell types and basement membrane components. The main cell types of the interstitium are fibroblasts, mast cells and myofibroblasts which are interwoven with tough connective tissue, mainly collagen and fibronectin. The main purposes of the interstitium are to provide a substratum support for both alveolar epithelial and endothelial cells and to regulate compliance within the whole lung.

# Alveolar epithelium as a pharmaceutical barrier

The permeability characteristics of the lung and recent advances in inhalational aerosol device technology have led to an increasing interest in exploiting the pulmonary route for the systemic delivery of macromolecule therapeutics, particularly recombinant proteins and polypeptides. Numerous reviews are available on the subject of pulmonary delivery of proteins and polypeptides and the reader is directed to these (Byron and Patton, 1994; Edwards *et al.*, 1998; Patton, 1996; Patton, 1998; Patton *et al.*, 1999; Smith, 1997; Wall, 1995; Yu and Chien, 1997). Nevertheless, it is of note that absorption studies, such as those of Colthorpe and colleagues (1992, 1995), have elegantly demonstrated that the extent of systemic absorption of proteins, such as insulin or growth hormone, following lung administration positively correlate with the depth of their deposition within the lung. Recently a proof-of-concept study has been reported upon the efficacy of inhaled human insulin in type-I diabetic patients (Skyler *et al.*, 2001). Anatomical determinants would also support the view that the lung periphery, and the alveolar epithelium in particular, is the appropriate lung surface to target when aiming to systemically deliver macromolecules. Furthermore, in the transport of macromolecules across the pulmonary alveolar epithelial-capillary endothelial barrier, evidence indicates that it is the alveolar epithelium that possesses a more restrictive paracellular pathway than that provided by the capillary endothelium (Schneeberger, 1977). As a corollary, the mechanisms of transport of drug molecules within alveolar epithelium are the subject of genuine interest, and *in vitro* models of the alveolar epithelium represent valuable research tools. Some of the anatomical determinants that may favor drug absorption in the pulmonary alveolar region include:

1. The location of alveolar epithelium beyond the mucociliary escalator clearance mechanism of the conducting airways. In the alveolus, macrophages are the first line of defence against particulates that have escaped the mucociliary escalator.
2. The pulmonary region of the lung receiving a high blood perfusion close to that of total cardiac output.
3. The relatively large alveolar epithelial surface area estimated at 100–140 m$^2$, although not all of this surface is likely to be concurrently available for the absorption of the inhaled drug. In constrast, the surface area of the conducting airways from main bronchi to terminal bronchioles approximates 2.5 m$^2$, whereas the total surface area of the small intestinal mucosa including villi and microvilli is approximately 200 m$^2$ in an adult.
4. A low alveolar fluid volume relative to the alveolar epithelial surface area (7–20 ml per 100 m$^2$) which minimizes dilution of dissolved drug.
5. A thin cellular barrier from airspace to capillary blood presented by the alveolar epithelial and pulmonary capillary cells.

# *Ex vivo* models for alveolar epithelial transport studies

Organ culture involving the removal and maintenance of slices of rat lung in flux chambers has been previously used as a model to investigate the accumulation and cellular localization of certain drugs and peptides, such as pentamidine (Jones *et al.*, 1992). However, the altered metabolism of a lung slice as a result of the slicing procedure and the need for oxygen and test compound to diffuse through the tissue will severely limit the widespread use of this technique in drug transport studies.

Excised intact bullfrog (Kim and Crandall, 1983), or *Xenopus* (Kim, 1990) alveolar epithelium has been successfully exploited in studying pulmonary transport processes. Amphibian alveolar epithelium has been shown to be similar to mammalian alveolar epithelium with respect to certain anatomical and biochemical considerations, such as alveolar cell morphology, dimensions of the air–blood barrier and surfactant composition (Meban, 1973). Kim *et al.* (1985) studied the polarized trans-alveolar trafficking of albumin across bullfrog alveolar epithelium. Wall and Pierdomenico (1996) demonstrated the presence in *Xenopus* pulmonary membrane of both active and passive transport mechanisms for amino acids and small peptides. Okumura *et al.* (1997) examined the transport of a number of different solutes of varying physico-chemical characteristics across *Xenopus* pulmonary membrane following intrapulmonary administration. The results of such studies suggest the use of amphibian lung as a suitable model to predict the pulmonary absorption of drugs across mammalian lung. However, a number of issues should be considered including the relatively few diffusional replicates that can be obtained from a single amphibian alveolar membrane and the need for additional animal holding facilities beyond that normally available as standard for rodent or rabbit species.

Experiments utilizing an isolated perfused lung (IPL) for the examination of various hemodynamic parameters such as pulmonary blood flow and arterial pressure–flow relationships have been reported for over 80 years (for the historical perspective see Fisher, 1985). Experimental conditions such as temperature, pH, hydrostatic and osmotic pressures within the pulmonary perfusate can be readily controlled. However, it is only comparatively recently that an IPL preparation that preserves tissue integrity for several hours has been developed (Niemeier and Bingham, 1972). The most common species used in IPL models is the rat, although studies with other species have been reported. If the lung is totally isolated from the animal then the vascular system that remains intact is that which serves only the pulmonary region, and hence airway to perfusate transfer of solute will reflect predominantly the pulmonary transport properties and less that of the tracheobronchial tree. IPL models have been used to study the transport of small solutes, e.g. $Na^+$, sucrose, mannitol, and large solutes, e.g. dextran, albumin, across the alveolar epithelial barrier (Basset *et al.*, 1987; Crandall *et al.*, 1986; Effros *et al.*, 1988).

The IPL model has also be used to evaluate the transport of various pharmaceutically relevant compounds. For example, a series of closely related papers by Ralph W. Niven and Peter R. Byron describe methods for using a rat IPL preparation to screen aerosol formulations (Niven and Byron, 1988), surfactant absorption enhancers (Niven and Byron, 1990) and the absorption of synthetic polypeptides (Niven *et al.*, 1990), within the context of macromolecular delivery to the systemic circulation. Additionally, the work of Byron and Patton (1994) demonstrated the use of a rat IPL model in the investigation of the kinetics of pulmonary absorption of hydrophilic biocompatible synthetic polypeptides. These investigators examined absorption as a function of polymer concentration, dose and formulation vehicle. The IPL model has also been shown to be of value when investigating peptidase activity within the pulmonary circulation with relevance to therapeutic peptide delivery (Forbes *et al.*, 1995). There are, however, limitations with the IPL model in terms of its exploitatation for studying the pulmonary absorption of pharmaceuticals. One of the key issues is the relatively short tissue viability of two–three hours; beyond this time functional breakdown of the epithelium becomes evident (Salidas *et al.*, 1998). This breakdown is a particular limitation when studying the alveolar to perfusate transport of macromolecules which will generally display low absorption rate constants. Additionally, the IPL model, like *in vivo*, lung dosing, does not overcome the difficulty of obtaining a reproducible and quantitative delivery of drug down to the alveolar membranes.

## Cell culture models of the alveolar epithelium

The complex nature of the lung architecture means that the alveolar epithelium is not readily accessible. Therefore, the use of cultures of alveolar cells as a reliable *in vitro* experimental model for the prediction of the extent, rate and mechanism of alveolar absorption of pharmaceuticals has gained acceptance amongst investigators (for review see Mathias *et al.*, 1996). Specifically, the use of primary cultures of isolated AT II cells, when grown over a five–eight day period on semi-permeable polycarbonate membranes, generate monolayers with high transepithelial resistance (>1000 $\Omega$ cm$^2$) resembling that of the *in vivo* pulmonary barrier. Many lines of evidence indicate that isolated AT II cells in culture transdifferentiate with time, losing their characteristic alveolar type-II phenotype, coupled with the acquisition of certain morphological and biochemical markers distinctive of the *in vivo* AT I cell. For example, there are morphological changes that include a loss of microvilli, an increase in cell surface area and the development of thin cytoplasmic attentuations extending away from a protruding nuclei (Cheek *et al.*, 1989) or biochemical changes including reactivity to AT I specific membrane components (Danto *et al.*, 1992; Dobbs *et al.*, 1988). However, the parallels between *in vitro–in vivo* AT II transdifferentiation remain to be fully defined, and the term AT I-'like' cell to represent the *in vitro*-derived AT I phenotype is often adopted.

Cell lines of alveolar epithelium derived from tumors (Giard *et al.*, 1973) or from immortalization strategies (Discroll *et al.*, 1995; Matsui *et al.*, 1994; Pasternack *et al.*, 1996) often lack the full differentiated phenotype of the *in vivo* cell. One of the features that appear to be universally lacking in the cell line models of alveolar epithelium is the ability of the cultures to generate a restrictive barrier, exemplified in such cells by low expression of tight junctional proteins such as zonula occludins (ZO)-1 or low transepithelial electrical resistance (TEER) and correspondingly high permeability to hydrophilic solutes that traverse cell monolayers via the paracellular route.

The most common example of an alveolar epithelial cell line is the human cell A549, isolated from a lung adenocarcinoma (Giard *et al.*, 1973). This cell shows similarities to the AT II cell, including the synthesis of phospholipids, the presence of lamellar bodies and microvilli. While A549 cells have been used as a model of the type-II cell in a wide range of applications (Asano *et al.*, 1994; Rahman *et al.*, 1996; Robbins *et al.*, 1994), for drug transport studies they lack many of the desired properties to serve as a model of the alveolar permeability barrier. For example, A549 cells are morphologically distinct from the AT I-cell phenotype and as stated above do not form a sufficiently restrictive paracellular pathway (Godfrey, 1997). Other cell lines which are used for transport studies like Calu-3 or 16HBE14o⁻ correspond to a more bronchiolar cell type (see Chapter 13 within this textbook by Ehrhardt and colleagues). In summary, therefore, the most appropriate permeability model for alveolar epithelium is the primary alveolar epithelial cultures derived from isolation of the AT II cell from rat (most commonly) also from rabbit and significantly from human tissue, and which over time transdifferentiates to an ATI-'like' phenotype.

## PROTOCOLS FOR ATII CELL ISOLATION

### Isolation and culture of rat type-II alveolar cells

The following procedure for rat type-II cell isolation is based on the method by Richards *et al.* (1987) with slight modifications.

### *Lung isolation*

Male pathogen-free rats (120–180 g) are used. Following anesthesia of the animal with sodium phenobarbital (60 mg/kg intraperitoneal), the animal is placed on its back and the abdomen and thorax doused liberally with 70 per cent alcohol for disinfection. A small incision in the neck is made to allow access to the trachea, which is then carefully separated from the surrounding tissue. The trachea is snipped near the top and a 20 G catheter is inserted and positioned so that it rests above the major bifurcation, held in place using thick linen thread. Access to the peritoneal cavity is then gained by a midline incision of the abdomen. From here, access to the heart and lungs is achieved by cutting through the left and right sides of the ribcage and peeling back the sternum. To expose the pulmonary artery at the top of the heart, the thymus is removed. A hemostat is then used to grip the bottom of the heart, which is then displaced to the left side of the thoracic cavity. Entry to the right ventricle, and hence to the pulmonary artery, is gained by making a small snip approximately halfway along the length of the heart on the right-hand side (look for a crimpled region of heart tissue which identifies the location for catheter insertion). A 20 G catheter, attached to a 500 ml saline bag suspended approximately 1 m above the rat, is carefully inserted through the right ventricle into the pulmonary artery, and the flow of saline is started at a rate equivalent to cardiac output (40–50 ml/min). To release blood and saline after passage through the lungs, the left atrium is punctured. To aid blood removal from the lung capillaries, the lungs are inflated several times by attaching a 10 ml syringe to the catheter in the trachea and inflating the lungs with 3–4 ml volumes, being careful not to over-inflate leading to damage of the alveolar epithelium. Inflations are ceased when the lungs appear white. Following removal of the catheter from the pulmonary artery, the lungs are then carefully dissected out of the animal and placed in a sterile Petri dish.

## Lung lavage

To remove resident leukocytes and free alveolar macrophages from the airways, the lungs are lavaged with 0.9 per cent saline warmed to 37 °C. The lavage is achieved by attaching a 20 G cannula to a 10 ml syringe held in a retort stand. The end of the cannula is carefully inserted into the trachea of the lungs, and the lungs are allowed to fill with saline under the influence of gravity. After the lungs have filled, they are gently removed from the cannula and inverted to allow the saline to run freely out of the lungs. This is repeated until the lavage fluid is clear.

## Isolation and purification of ATII cells

After lavage a 20 G cannula is placed into the trachea and tied into place using thick linen thread. The cannula is then attached to a 10 ml syringe held in a retort stand. The suspended lungs are lowered into a beaker containing buffer A (133 mM NaCl, 5.2 mM KCl, 6 mM $Na_2HPO_4 \cdot 2H_2O$, 1 mM $NaH_2PO_4 \cdot 2H_2O$, 10.3 mM HEPES, 5.6 mM glucose) pH 7.4 at 37 °C. Over a 20 minute period, 80 units of elastase solution at 37 °C (2 units/ml in buffer A) is instilled into the lungs via the tracheal cannula. The elastase that is used by most investigators is aqueous porcine elastase from Worthington Biochemical Corporation (New Jersey, USA). Following this procedure, the lungs are carefully removed and placed in a sterile Petri dish to allow removal of trachea and esophagus. The lungs are then finely chopped (1 mm$^3$), and 5 ml of fetal bovine serum (FBS) (37 °C) is added to terminate elastase digestion. Following this, 10 ml of DNAse IV (250 µg/ml) (37 °C) prepared in buffer B (same as buffer A but includes 1.9 mM $CaCl_2 \cdot 2H_2O$ and 1.3 mM $MgSO_4 \cdot 7H_2O$) is then added to prevent cell clumping. The tissue homogenate is then sequentially filtered through gauze webbing and nylon membrane filters (150 µ and 30 µ pore size). The resultant filtrate is then layered on to a one-step discontinuous Percoll gradient composed of layers at 1.040 g/ml and 1.089 g/ml density made up as in Table 12.1.

Following centrifugation (250 g for 20 minutes at 4 °C in a swing-out rotor), an ATII enriched band is located at the layered interface (Figure 12.4.). To harvest this band, 75 per cent of the 1.040 g/ml layer containing debris is removed and discarded. The white fluffy ATII-containing cell band is then carefully harvested and added to 35 ml of buffer A containing DNAse IV (50 µg/ml). To pellet the AT II cells, centrifugation at 250 g for 20 minutes at 4 °C is carried out, after which the

Table 12.1 Composition of layers in one-step discontinuous Percoll gradient

|  | 1.040 g/ml layer (Light gradient) | 1.089 g/ml layer (Heavy gradient) |
| --- | --- | --- |
| Percoll | 2.72 ml | 6.49 ml |
| FBS | 50 µl | 50 µl |
| Sterile water | None | 2.5 ml |
| Sterile water with phenol red indicator | 6.28 ml | None |
| 10 × buffer A | 1 ml | 1 ml |

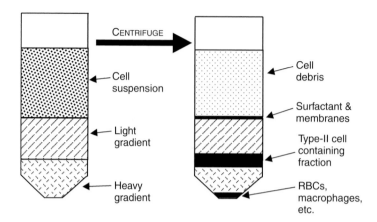

*Figure 12.4* Location of type-II cell band following density gradient centrifugation.

cells are resuspended in 10 ml of culture medium comprising of Dulbecco's modified Eagle medium (DMEM) supplemented with 10 per cent FBS and the antibiotics penicillin G (100 units/ml) and gentamicin (50 μg/ml) and the steroid hormone dexamethasone (0.1 μM). To increase the purity of the ATII isolation, differential attachment is undertaken by addition of the cell suspension to a sterile Petri dish for 1 hour at 37 °C in a humidified atmosphere (5 per cent $CO_2$/95 per cent air). Following this, the cell suspension is carefully removed from the Petri dish and placed in a 50 ml sterile centrifuge tube. After cell counting and resuspension of the cells in appropriate volumes of culture medium, ATII cells can be plated out at between 0.9 and $1.5 \times 10^6$ cells/cm$^2$ on tissue culture-treated plastic. Cultures were maintained in a humidified atmosphere (5 per cent $CO_2$/95 per cent air) and replenished with fresh media every 48 hours.

## Isolation and culture of human type-II alveolar cells

The isolation protocol is modified after the protocol of Elbert *et al.* (1999). The schematic isolation process is depicted in Figure 12.5.

### Materials

Sterile cell culture pipettes, beaker (100 ml and 400 ml), Erlenmeyer flask, Eppendorf tubes, Petri dishes (10 cm diameter), cell strainer 40 μm and 100 μm pore size (Falcon), tissue-choper, centrifuge tubes (15 ml and 50 ml), scissors and tweezers, counting chamber, gauze filters, Transwell clear Corning Costar (6 mm diameter).

### Reagents

1. Trypsin type-I (Sigma T-8003) in sterile Balanced Salt Solution A (BSSA), 150 mg aliquots.
2. Elastase LS022795, Worthington, 0.641 mg aliquots.

3. DNAse I (Sigma, D-5025, 10,000 units per aliquot).
4. FBS aliquots.
5. Amphotericin, 250 µg/ml.
6. Penicillin-Streptomycin (10,000 units/ml Penicillin, 10 mg/ml Streptomycin), 10 ml aliquots.
7. Fibronectin 1 mg/ml reconstitued in sterile water, 100 µg portions.
8. Rat tail collagen type-I in 0.01 N acetic acid, 3–5 mg/ml, store at 4 °C.
9. Magnetic CD 14 M-450 dynabeads (Dynal, Hamburg), store at 4 °C.
10. Percoll (P-1644, 100 per cent stem solution, Sigma, Deisenhofen), autoclaved, store at 4 °C.

## Buffers and media

1  BSSA: 137 mM NaCl, 5 mM KCl, 0.7 mM and 10 mM HEPES, 5.5 mM glucose, pH 7.4, Penicillin-Streptomycin 10 ml, sterile-filtrate, store up to three weeks at 4 °C.
2. Balanced Salt Solution B (BSSB): prepare according to BSSA; add $MgSO_4$ ·$7H_2O$ (0.7 mM), $CaCl_2$·$2H_2O$ (1.8 mM).
3. Small Airway Growth Medium Bulletkit (SAGM): basic medium with supplements CC-3118, prepare according to the manufacturer's recommendations.
4. DME/F12 powdered medium (Sigma): 1.2 g $NaHCO_3$ (Merck), Penicillin-Streptomycin stem solution, 5 ml gentamicin stem solution (10 mg/ml), 4.0 ml amphotericin.
5. 10×phosphate buffered saline (PBS): 1.3 M NaCl, 54 mM KCl, 110 mM glucose, 106 mM HEPES, 26 mM $Na_2HPO_4 \cdot 7H_2O$ (Merck, Darmstadt).
6. Percoll solutions
   a   low density (1.040 g/ml): 4 ml 10×PBS, 10.88 ml Percoll (sterile), 25.12 ml $H_2O$
   b   high density (1.089 g/ml): 4 ml 10×PBS, 25.96 ml Percoll (autoclaved) 10.04 ml $H_2O$. Add some neutral red to one of the Percoll solutions.
7. Washing buffer for the magnetic beads (sterile): PBS, 0.5 per cent bovine serum albumin (BSA), 2 mM ethylenediaminetetraacetic acid (EDTA).
8. Enzyme inhibition medium: 40 ml DME/F12, 10 ml FBS, DNAse 1 aliquot.
9. Adhesion medium: 22.5 ml SAGM, 22.5 ml DME/F12, 1 aliquot of DNAse.
10. Coating solution: mix on ice 10 ml SAGM, 100 µl fibronectin, 100 µl collagen-I in a centrifuge tube.

## Filter coating

Filters should be coated freshly for each isolation, add 200 µl of the coating solution on the Transwells (No. 3470, Corning Costar) and incubate the filters in the incubator for at least two hours.

## Isolation protocol

For the isolation of human cells out of patient tissue, at least 2 g of normal lungparenchyma is needed. The isolation should start on the same day of the lung resection to avoid decay of the tissue. After the lung resection the tissue should be stored in sterile Ringer-solution or DMEM. All of the isolation steps should be performed under sterile conditions under a laminar flow bank. Prepare all the components under a laminar flow bank and sterilize all equipment such as scissors, tweezers, razor blade, tissue-chopper with 70 per cent isopropanol. Document the age and the sex of the patient, weigh the piece of tissue, wet the tissue with BSSA in a sterile Petri dish and cut the tissue into small pieces. Remove the visible bronchi. Chop the tissue with the tissue-chopper into small pieces (blade cuts of 0.6 mm thickness), and wash the chopped tissue three times using 30 ml BSSA until the solution remains clear. After each wash-step, filter the material through 100 µm cell strainers to collect all the cells.

In the meantime, pre-warm the BSSB solution for the enzymatic digestion. Add the rinsed tissue into an Erlenmeyer flask containing 30 ml BSSB and add one aliquot of trypsin and elastase. Incubate the tissue in a 37 °C water-bath under moderate shaking for 40 minutes. During the digestion prepare the inhibition solution. After the enzymatic digestion the enzymes are inactivated by adding the inhibition solution and the suspension is pipetted up and down to loosen the cells out of the tissue. The cell suspension is passed through 40 µm cell strainer and spin the cells down in centrifuge tubes. The cell pellet is resuspended in 45 ml adhesion medium, and the macrophages are removed by differential adhesion on three plastic Petri dishes for 90 minutes. To prepare the Percoll gradient, 10 ml of the low-density Percoll is added in a centrifuge tube and underlaid carefully by 10 ml of the high-density Percoll. The integrity of the Percoll gardient is checked. Two seperate phases of the Percoll should be clearly visible. After the differential seeding, the cell suspension is collected into centrifuge tubes, cells are spinned down (1400 rpm, 10 minutes) and resuspended in 3 ml DME/F12 in total and the suspension is carefully added on the top of the Percoll gradient. The gradients are spinned by 1300 rpm for 20 minutes without break. Then the intermediate phase containing the alveolar cells is taken off using a pasteur pipette and the cells are transferred into centrifuge tubes containing BSSA to wash away the Percoll residues. Cells are pelleted by centrifugation and the wash-step is repeated.

In the spare time, a suitable amount of magnetic beads ($10\,\mu l/10^6$ cells) is washed twice using two times 2 ml of the washing buffer and a magnet to bind the dyna-beads. After centrifugation the cells are resuspended in the remaining supernatant of the pellet (maximum 500 µl) and the suspension is added to the beads. The suspension is rotated for 20 minutes on an overhead shaker and then the macro-phages are separated and bound to the magnetic beads by a magnet. The remaining cell suspension containing the alveolar cells is transferred into 10 ml of prewarmed SAGM and mixed well. For cell counting 50 µl are taken out. The cells in the medium are centrifuged again and the cell density is adjusted to 0.6 million cells/milliliter. If 0.33 mm inserts are used, seed 200 µl of the cell suspension per filter on the apical surface (resulting seeding density $3.6\times10^6$ cells per $cm^2$). Fill the basolateral compartment with 800 µl of SAGM. The purity of the isolation can be determined by alkaline phosphatase staining.

For the culture of human alveolar SAGM medium (Cellsystems) should be used. Feeding the cells every day starting from day two of the culture results in high TEER values. For the small filters use 200 µl medium for the apical and 800 µl for the basolateral chamber. The cells should be fed with prewarmed medium taking care not to damage the monolayer, which will result in a decline of the TEER.

## Characterization of alveolar epithelial cell cultures

At this point we will concentrate on the basic methods of characterization of alveolar cell cultures, including features which are important for drug transport studies.

The first issue is to ensure that the isolation has yielded a high purity of ATII cells. This purity is readily addressed morphologically in the first instance, and routine use of light microscopy is expected, with electron microscopy undertaken periodically on the isolations to ensure validation of the isolation technique. For rat ATII cell isolations the investigator would be looking at the following aspects of the culture under the microscope.

Upto 48 hours post-seeding, the isolated type-II cells retain a largely cuboidal shape with the presence of multivesicular bodies, apical microvilli, and numerous lamellar bodies. The cuboidal shape and lamellar bodies can be assessed under light microscopic conditions. After three days in culture the isolated rat type-II cells start to spread and flatten to fill the free space of the culture well, and the number of cells possessing lamellar bodies reduces. Lamellar bodies are even less evident by day five in culture. By days seven and eight post-seeding, the appearance of the cultures correspond to the morphology expected of type-I cells rather than ATII cells, with a more squamous and elongated thin cytoplasm.

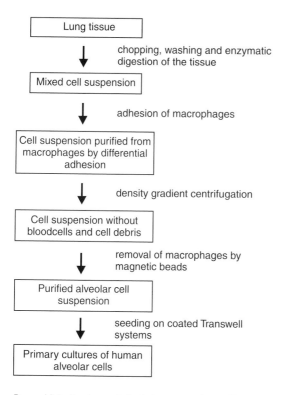

*Figure 12.5* A schematic isolation procedure of human alveolar cells.

## Tannic acid staining of lamellar bodies in rat ATII cells

For the rat cultures tannic acid status can be undertaken in the alveolar type-II cells to enable constrast staining of the lamellar bodies. Cells are bathed in PBS (pH 7.4)×3 to remove excess culture media, and then fixed with 1.5 per cent glutaraldehyde in PBS for 15 minutes. The cultured cells are then bathed twice with PBS to remove excess glutaraldehyde prior to post-fixation with 1 per cent osmium tetroxide in PBS for 1.5 hours. Following fixation, the cells are again bathed twice with PBS and then incubated overnight in freshly prepared 1 per cent tannic acid in PBS (pH 6.8). The cells are then bathed twice with PBS and twice with water before examination under the light microscopy (Mason *et al.*, 1985). The purity of type-II cell isolation can be assessed by the number of the cells containing blackened (tannic acid stained) surfactant granules within the cytoplasm. The purity of the cultures for rat ATII cells should be equal to or greater than 95 per cent. This technique is not applicable to human isolations.

## Differentiation markers

Cytokeratins belong to a group of intermediate filament proteins commonly used as markers for epithelial origin depending on the source of the tissue and the differentiation state. In the lung, a set of cytokeratines 8, 18, 19 has been described (Paine *et al.*, 1998), which are selectively, but not specifically, expressed depending on the state of differentiation of the alveolar cells. For rat alveolar epithelial cells, a shift from cytokeratin 19 to cytokeratin 18 has been seen in the early stages of type-II to type-I-'like' cell differentiation.

Differentiation of alveolar cells may also be monitored by their differential affinity for the lectins RCA (ricinus communis agglutinin) and MPA (maclura promifera agglutinin). Type-II cells bind the MPA, while type-I cells display a higher affinity to the RCA. The lectins can be coupled with fluorescent dyes to allow fluorescent microscopy or flow cytometry analysis. The production and secretion of surfactant proteins A, B, C, D by alveolar type-II cells can also be used as markers to the alveolar cells. The marker most specific to type-II alveolar cells is surfactant protein C (Kalina *et al.*, 1992). In contrast to other surfactant proteins it shows the highest specifity to alveolar type-II cells. Another surfactant protein SP-D (Madsen *et al.*, 2000), a collectin, is not only restricted to type-II cells but also expressed by alveolar macrophages or Clara cells in the lung and also in other mucosal tissues like the gastric mucosa in rats. Immunoreactivity for surfactant proteins A is also found in type-II and in Clara cells. Other proteins used as type-I cell marker, especially in rat cells, are the water channel proteins aquaporin 5 and T1-alpha (Nielsen *et al.*, 1997; Dobbs *et al.*, 1988).

## Competancy of tight junctional complex (ZO)

For transepithelial transport investigations the alveolar epithelial cultures need to generate a restrictive paracellular pathway. This pathway is established primarily and most readily by measurement of the TEER across the cultured monolayers.

The TEER is expressed in $\Omega\,cm^2$ and measured using an EVOM™ (World Precision Instruments, Berlin) meter and either chopstick-electrodes or ENDOhm™ chambers. The TEER provides an initial estimate of the paracellular restrictiveness of the alveolar

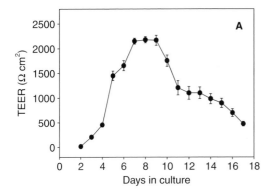

*Figure 12.6* Development of transepithelial electrical resistance (TEER) values in human alveolar cells with ongoing culture.

cell preparation, but additional solute permeability data for paracellular probes such as mannitol also need to be undertaken. The resistance can be measured after day two of culture, where the resistance is always measured immediately before a change of culture medium and not immediately after. The measured resistance in ohms needs to be corrected for background resistance using blank filters (coated as appropriate).

For rat cells the TEER should reach $>1000\,\Omega\,cm^2$ by day five–six of culture and remain greater, $500\,\Omega\,cm^2$, through to day eight–nine post-seeding. For human cells a plateau in TEER should be achieved after the sixth or seventh day. The TEER plateau maintains only for a short time period (two–three days) within which transport experiments should be carried out. To use the cells for transport studies the TEER values should be at least $1000\,\Omega\,cm^2$, which corresponds to a resistance value of about $3000\,\Omega$ on a filter with an area of $0.33\,cm^2$. In the same way the activity of ionic transporters can be determined by measuring the potential difference caused by different distribution of ionic transporters in apical and basolateral compartments of the cells. A time course of the TEER development in a human alveolar cell cultures is shown in Figure 12.6.

Together with electrical measurements and solute permeability data, the competance of the tight junctional complexes within the alveolar cultures can be assessed by staining for tight junctional proteins; for example, staining for structural proteins such as ZO-1, ZO-2, occludin or claudins. Commonly used methods to investigate the expression of such proteins are immunofluorescence microscopy or Western-blot. An immunofluorescence staining of the ZO-1 protein in human alveolar cells is depicted in Figure 12.7.

## SPECIFIC APPLICATIONS

### Transport studies in primary alveolar cells

The ability of isolated primary ATII cells to form a polarized monolayer and acquire characteristics of the *in vivo* ATI cell when grown on permeable membrane

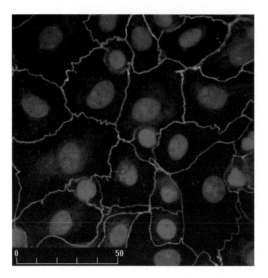

*Figure 12.7* Zona occluden (ZO)-1 staining on human alveolar cell culture, cell nuclei are stained by propidium iodide depicted in red, the tight junction protein ZO-1 is depicted in green, bar 50 µm. (*See Color plate 6*)

supports has allowed various research groups to study vectorial electrolyte and drug transport processes present in alveolar epithelium. Mason *et al.* (1982) was the first to describe the active transport of sodium in cultured ATII cells. Transport was found to occur predominantly from the apical to basolateral surface and is consistent with the concept that the alveolar epithelium has a direct role in keeping the alveoli relatively dry of fluid. Later reports (Goodman *et al.*, 1984) suggest that the active transport of sodium can be modulated by endogenous and exogenous chemical agents that exert their effect by elevating cAMP levels. The studies of Kim *et al.* (1991, 1992) investigated the mechanisms of active ion flux across alveolar epithelial monolayers and presented evidence to suggest that the active transport of both sodium and chloride ions is mediated by a $Na^+,K^+$-ATPase pump located in the basolateral membrane.

As previously mentioned, primary cultured rat ATII epithelial cell monolayers are the most useful *in vitro* system for the evaluation of mechanisms and pathways of alveolar solute and drug transport. Investigations using $\beta$-adrenergic drugs have revealed how the lipophilicity of a model drug affects the drug transport (Saha *et al.*, 1994). The influence by molecular size of solutes and permeability of the alveolar monolayers have been studied. More significantly, however, due to the recent exploitation of the pulmonary route for the systemic delivery of genetically engineered macromolecules, significant interest has centered on the permeability characteristics and transport rates of peptide and protein drugs across such monolayers.

The paracellular transport of peptides and the influence of their molecular structures have recently been described by Dodoo *et al.* (2000). A recent report of

Matsukawa *et al.* (1996a) evaluated the flux of five select radiolabeled proteins of varying molecular weight across cultured rat alveolar epithelial cell monolayers. The transport rates of all proteins investigated were rapid and found to be asymmetric. For example, the apparent permeability of $^{14}$C-BSA (molecular weight 67,000) was $0.768 \times 10^{-7}$ cm/sec and $0.39 \times 10^{-7}$ cm/sec in the apical-to-basolateral (A→B) and basolateral-to-apical (B→A) directions, respectively, while the apparent permeability of $^{14}$C-ovalbumin (molecular weight ~ 43,000) was found to be $1.09 \times 10^{-7}$ cm/sec in the A→B direction and $0.58 \times 10^{-7}$ cm/sec in the B→A direction. The results demonstrated that the rates of protein transport across alveolar epithelium are much faster than those predicted for simple restricted diffusion. This led the authors to speculate that the probable route of transport was via receptor-mediated transcytotic pathways. Indeed, a number of apically located albumin binding sites have been identified in cultured rat ATII cells (Kim *et al.*, 1995). The work of Deshpande *et al.* (1994) studied the endocytosis and transcytosis of Horseradish peroxidase (HRP)-transferrin conjugates, which also provided evidence of receptor-mediated trafficking pathways in the above cells.

Two studies undertaken by Morimoto *et al.* (1993) and Meredith and Boyd (1995) strongly suggest the presence of a proton-coupled transport protein in the apical surface of cultured rat ATII cells. Apart from facilitating the transport of dipeptides, such a protein transporter is of pharmacological interest since it may provide a transport route for other biologically active peptides, such as the anticancer agent bestatin and the angiotensin-converting enzyme inhibitor captopril.

Although the pulmonary route is generally considered to have a low protein metabolizing capacity, several reports indicate that cultured rat ATII cells do possess a significant enzymatic barrier to certain important pharmacological peptides. Degradation of insulin (Yamahara *et al.*, 1994) and two enkephalin peptides, Met-enkephalin and enkephalinamide (Wang *et al.*, 1993), have been observed in transport studies utilizing the above model. Furthermore, Forbes *et al.* (1999) followed the temporal expression of various peptidases as the cells progressed from ATII to ATI phenotype. The enzyme activity of aminopeptidase N was higher at day seven (ATI-like) than at day two, whereas there was a twofold reduction in dipeptidyl peptidase IV activity which was accompanied with a complete loss in activity of the angiotensin-converting enzyme. However, the transport of certain polypeptides, such as intact thyrotropin-releasing hormone across rat alveolar epithelial cell monolayers, is shown to be largely unaffected by peptidase activity (Morimoto *et al.*, 1994). These results have implications for the use of cultured ATII cells for the study of protein absorption across alveolar epithelium.

Cultures of rat alveolar epithelial cell monolayers have also been utilized to investigate the pinocytic capacity of the alveolar epithelial barrier in relation to the transport of therapeutic proteins. Using non-specific fluid phase markers, Matsukawa *et al.* (1996b) found the pinocytic activity across rat ATII monolayers to be symmetrical. The permeability coefficients of HRP in both the A→B and B→A directions were calculated to be approximately $7.0 \times 10^{-9}$ cm/sec. At $4 \,^{\circ}C$ this was decreased by 70 per cent suggesting that translocation of HRP across the alveolar epithelial barrier does not take place via paracellular routes but most probably via non-saturable vesicular pathways (pinocytosis).

The two vesicular transport systems that may be functional in alveolar epithelium are the clathrin coated pits and the caveolae. Despite much speculation upon the role of alveolar epithelial vesicles serving as a transcytotic pathway for the absorption of therapeutic proteins (reviewed in Gumbleton *et al.*, 2000; Gumbleton, 2001), very limited supporting experimental data in organized lung tissue are available. A key structural protein for caveolae is caveolin, and recently it has been shown in rat ATI-like cell monolayers that the expression of caveolin-1 and the biogenesis of caveolae is evident as a function of the transdifferentiation process, and indeed such expression and caveolae biogenesis is upregulated during the transdifferentiation process by the presence of dexamethasone in the culture media (Hollins *et al.*, 1999).

## Biomedical and toxicological aspects

The integrity of the alveolar barrier can be disturbed by cell death caused by hypoxia in a wide range of diseases. Cell culture models can help to elucidate how the cells can maintain their functionality under hypoxic conditions. Changes in the transport characteristics of, e.g. impairment of cations transport or upregulation (Mairbäurl *et al.*, 1997) of glucose transporters (Ouddir, 1999) and other transport proteins (Clerici and Matthay, 2000), have been already shown during hypoxia using cell culture models. Cell lines like A549 are commonly used in such investigations, but due to their missing ability to differentiate into type-I cells, primary cells are still needed to check effects of hypoxia on both alveolar cell types. Furthermore, the injury of the alveolar cells during lung transplantation by ischemia and reperfusion is currently under investigation (Cardella *et al.*, 2000). Cell culture models may help to understand mechanisms like apoptosis or necrosis of lung injury and to develop suitable prevention methods. In summary, alveolar cell cultures also offer, in addition to pharmaceutical applications, useful tools in the investigation of biomedical aspects.

## REFERENCES

Asano, K., Chee, C. B. E. *et al.* (1994) Constitutive and inducible nitric oxide synthase gene expression, regulation and activity in human lung epithelial cells. *Proc. Natl. Acad. Sci. USA.*, **91**, 10089–10093.

Basset, G., Crone, C. and Saumon, G. (1987) Significance of active ion transport in trans-alveolar water absorption: a study on isolated rat lung. *J. Physiol.*, **384**, 311–324.

Bastacky, J., Lee, C. Y., Goerke, J., Koushafar, H., Yager, D., Kenaga, L. *et al.* (1995) Alveolar lining layer is thin and continuous: low temperature scanning electron microscopy of rat lung. *J. Appl. Physiol.*, **79**, 1615.

Byron, P. R. and Patton, J. S. (1994) Drug delivery via the respiratory tract. *J. Aerosol. Med.*, **7**, 49–75.

Byron, P. R., Sun, Z., Katayama, H. and Rypacek, F. (1994) Solute absorption from the airways of the isolated rat lung. IV. Mechanisms of absorption of fluorophore-labeled poly-alpha, beta-[N(2-hydroxyethyl)-DL-aspartamide]. *Pharm. Res.*, **11**, 221–225.

Cardella, S., Keshavjee, E., Mourgeon, S. D. *et al.* (2000) A novel cell culture model for studying ischemia-reperfusiom injury in lung transplantation. *J. Appl. Physiol.*, **89**(4), 1553–1560.

Cheek, J. M., Evans, M. J. and Crandell, E. D. (1989) Type-I cell-like morphology in tight alveolar epithelial monolayers. *Exp. Cell Res.*, **184**, 375–387.

Clerici, C. and Matthay, M. A. (2000) Hypoxia regulates gene expression of alveolar epithelial transport proteins. *J. Appl. Physiol.*, **88**(5), 1890–1896.

Colthorpe, P., Farr, S. J., Taylor, G., Smith, I. J. and Wyatt, D. (1992) The pharmacokinetics of pulmonary delivered insulin: a comparison of intra-tracheal and aerosol administration to the rabbit. *Pharm. Res.*, **9**, 765–768.

Colthorpe, P., Farr, S. J., Smith, I. J., Wyatt, D. and Taylor, G. (1995) The influence of regional deposition on the pharmacokinetics of pulmonary-delivered human growth hormone in rabbits. *Pharm. Res.*, **12**, 356–359.

Crandall, D., Heming, T. A., Palombo, R. L. and Goodman, B. E. (1986) Effects of terbutaline on sodium transport in isolated perfused rat lung. *J. Appl. Physiol.*, **60**, 289–294.

Crapo, J. D., Barry, B. E., Gehr, P., Bachofen, M. and Weibel, E. R. (1982) Cell number and cell characteristics of normal human lung. *Am. Rev. Respir. Dis.*, **125**, 332–337.

Danto, S. I., Zabski, S. M. and Crandell, E. D. (1992) Reactivity of alveolar epithelial cells in primary culture with type-I cell monoclonal antibodies. *Am. J. Respir. Cell Mol. Biol.*, **6**, 296–306.

Deshpande, D., Velasquez, D. T., Wang, L. Y., Malanga, C. J., Ma. J. K. H. and Rojanasakul, Y. (1994) Receptor-mediated peptide delivery in pulmonary epithelial monolayers. *Pharm. Res.*, **11**, 1121–1126.

Discroll, K. E., Carter, J. M. *et al.* (1995) Establishment of immortalized alveolar type-II epithelial cell lines from adult rats. *In Vitro Cell: Dev. Biol.*, **31**, 516–527.

Dobbs, L. G., Williams, M. C. and Gonzales, R. (1988) Monoclonal antibodies specific to apical surfaces of rat alveolar type-I cells bind to surfaces of cultured, but not to freshly isolated, type-II cells. *Biochim. Biophys. Acta.*, **970**, 146–156.

Dodoo, A. N., Bansal, S., Barlow, D. J. *et al.* (2000) Systematic investigations of the influence of molecular structures on the tranport of peptides across cultured alveolar monolayers. *Pharm. Res.*, **17**, 7–14.

Edwards, D. A., Ben-Jebria, A. and Langer, R. (1998) Recent advances in pulmonary drug delivery using large, porous inhaled particles. *J. Appl. Physiol.*, **84**, 379–385.

Effros, R. M., Mason, G. R., Hukkanen, J. and Silverman, P. (1988) New evidence for active sodium transport from fluid-filled rat lungs. *J. Appl. Physiol.*, **66**, 906–919.

Elbert, K. J., Schäfer, U. F., Kim, K.-J., Lee, V. H. L. and Lehr, C.-M. (1999) Monolayers of human alveolar epithelial cells in primary culture for pulmonary drug absorption and Transportstudies. *Pharm. Res.*, **16**, 601–608.

Fisher, A. B. (1985) The isolated perfused lung. In (H. P. Witschi and J. D. Brain eds), 'Handbook of Experimental Pharmacology', Vol. 75, pp. 149–173.

Forbes, B. J., Wilson, C. G. and Gumbleton, M. (1995) Extraction of peptidase substrates by the isolated perfused rat lung. *Pharm. Sci.*, **1**, 569–572.

Forbes, B. J., Wilson, C. G. and Gumbleton, M. (1999) Temporal dependence of ectopeptidase expression in alveolar epithelial cell culture: implications for study of peptide absorption. *Int. J. Pharm.*, **180**, 225–234.

Gehr, P., Bachofen, M. and Weibel, E. R. (1978) The normal human lung: ultrastructure and morphometric estimation of diffusion capacity. *Respir. Physiol.*, **32**, 121–140.

Giard, D. J., Aaronson, S. A. *et al.* (1973) *In vitro* cultivation of human tumors: establishment of cell lines derived from a series of solid tumors. *J. Natl. Cancer Inst.*, **51**, 1417–1423.

Godfrey, R. W. A. (1997) Human airway epithelial tight junctions. *Microsc. Res. Techn.*, **38**, 488–499.

Goodman, B. E., Brown, S. E. S. and Crandell, E. D. (1984) Regulation of transport across pulmonary alveolar epithelial cell monolayers. *J. Appl. Physiol. Respir. Environ. Exercise Physiol.*, **57**, 703–710.

Gumbleton, M. (2001) Caveolae as potential macromolecule trafficking compartments within alveolar epithelium. In M. Gumbleton (ed), Caveolae mediated membrane transport', *Adv. Drug Deliv. Rev.*, theme issue, summer 2001.

Gumbleton, M., Abulrob, A. G. and Campbell, L. (2000) Caveolae: an alternative membrane transport compartment. *Pharm. Res.*, **17**, 1035–1048.

Haies, D. M., Gil, J. and Weibel, E. R. (1981) Morphometric study of rat lung cells: numerical and dimensional characteristics of parenchymal cell populations. *Am. Rev. Respir. Dis.*, **123**, 533–541.

Hollins, A. J., El-Aid, A., Campbell, L. and Gumbleton, M. (1999) Effect of glucocorticoid upon cell differentiation within an alveolar epithelial cell culture used to model pulmonary absorption in-vitro. *Pharm. Sci.*, **1**, (abtsract suppl.) 2629, S-257.

Jones, H. E., Blundell, G. K., Wyatt, I., John, R. A., Farr, S. J. and Richards, R. J. (1992) The accumulation of pentamidine into rat lung slices and its interaction with putrescine. *Biochem. Pharmacol.*, **43**, 431–437.

Kalina, M., Mason, R. J. and Shannon, J. M. (1992) Surfactant protein C is expressed in alveolar type-II cells but not in Clara cells of rat lung. *Am. J. Respir. Cell Mol. Biol.*, **6**(6), 594–600.

Kim, K. J. and Crandall, E. D. (1983) Heteropore populations of bullfrog epithelium. *J. Appl. Physiol.*, **54**, 140–149.

Kim, K. J., LeBon, R. R., Shinbane, J. S. and Crandall, E. D. (1985) Asymmetric 14C-albumin transport across bullfrog epithelium. *J. Appl. Physiol.*, **59**, 1290–1297.

Kim, K.J. (1990) Active $Na^+$ transport across Xenoplus lung alveolar epithelium. *Respir. Physiol.*, **81**(1) 29–39.

Kim, K. J., Cheek, J. M. and Crandell, E. D. (1991) Contribution of active $Na^+$ and Cl fluxes to net ion transport by alveolar epithelium. *Respir. Physiol.*, **85**, 245–256.

Kim, K. J., Suh, D. K., Lubman, R. L., Danto, S. P., Borok, Z. and Crandell, E. D. (1992) Studies on the mechanisms of active ion fluxes across alveolar epithelial cell monolayers. *J. Tissue Cult. Methods*, **14**, 187–194.

Kim, K. J., Ratton, V., Oh, P., Schnitzer, J. E., Kalra, V. K. and Crandell, E. (1995) Specific albumin-binding protein in alveolar epithelial cell monolayers. *Am. J. Respir. Crit. Care Med.*, **151**, A190.

Mackein C. C. (1955) *Acta Anat.*, **23**, 449–463.

Madsen, J., Kliem, A. *et al.* (2000) Localization of lung surfactant protein D on mucosal surfaces in human tissues. *J. Immunol.*, **164**(11), 5866–5870.

Mairbäurl, H., Wodopia, R., Eckes, R. S. *et al.* (1997) Impairment of cation transport in A549 cells and rat alveolar epithelial cells by hypoxia. *Am. J. Physiol.*, **273**(4 Pt 1), L797–L806.

Mason, R. J., Walker, S. R., Shields, B. A. and Henson, J. A. (1985) Identification of rat alveolar type-II epithelial cells with a tannic acid and polychrome stain. *Am. Rev. Respir. Dis.*, **131**, 786–788.

Mason, R. J., Williams, M. C., Widdicombe, J. H., Sanders, M. J., Misfield, D. S. and Berry, L. C. (1982) Transepithelial transport by pulmonary alveolar type-II cells in primary culture. *Proc. Natl. Acad. Sci. USA*, **79**, 6033–6037.

Mathias, N. R., Yamashita, F. and Lee, V. H. L. (1996) Respiratory epithelial cell culture models for evaluation of ion and drug transport. *Adv. Drug Deliv. Rev.*, **22**, 215–249.

Matsui, R., Goldstein, R. H. *et al.* (1994) Type-I collagen formation in rat type-II alveolar cells immortalized by viral gene products. *Thorax*, **49**, 201–206.

Matsukawa, Y., Yamahara, H., Lee, V. H. L., Crandell, E. D. and Kim, K. J. (1996a) Rates of protein transport across rat alveolar epithelial cell monolayers. *Proc. Int. Symp. Control Rel. Bioact. Mater.*, **23**, 491–492.

Matsukawa, Y., Yamahara, H., Lee, V. H. L., Crandell, E. D. and Kim, K. J. (1996b) Horse-radish peroxidase transport across rat alveolar epithelial cell monolayers. *Pharm. Res.*, **13**, 1331–1335.

Meban, C. (1973) The pneumonocytes in the lung of Xenopus lavevis. *J. Anat.*, **114**, 235–244.

Meredith, D. and Boyd, A. C. R. (1995) Dipeptide transport characteristics of the apical membrane of rat lung type-II pnemonocytes. Am. *J. Physiol.*, **269** (*Lung Cell Mol. Physiol. 13*), L137–L147.

Morimoto, K., Yamahara, H., Lee, V. H. L. and Kim, K. J. (1993) Dipeptide transport across rat alveolar epithelial cell monolayers. *Pharm. Res.*, **10**, 1668–1674.

Morimoto, K., Yamahara, H., Lee, V. H. L. and Kim, K. J. (1994) Transport of thyrotropin-releasing hormone across rat alveolar epithelial cell monolayers. *Life Sci.*, **54**, 2083–2092.

Newman, G. R., Campbell, L., von Ruhland, C., Jasani, B. and Gumbleton, M. (1999) Caveolin and its cellular and subcellular immunolocalisation in lung alveolar epithelium: implications for alveolar type-I cell function. *Cell Tissue Res.*, **295**, 111–120.

Nielsen, S., King, L. S. *et al.* (1997) Aquaporins in complex tisues: II. Subcellular distribution in respiratory and glandular tissues of rat. *Am. J. Physiol.*, **42**, C1549–C1561.

Niemeier, W. R. and Bingham E. (1972) An isolated perfused lung preparation for metabolic studies. *Life Sci.*, **11**, 807–820.

Niven, R. W. and Byron, P. R. (1988) Solute absorption from the airways of the isolated rat lung. I. The use of absorption data to quantify drug dissolution or release in the respiratory tract. *Pharm. Res.*, **9**, 574–579.

Niven, R. W. and Byron, P. R. (1990) Solute absorption from the airways of the isolated rat lung. II. Effect of surfactants on absorption of fluorescein. *Pharm. Res.*, **7**, 8–13.

Niven, R. W., Rypacek, F. and Byron, P. R. (1990) Solute absorption from the airways of the isolated rat lung. III. Absorption of several peptidase-resistant, synthetic polypeptides: poly-(2-hydroxyethyl)-apartamides. *Pharm. Res.*, **7**, 990–994.

Okumura, S., Tanaka, H., Shinsako, K., Ito, M., Yamamoto, A. and Muranishi, S. (1997) Evaluation of drug absorption after intrapulmonary administration using *Xenopus* pulmonary membranes: correction with in vivo pulmonary absorption studies in rats. *Pharm. Res.*, **14**(9), 1282–1285.

Ouddir, A., Planes, C., Fernandes, I., Van Hesse, A. and Clerici, C. (1999) Hypoxia upregulates activity and expression of the glucose transporter GLUT1 in alveolar epithelial cells. *Am. J. Cell. Mol. Biol.*, **21**, 710–718.

Paine, R., Zeev, A. B., Farmer, S. and Brody, J. (1998) The pattern of cytokeratin synthesis is a marker of type-II cell differentiation in adult and maturing fetal lung alveolar cells. *Dev. Biol.*, **129**, 505–515.

Pasternack, M., Floerchinger, C. S. and Hunninghake, G. W. (1996) E1A-induced immortalization of rat type-II alveolar epithelial cells. *Exp. Lung Res.*, **22**, 525–539.

Patton, J. S. (1996) Mechanisms of macromolecule absorption by the lungs. *Adv. Drug Deliv. Rev.*, **19**, 3–36.

Patton, J. S. (1998) Breathing life into protein drugs. *Nature Biotechnol.*, **16**, 141–143.

Patton, J. S., Bukar, J. and Nagarajan, S. (1999) Inhaled Insulin. *Adv. Drug Deliv. Rev.*, **35**, 235–247.

Rahman, J., Bel, A., Mulier, B. *et al.* (1996) Transcriptional regulation of gamma-glutamyl-cystein synthetase-heavy subunit by oxidants in human alveolar epithelial cells. *Biochem. Biophys. Res. Commun.*, **229**, 832–837.

Richards, R. J., Davies, N., Atkins, J. and Oreffo, V. I. C. (1987) Isolation, biochemical characterization, and culture of lung type-II cells of the rat. *Lung*, **165**, 143–158.

Robbins, R. A., Barnes, P. J. *et al.* (1994) Expression of inducible nitric oxide in human lung epithelial cells. *Biochem. Biophys. Res. Comm.*, **2**, 209–218.

Saha, P., Kim, K.-J., Yahamara, H. *et al.* (1994) Influence of lipophilicity on b-blocker permeation across rat alveolar epithelial cell monolayers. *J. Appl. Physiol.*, **54**, 140–146.

Saldias, F. J., Comellas, A., Guerrero, C. *et al.* (1998) Time course of active and passive liquid and solute movement in the isolated perfused rat lung model. *J. Appl. Physiol.*, **85**(4), 1572–1577.

Schneeberger, E. E. (1977) The integrity of the air-blood barrier. In J. D. Brain, D. F. Proctor and L. M. Reid (eds), Respiratory defense mechanisms, Dekker Inc. New York, pp. 687–708.

Schnitzer, J. E., Oh, P., Pinney, E. and Allard, J. (1994) Fillipin-sensitive caveolae-mediated transport in endothelium: reduced transcytosis, scavenger endocytosis and capillary permeability of select macromolecules. *J. Cell. Biol.*, **127**, 1217–1232.

Skyler, J. S., Cefalu, W. T., Kourides, I. A., Landschulz, W. H., Balagtas, C. C. and Gelfand, R. A. (2001) Efficacy of inhaled human insulin in type-I diabetes mellitus: a randomised proof-of-concept study. *Lancet*, **357**, 331–335.

Smith, P. L. (1997) Peptide delivery via the pulmonary rout: a valid approach for local and systemic delivery. *J. Control. Rel.*, **46**, 99–106.

Uhal, B. D. (1997) Cell cycle kinetics in the alveolar epithelium. *Am. J. Physiol.*, **272**, L1031–L1045.

Wall, D. A. (1995) Pulmonary absorption of peptides and proteins. *Drug Deliv.*, **2**, 1–20.

Wall, D. and Pierdomenico, D. (1996) Drug transport across *Xenopus* alveolar membranes *in vitro*. In R. T. Borchardt (ed), Models for assessing drug absorption and metabolism, Plenum Press, New York, 1996.

Wang, L. Y., Toledo-Velasquez, D., Schwegler-Berry, D., Ma, J. K. H. and Rojanasakul, Y. (1993) Transport and hydrolysis of enkephalins in cultured alveolar epithelial monolayers. *Pharm. Res.* **11**, 1662–1667.

Weibel, E. R. (1963) Morphometry of human lung Academic Press, New York.

Yamahara, H., Lehr, C.-M., Lee, V. H. L. and Kim, K. J. (1994) Fate of insulin during transit across rat alveolar epithelial cell monolayers. *Eur. J. of Biopharm.*, **40**, 294–298.

Yu, J. and Chien, Y. W. (1997) Pulmonary drug delivery: physiologic and mechanistic aspects. *Crit. Rev. Ther. Drug Carrier Sys.*, **14**, 395–453.

# Chapter 13

# Bronchial epithelial cell cultures

*Claire Meaney, Bogdan I. Florea, Carsten Ehrhardt,*
*Ulrich F. Schäfer, Claus-Michael Lehr,*
*Hans E. Junginger and Gerrit Borchard*

## INTRODUCTION

### Physiology of the bronchial epithelium

The conducting airways consist of the trachea, bronchi (contain cartilage), and bronchioles (lack cartilage). The composition of the conducting airways includes an epithelial lining, basement membrane (basal lamina; 80–90 nm, rich in laminin and type-IV collagen), subepithelial connective tissue (submucosa or lamina propria), smooth muscle, and adventitia (Kuhn, 1988). The epithelium of the conducting airways varies from a pseudo-stratified columnar type, mainly consisting of three cell types (ciliated, basal, and secretory cells), interconnected by tight junctions in the proximal bronchi, to a progressively more cuboidal, non-ciliated epithelium in the distal bronchioles. The thickness of the epithelium decreases from approximately 60 μm in the trachea to 10 μm in the bronchioles. The total surface area in humans averages at about 2.5 m$^2$, and during breathing the airways contain merely 4 per cent of the total volume of air inhaled.

Bronchial glands and cartilage progressively decreases, while the proportion of smooth muscles increases from proximal to distal airways. The smooth muscle layer in the trachea and mainstem bronchi is discontinuous and occupies a mere 5 per cent of the total wall thickness, while in bronchioles a spiral network of smooth muscles exist, which completely encircles the airways. This network occupies about 20 per cent of the total wall thickness. Ciliated cells (20–60 μm tall) make up about 50 per cent of the epithelial surface. With an average number of 250 cilia on the apical surface, their major function is the propulsion of mucus in the tracheal direction and out of the lung. Mucus is a mixture of mucus gland, goblet cell and epithelial cell secretions, mainly produced by secretory (mucus, goblet or Clara) cells. These cells can be characterized by electron dense secretory granules in the cytoplasm. Secretory cells are mainly present in the distal airways, but they can also be found in the larger airways. Basal cells found in the airway epithelium are of a flattened or spindle shape, and are thought to be the progenitor cells of ciliated and Clara cells (Mariassy, 1992).

Drug transport at tracheal/bronchial epithelium is mediated via passive paracellular diffusion (Mathias *et al.*, 1996), transcytosis in intracellular vesicles (Richardson *et al.*, 1976), or specific receptor binding (Curiel *et al.*, 1992; Ferkol *et al.*, 1993). The paracellular permeability ($P_{app}$), assessed as transepithelial electrical resistance

(TEER) or flux of marker substances, is regulated by tight junction opening under treatment with anesthetic ether (Richardson *et al.*, 1976), cytochalasin B (Ma *et al.*, 1993) and poly-L-lysine (Ferkol *et al.*, 1993). In gene delivery (e.g. the gene encoding the cystic fibrosis transmembrane regulator protein in cystic fibrosis), receptor-mediated transport via the transferrin (Curiel *et al.*, 1992) and the polymeric immunoglobulin receptor (pIgR) (Ferkol *et al.*, 1993) has been reported.

Most of the studies mentioned here were performed *in vivo* by instillation or aerosolization, often without distinction between drug absorption at the alveolar or tracheal epithelium. In order to examine to what extent the tracheal/bronchial epithelium contributes to total drug absorption, cell culture models of the tracheal epithelium should be employed. However, very few *in vitro* studies on absorption mechanisms, permeation pathways, and individual absorption sites using cultured bronchial/ tracheal cells are to be found in literature to date.

## *In vitro* models of the bronchial epithelium

Data on the $P_{app}$ of drugs at the airway epithelium from *in vivo* studies are scarce, because of the inaccessibility and the delicate structure of the lung. In order to predict absorption *in vivo*, cell cultures' modeling absorptive epithelia were shown to be useful tools. Only recently, some drug transport studies involving airway epithelial cell cultures appeared in literature (Forbes *et al.*, 1998). These studies involved both established cell lines, such as human 16HBE14o- (Forbes *et al.*, 1998) and Calu-3 (Cavet *et al.*, 1997; Meaney *et al.*, 1999), and primary cell cultures derived from a number of species, such as rabbits (Mathias *et al.*, 1996), guinea-pig (Adler *et al.*, 1990), rat (Clark *et al.*, 1995), hamster (Lee *et al.*, 1984) and human (Gruenert *et al.*, 1995). While cell lines do not afford painstaking isolation procedures and suffer from occasionally low availability, they may not present features (receptor expression, gene products, enzyme production, etc.) of the epithelium *in vivo*. Primary cell cultures of the airway epithelium were shown to better resemble the *in vivo* situation, featuring tight cell layers, apical cilia, and mucus production under certain culture conditions (Mathias *et al.*, 1996). However, as primary cultures are prone to infection with bacteria or mycoplasms, specific pathogen-free animals should be the preferable source of material for isolation of primary cells.

In the following section, the culture of two cell lines, Calu-3 (Fogh *et al.*, 1977) and 16HBE14o- (Cozens *et al.*, 1994), will be described. Also, the primary culture of porcine tracheal epithelial cells as models for the bronchial epithelium will be examined.

## CALU-3: A HUMAN SUBMUCOSAL GLAND CELL LINE

Calu-3 cells, a human submucosal gland cell line, were obtained from the American Type Culture Collection (HTB-55; ATCC, Manassas, USA). Under the culture conditions described herein, Calu-3 cells grow in tight monolayers (Figure 13.1) develop apical cilia (Figure 13.2) and produce mucus (Figure 13.3). Next to the culture methods, measurement of Transepithelial electrical resistance (TEER),

(A)

(B)

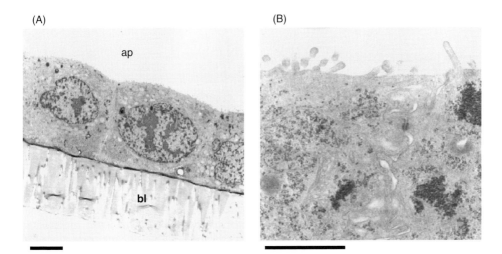

*Figure 13.1* Transmission electron micrographs of Calu-3 cells. Calu-3 cells grow as tight monolayers of columnar-shaped cells (A bar indicates 10 μm). Tight junctional complexes (tj) and desmosomes (D) can be observed at larger magnifications (B bar indicates 5 μm). ap: apical membrane, bl: basolateral membrane.

(A)

(B)

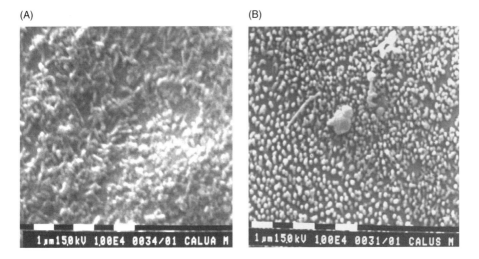

*Figure 13.2* Scanning electron micrograph of Calu-3 cells grown at an air interface (A), and under submerged conditions (B). One bar on the bottom of each of the micrographs indicates 1 μm length.

measurement of the flux of a hydrophilic marker compound, electron microscopy techniques and specific mucus staining are described as quality controls for the obtained culture. All cell culture compounds were purchased from Sigma-Aldrich Chemie B. V. (Zwijndrecht, The Netherlands), if not stated otherwise.

(A)

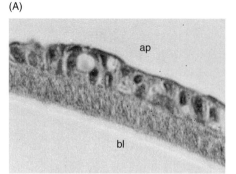

(B)

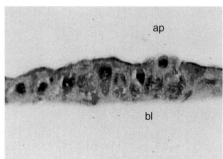

*Figure 13.3* Staining of mucus produced by Calu-3 cells cultured under submerged conditions (A) and at an air interface (B). Staining was done with periodic Schiff's reagent and counterstained with Alcian blue after fixation in paraformaldehyde and embedding in paraffin. Magnification 40×, ap: apical membrane, bl: basolateral membrane. (*See Color plate 7*)

## Materials needed

- Vitrogen 3 mg/ml (Cohesion, Palo Alto, USA)
- 0.25 per cent trypsin ethylenechamine tetraacetic acid (EDTA) solution
- Sterile pipettes 5 and 10 ml
- Sterile Pasteur pipettes
- 50 ml sterile centrifugation tubes
- Culture flasks 75 cm$^2$ (Costar, Schiphol, The Netherlands)
- Transwell® 24 mm diameter (Costar)
- Dulbecco's minimal Eagle's Medium (DMEM) supplemented with 10 per cent fetal calf serum (FCS), benzyl-penicillin G (160 U/ml) and streptomycin sulfate (100 µg/ml)
- Sterile phosphate buffered saline (PBS, pH 7.4)
- 0.1 per cent (Weight/Volume, w/v) Trypan blue solution in PBS

## Thawing of cells

1  Pipette 45 ml culture medium (DMEM/FCS) in a 50 ml tube.
2  Take one cryo vial with cells from the liquid nitrogen and thaw it in a water-bath.
3  Pipette the content of the cryo vial into the tube containing 45 ml culture medium.
4  Centrifuge the cells at 1300 rpm for eight minutes.
5  Aspirate the supernatant with a sterile needle.
6  Resuspend the pellet in 5 ml culture medium.
7  Pipette the content to a 25 cm$^2$ culture flask and mark the passage number and date.
8  Change the medium every other day and wait until the cells reach 80 per cent confluency for passaging (trypsinization).

## Collagen coating of Transwell® inserts

1   Calu-3 cells must be seeded on collagen coated Transwells®. Therefore, it is necessary to pre-coat the wells before starting the trypsinization procedure.
2   Vitrogen solution (3 mg/ml) is diluted to 30 μg/ml with PBS.
3   660 μl of this solution is applied to each Transwell® insert (4.71 cm²).
4   This solution is then allowed to dry for at least two hours in the laminar flow hood.

## Trypsinization procedure

1   Check the flasks under the microscope for confluency and choose the one which has a confluency of about 80 per cent.
2   Warm the culture medium, PBS and the trypsin/EDTA vial in a water-bath at 37 °C for about 15 minutes.
3   Transfer the flask, the warm PBS and the trypsin/EDTA vial into the laminar flow hood.
4   Aspirate the medium from the flask and add 10 ml PBS. Close the flask and rinse gently. Aspirate the PBS.
5   Add 3 ml trypsin/EDTA solution and incubate for 10 minutes.
6   Check microscopically if all cells are detached from the flask surface and place the flask into the laminar flow hood.
7   If some cells still remain attached, use the cell scraper to detach them gently.
8   Transfer the warm culture medium to the laminar flow hood.
9   Add 10 ml culture medium to the flask to stop trypsin activity.
10  Transfer the cell suspension into a 50 ml tube, rinse the flask with an additional 10 ml of culture medium, close the tube and transfer it to the centrifuge.
11  Centrifuge at 1000 rpm for eight minutes.
12  Transfer the tube back to the laminar flow hood and aspirate the supernatant gently.
13  Add 5 ml culture medium and loosen the cell pellet with a sterile Pasteur pipette. Triturate until a cell suspension is formed.

## Seeding of cells

1   When seeding new flasks add 10 ml of culture medium to each of three 75 cm² flasks and divide the cell suspension equally between the three flasks. Ensure that the medium covers the entire lower surface of the flasks. Note the date, passage number and cell type on the flask. Check the flasks microscopically and then place them in the incubator.
2   For seeding of Transwell® filters it is first necessary to count the number of cells in the cell suspension. Pipette 100 μl trypan blue solution into an Eppendorf vial. Add 50 μl cell suspension and mix gently. Count the cells in a hemocytometer.
3   Seeding of Transwell® filters of a surface area of 4.71 cm².
    Seeding density: $10^5$ cells/cm²
    $4.71 \times 10^5$ cells/filter → $4.71 \times 10^5$ cells/1.5 ml → $3.16 \times 10^5$ cells/ml.

Dilute the cell suspension to $3.16 \times 10^5$ cells/ml for seeding onto six well filters. Dilution factor: number of cells/ml in cell suspension/$3.16 \times 10^5$ cells/ml.

4   Aspirate the apical medium of the pre-equilibrated collagen coated Transwell® inserts and then add 1.5 ml of the diluted cell suspension to each filter. Transfer the plates back to the incubator.

5   On the day after seeding, aspirate the culture medium of the apical side of the filters for those cells which are to be grown at an air interface.

6   Subsequently, the medium is changed every other day and Calu-3 cell monolayers are cultured up to 16–17 days before being used for transport experiments. Cell monolayers being cultured at an air interface receive 2 ml culture medium in the basolateral chamber only. Cells grown under submerged conditions receive also 2 ml at the apical side.

## Quality control

### TEER

TEER is a measure of the tightness of the cell layers which forms due to formation of intercellular tight junctions. TEER of the cell layers is usually monitored by means of a Millicell®-ERS (Millipore Corp., Bedford, USA), connected to a pair of chopstick electrodes, as described previously (Borchard *et al.*, 1996). Average TEER values for Calu-3 cell monolayers have been found to be in the range of 350–450 $\Omega \text{cm}^2$ after 16 days in culture.

### $^{14}$C-mannitol transport

Information on the overall integrity of the cell monolayer is obtained from the measurement of the permeation of a hydrophilic marker such as radiolabeled mannitol across cell monolayers.

1   Cell monolayers are pre-incubated for one hour with transport buffer, consisting of DMEM/HEPES (40 mM, pH 7.2).

2   Replace the transport buffer on the apical side with 1.5 ml of a $^{14}$C-mannitol (Amersham Life Sciences, Little Chalfort, UK) solution (4 µM, specific activity: 0.2 µCi/ml) in transport buffer.

3   Take samples of 200 µl from the basolateral side every 30 minutes for three hours, and substitute with 200 µl transport buffer.

4   Mix the samples with 3 ml Ultima-Gold scintillation cocktail (Packard Instruments, Groningen, The Netherlands) in scintillation vials, and radioactivity is counted in a liquid scintillation counter. Transport buffer (200 µl) mixed with 3 ml scintillation cocktail serves as background sample.

5   Calculate the apparent permeability ($P_{app}$) of mannitol by the following equation:

$$P_{app} = \frac{\frac{(dQ)}{(dt)}}{(60 \times A \times C_0)} \tag{13.1}$$

$dQ/dt$: amount marker transported with time; 60: conversion factor (minutes $\rightarrow$ seconds); $A$: surface area of filter; $C_0$: initial mannitol concentration in donor.

Typical values for $P_{app}$ in Calu-3 cells are in the range of $10^{-7}$ cmsec$^{-1}$.

## Electron microscopy

*Transmission electron microscopy (TEM)*

1   For TEM of Calu-3 cell monolayers, cell layers grown for 17 days are fixed in 0.1 per cent (w/v) glutaraldehyde in 0.14 M cacodylate buffer (pH 7.4, 300 mOsmol) for 60 minutes at room temperature.
2   Rinse the samples twice in Ringer solution and postfix in 1 per cent (w/v) osmium tetroxide in Millonig phosphate buffer (pH 7.3, 300 mOsmol) for 60 minutes at 4 °C. This is according to a procedure as described by Koerten *et al.* (1990).
3   After rinsing, dehydrate the samples in a graded series of ethanol.
4   The specimens are embedded in epoxy resin LX-112 (Ladd Research Industries, Burlington, USA) and polymerized for 72 hours at 60 °C.
5   Cut ultrathin sections (60 nm) on an ultramicrotome, collect on copper grids, stain with uranyl acetate and lead hydroxide and examine using a TEM.

*Scanning electron microscopy (SEM)*

1   Cell layers for SEM are fixed in 1.5 per cent (w/v) glutaraldehyde in 0.14 M cacodylate buffer (pH 7.4, 300 mOsmol) for 60 minutes at room temperature.
2   Samples are then gold sputtered and examined using a SEM.

## Mucus staining

The mucus layer on the surface of the airway epithelium may represent a diffusion barrier in drug delivery to this particular site. Production of mucus is a unique property of Calu-3 cells, contributing to their predictive potential. In order to detect mucus, Calu-3 cell monolayers are fixed in paraformaldehyde, embedded in paraffin, stained with periodic acid–Schiff's stain (PAS) and examined by light microscopy. These techniques have been used to detect mucins in cultured human lung adenocarcinomas (Albright *et al.*, 1991).

## 16HBE14o-: AN IMMORTALIZED HUMAN BRONCHIAL CELL LINE

The 16HBE14o- cell line was generated by transformation of normal bronchial epithelial cells of a one-year-old heart-lung patient with SV40 large T antigen using the replication-defective pSVori-plasmid (Cozens *et al.*, 1994). The cells were generously gifted by Dieter C. Gruenert, Cardiovascular Research Institute at the University of California, San Francisco, USA. The 16HBE14o- cells are able to form confluent cell layers with functional tight and adherens junctions

(A)                                           (B)

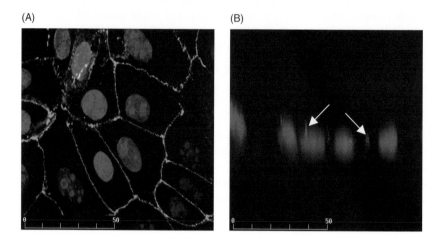

*Figure 13.4* Expression of zonula occludens (ZO)-1 in 16HBE14o- cell cultures. Cells were seeded at a density of $10^5$ cells/cm$^2$ on Transwell® inserts and cultured under submerged conditions. Cells were fixed one week after seeding and stained for the tight junctional protein ZO-1 (green) as described. Top view (A) and cross section (B). Nuclei were counter stained with propidium iodide (red). Bars represent μM. (*See Color plate 8*)

(A)                                           (B)

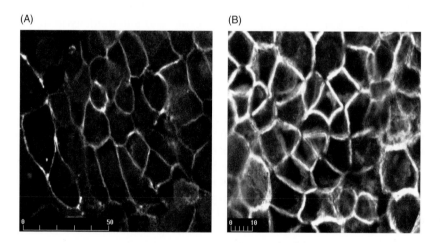

*Figure 13.5* Expression of occludin and E-cadherin in 16HBE14o- cell cultures after one week. Cells were seeded at a density of $10^5$cells/cm$^2$ on Transwell® inserts and cultured under submerged conditions. Cultures were stained for proteins occludin (A) and E-cadherin (B) as described. Bars represent μM.

(Figures 13.4 and 13.5). Like for the Calu-3 cells, measurement of TEER and flux of a hydrophilic marker substance is described. All cell culture components were purchased from Sigma-Aldrich Chemie GmbH (Deisenhofen, Germany), if not stated otherwise.

## Materials needed

- 0.25 per cent trypsin/EDTA solution
- Sterile pipettes 5 and 10 ml
- Sterile Pasteur pipettes
- 15 and 50 ml sterile centrifugation tubes
- Culture flasks 75 cm$^2$ (Greiner, Frickenhausen, Germany)
- Transwell-Clear® 12 mm diameter, pore size 0.4 μm (Corning Costar Corp., Wiesbaden, Germany)
- Minimum essential medium (MEM) supplemented with 10 per cent FCS, penicillin G (100 U/ml), streptomycin sulfate (100 μg/ml), glucose (0.6 g/l) and non-essential amino acids (1 mM)
- Sterile PBS (pH 7.4)
- 0.1 per cent (w/v) trypan blue solution in PBS

## Thawing of cells

1  Pipette 15 ml culture medium (MEM/FCS) in a 75 cm$^2$ culture flask.
2  Take one cryo vial with cells from the liquid nitrogen and thaw it in a water-bath.
3  Pipette the content of the cryo vial into the flask containing the culture medium.
4  Mark the passage number and date.
5  Change the medium every other day, beginning the next day after the thawing and wait until the cells are grown to 80–90 per cent confluency for passaging.

## Trypsinization procedure

16HBE14o- cells are passaged like described for the Calu-3 cells with the following changes:

1  Add only 7 ml of warmed culture medium to the flask to stop trypsin activity.
2  It is not necessary to centrifuge the cells after trypsinization. Instead, just give the cell suspension directly into the newly prepared flask or use it for counting before seeding onto Transwell® filter inserts.

## Seeding of cells

1  When seeding new flasks add 13 ml of culture medium to each of five 75 cm$^2$ flasks and divide the cell suspension equally between the five flasks. Ensure that the medium covers the entire lower surface of the flasks. Note the date, passage number, and cell type on the flask. Check the flasks microscopically and then place them in the incubator.
2  For seeding of Transwell® filters it is first necessary to count the number of cells in the cell suspension. Pipette 50 μl trypan blue solution into an Eppendorf vial containing 400 μl medium. Add 50 μl cell suspension and mix gently. Count the cells in a Fuchs-Rosenthal-chamber. Multiply the result by 50,000 to assess the cell number per milliliter.

3   Seeding of Transwell® filters of a surface area of $1.13\,cm^2$.
    Seeding density: $10^5\,cells/cm^2$.
    $1.13\times10^5\,cells/filter \rightarrow 1.13\times10^5\,cells/0.5\,ml \rightarrow 2.26\times10^5\,cells/ml$.
    Dilute the cell suspension to $2.26\times10^5\,cells/ml$ for seeding onto 12 well filters.
    Dilution factor: number of cells/ml in cell suspension/$2.26\times10^5\,cells/ml$.
4   Add 0.5 ml of the diluted cell suspension to each uncoated Transwell® filter insert and then add 1.5 ml of prewarmed medium to the basolateral compartment. Transfer the plates back to the incubator.
5   On the day after seeding, aspirate the culture medium of the apical side of the filters for those cells which are to be grown at an air interface.
6   Subsequently, the medium is changed every day and 16HBE14o- cell layers are cultured up to eight to ten days before being used for transport experiments. Cell monolayers being cultured at an air interface receive 700 µl culture medium in the basolateral chamber only. Cells grown under submerged conditions receive 500 µl at the apical side and 1.5 ml at the basolateral.

## Quality control

### TEER

Measure the TEER as described in TEER with a chopstick electrode connected to an EVOM voltohmmeter (WPI, Berlin, Germany). We found average TEER values for 16HBE14o- cell layers to be in the range of $600\text{--}800\,\Omega cm^2$ for cells grown under submerged conditions after eight to ten days in culture and of $100\text{--}200\,\Omega cm^2$ for air-interfaced grown cells after two-weeks of culturing.

### Fluorescein-Na transport

Besides $^{14}$C-mannitol, fluorescein-sodium can be employed as a hydrophilic marker to gain information on the overall integrity of the cell layer.

1   Cells are pre-incubated for one hour with transport buffer, consisting of HEPES buffered Krebs-Ringer solution (KRB; pH 7.4).
2   Replace the KRB on the apical side with 520 µl of a fluorescein solution (50 µM) in KRB.
3   Immediately, take a 20 µl sample from the apical side to determine the initial concentration.
4   Take samples of 200 µl from the basolateral side at intervals of 30, 60, 120, 180 and 240 minutes, and substitute with 200 µl transport buffer.
5   The samples are collected in a 96-well plate (Greiner) and fluorescence is measured in a fluorescence plate reader (Cytofluor II, PerSeptive Biosystems, Wiesbaden, Germany) at excitation and emission wavelengths of 485 and 530 nm, respectively. Samples should be diluted with KRB whenever appropriate.
6   Calculate the apparent $P_{app}$ of fluorescein by the following equation:

$$P_{app} = \frac{\dfrac{dQ}{dt}}{(60 \times A \times C_0)} \tag{13.2}$$

d$Q$/d$t$: amount marker transported with time; 60: conversion factor (minutes $\rightarrow$ seconds); $A$: surface area of filter; $C_0$: initial fluorescein concentration in the donor compartment.

Typical values for $P_{app}$ in 16HBE14o- cells are in the range of $10^{-7}$ cm sec$^{-1}$.

7  Measure the TEER before and after the experiment to prove the integrity of the cell layer.

### Immunocytochemical staining

1  Cells are prepared for staining by ten minutes of fixation with 2 per cent (w/v) paraformaldehyde.
2  After washing with PBS the remaining paraformaldehyde is blocked for ten minutes in 50 mM NH$_4$Cl solution in PBS.
3  This step is followed by another washing step and permeabilization with 0.1 per cent Triton X-100 for seven minutes.
4  After thoroughly washing the cells with PBS, 200 µl of the primary antibody solution is added. Therefore, all antibodies have to be diluted 1:100 in PBS containing 1 per cent bovine serum albumin (BSA).
5  Rabbit polyclonal Zonula occludens (ZO)-1 antibody was obtained from Zymed (South San Francisco, USA), mouse monoclonal occludin and E-cadherin immunoglobulin (IgG1) antibodies were purchased from BD Transduction Laboratories (Heidelberg, Germany). Mouse IgG1$\kappa$ (Sigma, Deisenhofen, Germany) was used as isotype control.
6  After 60 minutes of incubation with the primary antibody, the cells are washed at least three times with PBS before being reacted with a 1:100 dilution of a fluorescein isothiocyanate (FITC)-labeled goat anti-mouse F(ab')$_2$ fragment or swine anti-rabbit F(ab')$_2$ fragment in PBS + 1 per cent BSA (DAKO, Hamburg, Germany).
7  For counterstaining of the nuclei, 1 µg/ml of propidium iodide (PI) is added.
8  After 30 minutes of incubation with the secondary antibody and the PI, the samples are washed again three times with PBS.
9  The filter is cautiously cut out with a scalpel and transferred to a slide, where it is embedded in FluorSave anti-fade medium (Calbiochem, Bad Soden, Germany).
10 Images are obtained by fluorescence microscopy or with confocal laser scanning microscopy with the instrument settings adjusted in a way that no positive signal can be observed in the channel corresponding to green fluorescence of the isotype controls.

## DEVELOPMENT OF A PRIMARY PORCINE TRACHEAL EPITHELIAL CELL CULTURE (PTC)

In this section, the development of a primary cell culture as a model for the bronchial epithelium is described. Primary cell cultures derived from tracheal and bronchial tissue of hamster, rats, rabbits or humans have been described previously (Wu, 1986). Because of the close resemblance of human and porcine physiology and morphology, porcine tracheas from specific pathogen-free pigs have been

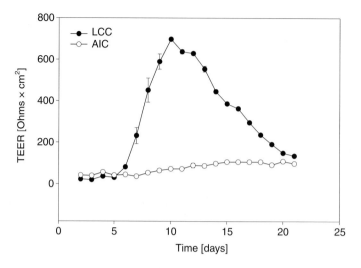

*Figure 13.6* Development of transepithelial electrical resistance (TEER) in 16HBE14o- cells. Cells were seeded at a density of $10^5$ cells/cm$^2$ on Transwell® inserts and cultured either under liquid covered (LLC) or air-interfaced culture (ALC) conditions.

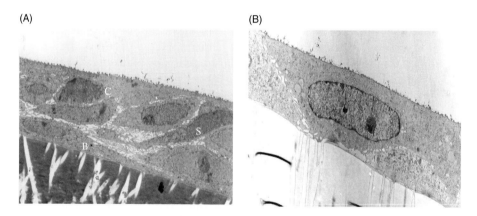

*Figure 13.7* Transmission electron micrograph of percine tracheal epithelial cell culture (PTC) (PN 1) cultured at air interface and in standard culture medium for four days (A), showing a multilayer of different cell types (B: basal cell, C: ciliated cell, S: serosal cell). After passaging, the thickness of the PTC layer is significantly reduced (B, PN 2).

used as the source for a primary cell culture. Under the cell culture conditions described herein, PTC form multilayers of different cell types (Figure 13.6), express intercellular tight junctions and apical cilia (Figure 13.7), and are able to produce mucus (Figure 13.8). Air interface and submerged culture conditions were also compared, which appeared to have an influence on the morphology of PTC.

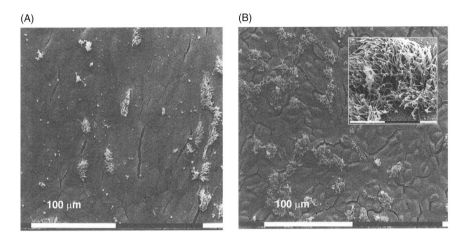

*Figure 13.8* Scanning electron micrograph of percine tracheal epithelial cell culture (PTC) grown under submerged conditions (A), and at air interface (B) in culture medium (Dulbecco's minimal Eagle's medium, DMEM:F12, 2 per cent Ultroser G), supplemented with retinoic acid (RA) at a concentration of 25 ng/ml. Cilial development in approximately 25 per cent of epithelial cells was achieved (see insert, bar: 10 μm).

## Materials needed

- Collagen coated 12-well filter plates
- Cell strainer
- Sterile blade
- Sterile PBS (pH 7.4)
- 0.4 mg/ml solution of pronase type XIV in PBS
- DNAse I
- DMEM:F12 (1:1) medium with 10 per cent FCS
- DMEM:F12 (1:1) medium with 2 per cent Ultroser G.

(All media have been supplemented with Penicillin and streptomycin and fungizone).

## Isolation and culturing of primary PTC

All cell culture materials were purchased from Sigma-Aldrich, unless otherwise stated.

1   Tracheas from specific pathogen-free pigs were obtained from the Institute for Animal Health (ID-DLO, Lelystad, The Netherlands) after sacrificing the animals and transported in ice-cold DMEM.
2   Rinse tracheas with ice-cold PBS to remove mucus, blood and debris.
3   Open the tracheas longitudinally along the anterior surface and cut into 1–2 cm pieces.
4   Incubate the pieces overnight in a 0.4 mg/ml solution of pronase type XIV at 4 °C.

5   Place the tracheas in cold isolation medium, consisting of DMEM:F12, supplemented with 10 per cent FCS, (HyClone, Logan, USA), benzylpenicillin G (160 U/ml) and streptomycin sulfate (100 μg/ml), to stop enzymatic activity.
6   Gently scrape off surface epithelial cells using a new sterile blade and pool cells in DMEM:F12.
7   Pellet cells by centrifugation at 1000 rpm at room temperature for ten minutes.

## Seeding of cells

1   Resuspend cells in isolation medium and filter through a 40 μm cell strainer.
2   Re-pellet by centrifugation at 1000 rpm at room temperature for ten minutes. The resulting cell pellet is re-suspended in isolation medium.
3   Isolation yield is counted in a haemocytometer, and viability is estimated by light microscopy using 50 μl cell suspension mixed with 100 μl of a 0.1 per cent (w/v) trypan blue solution in PBS.
4   Seed the isolated PTC onto collagen coated Transwell® inserts (for procedure see 2.3.) at a density of $1.3 \times 10^{6}$ cells/cm$^2$.
5   Place 1 ml of DMEM:F12 medium (1:1) containing 10 per cent FCS in the basolateral chambers of a Costar 12-well plate, and 200 μl of cell suspension of the same medium to the apical sides.
6   Remove the medium from the plates the following day and replace with DMEM:F12 (1:1) medium containing 2 per cent Ultroser G. In the case of cells cultured at the air interface, 800 μl of this medium is added to the basolateral chamber and no medium is added to the apical chamber. For cell monolayers cultured in the submerged state 200 μl medium is added to the apical side and 1 ml is added to the basolateral side.

## Quality control

In addition to the measurement of TEER (see TEER for details), $^{14}$C-mannitol flux ($^{14}$C-mannitol transport), and mucus production (Figure 13.9), we also determine the epithelial character of the cell layer by cytokeratin staining. Typical staining obtained is depicted in Figure 13.10.

### Cytokeratin staining

1   PTC are grown on glass cover slips of a diameter of 25 mm in a 6-well plate for eight days.
2   Wash cells with PBS and fix in 3.7 per cent formaldehyde for ten minutes.
3   Wash again three times with PBS and block unspecific binding by incubation with TBP (0.1 per cent Triton/0.5 per cent BSA in PBS, pH 7.4) for 60 minutes.
4   Incubate the cells with an FITC-labeled monoclonal anti-cytokeratin peptide 18 antibody F4772 (20 μl in 1 ml TBP) for two hours. In order to protect the FITC label from bleaching, the specimens have to be kept in the dark.
5   Wash cells twice with TBP and once with PBS and post-fix with 3.7 per cent (w/v) formaldehyde for five minutes.

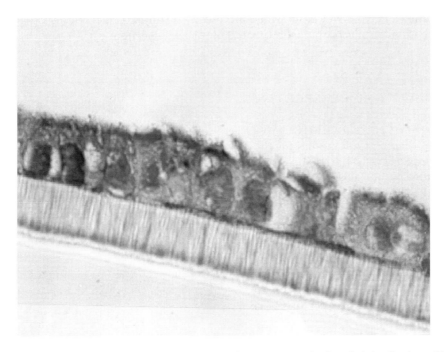

*Figure 13.9* Specific staining of mucin produced in percine tracheal epithelial cell culture (PTC) grown at an air interface by periodic acid–Schiff's (PAS) reagent and counterstained with Alcian blue. Note the presence of cilia at the apical cell membranes. microporous membrane (×100). (*See Color plate 9*)

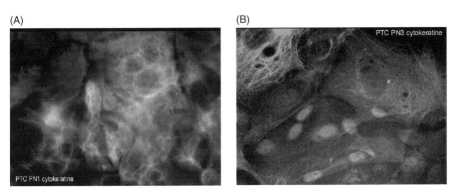

(A)

(B)

*Figure 13.10* Staining of cytokeratin in percine tracheal epithelial cell culture (PTC) of passage numbers 1 (A) and 3 (B). Cytokeratin was stained with an fluorescein isothiocyanate (FITC)-labeled monoclonal anti-cytokeratin 18 antibody (green fluorescence), nuclei were counterstained with Hoechst 33258 (red). (*See Color plate 10*)

6  Wash cells with PBS and incubate with a $2\,\mu g/ml$ solution of Hoechst 33258 in PBS for 15 minutes to counterstain cell nuclei.
7  Wash with PBS, mount the coverslips on microscopical slides and let to dry overnight at room temperature.

8   Store at −20 °C for further use.
9   Samples may be viewed by videomicroscopy or fluorescence microscopy, and the number of cells stained estimated by visual inspection.

## SUMMARY AND PERSPECTIVES

Comparing Calu-3 and 16HBE14o- cell cultures, one can draw the conclusion that both may be relevant models for the airway epithelium. Whereas Calu-3 (like primary tracheal cells) have to be grown at an air interface to produce apical cilia, this condition is not favorable for HBE cells, indicating the basal cell origin of 16HBE14o-. P-glycoprotein has been shown to be present in both cell cultures; however, metabolic activity in 16HBE14o- cells has been characterized to a lesser extent than in Calu-3. Considering the $P_{app}$ of both cell cultures, 16HBE14o- cell layers show values which are comparable to those measured *ex vivo*. Calu-3 cells show a ten-fold lower $P_{app}$ for mannitol than found *in vivo*, which is actually the same ratio found for mannitol $P_{app}$ differences between the intestinal epithelium and Caco-2 cells. The decision for which model to use, therefore, depends on the aim of the studies to be performed.

## REFERENCES

Adler, K. B., Cheng, P. W. and Kim, K. C. (1990) Characterization of guinea pig tracheal epithelial cell maintained in biphasic organotypic culture: cellular composition and biochemical analysis of released glycoconjugates. *Am. J. Respir. Cell Mol. Biol.*, **2**, 145–154.

Albright, C. D., Keenan, K. P., Colombo, K. L. and Resau, J. H. (1991) Morphologic identification of epithelial cell types of the human tracheo-bronchus in cell culture. *J. Tissue Cultur. Methods*, **13**, 5–12.

Borchard, G., Lueben, H. L., de Boer, A. G., Verhoef, J. C., Lehr, C.-M. and Junginger, H. E. (1996) The potential of mucoadhesive polymers in enhancing intestinal peptide drug absorption. III. Effects of chitosan glutamate and carbomer on epithelial tight junctions *in vitro*. *J. Control Rel.*, **39**, 131–138.

Cavet, M. E., West, M. and Simmons, N. L. (1997) Transepithelial transport of the fluoroquinolone ciprofloxacin by human airway epithelial Calu-3 cells. *Antimicrob. Agents Chemother.*, **41**, 2693–2698.

Clark, A. B., Randell, S. H., Nettesheim, P., Gray, T. E., Bagnell, B. and Ostrowski, L. E., (1995) Regulation of ciliated cell differentiation in cultures of rat tracheal epithelial cells. *Am. J. Respir. Cell Mol. Biol.*, **12**, 329–338.

Cozens, A. L., Yezzi, M. J., Kunzelmann, K., Ohrui, T., Chin, L., Eng, K. *et al.* (1994) CFTR expression and chloride secretion in polarized immortal human bronchial epithelial cells. *Am. J. Respir. Cell Mol. Biol.*, **10**, 38–47.

Curiel, D. T., Agarwal, S., Romer, M. U., Wagner, E., Cotten, M., Birnstiel, M. L. *et al.* (1992) Gene transfer to respiratory epithelial cells via the receptor-mediated endocytosis pathway. *Am. J. Respir. Cell Mol. Biol.*, **6**, 247–252.

Ferkol, T., Kaetzel, C. S. and Davis, P. B. (1993) Gene transfer into respiratory epithelial cells by targeting the polymeric immunoglobulin receptor. *J. Clin. Invest.*, **92**, 2394–2400.

Ferkol, T., Perales, J. C., Eckman, E., Kaetzel, C. S., Hanson, R. W. and Davis, P. B. (1995) Gene transfer into the airway epithelium of animals by targeting the polymeric immunoglobulin receptor. *J. Clin. Invest.*, **95**, 493–502.

Fogh, J., Fogh, J. M. and Orfeo, T. (1977) One hundred and twenty-seven cultured human tumor cell lines producing tumors in nude mice. *J. Natl. Cancer Inst.*, **59**, 221–226.

Forbes, B. and Lansley, A. B. (1998) Transport characteristics of formoterol and salbutamol across a bronchial epithelial drug absorption model. *Eur. J. Pharm. Sci.*, **6**, S24.

Gruenert, D. C., Finkbeiner, W. E. and Widdicombe, J. H. (1995) Culture and transformation of human airway epithelial cells. *Am. J. Physiol.*, **286**, L347–L360.

Koerten, H. K., Hazekamp, J., Kroon, M. and Daems, W. T. (1990) Asbestos body formation and iron accumulation in mouse peritoneal granulomas after the introduction of asbestos fibres. *Am. J. Pathol.*, **136**, 141–157.

Kuhn, C. (1988) Normal anatomy and histology. In W. M. Thurlbeck (ed), Pathology of the lung. Thieme, New York, 11–50.

Lee, T. C., Wu, R., Brody, A. R., Barrett, J. C. and Nettesheim, P. (1984) Growth and differentiation of hamster tracheal epithelial cells in vitro. *Exp. Lung Res.*, **6**, 27–45.

Ma, T. Y., Hollander, D., Riga, R. and Bhalla, D. (1993) Autoradiographic determination of permeation pathway of permeability probes across intestinal and tracheal epithelia. *J. Lab. Clin. Med.*, **122**, 590–600.

Mariassy, A. T. (1992) Epithelial cells of trachea and bronchi, In R. Parent (ed), Comparative biology of the normal lung, CRC Press, London, 63–76.

Mathias, N. R., Kim, K.-J. and Lee, V. H. L. (1996) Targeted drug delivery to the respiratory tract: solute permeability of air-interface cultured rabbit tracheal epithelial cell monolayers. *J. Drug Target.*, **4**, 79–86.

Meaney, C., Florea, B. I., Borchard, G. and Junginger, H. E. (1999) Characterization of a human submucosal gland cell line (Calu-3) as an in vitro model of the airway epithelium. *Proc. Int. Symp. Cont. Rel. Bioact. Mater.*, **26**, 198–199.

Richardson, J., Bouchard, T. and Ferguson, C. C. (1976) Uptake and transport of exogenous proteins by respiratory epithelium. *Lab. Invest.*, **35**, 307–314.

Wu, R. (1986) *In vitro* differentiation of airway epithelial cells, In L. J. Schiff (ed.), *In vitro* models of respiratory epithelium, CRC Press Inc., Boca Raton, 1–26.

Yu, X. Y., Schofield, B. H., Croxton, T., Takahashi, N., Gabrielson, E. W. and Spannhake, E. W. (1994) Physiological modulation of bronchial epithelial cell barrier function by polycationic exposure. *Am. J. Respis. Cell. Mol. Biol.*, **11**, 188–189.

# Chapter 14

# *In vitro* methodologies to study nasal delivery using excised mucosa

*Annette M. Koch, M. Christiane Schmidt and Hans P. Merkle*

## INTRODUCTION

Nasal delivery of therapeutic proteins and peptides is an established alternative to the parenteral route. In contrast to preoral delivery, nasal application is less prone to dilution and metabolic degradation typical for the gastrointestinal (GI) tract. Moreover, in light of promising results with nasal immunization, and the recent introduction of a nasal influenza vaccine (Glück *et al.*, 1999), nasal mucosal vaccination may gain in importance. Although bioavailability and immune response *in vivo* are the gold standards for the assessment of the nasal route, *in vitro* studies are helpful and important tools. They may be applied to examine principal mucosal permeabilities and pathways, or mechanisms of permeation. Moreover, metabolism and cellular toxicity issues can be addressed. A general review covering the existing nasal *in vitro* models was previously given by Schmidt *et al.* (1998a) and forms the major basis for large parts of this chapter. Briefly, cell line cultures from nasal epithelial cells, due to difficulties in forming confluent monolayers, have shown to be impracticable for routine transport studies. In particular, they are unlikely to serve as a complete model for the rather complex nasal epithelium, with its various region-dependent cell types and all its carriers and enzymes. Moreover, cell line cultures often lack the phenotype typical for the nasal epithelium, i.e. a pseudostratified columnar epithelium. Primary cell cultures of human origin are highly differentiated but difficult and costly to cultivate, and allow only limited passaging. Finally, primary cultures cannot reflect the heterogeneous cellular coexistence typical for the nasal mucosa, consisting of ciliated and non-ciliated cells, basal cells, goblet cells etc. By contrast, excised mucosa, as described here in detail, represents the most relevant model available to date for the respiratory nasal epithelium (Figure 14.1).

Excised nasal mucosae of different species are frequently used tissues to study nasal transport and metabolism. Taking the difficulties into account of obtaining human tissue from nasal biopsies, it becomes obvious that most studies were performed with epithelia excised from animals. For obvious reasons studies using human tissue are particularly rare (De Fraissinette *et al.*, 1995; Peter *et al.*, 1992). For a majority of studies rabbit tissue has been used (Bechgaard *et al.*, 1992; Cremaschi *et al.*, 1991a; Maitani *et al.*, 1997). In addition, mucosae from ovine (Reardon *et al.*, 1993; Wheatley *et al.*, 1988), canine (Hersey and Jackson, 1987), porcine (Wadell *et al.*, 1999) and human origin (De Fraissinette *et al.*, 1995) were selected. The

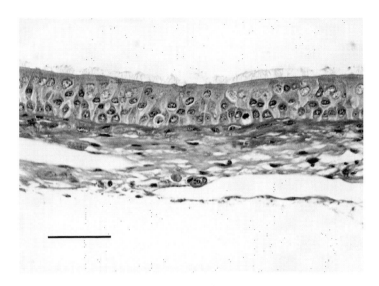

*Figure 14.1* Cross-section of excised bovine nasal epithelium from frontal part of nasal conch. It shows a typical, respiratory type of epithelium (ciliated, pseudostratified and columnar). Respiratory epithelium usually consists of four cell types: columnar cells, both ciliated and non-ciliated (mainly seen in this figure), goblet cells for mucus production, and basal cells. Below the epithelium is the basement membrane, consisting of collagen fibrils, and the lamina propria which is rich in blood vessels and mucus glands. The capillary vessels are situated just below the epithelial layer. Nasal mucosa is of high permeability and referred to as leaky-type epithelium. Bar length is 30 μm (Schmidt *et al.*, 1998a).

focus of this review is on bovine nasal mucosa which can be obtained directly from slaughterhouses. It is readily and economically available in sufficient quantity and of reproducible quality. This model has been shown to be well-suited for studies on nasal permeation and nasal metabolism of peptides (Ditzinger, 1991; Lang *et al.*, 1996a,b).

## MATERIALS AND METHODS

### Tissue preparation

The techniques for dissecting animal nasal epithelium may widely depend on the species selected. In case of bovine tissue, dissection should take place immediately after slaughter (see Protocol 14.1) to preserve viability. Preservation by freezing may impair the integrity of this tissue.

To obtain rabbit nasal mucosal tissue, animals are sacrificed, and a longitudinal incision through the lateral wall is made and the nasal cavity fully opened. After removing the lateral wall, the entire nasal septum or mucosa from the roof of the nostrils (upper and lower nasal conch) (Porta *et al.*, 2000) is isolated using operation

## Protocol 14.1

---

*Preparation of excised bovine nasal mucosa*

Nasal tissue may be obtained from snouts of freshly slaughtered cattle. Suitable tissue is obtained by removing the skin covering the nose and cutting out the frontal part of the nasal conch (*conchae nasales dorsales*) above the *os incisivum* with a sharp knife starting from the *incisura nasoincisiva* (see Figure 14.2). Keep tissue specimens on ice for transport and storage.

As soon as possible, but at the latest one hour after tissue dissection, mucosa of a thickness of about 100 μm, consisting of the epithelium and some of the connective tissue, is carefully stripped from the lateral cartilage of the specimen. For this purpose the use of a small scalpel at little pressure is suggested. Cut out a section of about 3–4 cm$^2$ (as indicated in Figure 14.2), which is then carefully stripped off with a pair of tweezers and the help of a blunt scalpel. Stripping direction is as indicated in Figure 14.2.

With the stripped section still attached to the specimen, the excised mucosa is carefully transferred onto a microscope slide and finally truncated at the end still connected to the specimen. From the slide the tissue is directly transferred into preheated diffusion chambers for equilibration and subsequent permeation or metabolism studies.

For microscopy studies this procedure should be done in a Petri dish containing constantly oxygenated and glucose-containing buffer medium. In this way the tissue can be permanently kept in a humid atmosphere and perfectly rinsed.

---

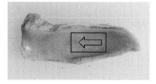

*Figure 14.2* Excision of bovine nasal mucosa. Left panel: bovine snout. Middle panel: bovine snauze after cutting out specimen. Right panel: nasal specimen, indicating area and direction for subsequent stripping of mucosa. (*See Color plate 11*)

scissors or a scalpel (Jorgensen and Bechgaard, 1993; Maitani *et al.*, 1991). Porcine nasal mucosa (Wadell *et al.*, 1999) is obtained in a similar manner as bovine mucosa. After slaughter the septum is fully exposed by a longitudinal incision through the lateral wall of the nose. Septal mucosa is removed from the underlying bone by cutting along the septum and pulling the mucosa off the septum with hemostatic forceps. Alternatively, the connective tissue is removed using an electro-dermatome. Finally, cavity tissue is removed from the conchae and the lower cavity using hemostatic forceps.

# Physiologic test media, oxygenation and temperature control

A list of buffer systems used for permeation studies is given by Reardon (1996). Although the pH of the human nasal mucus layer is around 5.5–6.5 (Chien *et al.*, 1989), most studies with excised nasal tissue are performed at pH 7.4. Studies by Maitani *et al.* (1997) using excised rabbit mucosa show that enzyme activity decreases when the pH is dropped to 5.2. Problems resulting from pH-dependent solubility, or from physical instabilities, e.g. fibrillation and aggregation of large peptides and proteins, can occur with several drugs, e.g. human calcitonin (hCT) and insulin. In the case of hCT low percentages of methylhydroxypropylcellulose may be added to inhibit fibrillation (Schmidt *et al.*, 1998a).

Most nasal studies are conducted at pH 7–7.4. At this pH, the use of a phosphate buffer, e.g. Dulbecco's modified PBS (see Protocol 14.2) is preferred because of its higher pH stability. This medium does not contain hydrogen carbonate ions ($HCO_3^-$), therefore, pure oxygen ($O_2$) without a carbon dioxide ($CO_2$) atmosphere can be used for the oxygenation of the buffer. For oxygenation in $CO_2$-containing incubators, e.g. for preparing samples for confocal microscopy, usually Krebs-Henseleit buffer (KHB), a bicarbonate physiological saline (pH 7.4), is suggested.

Electrophysiological studies show the importance of exogenous glucose in the buffer medium (Cremaschi *et al.*, 1991b; Kubo *et al.*, 1994). In the absence of glucose, the transepithelial potential difference between the mucosal and serosal side ($V_{ms}$) is reduced by 1–2 mV after 30 minutes, indicating that the nasal mucosa depends on transport processes powered by exogenous glucose. Similarly, experiments done under $O_2$-free conditions result in a reduction of $V_{ms}$ to zero after 10–30 minutes (Kubo *et al.*, 1994). It can be concluded that exogenous glucose and oxygenation of the buffer systems are essential to retain tissue viability.

Due to its close contact with the inspired air, the temperature of the nasal mucosa is expected to be subject to climate- and exercise-related variations. In light of this fact, it is surprising to see that most authors prefer an experimental temperature of 37 °C. A more suitable temperature could be 28–30 °C. Chien *et al.* (1989) reported that the optimum temperature reported for the mucociliary activity is in

## Protocol 14.2

*Buffer media for transport studies described in Protocols 14.3 and 14.4*

Buffer media most recommended for transport studies in Protocols 14.3 and 14.4 and metabolism studies are:

- Dulbecco's modified phosphate-buffered saline (PBS), (Cat. No. 14040, Life Technologies Inc., GIBCO BRL, Rockville, MD), with the addition of 1.0 g/l glucose. This physiologic saline solution shows high pH-stability and is suitable in combination with pure oxygen ($O_2$) (medicinal grade) for gassing.
- Krebs-Henseleit buffer (KHB), a bicarbonate physiological saline (regularly described in the literature) in combination with 5 per cent carbon dioxide ($CO_2$) containing $O_2$ (medicinal grade).

this range. The temperature selected by Cremaschi *et al.* (1991b) is 27 °C, mainly because of viability and stability considerations. Hersey and Jackson (1987) even made all their permeation experiments at room temperature (22–24 °C).

## Tissue equilibration

As monitored by electrophysiological measurements (see next section), the time periods necessary to reach equilibration of nasal tissue after dissection ranges from about 20–30 minutes (Cremaschi *et al.*, 1991a) up to 60–120 minutes (Bechgaard *et al.*, 1992; Hosoya *et al.*, 1993; Kubo *et al.*, 1994; Reardon *et al.*, 1993; Wheatley *et al.*, 1988). In our bovine model, equilibration is reached after 20–30 minutes and the electrophysiological parameters remain stable for a period of at least 120 minutes (Lang *et al.*, 1996a). In ovine tissue, Reardon *et al.* (1993) and Wheatley *et al.* (1988) observed simultaneous increases in both the short-circuit current ($I_{sc}$) and the $V_{ms}$ during this period (see next section). The resulting transepithelial resistance ($R_m$) rose from 25 to 100 $\Omega$ cm$^2$. The increase in $I_{sc}$ was explained by the re-establishment of ion gradients after the tissue was placed in warmed buffer with oxygenation (Wheatley *et al.*, 1988; Reardon *et al.*, 1993). In other cases the resulting $R_m$ remained approximately constant (Cremaschi *et al.*, 1991a).

## PROBLEM SOURCES AND QUALITY CONTROL

For *in vitro* permeation studies with excised mucosae, tissue viability and integrity during the time course of the experiment are the two main issues to assure quality and reproducibility.

## Viability tests

The principal tools to test the viability of excised nasal tissue are electrophysiological measurements and cell staining assays. Viability tests are efficient tools to detect cytotoxicity-related damages by drugs, drug carriers, permeation enhancers and other biological modifiers. The features to be tested include (i) active ion transport, (ii) activity of typical enzymes, and (iii) functional integrity of cellular and subcellular membranes. To unequivocally assure the viability of an excised nasal model always more than one test should be considered.

*Electrophysiological measurements* – Electrophysiology is a powerful tool to monitor the viability and the physical integrity of excised tissue. Because of the highly leaky character of excised nasal mucosa, however, the significance of electrophysiology for this tissue is lower than for tissues whose permeabilities are under strict tight junctional control, e.g. the intestinal epithelium. Thus, for the nasal mucosa the exact distinction of intact from damaged tissue is more difficult than with tighter tissues, e.g. intestinal or endothelial cell lines.

For measurements under short-circuit conditions, two pairs of silver/silver chloride electrodes mounted in glass capillaries and closed with a diaphragm are inserted into the mucosal and serosal bathing solutions of a donor–receiver set-up, i.e.

one pair to pass the current across the tissue and a second pair to measure the corresponding voltage. For calibration, the resistance of the buffer solution has to be compensated in the absence of the mucosa. For the measurement of the $V_{ms}$, the mucosa is then inserted between the two half-cells of the diffusion chamber.

As $V_{ms}$ largely depends on the flux of ions, current is a better measure of tissue viability. To nullify $V_{ms}$, an external current, i.e. the $I_{sc}$, is applied with the second pair of electrodes leading to short-circuit conditions. $I_{sc}$ depends on the ionic permeability of the tissue and serves as an indicator of viability. $I_{sc}$ arises from the active $Na^+$ transport across the mucosal side via an amiloride-sensitive conductive pathway and the exit across the serosal side via a $Na^+/K^+$-ATPase (Reardon, 1996). Under short-circuit conditions, charged species cannot passively permeate the mucosa. $V_{ms}$ and $I_{sc}$ are related by the transepithelial resistance $R_m$:

$$R_m = \frac{V_{ms}}{I_{sc}}. \tag{14.1}$$

Tissue damage and cell death causes decrease of the $R_m$. In the course of an experiment, short-circuiting is usually performed in a pulsed mode using a multi-channel voltage-current clamp. Under the pulsed voltage-clamp mode, Equation 14.1 transforms into:

$$R_m = \frac{\Delta V_{ms}}{\Delta I_{vc}}, \tag{14.2}$$

where $\Delta V_{ms}$ is the potential difference during a current pulse, and $\Delta I_{vc}$ the current pulse difference generated from the voltage clamp.

Experimentally more simply, experiments may also be carried out under open-circuit conditions. In this case, the voltage across the membrane is measured and no current is applied. The measured voltage reflects the spontaneous $V_{ms}$ of the mucosa. Thus, the determination of $V_{ms}$ requires only two electrodes and can be used for routine checks of tissue integrity and viability, e.g. at the beginning and/or the end of an experiment (Cremaschi *et al.*, 1996b).

According to Bechgaard *et al.* (1992), using rabbit nasal mucosa mounted on Ussing chambers, a reasonable definition of the window of viability may be the time where $I_{sc}$ is more than 80 per cent of the mean $I_{sc}$ value between one and ten hours. The $R_m$ was constant for more than ten hours and they concluded that full viability during this time period can be preserved. In bovine mucosa, constant $I_{sc}$ values were found for at least two hours after preparation of the tissue (Schmidt *et al.*, 2000b). Rate and extent of recovery to normal electrophysiological parameter levels, e.g. after exposure to absorption enhancers and subsequent wash-outs, may be another aspect of monitoring tissue viability.

Electrophysiological measurements may be combined with functional tests of transmembrane ion channels. Wheatley *et al.* (1988) reported on tests involving several agents known to affect ion transport across viable nasal tissue. The agents suggested include amiloride (apical $Na^+$ channel blocker), ouabain (inhibitor of $Na^+/K^+$-ATPase), and epinephrine (stimulation of $Cl^-$ secretion). It was demonstrated that in ovine nasal mucosa viability is maintained for up to eight hours, even after the biological response to the ion channel modifier amiloride was tested.

*Table 14.1* Transepithelial electrical resistances (TEERS), thicknesses and mannitol permeabilities of excised nasal tissues and confluent cell culture monolayers

| Species | $R_m$, $\Omega\,cm^2$ | Thickness, $\mu m$ | $P_{eff}$ mannitol, $10^6$ cm/s | References |
|---------|-------|-----------|----------------|------------|
| *Excised mucosae* | | | | |
| Cattle, cavity | 42 | 100 | 90 | (Schmidt et al., 2000b) |
| Pig, septum | 52 | 250 | 8.5 | (Wadell et al., 1999) |
| Pig, cavity | 68 | 400 | 5.7 | (Wadell et al., 1999) |
| Cattle, cavity | | 800 | 11 | (Koch et al., unpublished data) |
| Pig, septum | 74 | 850 | 3.9 | (Wadell et al., 1999) |
| Sheep, cavity | 100 | | 2.2 | (Wheatley et al., 1988) |
| Rabbit, septum | 45 | | | (Kubo et al., 1994) |
| Rabbit, septum | 70 | | | (Bechgaard et al., 1992) |
| Rabbit, region | 45 | 335 | | (Cremaschi et al., 1991b) |
| upper concha | | | | (Cremaschi et al., 1998) |
| Human | 65 | | | (Boucher et al., 1986) |
| *Cell culture monolayers* | | | | |
| RPMI 2650 | 114 | | | (Peter, 1996) |
| Human nasal primary cell cultures | 360–640 | | | (Werner and Kissel, 1995) |

However, electrophysiological data can only cover selected aspects of tissue viability, i.e. the function of ion channels, or tight junction control. For a comparison of transepithelial electrical resistances of excised nasal tissue and confluent cell culture monolayers see Table 14.1. This table shows that species, site of excision and thickness of isolated mucosa may have some influence on the transepithelial electrical resistance. Typically, $R_m$ is constant within a reasonable range.

*Viability staining* – Other aspects of cell viability are the enzymatic activity in the various subcellular compartments and/or the functional integrity of cellular and subcellular membranes. Trypan blue is a widely used dye (Kotze et al., 1999) to detect damaged or dead cells. Typically, it is excluded from viable cells, but taken up by non-viable ones. This test is mainly used in cell suspensions.

The current methodology is a two-component staining assay for excised tissue viability testing (see Protocol 14.3) (Schmidt et al., 2000b). One element of this assay is the testing of ubiquitous intracellular esterase activity. In enzymatically active viable cells, the virtually non-fluorescent and cell-permeable calcein AM ester is converted to the strongly fluorescent calcein. According to its polyanionic character, the free calcein is retained in viable cells only, thereby producing an intense green fluorescence. In non-viable cells or in cells of reduced viability some esterase activity may remain and also metabolize calcein AM. In such leaky cells, fluorescent calcein cannot be retained and is released depending on the extent of the damage. Active secretion might be a complication of this test. The second element of this assay is the staining of the cell nucleus by ethidium homodimer (EthD-1). Based on its strongly positive charge this marker only enters non-viable cells or cells with compromised cell membranes. By binding to the nucleic acids in the nucleus it forms a bright red fluorescence complex, labeling the nuclei of dead cells. The advantage of this two-component assay is the simultaneous testing of enzymatic activity and membrane function.

## Protocol 14.3

*Viability staining assay and visualization of permeation pathways*

Periodic viability staining is recommended to ensure the experimental significance of permeation or metabolism studies with excised mucosa, as well as for studies involving fluorescence or confocal microscopy of the nasal tissue, e.g. for pathway visualization. Depending on the experimental set-up, viability staining may be simultaneous or subsequent to these experiments. The protocol can also be applied to assess cellular permeation pathways by means of suitable fluorescence markers or biochemical assays. Adjustment of incubation time may be necessary for optimization.

- Transfer excised mucosa on a coverslip and place coverslip on tissue-holder (see Figure 14.4).
- Wash excised mucosa gently three times with preheated Krebs-Henseleit buffer (KHB). For this purpose, add about 200 µl of KHB and carefully remove buffer by aspiration using a Pasteur pipette.
- For equilibration incubate mucosa with KHB in a cell culture incubator (5 per cent $CO_2$), e.g. during 30 minutes at 37 °C. To avoid evaporation and subsequent dessication of the mucosal tissue such experiments should be performed in a closed Petri dish.
- Replace pure KHB by a mixture of 16 µM calcein AM (4 mM stock solution in DMSO) and 4 µM ethidium homodimer (EthD-1, 2 mM stock solution in DMSO) in KHB and incubate mucosa again for 30 minutes at 37 °C (viability assay: LIVE/DEAD Euko Light, Molecular Probes Europe, Leiden, The Netherlands).
- Remove staining solution and wash five times with KHB.
- Observe fluorescence immediately using fluorescence (or confocal) microscopy. Typical settings, for example for a Zeiss Axiovert inverted microscope (Carl Zeiss, Oberkochen, Germany) equipped for fluorescence microscopy are: 40× lens, excitation filter 450–490 nm, emission filter 520 nm longpass. For pathway visualization studies, the tissue may be fixed for preservation.
- Intense green cytosolic fluorescence indicates practically full tissue viability through intracellular esterase activity converting the non-fluorescent calcein AM ester into the fluorescent polycationic calcein.
- Red fluorescence, as generated by intercalation of nuclear DNA with EthD-1, indicates dead cells. In viable tissue, red fluorescence is seen occasionally, indicating isolated dead cells. Red fluorescence in deeper sections and out of the focused plane indicates dead cells that are part of the connective tissue below the basement membrane, which may have been caused by the separation of the mucosa from the underlying connective tissue.
- The experimental set-up is given in Figure 14.4.

Another assay involving the combination of fluorescein diacetate and propidium iodide is based on a similar mechanism (Jones and Senft, 1985). However, the relatively strong fluorescence of the unmetabolized fluorescein diacetate may be a disadvantage. A quantitative method to test potential cytotoxic effects of substrates and metabolites during permeation or metabolism studies is based on the production of mitochondrial succinic dehydrogenase in viable cells. The principle reaction of the test is the reduction of the soluble yellow tetrazolium salt 3-(4,5-dimethylthiazol-2-yl)-

2, 5-diphenyltetrazolium bromide (MTT) to a blue insoluble MTT formazan product by succinic dehydrogenase (Mosmann, 1983). Decrease of MTT formazan, relative to a standard, is indicative of a cytotoxic effect.

## Integrity tests using markers

Similar to viability testing, the control of the physical integrity must be an integral part of *in vitro* experimentation with excised nasal mucosae. In addition to the monitoring of the electrophysiological parameters, $V_{ms}$, $I_{sc}$ and $R_m$ (see above), the fluxes of passive permeability markers across the excised tissue are often used to assess epithelial integrity. For a suitable experimental set-up see Protocol 14.4. Typical markers are inulin, lucifer yellow, mannitol, polyethyleneglycol (PEG) and sucrose. With passive and metabolically stable permeability markers (mannitol, sucrose), mucosal-to-serosal and serosal-to-mucosal fluxes should not differ significantly. Using a donor–receiver diffusion system, the $^3$H-mannitol fluxes (initial donor concentration at 0.5 μCi/ml) correlated closely with the $R_m$ of ovine nasal tissue (Wheatley *et al.*, 1988). Typical permeability coefficients ($P_{eff}$) of mannitol in bovine nasal mucosa are in the order of $10^{-4}$ cm s$^{-1}$ (Schmidt *et al.*, 2000a).

**Protocol 14.4**

---

*Permeation experiment using mannitol as model compound*

- Prepare mucosa as described in Protocol 14.1. Carefully transfer excised tissue from the microscope slide onto the respective opening of the prewarmed and prehumidified half-cell of a diffusion set-up.
- Close both half-cells of the diffusion set-up.
- Fill in equilibration medium, e.g. preheated Dulbecco's modified phosphate-buffered saline (PBS) or Krebs-Henseleit saline (see Protocol 14.2). For tissue protection, fill both half-cells simultaneously.
- Equilibrate tissue for 30 minutes.
- Simultaneously remove equilibration media, and simultaneously add donor and receiver solutions, the donor solution containing 0.5 μCi/ml $^3$H-mannitol (e.g. D-(1–3H(*N*)(-Mannitol, NET 101, NEN Life Science, Boston, MA).
- At evenly spaced time-points, e.g. after 15, 30, 45, 60 minutes, take samples from receiver solution, add 2 ml scintillation cocktail (e.g. Ultima Gold, high flash point LSC cocktail from Packard Instruments, Packard BioScience, Meriden, CT) and count $^3$H-mannitol activity using a liquid scintillation counter (e.g. Beckmann Instruments Inc., Fullerton, CA). Samples are instantly replaced by pure preheated buffer solution.
- At time zero and at the end of the experiment, samples are taken from the donor solution as controls for donor depletion.
- When testing the permeability of compounds, simultaneous determination of mannitol permeability may serve as control for tissue integrity. It is of advantage to take the samples for mannitol and the examined substance from the receiver compartment at identical time-points. This simultaneous sample-testing allows for calculation of loss of compound owing to mannitol sampling.

---

In this context, it is worth mentioning that the thickness of the connective tissue underlying the epithelial layer highly influences the permeability of passive permeability markers. This influence results from the highly leaky character of this tissue. Wadell *et al.* (1999) found that when the thickness of excised porcine nasal mucosa was increased from around 200 µm to 800 µm by changing the dissecting technique, the effective mannitol permeability decreased by approximately 50 per cent (see Table 14.1). Similarly, in bovine tissue, mannitol permeability decreased by almost 90 per cent when switching from a 100 µm sample, obtained with a stripping technique, to a sample of 800 µm, obtained with a dermatome (Koch, unpublished data). Nevertheless, the barrier properties of the epithelium can be assessed most accurately when the contribution of the connective tissue is as small as possible. It needs to be considered that *in vivo* the capillary network to absorb the drug is right under the basement membrane in the lamina propria. So, for systemic absorption only transport through the epithelial layer but not through the underlying connective tissue is typical. The thickness of excised rabbit nasal mucosa was reported to be 335 µm (Cremaschi *et al.*, 1998) and 346 µm (Maitani *et al.*, 1997).

Further to mannitol, high molecular weight molecules, e.g. inulin and PEG 4000 have been used as so-called non-absorbable markers (Bechgaard *et al.*, 1992; Wheatley *et al.*, 1988). In light of the measurable permeation of these compounds through the leaky-type nasal epithelium (Donovan and Huang, 1998), this terminology may be misleading and should not be taken literally. McMartin *et al.* (1987) concluded that in nasal tissue, high molecular weight hydrophilic molecules (>1000 Da) may passively permeate through aqueous channels. In our studies in bovine tissue, we found a linear relationship of log molecular weight versus effective permeability (Koch, unpublished data) when following the transport of fluorescein isothiocyanate (FITC) dextrans ranging from 4000–40,000 Da.

## SPECIFIC APPLICATIONS

### Transport studies

Transport across nasal mucosa or nasal cell cultures can be mediated through one or a combination of several routes. For instance, permeation may occur through a paracellular and/or a transcellular pathway. Transport through the paracellular route, i.e. the passive diffusion through the space between adjacent epithelial cells, is limited by the barrier function of the tight junctions. Alternatively, the transcellular pathway may be either passive or active. Transcytosis, an active transcellular transport process, is either receptor-mediated or non-specific (pinocytosis). To discriminate between passive and active transport, permeation studies performed in diffusion chambers are most useful. Passive diffusion is indicated by low energy dependency and should exhibit non-saturable kinetics and equal flux rates in either direction (mucosal-to-serosal versus serosal-to-mucosal), whereas direction specificity would suggest active transport. However, this conclusion would only apply for a metabolically stable permeant or under even distribution of the metabolic activity in the barrier. When the metabolic enzymes, are polarized at the mucosal surface, direction specificity would also occur under passive diffusion.

In contrast to passively transported permeants, active transepithelial transport should show saturation kinetics and direction specificity. Equally typical are substrate specificity and dependence on metabolic energy. The potential transport mechanisms most often cited for proteins and peptides in respiratory epithelia involve endocytosis and transcytosis processes (Johnson and Boucher, 1993). The molecular and cellular mechanisms involved in endocytotic processes were reviewed by Schaerer *et al.* (1991). Active transport of polypeptides, e.g. elcatonin and adrenocorticotropic hormone (ACTH), was studied by Cremaschi *et al.* (1991a). With increasing concentration the polypeptides showed saturable permeation kinetics. Endocytotic uptake of hCT was studied by Schmidt *et al.* (1998b). See below for a pathway visualization study of this polypeptide.

### Permeability studies

The effective $P_{eff}$ (cm/s) under steady-state conditions across excised mucosa can be calculated according to:

$$P_{eff} = \left(\frac{dC}{dt}\right)_{ss} \frac{V}{A \times C_D}, \tag{14.3}$$

where $(dC/dt)_{ss}$ is the time dependent change of concentration in the steady-state, $A$ is the permeation area, $V$ the volume of the receiver compartment and $C_D$ the initial donor concentration (Lang *et al.*, 1996a).

A selection of effective $P_{eff}$ for therapeutic oligo- and polypeptides across excised nasal tissue (human, cattle, rabbit) is given in Table 14.2, with permeabilities ranging from $10^{-7}$ to $10^{-5}$ cm/s at molecular weights from 362 up to 5600.

The data indicate that there is no obvious correlation between the molecular weight and the $P_{eff}$. Depending on the individual permeant various parameters may play the key role. Besides the molecular weight (or size), other physicochemical parameters and properties, such as lipophilicity, hydrogen bonding, interaction with mucus, charge and conformation may all have an influence on passive permeability. With ionic drugs, peptides and proteins in particular, the negative charge of the nasal mucosa might be of importance. This charge results from negatively charged lipids and the sialic acid of the glycoproteins in the mucus (Maitani *et al.*, 1991).

Susceptibility to metabolic cleavage in the nasal mucosa is another crucial factor. For instance, the rapid metabolism of gonadorelin (luteinizing hormone–releasing hormone, LHRH) in biological fluids and tissues explains its relatively low effective $P_{eff}$ in bovine nasal mucosa. Buserelin and Hoe013, on the other hand, are synthetically stabilized peptides and thus permeate at much higher rates (Ditzinger, 1991). Octreotide, a cyclic octapeptide, is metabolically stable and, therefore, of higher permeability in the mucosa. Even medium-size polypeptides, like human and salmon calcitonin, permeate at reasonable rates, partly because of their moderate susceptibility to mucosal metabolism.

The molecular sieving effect of nasal mucosa is clearly established by a reciprocal correlation between log $(P_{eff})$ and log (molecular weight) of FITC-dextrans (Kubo *et al.*, 1994; Uchida *et al.*, 1991). However, the characteristics of FITC-dextrans are quite different from peptides and proteins and, therefore, extrapolation to biopharmaceuticals is difficult.

*Table 14.2* Effective permeability coefficients ($P_{eff}$) of peptides in excised nasal mucosa (mucosal-to-serosal)

| Peptides | Molecular weight(MW) | $P_{eff}$ $10^6$ cm/s | Species | References |
|---|---|---|---|---|
| Thyrotropin-releasing hormone | 362.4 | 4.9 | Rabbit | (Jorgensen and Bechgaard, 1993) |
| Thymotrinan (TP3) | 417 | 15.7[a] | Cattle | (Schmidt et al., 2000a) |
| Thymocartin (TP4) | 517 | 17.0[b] | Cattle | (Lang et al., 1996a) |
| Thymopentin (TP5) | 680 | 9.2[c] | Cattle | (Lang et al., 1996a) |
| Octreotide | 1072 | 43 | Cattle | (Lang, 1995) |
| Sandostatin® | | 5.0 | Human | (De Fraissinette et al., 1995) |
| Angiopeptin | 1156 | 0.91 | Rabbit | (Jorgensen and Bechgaard, 1993) |
| Gonadorelin, LHRH | 1182 | 1.86 | Cattle | (Ditzinger, 1991) |
| Buserelin | 1239 | 14.7 | Cattle | (Ditzinger, 1991) |
| Hoe 013 | 1511 | 16.3 | Cattle | (Ditzinger, 1991) |
| [1,7 Asu]eel calcitonin | 3362 | 0.38 | Rabbit | (Cremaschi et al., 1991a) |
| Human calcitonin | 3418 | 20.1 | Cattle | (Lang, 1995) |
| Salmon calcitonin | 3432 | 19.2 | Cattle | (Lang, 1995) |
| Insulin | 5600 | 0.24 | Rabbit | (Bechgaard et al., 1992) |
| | | 0.72 | | (Carstens et al., 1993) |
| | | 0.95 | | (Maitani et al., 1992) |

Notes
a  At 1.0 μmol/mL; at 0.2 μmol/mL $P_{eff} = 0.37 \pm 0.5 \times 10^6$ cm/s.
b  At 0.85 μmol/mL; at 0.1 μmol/mL $P_{eff} = 0.15 \pm 0.1 \times 10^6$ cm/s.
c  At 0.85 μmol/mL.

*Experimental set-ups* – Donor–receiver experiments represent the typical set-up for *in vitro* permeation studies across cell cultured tissues and excised mucosae. For a selection of various diffusion chambers see Figure 14.3. Usually such chambers are well-defined by their volume, geometry and hydrodynamics. The main subjects of such permeation studies are (i) rates and pathways of nasal mucosal transport, (ii) transport and concurrent metabolism, and (iii) effects and mechanisms of nasal absorption enhancement. The typical volumes of the donor or receiver compartments range from about 1–12 ml. The areas available for diffusion are in the range of 1 cm² or lower. To protect the biological function of the tissues, gassing of the systems with carbogen gas (95 per cent $O_2$ and 5 per cent $CO_2$, in case of carbonate sensitive buffers) or $O_2$ alone is required. In conventional donor–receiver set-ups, convection of the buffer media is achieved by means of magnetic bars (see Figure 14.3d). In Ussing chambers circulation and mixing is achieved by a constant gas-lift in both chambers (see Figure 14.3a). Precisely designed ports to accommodate electrodes for electrophysiological measurements are standard features of such equipment. Today, there is a tendency to develop miniaturized equipment which is especially needed when only small amounts of the permeant are available or when detection limits are most critical. Such systems down to 0.2 ml in donor and receiver compartments are on the market now and could be valuable tools for routine permeation testing of nasal drug candidates.

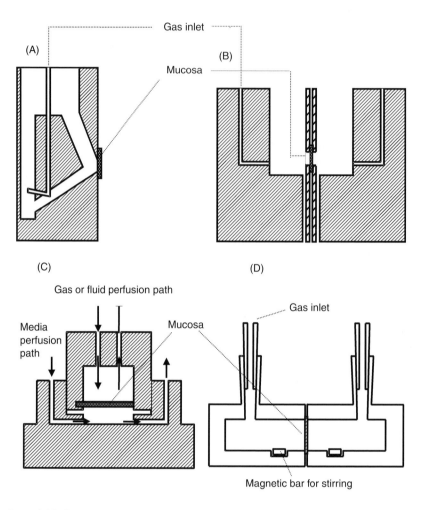

*Figure 14.3* Cross-sections of several donor–receiver systems. (A) Acrylic half-cell of vertical Ussing diffusion chamber system, Cornning Costar Corp., Cambridge, MA. Oxygen gassing is via lateral ducts. The tubular shape of the half-cell provides optimal gassing and mixing. Typically, half-cell volumes are from several ml to 0.2 ml. (B) Diffusion chamber system (donor and receiver), made of polytetrafluoroethylene (PTFE), as suggested by Porta *et al.* (2000) for nanospheres uptake and permeation studies. The use of PTFE is suggested to minimize adsorption of nanospheres. Tissue is mounted between two thin PTFE plates, sealed by poly(dimethylsiloxane) sheets. The set-up prevents nanosphere aggregation and sedimentation resulting from poor convection. (C) Acrylic horizontal diffusion chamber system, Corning Costar Corp., Cambridge, MA. Tissue can be exposed to an air interface to simulate nasal, bronchial or alveolar epithelia. Östh and Bjork (1997) were the first to use this system in combination with excised nasal mucosa (porcine). In this set-up, the serosal surface is constantly perfused with medium, leading to an increased receiver fluid volume which may limit analytical sensitivity. (D) Side-by-side diffusion chamber system, Crown Glass Co. Inc., Somerville, NJ. Set-up made of glass, stirring is provided by two magnetic bars.

### Visualization of pathways

Using confocal laser scanning microscopy (CLSM), the permeation pathways of self-fluorescent and fluorophore-labeled permeants may be localized. The principles of CLSM and its biological applications were reviewed by Shotton (1989). As a non-invasive technique CLSM allows the focusing of optical sections of thick specimens by eliminating the glare of defocused images from planes in front and behind the plane of interest. Conventional microscopy would require fixation of the specimen and time-consuming tissue sectioning procedures. Permeation of the permeants requires viable tissue. Fixation prior to permeation may result in an altered distribution of the permeant. Later on, the tissue may be fixed, and co-staining of cell structures may take place, e.g. nuclei by 4′,6-diamino-2-phenylindole dihydrochloride (DAPI) and actin by phalloidin (Schmidt *et al.*, 1998b). The fixative should ensure adequate and sufficiently rapid fixation. Some fixatives may cause fluorescence quenching. For embedding, an anti-bleaching agent, e.g. n-propyl gallate, should be added to the embedding medium to avoid fluorescence bleaching (Lang *et al.*, 1998). The use of CLSM for quantitative studies is largely limited by bleaching and quenching phenomena. Confocal microscopy of living cells is dependent on the properties and availability of suitable fluorescent probes. A fluorescent probe should produce a strong signal and be both slow to bleach and non-toxic (Terasaki and Dailey, 1995). An inherent problem of using fluorophore-labeled compound is that the label may alter the compound's original permeability and partitioning. Moreover, there should be sufficient information about the metabolic stability of the labeled compound in the tissue. Potential artifacts, related to the effects that polarity, pH, ionic strength etc. may have on the apparent fluorescence intensity and quenching in the various compartments of the cells, need to be carefully considered. Such unknowns can make the interpretation of confocal microscopy quite difficult and suggest the use of this technique only in combination with other experiments. As an alternative, localization of the compound in nasal tissue can also be visualized by immunohistochemical staining, provided that a suitable antibody is available.

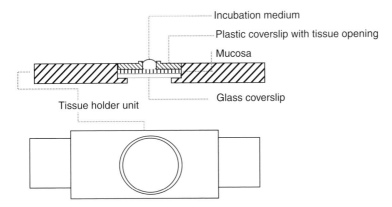

*Figure 14.4* Miniaturized set-up for viability testing and pathway visualization. Volume of incubation medium is 200 µl. Set-up may be directly used for microscopic inspection.

*Experimental set-up* – For pathway visualization studies of fluorophore-labeled compounds we developed techniques (see Figure 14.4 and Protocol 14.3) with miniaturized donor chambers. Volumes as low as 200 µl may be tested with this approach. To assure optimum conditions for the viability of the tissue, the experiment may be carried out in a cell culture incubator (five per cent $CO_2$, 37 °C). For pathway visualization after incubation with fluorophore-labeled permeants, the tissue-holder unit can be directly placed under a microscope (fluorescence microscope, CLSM). By CLSM optical sections of the tissue can be taken. After image processing, data can be illustrated as single layers and orthogonal cut-sections, or as 3D projection pictures.

### Identification of permeation pathways – calcitonin as an example

The nasal transfer of hCT in *bovine nasal mucosa* was shown to be direction-specific and highly temperature-sensitive. Higher rates of mucosal-to-serosal permeation versus lower serosal-to-mucosal rates at 37 °C, in combination with the equalized but much lower permeation rates at 4 °C, indicated that an active transport component may be involved (Lang *et al.*, 1998). The lower permeation rates at 4 °C, compared with the serosal-to-mucosal rate at 37 °C may be due to the lowered fluidity of the biological membranes at reduced temperature. CLSM visualization studies suggested endocytotic uptake of hCT (see Figure 14.5).

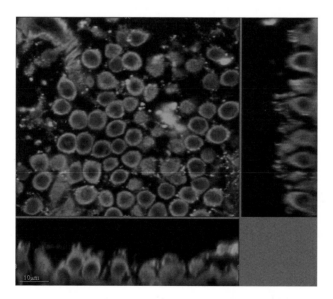

*Figure 14.5* Confocal laser scanning microscopy (CLSM) image for the intracellular uptake of human calcitonin (hCT). Mucosa was fixed (paraformaldehyde) and permeabilized (Triton X-100) for staining after hCT uptake experiment. Cell nuclei were stained by DAPI (blue), hCT in vesicular structures was labeled by immunofluorescence (green) (Schmidt *et al.*, 1998b). (*See Color plate 12*)

Using rabbit nasal mucosa and [1,7 Amino-suberic acid]eel calcitonin (elcatonin) similar observations were reported by Cremaschi *et al.* (1996a). The addition of endocytosis inhibitors, interacting with the cell cytoskeleton and inhibiting the formation and the trafficking of intracellular vesicles, completely inhibits the active net transport, i.e. the difference between mucosal-to-serosal and serosal-to-mucosal flux. Cytochalasin B and D, colchicine, aluminium fluoride and monensin, the inhibitors used in these studies, reduce $R_m$ drastically and shows an effect on the integrity of the tight junctions indicated by the higher paracellular permeability of the polypeptide. Furthermore, when adsorbed on microspheres with a diameter of 0.5 µm, elcatonin displays a positive net flux in contrast to uncovered particles (Cremaschi *et al.*, 1996a,b). In conclusion, endocytosis by a specific uptake mechanism is a typical feature of the nasal uptake of some peptides and may have a physiologic role for the sampling of antigens by the nose (Cremaschi *et al.*, 1991a; Porta *et al.*, 2000). How such findings may be implemented for drug delivery purposes needs to be further evaluated.

## Mechanisms and toxicity of nasal absorption enhancers

For efficacious nasal delivery of many therapeutic peptides and proteins, co-administration of absorption enhancers is imperative. Since the nasal epithelium represents an important defense system, e.g. against mucosal infections, the study of the underlying mechanisms and the local toxicology of absorption enhancers is highly important.

The mechanisms of action of absorption enhancers are not fully understood. The diversity of effects is widespread, among these are perturbation of lipid membrane fluidity, extraction of lipids and proteins, chelation of $Ca^{2+}$ ions, inhibition of metabolic enzymes, inhibition of aggregation, tight junction regulation, or combinations thereof.

*In vitro* nasal mucosa models are useful tools for obtaining information on (i) the effect of enhancers on the flux and the concentration dependency of this effect, (ii) the reversibility and recovery time of the enhancement, (iii) the extent of tissue damage by histological examination, (iv) the respective mechanisms involved, and (v) the effect on the ciliary beat frequency. Absorption promoters include enzyme inhibitors to overcome the enzymatic barrier (Sarkar, 1992) and permeation enhancers to overcome the physiological (physical) barrier (Sayani and Chien, 1996). Synergism of both effects is possible.

For instance, bacitracin is both a potent inhibitor of various proteolytic enzymes and may interact with membrane lipids. The enhancement effect of bacitracin and its reversibility on the permeability of buserelin and gonadorelin was studied using the bovine nasal mucosa model (Ditzinger, 1991). After addition of 0.05 mol/l bacitracin, buserelin showed a three-fold and gonadorelin a ten-fold increase in permeation. The same effect has also been shown *in vivo* in rats. Complete recovery of the epithelium was demonstrated. After a washing period of 2.5 hours in buffer; the enhancement effect of bacitracin on buserelin permeation was completely reversed and normal permeability re-established.

The effect of bile salts on the membrane potential was investigated by Maitani *et al.* (1991) and led to the suggestion that bile salts penetrate into the nasal mucosa and increase the negative membrane charge density. Hosoya *et al.* (1994) observed an effect of bile acids and surfactants on the leaching of proteins from rabbit nasal mucosa and found a good correlation between the changes of $R_m$, $P_{eff}$ and protein leaching activity.

Histological examinations of tissue treated with permeation enhancers may give further information about their effect on tissue morphology. Canine nasal epithelium incubated with the sodium deoxycholate showed extensive loss of the surface layer (Hersey and Jackson, 1987). Changes of the morphology of the mucosal surface after exposure to enhancers have also been observed by scanning electron microscopy. After an exposure of two hours, ethylenediaminetetraacetic acid (EDTA) shows no effect on the surface and sodium glycocholate shows a slight denudation of cilia. Sodium taurodihydrofusidate, sodium deoxycholate and polyoxyethylene-9-lauryl ether cause a drastic morphological change in the mucosal surface (Hosoya *et al.*, 1994). The local toxicity detected *in vitro* might be of minor importance *in vivo* because of the continuous re-epithelialization in the nasal epithelium. It needs to be considered that most *in vitro* models cannot reflect the influence of the enhancers on the viscosity of the mucus or on the mucociliary clearance. The presence of mucus on isolated tissue is expected to be minimal.

The ongoing search of a nasal insulin formulation nicely demonstrates the need to find other ways than the discussed absorption enhancers to increase bioavailability. The achieved progress in this field has been reviewed recently (Hinchcliffe and Illum, 1999). Absorption enhancers for insulin so far include bile salts, surfactants, phospholipids and cyclodextrins. Lipid emulsions also have been examined to improve nasal insulin delivery (Mitra *et al.*, 2000). Most of these absorption enhancers have the disadvantages of being more or less toxic and not being too effective if bioavailability is compared with subcutaneous injection. Therefore, recent studies focus on microspheres given as a nasal powder to increase nasal absorption of insulin. An enhancement of absorption is achieved by increasing the residence time using mucoadhesive polymers as drug carriers. These delivery systems are discussed in the section below.

## Uptake of nano- and microspheres

There is increasing interest in the investigation of nano- or microspheres as nasal drug delivery systems. It is well-established that microspheres may be optimized as mucoadhesive systems to adhere to the epithelial surface, form a gel and then release the drug. On the other hand, nasal mucosa has also been shown to be permeable to some particulates. Therefore, the function of micro- and nanospheres as nasal drug delivery systems may not be restricted to surface adhesion. Moreover, if appropriately designed, they may be envisaged to be taken up as a whole and deliver, for example, encapsulated antigens to specialized epithelial cells, from there to antigen-presenting cells, to the local lymph nodes of the nasal-associated lymphoid tissue (NALT) and/or to the systemic circulation. Microspheres as nasal drug delivery systems, with special emphasis on the mucoadhesive systems, have recently been discussed (Pereswetoff-Morath, 1998). Such microspheres may be prepared from starch, gelatin, chitosan, albumin, dextran, carbopol, hyaluronic acid and cellulose derivatives. Their size is typically in the range of 1–50 μm, sometimes reaching up to 200 μm. Usually such microspheres are water insoluble, but are able to take up water and swell. They have been shown to enhance the absorption of several proteins and peptides, like insulin (Bjork and Edman, 1988) and other drugs, e.g. pentazocine (Sankar *et al.*, 2001). This facilitation of absorption is considered not only to result from mucoadhesion, prolonging the residence time in the nasal cavity, but also by local dehydration of the mucosa (Bjork *et al.*, 1995). Gelation of the microspheres and the concomitant dehydration of adjacent epithelial cells is suggested to cause transient loosening of tight junctions, enhancing the paracellular flux of high molecular weight peptides.

In this case, the possiblity of an uptake of the intact microspheres is assumed to be of minor importance, mainly due to size restriction. Size and surface properties of particulates, in combination with the porosity of the epithelium, determine the ability of microspheres to cross the epithelial barrier. Nasal mucosa is known to be a leaky type epithelium and microspheres the size of 0.8 μm have been detected in the systemic ciruclation after nasal administration to rats (Alpar *et al.*, 1994). These findings, in combination with mucoadhesion, seem to be a promising approach in nasal drug formulation. Nanoparticles that can adhere to the epithelial surface and be taken up intact could serve as carriers for labile substances, including antigens from the nasal cavity into the underlying tissue or the systemic circulation. As an example, insulin-loaded chitosan nanospheres ranging between 300 and 400 nm were found to enhance the absorption of insulin after nasal administration in rabbits as opposed to insulin solution (Fernandez-Urrusuno *et al.*, 1999). The authors speculated that a concerted action of effects could have contributed to this success, namely bioadhesion, opening of tight junctions and also the uptake of intact microspheres. These mechanisms, especially the pathways of microspheres across nasal epithelium, remain to be further examined. By what is known so far, particles are mainly taken up by nasal epithelial M cells, having similar features as those M cells present in the intestinal epithelium. Nasal M cells represent specialized cells and are part of the NALT. Their physiological role is the sampling of antigenic material, e.g. microorganisms, from the nasal epithelium and the subsequent transfer to the underlying lymphoid tissue for antigen processing and presentation (Kuper *et al.*, 1992). Therefore, both a general concern and an appealing potential of such microspheres may be the induction of a mucosal immune response, e.g. in the form of an adjuvant effect. For nasal vaccination such effects would be favored. It is a matter of debate whether other cells, e.g. normal epithelial cells, can also take up particulate matter.

The nasal uptake of large molecules and microspheres has been nicely reviewed, taking *in vivo* as well as *in vitro* studies into account (Donovan and Huang, 1998). The exact assessment of the extent and the mechanisms of particulate uptake is still subject to investigation. Validated *in vitro* methodology to quantify the uptake of microspheres is currently being developed.

*Experimental set-up* – Commercially available Ussing chamber systems are unsuitable to study microsphere uptake. Depending on the size and polymer, microparticles tend to stick to polyacrylate or glass chambers. Moreover, under the conditions of gas-lift systems massive aggregation and sedimentation of particles can occur. Therefore, other set-ups and materials, e.g. teflon or polytetrafluoro-ethylene (PTFE) are preferred. Using the set-up shown in Figure 14.3B, Cremaschi *et al.* (1999) found differences in the amounts of particles crossing the excised rabbit mucosa depending on different proteins adsorbed on the surface of the polystyrene microspheres with 0.5 μm in diameter. In such experiments, all factors that could increase the amount of particles that appear in the receiver chamber need to be controlled as far as possible. Among these are the possibility of edge damage at the rim of the tissue opening and also pores that might appear in the exposed tissue area owing to points of cellular damage. In nasal tissue, such pores were made responsible for a serosal-to-mucosal flux of microspheres (Cremaschi *et al.*, 1996a). *In vivo*, loss of epithelial integrity has been suggested to originate from infection or air pollution (Kuper *et al.*, 1992). Recently a novel diffusion chamber set-up was developed with the tissue in a horizontal position (Turner, 2000). The set-up was inspected under a confocal microscope and the penetration depth of nanospheres of different types was determined. Based on this method, once more a size dependency for the uptake of nanospheres in nasal epithelium could be confirmed. Additionally, kinetic information could be extracted, suggesting a passive mechanism for particle uptake.

Polystyrene latex (PS) microspheres are frequently suggested models to study nasal uptake (Cremaschi *et al.*, 1999). They are commercially available in a wide size range. The particles may be coated with proteins or ligands, both by physical interaction or chemical conjugation. Thus, dependency of uptake and penetration on size, nature of polymer, zeta potential, vehicle, coating with lectins or other adhesion factors and proteins could be investigated. Also, they are available with covalently conjugated fluorescent markers. However, such particles are not biodegradable. Other potential particulates are biodegradable micro- and nanoparticles made of poly(lactide) (PLA) or poly(lactide-*co*-glycolide) (PLGA). At this point the surfaces of such particles are difficult to functionalize by other than physicochemical means, e.g. adsorption.

Experimental techniques to study qualitatively and quantitatively the uptake and the translocation of microspheres in excised nasal epithelium are currently under investigation. For qualitative analysis light, fluorescence or confocal microscopy may be appropriate. Embedding and staining procedures for histology should be carefully selected to avoid dissolution of the polymer.

Quantification of the number of microspheres passing the epithelium from the donor compartment into the receiver half-cell is experimentally challenging. Various methods have been suggested, for example by counting with a hemocytometer. This laborious method is reliable for particles between 0.5 and 1 μm. Another option would be automatic particle counting. The Coulter counter technique measures the change in electrical resistance across a small orifice as a diluted suspension of the particles pass through. Microspheres down to a size range of 0.4–1.2 μm may be characterized by this method. Other methods to assess particles in the receiver fluid are by flow cytometry or fluorimetry, provided that the microspheres carry a fluorescent label for quantification. Gel permeation chromatography (GPC), after dissolution of the polymer in a solvent, may be another technique completely independent of particle size. Finally, after its incubation with microspheres, tissue can be dissolved and then subsequently analyzed for a suitable marker. Nevertheless, there may be difficulties distinguishing between microspheres that were actually taken up by the tissue and those only adsorbed on the surface. A thorough discussion of these methods, in conjuction with intestinal epithelium, can be found in a review by Delie (1998).

## Metabolism studies

The importance and even rate-limiting effect of proteolytic activities on the nasal mucosal transfer, for example of peptides is obvious. Nasal metabolism represents a pseudo-first-pass effect, based on a broad variety of metabolic enzymes present, including cytochrome P-450 enzymes (oxidative phase-I enzymes), conjugative phase-II enzymes, non-oxidative enzymes and proteolytic enzymes. This field was previously reviewed by Sarkar (1992). It was concluded that among the exopeptidases and endopeptidases, which are active in the nasal mucosa, aminopeptidases play an important role. For instance, Ditzinger (1991) showed that pyroglutamate aminopeptidase activity leads to an efficient enzymatic cleavage of the *N*-terminal pyroglutamate group of LHRH and buserelin. In the bovine model, among the endopeptidases involved, chymotrypsin- and trypsin-like activity was found to be responsible for the initial cleavage of hCT (Lang *et al.*, 1996b). Previously, neutral endopeptidase mRNA was identified in human nasal mucosa (Baraniuk *et al.*, 1993).

*In vitro* models represent useful screening tools, for example to identify pathways and kinetics of nasal metabolism, to develop more stable derivatives, and to evaluate suitable additives such as enzyme inhibitors. On the other hand, *in vitro* models ignore the degradation owing to the presence of enzymes or bacteria in the nasal fluid and the changes in pH caused by several diseases. In principle, there are two ways to perform metabolism studies using excised mucosae, (i) permeation and concurrent metabolism studies and (ii) reflection kinetics studies (see Figure 14.6).

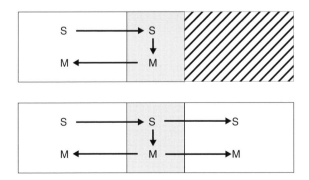

*Figure 14.6* Schematic views of set-ups for permeation and metabolism studies in cell sheets. Top panel: reflection kinetics approach, i.e. diffusion of substrate from donor into cell sheet, metabolite formation in cell sheet, and back-diffusion of metabolite by reflection at impermeable support interface replacing the receiver half-cell. Bottom panel: permeation and concurrent metabolism studies using a standard diffusion chamber set-up. In case of metabolization by mucosal enzymes, the metabolite will be released both into the receiver and donor half-cells.

## Permeation and concurrent metabolism studies

Excised tissue mounted in diffusion chambers allows the extent of metabolic degradation during permeation to be determined. For the procedure see Protocol 14.4. Metabolites need to be monitored in both the donor and receiver compartment. Because of the high initial substrate concentration in the donor, metabolite formation is supposed to be higher in the donor than in the receiver phase. High-performance liquid chromatography (HPLC) analysis possibly combined with mass spectroscopy is useful to identify and quantify the metabolites.

## Reflection kinetics studies

As an alternative to permeation and concurrent metabolism studies, metabolism in viable excised nasal tissue or in cell culture may be also studied using the reflection kinetics approach. The mathematics of such in-and-out experiments was first described by Yu *et al.* (1979) assuming first-order metabolism kinetics and has been more recently reflected by Steinsträsser *et al.* (1995, 1997) for Michaelis–Menten-type saturation kinetics. The reflection kinetics technique uses a single compartment set-up and consists of: (i) diffusion of substrate from a donor compartment (e.g. one half-cell of a side-by-side diffusion system) into a cell sheet, (ii) metabolism within the cell sheet, and (iii) back-diffusion of the metabolites into the donor compartment due to reflection at an impermeable support interface (e.g. an acrylic block replacing the second half-cell). Schematic views of both set-ups are given in Figure 14.6.

In order to better understand the influence of metabolism and of diffusion in this system, a physical model approach may be used for interpretation of the data. A suitable physical model was previously introduced for the (keratinocyte) HaCaT cell culture sheet model (Steinsträsser *et al.*, 1995, 1997) and was also used to explain data derived from the human nasal RPMI 2650 cell culture model (Peter, 1996). Simulations based on this model can be used to interpret the data in terms of the rate-controlling effects, e.g. the influence of Fickian diffusion versus Michaelis–Menten-type metabolism. Experiments under reflection kinetics conditions can be performed with the excised bovine nasal mucosa model (Lang *et al.*, 1996a,b). Metabolic rates ($V_M$, mmol/cm/s) can be calculated from the linear parts of the concentration-time profile $(dC/dt)_{ss}$ in the steady-state according to:

$$V_M = -\left(\frac{dC}{dt}\right)_{ss}\frac{V}{A},\qquad (14.4)$$

where $A$ (cm$^2$) is the surface area of the membrane and $V$ (ml) the volume of the bulk solution.

The experimental procedure of the reflection studies is more straightforward and technically easier to handle than the corresponding donor–receiver permeation studies. Moreover, reflection studies are regarded as more sensitive to metabolism.

As an example, the metabolism of hCT was previously presented. (Lang *et al.*, 1996b). The peptide sequences of the main nasal metabolites of hCT extracted from the bulk solution were analyzed by liquid secondary ionization mass spectrometry (LSIMS), following HPLC fractionation of the metabolites, and by matrix-assisted laser desorption ionization mass spectrometry (MALDI). The sites of primary cleavage of hCT in nasal mucosa were in the medium section of the molecule (cleavage sites: 16–17, 18–19, and 19–20) and owing to chymotryptic- and tryptic-like endopeptidases. There was no initial *N*- or *C*-terminal cleavage of the peptide taking place.

## CONCLUSIONS

Further improvement of the performance and significance of nasal *in vitro* models appears to be possible. Among the subjects to be studied are, above all, the permeabilities of given permeants, the localization of metabolic enzymes, the differences in metabolic activities as compared with intact *in vivo* human tissue, the cell biology of enhancer effects, and the relevance of active transport and endocytosis. Routes and pathways of permeation and metabolism will be subjects of continuing interest. Fluorophore-labeled markers and drugs and new fluorogenic enzyme substrates in combination with sophisticated microscopy techniques, e.g. CLSM, will play a major role in observing permeation pathways and enzyme activities in the nasal tissue. Finally, the uptake of particulates will be a prime subject for the development of nasal mucosal vaccines.

## REFERENCES

Alpar, H. O., Almeida, A. J. and Brown, M. R. (1994) Microsphere absorption by the nasal mucosa of the rat. *J. Drug Target.*, **2**, 147–149.

Baraniuk, J. N., Ohkubo, K., Kwon, O. J., Mak, J., Rohde, J., Kaliner, M. A. *et al.* (1993) Identification of neutral endopeptidase mRNA in human nasal mucosa. *J. Appl. Physiol.*, **74**, 272–279.

Bechgaard, E., Gizurarson, S., Jorgensen, L. and Larsen, R. (1992) The viability of isolated rabbit nasal mucosa in the Ussing chamber, and the permeability of insulin across the membrane. *Int. J. Pharm.*, **87**, 125–132.

Bjork, E. and Edman, P. (1988) Degradable starch microspheres as a nasal delivery system for insulin. *Int. J. Pharm.*, **47**, 233–238.

Bjork, E., Isaksson, U., Edman, P. and Artursson, P. (1995) Starch microspheres induce pulsatile delivery of drugs and peptides across the epithelial barrier by reversible separation of the tight junctions. *J. Drug Target.*, **2**, 501–507.

Boucher, R. C., Stutts, M. J., Knowles, M. K., Cantley, L. and Gatzy, J. R. (1986) $Na^+$ transport in cystic fibrosis respiratory epithelia. Abnormal basal rate and response to adenylate cyclase activation. *J. Clin. invest.*, **78**, 1245–1252.

Chien, Y. W., Su, K. S. E. and Chang, S.-F. (1989) *Nasal systemic drug delivery*, Marcel Dekker Inc., New York.

Cremaschi, D., Ghirardelli, R. and Porta, C. (1998) Relationship between polypeptide transcytosis and lymphoid tissue in the rabbit nasal mucosa. *Biochim. Biophys. Acta.*, **1369**, 287–294.

Cremaschi, D., Porta, C. and Ghirardelli, R. (1996a) The active transport of polypeptides in the rabbit nasal mucosa is supported by a specific vesicular transport inhibited by cytochalasin D. *Biochim Biophys. Acta.*, **1283**, 101–105.

Cremaschi, D., Porta, C. and Ghirardelli, R. (1999) Different kinds of polypeptides and polypeptide-coated nanoparticles are accepted by the selective transcytosis shown in the rabbit nasal mucosa. *Biochim. Biophys. Acta.*, **1416**, 31–38.

Cremaschi, D., Porta, C., Ghirardelli, R., Manzoni, C. and Caremi, I. (1996b) Endocytosis inhibitors abolish the active transport of polypeptides in the mucosa of the nasal upper concha of the rabbit. *Biochim. Biophys. Acta.*, **1280**, 27–33.

Cremaschi, D., Rossetti, C., Draghetti, M. T., Manzoni, C. and Aliverti, V. (1991a) Active transport of polypeptides in rabbit nasal mucosa: possible role in the sampling of potential antigens. *Pflugers. Archiv.*, **419**, 425–432.

Cremaschi, D., Rossetti, C., Draghetti, M. T., Manzoni, C., Porta, C. and Aliverti, V. (1991b) Transepithelial electrophysiological parameters in rabbit respiratory nasal mucosa isolated *in vitro. Comp. Biochem. Physiol. – A-Comp. Physiol.*, **99**, 361–364.

De Fraissinette, A., Kolopp, M., Schiller, I., Fricker, G., Gammert, C., Pospischil, A. *et al.* (1995) *In vitro* tolerability of human nasal mucosa: histopathological and scanning electron-microscopic evaluation of nasal forms containing sandostatin(TM). *Cell Biol. Toxicol.*, **11**, 295–301.

Delie, F. (1998) Evaluation of nano- and microparticle uptake by the gastrointestinal tract. *Adv. Drug Deliv. Rev.*, **34**, 221–233.

Ditzinger, G. (1991) Wirkung, Funktion und Verträglichkeit strukturanaloger zyklischer Peptide als Absorptionsförderer auf Schleimhäuten, *PhD-Thesis*, Johann Wolfgang Goethe-Universität, Frankfurt am Main, Germany.

Donovan, M. D. and Huang, Y. (1998) Large molecule and particulate uptake in the nasal cavity: the effect of size on nasal absorption. *Adv. Drug Deliv. Rev.*, **29**, 147–155.

Fernandez-Urrusuno, R., Calvo, P., Remunan-Lopez, C., Vila-Jato, J. L. and Alonso, M. J. (1999) Enhancement of nasal absorption of insulin using chitosan nanoparticles. *Pharm. Res.*, **16**, 1576–1581.

Glück, U., Gebbers, J. O. and Glück, R. (1999) Phase 1 evaluation of intranasal virosomal influenza vaccine with and without Escherichia coli heat-labile toxin in adult volunteers. *J. Virol.*, **73**, 7780–7786.

Hersey, S. J. and Jackson, R. T. (1987) Effect of bile salts on nasal permeability in vitro. *J. Pharm. Sci.*, **76**, 876–879.

Hinchcliffe, M. and Illum, L. (1999) Intranasal insulin delivery and therapy. *Adv. Drug Deliv. Rev.*, **35**, 199–234.

Hosoya, K. I., Kubo, H., Natsume, H., Sugibayashi, K. and Morimoto, Y. (1994) Evaluation of enhancers to increase nasal absorption using Ussing chamber technique. *Biol. Pharm. Bull.*, **17**, 316–322.

Hosoya, K. I., Kubo, H., Natsume, H., Sugibayashi, K., Morimoto, Y. and Yamashita, S. (1993) The structural barrier of absorptive mucosae: site difference of the permeability of fluorescein isothiocyanate-labelled dextran in rabbits. *Biopharm. Drug Dispos.*, **14**, 685–695.

Johnson, L. G. and Boucher, R. C. (1993) In K. L. Audus and T. J. Raub (eds), *Biological Barriers to Protein Delivery*, Vol. 7, Plenum Press, New York, pp. 161–178.

Jones, K. H. and Senft, J. A. (1985) An improved method to determine cell viability by simultaneous staining with fluorescein diacetate-propidium iodide. *J. Histochem. Cytochem.*, **33**, 77–79.

Jorgensen, L. and Bechgaard, E. (1993) Intranasal absorption of angiopeptin: *In vitro* study of absorption and enzymatic degradation. *Int. J. Pharm.*, **99**, 165–172.

Kotze, A. F., Thanou, M. M., Luebetaen, H. L., de Boer, A. G., Verhoef, J. C. and Junginger, H. E. (1999) Enhancement of paracellular drug transport with highly quaternized N-trimethyl chitosan chloride in neutral environments: in vitro evaluation in intestinal epithelial cells (Caco-2). *J. Pharm. Sci.*, **88**, 253–257.

Kubo, H., Hosoya, K. I., Natsume, H., Sugibayashi, K. and Morimoto, Y. (1994) *In vitro* permeation of several model drugs across rabbit nasal mucosa. *Int. J. Pharm.*, **103**, 27–36.

Kuper, C. F., Koornstra, P. J., Hameleers, D. M., Biewenga, J., Spit, B. J., Duijvestijn, A. M. *et al.* (1992) The role of nasopharyngeal lymphoid tissue [see comments]. *Immunol. Today*, **13**, 219–224.

Lang, S., Oschmann, R., Traving, B., Langguth, P. and Merkle, H. P. (1996a) Transport and metabolic pathway of thymocartin (TP4) in excised bovine nasal mucosa. *J. Pharm. Pharmacol.*, **48**, 1190–1196.

Lang, S., Rothen-Rutishauser, B., Perriard, J. C., Schmidt, M. C. and Merkle, H. P. (1998) Permeation and pathways of human calcitonin (hCT) across excised bovine nasal mucosa. *Peptides*, **19**, 599–607.

Lang, S. R., Staudenmann, W., James, P., Manz, H.-J., Kessler, R., Galli, B. *et al.* (1996b) Proteolysis of human calcitonin in excised bovine nasal mucosa: elucidation of the metabolic pathway by liquid secondary ionization mass spectrometry (LSIMS) and matrix assisted laser desorption ionization mass spectometry (MALDI). *Pharm. Res.*, **13**, 1679–1685.

Maitani, Y., Ishigaki, K., Takayama, K. and Nagai, T. (1997) In vitro nasal transport across rabbit mucosa: effect of oxygen bubbling, pH and hypertonic pressure on permeability of lucifer yellow, diazepam and 17b-estradiol. *Int. J. Pharm.*, **146**, 11–19.

Maitani, Y., Uchida, N., Nakagaki, M. and Nagai, T. (1991) Effect of bile salts on the nasal mucosa: membrane potential measurement. *Int. J. Pharm.*, **69**, 21–27.

McMartin, C., Hutchinson, L. E., Hyde, R. and Peters, G. E. (1987) Analysis of structural requirements for the absorption of drugs and macromolecules from the nasal cavity. *J. Pharm. Sci.*, **76**, 535–540.

Mitra, R., Pezron, I., Chu, W. A. and Mitra, A. K. (2000) Lipid emulsions as vehicles for enhanced nasal delivery of insulin. *Int. J. Pharm.*, **205**, 127–134.

Mosmann, T. (1983) Rapid colorimetric assay for cellular growth and survival: application to proliferation and cytotoxicity assays. *J. Immunol. Methods*, **65**, 55–63.

Östh, K. and Bjork, E. (1997) Characterization of pig nasal tissue in a new horizontal Ussing chamber. *Proc. Control. Release Soc. Issues.*, **24**, 417–418.

Pereswetoff-Morath, L. (1998) Microspheres as nasal drug delivery systems. *Adv. Drug Deliv. Rev.*, **29**, 185–194.

Peter, H. (1996) Cell culture sheets to study nasal peptide metabolism: the human nasal RPMI 2650 cell line model, *PhD-Thesis*, Swiss Federal Institute of Technology Zurich, Zurich, CH.

Peter, H., Wunderli-Allenspach, H., Gammert, C. and Merkle, H. P. (1992) Human nasal cell culture system to study biotransformation of peptides. *Eur. J. Pharm. Biopharm.*, **38**, 31S.

Porta, C., Dossena, S., Rossi, V., Pinza, M. and Cremaschi, D. (2000) Rabbit nasal mucosa: nanospheres coated with polypeptides bound to specific anti-polypeptide IgG are better transported than nanospheres coated with polypeptides or IgG alone. *Biochim. Biophys. Acta.*, **1466**, 115–124.

Reardon, P. M. (1996) In R. T. Borchardt, P. L. Smith and G. Wilson (eds), *Models for Assessing Drug Absorption and Metabolism*, Vol. 8, Plenum Press, New York, pp. 309–323.

Reardon, P. M., Gochoco, C. H., Audus, K. L., Wilson, G. and Smith, P. L. (1993) *In vitro* nasal transport across ovine mucosa: effects of ammonium glycyrrhizinate on electrical properties and permeability of growth hormone releasing peptide, mannitol, and lucifer yellow. *Pharm. Res.*, **10**, 553–561.

Sankar, C., Rani, M., Srivastava, A. K. and Mishra, B. (2001) Chitosan based pentazocine microspheres for intranasal systemic delivery: development and biopharmaceutical evaluation. *Pharmazie*, **56**, 223–226.

Sarkar, M. A. (1992) Drug metabolism in the nasal mucosa. *Pharm. Res.*, **9**, 1–9.

Sayani, A. P. and Chien, Y. W. (1996) Systemic delivery of peptides and proteins across absorptive mucosae. *Crit. Rev. Ther. Drug Carrier Syst.*, **13**, 85–184.

Schaerer, E., Neutra, M. R. and Kraehenbuhl, J. P. (1991) Molecular and cellular mechanisms involved in transepithelial transport. *J. Membr. Biol.*, **123**, 93–103.

Schmidt, M. C., Peter, H., Lang, S. R., Ditzinger, G. and Merkle, H. P. (1998a) In vitro cell models to study nasal mucosal permeability and metabolism. *Adv. Drug Deliv. Rev.*, **29**, 51–79.

Schmidt, M. C., Rothen-Rutishauser, B., Rist, B., Beck-Sickinger, A., Wunderli-Allenspach, H., Rubas, W. *et al.* (1998b) Translocation of human calcitonin in respiratory nasal epithelium is associated with self-assembly in lipid membrane. *Biochemistry*, **37**, 16582–16590.

Schmidt, M. C., Rubas, W. and Merkle, H. P. (2000a) Nasal epithelial permeation of thymotrinan (TP3) versus thymocartin (TP4): competitive metabolism and self-enhancement. *Pharm. Res.*, **17**, 222–228.

Schmidt, M. C., Simmen, D., Hilbe, M., Boderke, P., Ditzinger, G., Sandow, J. *et al.* (2000b) Validation of excised bovine nasal mucosa as in vitro model to study drug transport and metabolic pathways in nasal epithelium. *J. Pharm. Sci.*, **89**, 396–407.

Shotton, D. M. (1989) Confocal scanning optical microscopy and its applications for biological specimens. *J. Cell Sci.*, **94**, 175–206.

Steinsträsser, I., Koopmann, K. and Merkle, H. P. (1997) Epidermal aminopeptidase activity and metabolism as observed in an organized HaCaT cell sheet model. *J. Pharm. Sci.*, **86**, 378–383.

Steinsträsser, I., Sperb, R. and Merkle, H. P. (1995) Physical model relating diffusional transport and concurrent metabolism of peptides in metabolically active cell sheets. *J. Pharm. Sci.*, **84**, 1332–1341.

Terasaki, M. and Dailey, M. E. (1995) In J. B. Pawley (ed), *Handbook of biological confocal microscopy* Plenum Press, New York, pp. 327–346.

Turner, J. (2000) Development of a novel in vitro system for nasal drug delivery development, *PhD-Thesis*, Aston University, Birmingham.

Uchida, N., Maitani, Y., Machida, Y., Nakagaki, M. and Nagai, T. (1991) Influence of bile salts on the permeability of insulin through the nasal mucosa of rabbits in comparison with dextran derivatives. *Int. J. Pharm.*, **74**, 95–103.

Wadell, C., Bjork, E. and Camber, O. (1999) Nasal drug delivery – evaluation of an *in vitro* model using porcine nasal mucosa. *Eur. J. Pharm. Sci.*, **7**, 197–206.

Werner, U., Kissel, T. (1995) Development of a human nasal epithelial cell culture model and its suitability for transport and metabolism studies under *in vitro* conditions. *Pharm. Res.*, **12**, 565–571.

Wheatley, M. A., Dent, J., Wheeldon, E. B. and Smith, P. L. (1988) Nasal drug delivery: an in vitro characterization of transepithelial electrical properties and fluxes in the presence or absence of enhancers. *J. Control. Release*, **8**, 167–177.

Yu, C. D., Fox, J. L., Ho, N. F. and Higuchi, W. I. (1979) Physical model evaluation of topical prodrug delivery-simultaneous transport and bioconversion of vidarabine-5′-valerate II: parameter determinations. *J. Pharm. Sci.*, **68**, 1347–1357.

# Chapter 15

# Cell culture models of the corneal and conjunctival epithelium

*Pekka Suhonen, Jennifer Sporty, Vincent H. L. Lee and Arto Urtti*

## INTRODUCTION

Drug treatment of eye diseases can be carried out by local ocular drug application or by systemic drug administration. In the latter case the drug is distributed widely in the body after, for example, oral or intravenous administration. Only a very small fraction of the dose gains access to the eye. Therefore, most ocular drugs are given locally as eye drops. The eye drops are applied on the ocular surface and in most cases the drug absorbs into the inner tissues of the eye. Despite local administration, less than 5 per cent of the instilled dose absorbs into the eye (Maurice and Mishima, 1984). For this reason the drugs are often applied in high concentrations.

Figure 15.1 illustrates the basic factors in the ocular pharmacokinetics. After instillation the eye drop solution flows from the ocular surface to the nasolacrimal drainage system. The ocular contact of the solution is short, with a half-life often less than one minute (Lee and Robinson, 1979). A concentration gradient between tear fluid and anterior tissues is the driving force of drug absorption. Most small lipophilic drugs that are used in the clinic are absorbed into the eye via the cornea. Some hydrophilic or large molecules absorb into the eye through the conjunctiva and sclera (Ahmed and Patton, 1984). After corneal permeation the drug diffuses rapidly to the aqueous humor where it distributes easily to the anterior uvea (iris, ciliary body). However, drug distribution to the lens is very slow owing to its highly organized dense structure (Urtti *et al.*, 1990). Only a small fraction of the drug distributed from the aqueous humor reaches the vitreous humor. For example, when compared with the drug concentrations in the lacrimal fluid immediately after instillation, the peak concentrations in the cornea, aqueous humor, and vitreous are typically 1000, 10,000 and 100,000 times less, respectively (Urtti *et al.*, 1990).

Drugs are eliminated from the anterior segment in conjunction with aqueous humor turnover and by the venous blood flow of the anterior uvea (Maurice and Mishima, 1984). From the posterior segment the drugs may be eliminated via the anterior route using the mechanisms described above, or through the blood–retinal barrier to the blood circulation. Both mechanisms move drugs into the systemic blood circulation.

Ocular pharmacokinetics is usually investigated *in vivo* using rabbits as animal models. Some investigators use non-invasive fluorometric techniques to follow the

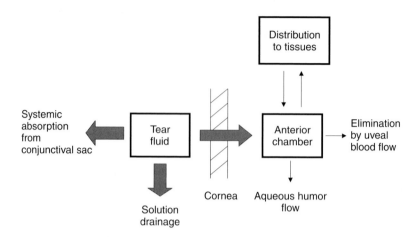

*Figure 15.1* Schematic model of the basic factors in the ocular pharmacokinetics.

concentration profiles in the aqueous humor. This method cannot be used to study the pharmacokinetics of commonly used drugs because it is applicable only to fluorescent model compounds that are not prescribed. Normally, ocular pharmacokinetics is studied using invasive experimental techniques. Rabbits must be sacrificed at each time-point to collect tissue samples. Because standard deviations are often quite substantial in ocular kinetic studies, large rabbit groups are needed for meaningful results. To generate concentration curves with six time-points for six experiments require 36 rabbits. Each drug used in comparative studies would require a minimum 36 rabbits. For example, a comparison of three drugs or formulations requires at least 108 rabbits. This kind of animal use is an ethical concern and the experiments are very expensive.

Isolated tissues from rabbit eyes are used to obtain relevant *in vitro* information about the absorption potential of the drugs. Corneal permeability data are often obtained by sacrificing a rabbit and excising the cornea or a piece of palpebral conjunctiva. The tissue is then placed into a diffusion chamber. There are some problems in the permeability studies with isolated corneas. Most notably, the corneal tissue retains its viability only for a few hours, so the interpretation of the findings on active transport and metabolism may vary with experiment timing. Also, differences in the structure of transporters and enzymes between species make predictions to a human model more difficult. Viable human cell culture models may avoid viability and species applicability problems.

Eye drops contain drugs and many excipients including preservatives, buffering agents, anti-oxidants, and solubilizers that can make a drug impractical for pharmacological use. Drugs and preservatives can be harmful to the cornea, and because the cornea is highly innervated any irritation or unpleasant sensation may hamper the compliance of the patient. Eye irritation from the use of such ocular products is studied *in vivo* in rabbits using the Draize test that is frequently criticized for its unreliability and unethical nature. A more reliable and morally acceptable substitute method is needed to screen such compounds and products for their irritation potential. Prescreening, for example could reduce the number of animal experiments by eliminating the most toxic compounds on the basis of cell tests.

New corneal and conjunctival cell culture models are needed for predicting the ocular absorption and tolerability of new drugs and excipients. This review describes some advances in the field of ocular cell cultures.

## ANTERIOR OCULAR BARRIERS FOR DRUG ABSORPTION

*Cornea*. The cornea is the main route of drug absorption from the tear fluid to the inner eye for lipophilic drugs like timolol (Ahmed and Patton, 1984), pilocarpine (Doane *et al.*, 1978) and hydrocortisone (Doane *et al.*, 1978). The main corneal penetration barrier usually is the epithelium (Sieg and Robinson, 1976). The lipophilicity of a compound determines the effectiveness of an epithelial layer as a barrier. In the case of hydrophilic atenolol, the entire epithelium behaves as the penetration limiting barrier, whereas drug permeability of compounds with intermediate lipophilicity, like timolol and levobunolol, is limited considerably by only the most superficial cell layers. The tight junctions in these outermost cell layers of corneal epithelium limit penetration of hydrophilic molecules (Wang *et al.*, 1991). In contrast, wing cells and basal cells are less tightly interconnected and allow intercellular penetration of macromolecules.

Corneal stroma is a loosely arranged hydrophilic layer without continuous cellular structure. Consequently, drug diffusion in the stroma is fast and in the typical range of values for drugs whose diffusion is not dependent on lipophilicity or molecular weight (Maurice and Mishima, 1984).

Many ophthalmic drugs show considerable lipophilicity, making transcellular drug penetration possible. Increasing the lipophilicity of a compound can enhance corneal permeability and a maximal permeability of beta-blockers was observed at log (octanol/water) partition coefficients of 2–3 (Huang *et al.*, 1983). Because this partition coefficient is in the same range as that of very lipophilic compounds (log P > 3), corneal permeability cannot be further improved by increasing the lipophilicity. Instead, steady (Wang *et al.*, 1991) or decreasing (Huang *et al.*, 1983; Schoenwald and Ward, 1978) permeabilities have been observed by further raising lipophilicity. This phenomenon is a result of impaired drug desorption from the lipophilic parts of the corneal epithelium to the hydrophilic stroma (Huang *et al.*, 1983). Therefore, the corneal stroma is the penetration rate-limiting barrier in cases with highly lipophilic drugs (log P > 3).

Data sets from more heterogeneous groups of compounds revealed substantial deviations from the simple partitioning effects resulting from varying molecular sizes and charges (Liaw *et al.*, 1992). At neutral pH the rabbit cornea behaves as if it were negatively charged. Consequently, positively charged L-lysine had ten times higher paracellular permeability than the negatively charged L-glutamic acid (Liaw *et al.*, 1992).

Two additional methods for increasing corneal permeability of drugs have been determined. Optimization of the formulation pH increases the fraction of unionized drug, allowing the drug to further penetrate the corneal route. Another more demanding approach is to prepare a prodrug derivative with improved corneal absorption characteristics that releases the active parent compound after enzymatic or chemical hydrolysis in the eye.

*Conjunctiva.* The conjunctiva covers most of the ocular surface and has a higher permeability than the cornea, especially when considering hydrophilic compounds (Wang *et al.*, 1991). Consequently, depending on dosing conditions and dosage form, 5–30 times more timolol (Chang and Lee, 1987; Urtti *et al.*, 1990) and pilocarpine (Urtti *et al.*, 1985) are absorbed through the conjunctiva to systemic circulation than transcorneally into the eye. The conjunctiva is also more permeable to large molecules than the cornea as a result of wider and more numerous paracellular spaces. The porosity caused by the paracellular spaces is much higher in the conjunctiva than in the cornea (Hämäläinen *et al.*, 1997). A favorable route of drug absorption is through the conjunctiva to the sclera and thereafter to the ciliary body (Ahmed and Patton, 1984). Drugs that are absorbed via this route do not usually gain access to the aqueous humor. Hydrophilic and large molecules, like insulin, that have poor corneal permeability prefer the conjunctiva to sclera route where penetration is more favorable.

## CONJUNCTIVA CELL CULTURES

The conjunctiva lines the eyelids and connects to the surface of the eyeball. Two folds in the conjunctiva form the conjunctival sacs that allow movement of the eyelids and eyeball. Applied drugs must first pass the bulbar conjunctiva to reach the uveal tract for treating uveitis and other inner eye diseases. The lower conjunctival sac is useful for drug administration via eye drops and suspensions and has been studied for its permeability characteristics.

The conjunctiva is more permeable to hydrophilic compounds than is the cornea (Wang *et al.*, 1991). Timolol, a hydrophilic compound, is systemically absorbed 2.5 times more efficiently by the conjunctiva than the cornea and, with a five–ten minutes contact time, ocular absorption is increased (Chang and Lee, 1987). Ocular absorption of timolol can be further increased four- to five-fold in concentration over systemic absorption by coadministering phenylephrine in concentrations between 0.8 and 8.2 mg/ml with 25 µl of 5 mg/ml timolol in eye drop form to pigmented rabbits (Kyyrönen and Urtti, 1990). Phenylephrine constricts capillaries in the conjunctiva, thereby preventing systemic absorption and further increasing ocular absorption.

These aforementioned studies relied on the use of excised tissue, requiring live animals, and raise ethical and financial concerns. Cell cultures, if able to precisely mimic the characteristics of the conjunctiva, would be an ethically and financially superior alternative. In the struggle to obtain a stable conjunctival cell line, better primary culturing methods have been produced. Unfortunately, a method for keeping the cells viable past three passages has eluded researchers. However, primary conjunctival cell cultures have proven useful for studying drug transport and mechanisms as well as for formulating new drugs. Much progress has been made from the first primary cell lines in liquid-covered conditions (LCC) to the newer, more biologically relevant air-interface condition.

## Liquid-covered cultures

The first functional primary conjunctival epithelial cell culture was developed by Saha *et al.* (1996a) and contained five different cell types. Fresh conjunctival tissue was isolated from pigmented rabbits and incubated in 0.2 per cent protease in S-MEM at 37 °C for 90 minutes to loosen cells from the tissue. The cells were scraped from the tissue with forceps and transferred into a 10 per cent fetal bovine serum (FBS) solution in minimum essential media for suspension culture (S-MEM) with 0.5 mg/ml DNAse-I to inhibit further protease action. Spinning the cells in a centrifuge twice at 210×g for ten minutes resulted in a cell pellet that was filtered through a 40 µm cell strainer and spun down again. The cell pellet was resuspended in PC-1 growth medium supplemented with 1 per cent FBS, 2 mM L-glutamine, 100 units/ml penicillin-streptomycin, 0.5 per cent gentamicin, and 0.4 per cent fungizone, and plated onto Transwells that were precoated with collagen types I and II. The Transwells were bathed in supplemented PC-1 medium and weaned off of 1 per cent FBS complementation by the third day. The resulting culture exhibited a transepithelial electrical resistance (TEER) of $1.9 \pm 0.2 \, k\Omega \, cm^2$, a potential difference (*PD*) of $14.2 \pm 1.6 \, mV$, and an equivalent short-circuit current ($I_{eq}$) of $8.0 \pm 0.4 \, \mu A/cm^2$. A typical TEER value for excised conjunctiva tissue is $1.3 \, k\Omega \, cm^2$. Typical *PD* and $I_{eq}$ values are $17.7 \, mV$ and $14.5 \, \mu A/cm^2$, respectively. Measurements from the primary culture closely match those found *in vivo*, proving the suitability of primary conjunctival cell cultures for drug transport studies. Unfortunately, the TEER for these cultures began to decrease after ten days of culturing. The short viability of these primary cultures still proves to be a problem and prevents stable cell lines from being produced.

Saha *et al.* (1996b) used this method to culture rabbit conjunctival epithelial cell layers to characterize the permeability of the culture to low molecular weight drugs varying in lipophilicity. The Transwells containing conjunctival epithelial layers were washed twice with modified Ringer's solution with 0.075 mM bovine serum albumin (BSA) to prevent the TEER from falling and supplemented with a low molecular weight drug. Samples were taken from the basolateral solution every half-hour, for four hours, to measure the concentration of the remaining drug in the solution using either a liquid scintillation counter or high performance liquid chromatography (HPLC). Apparent permeability coefficients ($P_{app}$) were determined for [3]H-mannitol, sotalol and atenolol (hydrophilic), metoprolol, timolol, and propranolol (moderately lipophilic), and betaxolol (highly lipophilic). It was found through these experiments that the permeability of tight conjunctival epithelial layers, created by culturing rabbit conjunctiva in LCC, to lipophilic beta-blockers (metoprolol, timolol, propranolol, and betaxolol) is of the same order of magnitude as freshly excised rabbit conjunctival tissue. The hydrophilic drugs, sotalol and atenolol, were 100 times more permeable across excised tissue than across cultures and may be a result of their preference for a paracellular transport pathway that does not require defeating hydrophobic forces in cell membranes over a transcellular pathway. Primary conjunctival epithelial cell cultured in LCC has proven to possess not only the same TEER, $I_{eq}$, and *PD*, but also very similar permeability profiles for lipophilic, low molecular weight drugs, making these primary cultures excellent models for the study of lipophilic drug transport.

Other studies have further supported the similarities between excised conjunctival tissue and primary conjunctival cultures. Cyclosporin A (CSA), a lipophilic drug, is prevented from entering mammalian cells by *p*-glycoprotein (*p*-gp), a drug efflux pump 170 kDa in size (gp170). Because CSA was being tested for the treatment of uveitis and allergic conjunctivitis, it was essential to characterize gp170, located by Western blots, in the conjunctiva (Saha *et al.*, 1998). Primary conjunctiva epithelial cultures were used as experimental models for assurance that gp170 from endo-thelial tissue was not affecting the uptake of CSA through the conjunctiva. Trans-port experiments were conducted by adding tritiated and unlabeled CSA in concentrations of 0.5 μm or 5 μm to Transwell inserts on either the apical side, for apical-to-basolateral transport, or the basolateral side, for basolateral-to-apical trans-port. Transport was terminated by the addition of unlabeled CSA to the receiving side of the insert (i.e. the basolateral side for apical-to-basolateral transport). A scintillation counter was used to determine the amount of $^3$H-CSA transported through the cell layers. The effects of blocking gp170 were also studied. A function-blocking, murine monoclonal antibody, or one of two *p*-gp inhibitors, verapamil or progesterone, was applied to the basolateral and apical sides before transport began. Tritiated CSA was applied as before and measured at various time inter-vals. The $P_{app}$ for CSA was found to be $0.83 \pm 0.09 \times 10^{-6}$ cm/sec for basal-to-apical transport and $0.09 \pm 0.0 \times 10^{-6}$ cm/sec for apical-to-basal transport, indicating that *p*-gp is an efflux pump (from basal-to-apical) rather than an influx pump. A similar pattern was also observed for verapamil, with and without the function blocking antibody, and dexamethasone. Because the functioning antibody could not completely block the transport of CSA, it was determined that approximately 30 per cent of the transport is *p*-gp-independent. Without a primary cell culture, it would have been questionable whether endothelial cells containing *p*-gp could have interfered with drug transport. The resulting data would have been useless to those studying CSA transport for disease treatment.

Another study relying on primary cultures resulted in the discovery and charac-terization of a proton-coupled dipeptide transporter (Basu *et al.*, 1998). Previous studies showed that L-carnosine, a naturally occurring hydrolysis-resistant di-peptide, could be transported across the conjunctiva via a carrier in a pH-depend-ent process. Primary conjunctiva epithelial cultures were grown in 12 mm Transwells coated with rat tail type-I collagen and fibronectin. Tritiated L-carnosine was added to either the basolateral or apical side of the insert to begin transport. Transport was terminated by removing the tritiated L-carnosine and submersing the inserts in an ice-cold bicarbonated Ringer's solution. The cell layers were solu-bilized and L-carnosine uptake was measured by liquid scintillation. Diffusional uptake of L-carnosine through the paracellular spaces was measured at 4 °C and subtracted from the overall uptake to determine that the carrier-mediated uptake constituted 94 per cent of the overall uptake. Further experimentation with the pH of the L-carnosine solutions showed that the uptake of L-carnosine through the transporter occurred through a proton-coupled dipeptide transporter. This trans-porter is most likely located on the apical side of the cell layers as determined by a five-fold higher uptake from the apical than the basolateral side. The discovery and characterization of this proton-coupled dipeptide transporter may lead to better ocular drug absorption and more effective application techniques. By isolating conjunctiva epithelial layers through primary culturing, specific interactions of drugs with the conjunctiva can be characterized to increase drug permeability and efficacy in the eye.

## Air-interface cultures (AIC)

A seemingly more accurate conjunctival epithelial model arose with the advent of AIC. Tracheal cells were the first to be grown as AIC and were found to have better drug transport characteristics since AIC more closely mimic the *in vivo* situation (Mathias *et al.*, 1995). Because conjunctival epithelial cells also occur at an air-interface *in vivo*, the AIC technique was applied to conjunctiva primary cultures (Yang *et al.*, 2000). The cells were isolated according to Saha *et al.*, 1996a and grown in LCC on Transwells until day four when the cells were transferred to an air-interface condition by removing media on the apical surface of the insert. Electrical and transport properties were measured for cells grown in this manner. Transferring cells to air-interface conditions on day four produced a peak TEER and *PD* of $1.06 \pm 0.06\,\mathrm{k\Omega cm^2}$ and $17.0 \pm 0.5\,\mathrm{mV}$, respectively. TEER and *PD* values decreased when transferred to an air-interface on day two or three. When compared with LCC, AIC on day four exhibited a higher $I_{eq}$ (181 per cent) and *PD* (130 per cent), but lower TEER (73 per cent), bringing the values of the AIC closer to those of the excised tissue. The presence of inhibitors reduced the $I_{eq}$ across the AIC by 63 per cent of the control for NPAA, 35 per cent for bumetanide, 66 per cent for ouabain, and 46 per cent for barium chloride. The findings correlate well with the inhibition found in excised tissue and suggest that the transporters tested in AIC are comparably functional to those *in vivo*. An additional study on the permeability of rabbit conjunctival epithelial AIC to hydrophilic solutes indicated that transport is dependent on molecular weight with a maximal permeating size of 20,000 Da. Conjunctival epithelial air-interface cultures produced a three-fold lower $P_{app}$ for lipophilic solutes than cultures grown under LCC and more closely mimicked the numbers found in excised tissue. Characterization of the rabbit conjunctiva AIC resulted in cell layers with electrical and transport properties more akin to excised tissue. This superior model furthers the applicability of cultures for drug transport studies.

## CORNEAL CELL CULTURE MODELS

The current method for culturing corneal epithelial cells is based on that developed by McPherson *et al.* (1956) in the 1950s while studying the cellular nutritional requirements for the preservation of corneal tissues. Since then, corneal epithelial cells have been cultured from different species. Cultured corneal cells have been used to evaluate intrinsic and extrinsic cell differentiation and regulation (Beebe and Masters, 1996; Doran *et al.*, 1980) as well as cell attachment to extracellular matrix proteins (Maldonado and Furcht, 1995).

## Primary cultures of corneal epithelium

Several different methods for culturing cells are routinely performed. These can be roughly divided into primary, established, or immortal cell line cultures. A primary culture consists of a complex organ or tissue's slice, a defined mixture of cells, or highly purified cells isolated directly from the organism. Techniques to purify the cell type of interest are usually employed to develop a homogenous primary culture. Primary cell cultures have an advantage in that they have been recently removed from the *in vivo* condition and are, therefore, expected to more closely resemble the function of the cell type *in vivo*. The disadvantage is that these cultures react to a constantly changing environment over the first days or weeks *in vitro*, including the damage sustained during the removal of cells from the animal, the change in environment from the animal to the *in vitro* culture, and changing composition of the culture as some cells in the mixed cultures die and others proliferate or differentiate. Primary corneal epithelial cells from humans and rabbits have been successfully cultured (Chan and Haschke, 1983; Chang *et al.*, 2000; Doran *et al.*, 1980; Jumblatt and Neufeld, 1983; Sun and Green, 1963).

## Secondary cultures of corneal epithelium

Primary cultures of human corneal epithelial cells usually cease to grow after one or two passages and cannot revive well from storage in liquid nitrogen. This short viability results in small cell yields for experiments. It is possible to avoid this kind of problem by using continuously growing cell lines. There are several major strategies to produce these cells. First, cells can be immortalized with oncogenes introduced by transfection (Houweling *et al.*, 1980), viral infection (Reddel *et al.*, 1988) or retroviral vectors (Cone *et al.*, 1988). Most authors have obtained epithelial cell lines through infection with SV40 (Hronis *et al.*, 1984; Steinberg and Defendi, 1983), the expression of transfected SV40-LT antigen encoding sequences (Agarwal and Eckert, 1990), or oncogenes from epitheliotropic viruses such as HPV16 (Halbert *et al.*, 1991) or adenoviruses (Kuppuswamy and Chinnadurai, 1988; Cone *et al.*, 1988). Although immortalization frequency is increased, results from this approach are quite variable and dependent on cell type, and in most cases the expression of differentiated phenotypes is altered or absent. A second method used to produce continuously growing cell lines is to generate spontaneous cell lines by extended subculture. This technique has proven particularly successful with rodent tissues such as small intestine (Quaroni *et al.*, 1979) or 3T3-fibroblasts (Todaro and Green, 1963). Although spontaneous establishment of human epithelial cell lines has been obtained by serial passage or by enhancing proliferation and delaying terminal differentiation, in general, spontaneous human cell lines have been difficult to obtain. However, the immortalization of corneal epithelial cells from rabbits and rats using a recombinant SV40 adenovirus has been published (Araki *et al.*, 1993, 1994; Aizawa *et al.*, 1991; Van Doren and Gluzman, 1984). Kahn *et al.* (1993) subsequently developed an *in vitro* model of human corneal epithelium using an Ad12-SV40 hybrid vector. Later, Araki-Sasaki *et al.* (1995) published an SV40-immortalized human corneal epithelial cell line with properties similar to normal corneal epithelial cells. Their SV40-adenovirus recombinant vector lacked the origin of SV40 viral replication and did not produce any viral particles. This cell line continued to grow for more than 400 generations exhibiting a cobble-stone-like appearance similar to normal corneal epithelial cells in culture. Recently, telomerase-immortalized corneal cells were introduced for transplantation purposes (Griffith *et al.*, 1999).

The human corneal epithelial culture produced by Araki-Sasaki *et al.* (1995) was grown using an air–liquid interface. This method has been popular among the investigators of dermatology and can produce proliferation and differentiation of epidermal cells (Krejci *et al.*, 1991; Mak *et al.*, 1991; Ponec *et al.*, 1988; Prunieras *et al.*, 1983), and has also been attempted for the conjunctiva (Yang *et al.*, 2000) and airway (Robinson and Kim, 1994). Minami *et al.* (1993) published a reconstruction of the cornea using this method. The air–liquid interface also seems to be critical for the differentiation of HCE cells in culture. Under this condition, HCE cells proliferated to form two or three layers within seven days (Araki-Sasaki *et al.*, 1995).

## Permeability models

After topical instillation, drugs are absorbed into the inner eye through either the cornea or the conjunctiva and sclera. The cornea is currently the main route of absorption for clinically used ocular drugs (Doane *et al.*, 1978; Lee and Robinson, 1986; Maurice and Mishima, 1984). Corneal permeability studies of drugs are usually performed *in vitro* using isolated rabbit corneas mounted on modified Ussing chambers (Huang *et al.*, 1983; Morimoto *et al.*, 1987; Schoenwald and Ward, 1978). There are many negative aspects surrounding this type of experiment. For example, in these studies numerous rabbits are sacrificed and the isolated corneas are viable for only six hours after dissection. In addition, metabolic enzyme and active transporter differences between species may impair the predictability of drug absorption in humans. New models for measuring drug permeability that require fewer rabbits, that are viable for longer periods of time, and that better mimic human models are needed.

Corneal cell culture models could be useful in testing the permeability of ocular drugs and formulations. The chosen model should be based on corneal epithelial cells because the epithelium is the rate-limiting barrier for drug permeation (Sieg and Robinson, 1976). The published epithelial models are based on primary cells (Chang *et al.*, 2000; Kawazu *et al.*, 1998, 1999) that usually stop growing after one or two passages, and revive weakly or not at all after storage in liquid nitrogen. Fresh rabbit cells must be isolated frequently and, therefore, may not be optimal for larger scale screening of new compounds and formulations.

Immortalized cell lines can be grown continuously and should be more practical for testing of permeability. SIRC cells, a corneal cell line from rabbits, were recently described for permeability testing (Goskonda *et al.*, 1999; Hutak *et al.*, 1997). Unfortunately, these cells exhibit a fibroblast phenotype that decreases the value of SIRC cells as a model (Niederkorn *et al.*, 1990). The HCE-T *in vitro* model of human corneal epithelium (Kahn *et al.*, 1993; Kruszewski *et al.*, 1995, 1997) represents a 3D culture for HCE-T cells and is grown on a collagen membrane to provide a species-and tissue-specific equivalent of the human corneal surface *in vivo*. Araki-Sasaki *et al.* (1995) established an immortalized HCE cell line that exhibits properties of normal corneal epithelial cells. This cell line continues to grow for more than 400 generations through 100 passages and the cells can be frozen and revived.

Toropainen et al. (2000) developed a model for drug permeability studies using immortalized epithelial cells from the human cornea. Typically, corneal epithelial cells used for permeability studies are cultured on a plastic support that is submerged in growth medium containing FBS (2–20 per cent), fibroblasts, or growth supplements such as epidermal growth factor (EGF) (10 ng/ml), insulin (5 μg/ml), or cholera toxin (0.1 μg/ml). Toropainen et al. (2000) used 15 per cent of FBS in the feeding medium of the cells. However, high concentrations of FBS may disturb cell proliferation and differentiation (Kruse and Tseng, 1993), and some groups have reported improved differentiation of the corneal epithelium in serum-free medium (Castro-Munozledo et al., 1997; Kruszewski et al., 1997). Toropainen et al. (2000) also cultivated the cells at lower FBS concentrations (2, 5, 10 per cent), but the cells did not survive. As described for the primary corneal epithelial cells (Minami et al., 1993), the air–liquid interface was critical for the differentiation of HCE cells in culture. No flat apical cells or proper barrier was obtained without air-lifting (Toropainen et al., 2000). Other important factors in the development of a culture model are the filter material, its pore size, and the coating components on the filter (i.e. the extracellular matrix).

Cultured corneal epithelial cell layers grown on a permeable membrane allow for the investigation into drug transport mechanisms; as well as into drug formulation factors influencing corneal epithelial drug transport. One advantage of such a model system is that it is possible to access both the apical and basolateral sides directly in mechanistic transport studies. Ideally, the microporous membrane should be sufficiently transparent so that development of the cell monolayer can be verified by microscopic techniques, and readily permeable to hydrophilic and lipophilic solutes and low and high molecular weight solutes. Polycarbonate, with or without collagen, is used in most cultures of corneal cells (Kawazu et al., 1998, 1999; Zieske et al., 1994). For barrier formation with immortalized HCE cells, however, polyester filters were found to provide better cell visibility using light microscopy or phase-contrast microscopy than polycarbonate filters (Toropainen et al., 2000).

Collagen facilitates the growth and differentiation of corneal epithelial cells in culture (Geggel et al., 1985; He and McCulley, 1991; Kawazu et al., 1998; Minami et al., 1993; Ohji et al., 1994; Zieske et al., 1994). Collagen coating helps cells to attach to the cultivating bed by stimulating their proliferation and differentiation. Human corneal epithelium cultured on collagen gels can synthesize and deposit basement membrane components like laminin and type-IV collagen (Fukuda et al., 1999; Ohji et al., 1994). The mixture of collagen and laminin can further improve the culture model (Toropainen et al., 2000).

3T3-fibroblast cells from mice are used as a feeder layer for differentiation of corneal epithelial cells. The interactions between the fibroblasts and corneal epithelial cells stimulate the differentiation of primary corneal epithelial cells (Castro-Munozledo, 1994; Zieske et al., 1994). For culturing immortalized HCE cells, Toropainen et al. (2000) coated the filters with a mixture of collagen and fibroblasts to provide a substrate resembling a corneal stroma. The fibroblasts did not divide in the collagen matrix, but they functioned as feeder cells for HCE cells. Based on morphology and on the barrier properties, a polyester filter with collagen and fibroblasts appeared to be the best condition for HCE cell culture.

The apical surface of corneal epithelium contributes to over half of the total electrical resistance of the cornea (Klyce and Crosson, 1985). The top two layers are the most important part of the cornea in limiting the permeability of hydrophilic drugs (Sieg and Robinson, 1976). Therefore, the most apical cell layers are the most important part of the HCE permeability model. This model resembles intact cornea with morphologically identifiable desmosomes, tight junctions, microvilli, and cell layers with apical flat cells (Toropainen *et al.*, 2000). As an *in vitro* model of permeability, the cultured corneal epithelium should predict the permeability of the cornea, i.e. the barrier of the culture should resemble that of the cornea and should give clearly different permeabilities for hydrophilic and lipophilic solutes. Toropainen *et al.* (2000) determined the permeabilities of $^3$H-mannitol and 6-carboxyfluorescein to evaluate the intercellular spaces of their cultured corneal epithelium model. Rhodamine B was used as a lipophilic marker of transcellular permeability. The authors compared the permeability differences between the cultured cells and excised rabbit corneas. The hydrophilic marker, 6-carboxyfluorescein, and the lipophilic marker, rhodamine B, differed in lipophilicity by five orders of magnitude (based on logP). Furthermore, the difference in permeability between 6-carboxyfluorescein and rhodamine B in the HCE cell culture model was substantial (21-fold), although smaller than in the isolated cornea (39-fold). In the culture model of Kawazu *et al.* (1998) only a 3.3-fold difference was observed between the permeabilities of hydrophilic atenolol and lipophilic propranolol.

The results suggest that the immortalized HCE cell culture model can discriminate between the permeabilities of lipophilic and hydrophilic drugs, and that the permeability values are close to those obtained from isolated rabbit corneas. It appears that the physical barrier of the HCE culture is comparable with that of the rabbit cornea. Therefore, this model can be used to predict ocular absorption of the drugs that permeate through the cornea by passive diffusion. The extent of this model's applicability is also dependent on the expression of the active transporters, efflux proteins, and metabolic enzymes in the cultured cells. These aspects of the cultured model should be elucidated in the future since it is possible that the expression profiles of the immortalized cell line, primary cells, and animal model *in vivo* are different. A general disadvantage of using *in vitro* permeation studies, however, is that an *in vitro* model does not take into account the composition of aqueous humor and tear fluid, or the mechanical stress of the eyelids and tear flow.

Cell culture models offer many potential advantages in the analysis of drug transport and drug metabolism (Audus *et al.*, 1990). Importantly, they offer the possibility of decreasing the number of animal experiments needed for research. In addition, these systems offer the potential to manipulate the environment or cellular properties as a means of addressing the mechanisms of drug permeation in living cells. From the standpoint of drug discovery and drug formulation, cell culture models can be used to speed up the identification of compounds or formulations with favorable pharmacokinetic properties, and to evaluate structure-absorption and structure–metabolism relationships on a large scale. Development of methods for testing *in vitro* absorption is of the utmost importance in current drug discovery because the modern combinatorial and automated methods of drug synthesis produce numerous compounds but only in minute quantities. Therefore, maximal predictive information should be gained from the *in vitro* testing of small quantities of the test substances.

## Toxicity models

Eye irritation tests in rabbits (i.e. Draize test) are the standard procedure for evaluating topical ocular safety (Draize *et al.*, 1944). The currently accepted Draize test employs rabbits and involves placing a foreign compound directly into the conjunctival sac of the rabbit eye. This assay is simple to perform, provides a conservative model for human ocular safety testing, allows fast economical results, and uses a laboratory animal that is easy to breed and maintain. There are, however, morphologic and biochemical differences between the rabbit eye and the human eye that have led this animal model to be questionable. Several studies have focused on determining the validity of alternative *in vitro* methods as replacements of the Draize test (Balls *et al.*, 1995; Ohno *et al.*, 1995; Sina *et al.*, 1995; Spielmann *et al.*, 1995). The Draize test has been widely criticized for ethical reasons, as well as for the questionable accuracy in predicting the human eye response. Reasons for doubting the Draize model include the high variability found in the *in vivo* data used as a reference (Balls *et al.*, 1995) and the lack of accuracy of the presented *in vitro* methods to mimic processes associated with eye irritation such as damage, inflammation, and repair (for reviews, see Earl *et al.*, 1997). Alternative methods to *in vivo* animal models have been proposed and are very useful and predictive when used as screening or adjunct tests for restricted classes of test substances acting through similar toxic mechanisms (Balls *et al.*, 1995). One alternative is to use isolated target organs (e.g. the enucleated rabbit or chicken eye, the isolated cornea) for toxicology studies (Prinsen, 1996; Igarashi and Northover, 1987). In this case, however, animals are needed as a source of the tissues and the viability of the tissues is questionable. Another alternative is the use of cell-based *in vitro* methods in safety evaluations. Cell-based tests are the most widely used alternative methods at the moment (for reviews, see Herzinger *et al.*, 1995). Many different cell culture systems and endpoints have been proposed for cytotoxicity screening of drugs, surfactants, solvents, and various chemicals (Herzinger *et al.*, 1995). Cytotoxicity has been evaluated using primary cells (Grant *et al.*, 1992; Sina *et al.*, 1992) and continuous cell lines from rabbit cornea (Pasternak and Miller, 1995) and human cornea (Saarinen-Savolainen *et al.*, 1998). Other models that use continuous cell lines have been proposed for ocular toxicology studies. These models include the SIRC cell line (rabbit origin) that, in fact, has fibroblast morphology (Niederkorn *et al.*, 1990) and the Madin–Darby canine kidney (MDCK) cell line (canine origin) (Botham *et al.*, 1997), which is derived from kidney. Neither of these cell lines provides species and tissue specificity.

Primary corneal epithelial cells can be used for rapid screening of acute topical ocular toxicities of a large number of compounds. The disadvantage of primary cultures is their restricted lifespan, which necessitates seeding of the fresh primary cells from animal or human eyes frequently. Immortalized cell lines can be grown continuously and should be more practical for screening than primary cells. Saarinen-Savolainen *et al.* (1998) evaluated the suitability of an immortalized HCE cell line for predicting eye irritation and toxicity potential of some commonly used ophthalmic drugs (dipivefrin, timolol, pilocarpine, dexamethasone) and pharmaceutical excipients (benzalkonium chloride, sodium edetate, polyvinyl alcohol, methylparaben, some cyclodextrins). Cytotoxic rankings for the test substances were made based on the results obtained from the MTT test (Hansen *et al.*, 1989) and the propidium iodide internalization test (Nieminen *et al.*, 1992). Toxic effects of the tested substances were detected at concentrations higher than typically used in clinical practice and, in some cases, only after longer exposure times. The authors found that the cytotoxic agents used in these *in vitro* assays were irritating or toxic *in vivo*, proving that the human corneal epithelial cell line is useful for evaluating ocular irritation of eye drop components. The toxic effect of a chemical compound *in vivo* depends on its concentration at the site of action and on the time for which the active concentration is maintained there. *In vivo* residence of topical ocular drugs and excipients on the cornea is short. Although *in vitro* cell culture is different from the *in vivo* situation, it could be useful in predicting corneal toxicity and in evaluating its mechanisms.

Comparison of the results from HCE cells with other cells is difficult owing to the different experimental conditions (e.g. cell densities, exposure times, *in vitro* tests, different test substances). It is, however, possible to compare the results of HCE cells in MTT assay (Saarinen-Savolainen *et al.*, 1998) with those of primary rabbit corneal epithelial cells (Grant *et al.*, 1992). Immortalized HCE cells seem to be more resistant to toxic effects of benzalkoium chloride (BAC) than the primary cells. In contrast, the cells of the 3D human-based tissue culture model (SKIN$^2$ZK1200) for assessment of eye irritation *in vitro* were less sensitive to the toxicity of BAC (Espersen *et al.*, 1997). Based on these findings, HCE cells approximately predict the concentrations that are tolerated *in vivo*. Sensitivity of HCE cells is intermediate between primary corneal epithelium cells and SKIN2ZK1200.

## CONCLUDING REMARKS

Primary rabbit conjunctival epithelial cell cultures have advanced greatly since the first culture grown under liquid conditions. With the aid of electrical profiles and transport studies on excised tissue and transport studies, conjunctival cultures could be characterized as closely mimicking excised tissues. Although still requiring the sacrifice of rabbits, these primary cultures can form uncontaminated tight conjunctival epithelial barriers that are convenient to work with and study. Excised tissue is not always appropriate to work with as it can contain cells from other conjunctival layers that can affect drug transport. This problem is most clearly seen in the *p*-gp transporter study (Saha *et al.*, 1998). Another disadvantage of excised tissue is its short viability period of about six hours after excision. At this point, the tissue becomes 'leaky' and the TEER drops making the tissue useless for drug transport and electrical studies. Comparatively, air-interface primary cultures obtain peak TEER values around four days and can be functional up to ten days. Disadvantages of primary cultures include a culturing lag time of four days before the cells are ready for experimentation and the cost of culturing media. Once fully characterized, conjunctival primary cultures may enable researchers to discover and improve methods for ocular drug delivery.

## REFERENCES

Agarwal, C. and Eckert, R. L. (1990) Immortalization of human keratinocytes by simian virus 40 large T-antigen alters keratin gene response to retinoids. *Cancer Res.*, **50**, 5947–5953.

Ahmed, I. and Patton T. F. (1984) Importance of the noncorneal absorption route in topical ophthalmic drug delivery. *Invest. Ophthalmol. Vis. Sci.*, **26**, 584–587.

Aizawa, S., Yaguchi, M., Nakano, M., Inokuchi, S., Handa, H. and Toyama, K. (1991) Establishment of a variety of human bone marrow stromal cell lines by the recombinant SV40-adenovirus vector. *J. Cell. Physiol.*, **148**, 245–251.

Araki, K., Ohashi, Y., Sasabe, T., Kinoshita, S., Hayashi, K., Yang, X. Z. *et al.* (1993) Immortalization of rabbit corneal epithelial cells by a recombinant SV40-adenovirus vector. *Invest. Ophthalmol. Vis. Sci.*, **34**, 2665–2671.

Araki, K., Sasabe, T., Ohashi, Y., Yasuda, M., Handa, H. and Tano, Y. (1994) Immortalization of rat corneal epithelial cells by SV40-adenovirus recombinant vector. *Nippon Ganka Gakkai Zasshi*, **98**, 327–333.

Araki-Sasaki, K., Ohashi, Y., Sasabe, T., Hayashi, K., Watanabe, H., Tano, Y. *et al.* (1995) An SV40-immortalized human corneal epithelial cell line and its characterization (see comments). *Invest. Ophthalmol. Vis. Sci.*, **36**, 614–621.

Audus, K. L., Bartel, R. L., Hidalgo, I. J. and Borchardt, R. T. (1990) The use of cultured epithelial and endothelial cells for drug transport and metabolism studies. *Pharm. Res.*, **7**, 435–451.

Balls, M., Botham, P., Bruner, L. and Spielmann, H. (1995) The EC/HO international validation study on alternatives to the Draize eye irritation test. *Toxicol. In vitro*, **6**, 871–929.

Basu, S. K., Haworth, I. S., Bolger, M. B. and Lee, V. H. (1998) Proton-driven dipeptide uptake in primary cultured rabbit conjunctival epithelial cells. *Invest. Ophthalmol. Vis. Sci.*, **39**, 2365–2373.

Beebe, D. C. and Masters, B. R. (1996) Cell lineage and the differentiation of corneal epithelial cells. *Invest. Ophthalmol. Vis. Sci.*, **37**, 1815–1825.

Botham, P., Osborne, R., Atkinson, K., Carr, G., Cottin, M. and van Buskirk, R. G. (1997) IRAG working group 3. Cell function-based assays. Interagency regulatory alternatives group. *Food. Chem. Toxicol.*, **35**, 67–77.

Castro-Munozledo, F. (1994) Development of a spontaneous permanent cell line of rabbit corneal epithelial cells that undergoes sequential stages of differentiation in cell culture. *J. Cell Sci.*, **107**(8), 2343–2351.

Castro-Munozledo, F., Valencia-Garcia, C. and Kuri-Harcuch, W. (1997) Cultivation of rabbit corneal epithelial cells in serum-free medium. *Invest. Ophthalmol. Vis. Sci.*, **38**, 2234–2244.

Chan, K. Y. and Haschke, R. H. (1983) Epithelial–stromal interactions: specific stimulation of corneal epithelial cell growth *in vitro* by a factor(s) from cultured stromal fibroblasts. *Exp. Eye Res.*, **36**, 231–246.

Chang, S. C. and Lee, V. H. (1987) Nasal and conjunctival contributions to the systemic absorption of topical timolol in the pigmented rabbit: implications in the design of strategies to maximize the ratio of ocular to systemic absorption. *J. Ocul. Pharmacol.*, **3**, 159–169.

Chang, J.-E., Basu, S. K. and Lee, V. H. L. (2000) Air-interface condition promotes the formation of tight corneal epithelial cell layers for drug transport studies. *Pharm. Res.*, **17**, 670–676.

Cone, R. D., Grodzicker, T. and Jaramillo, M. (1988) A retrovirus expressing the 12S adenoviral E1A gene product can immortalize epithelial cells from a broad range of rat tissues. *Mol. Cell. Biol.*, **8**, 1036–1044.

Doane, M. G., Jensen, A. D. and Dohlman, C. H. (1978) Penetration routes of topically applied eye medications. *Am. J. Ophthalmol.*, **85**, 383–386.

Doran, T. I., Vidrich, A. and Sun, T. T. (1980) Intrinsic and extrinsic regulation of the differentiation of skin, corneal and esophageal epithelial cells. *Cell*, **22**, 17–25.

Draize, J. H., Woodard, G. and Calvery, H. (1944) Methods for the study of irritation and toxicity of substance applied topically to the skin and mucous membranes. *J. Pharmacol. Exp. Ther.*, **82**, 377–390.

Earl, L., Dickens, A. and Rowson, M. (1997) A critical analysis of the rabbit eye irritation test variability and its impact on the validation of alternative methods. *Toxicol. In Vitro.*, **11**, 295–304.

Espersen, R., Olsen, P., Nicolaisen, G., Jensen, B. and Rasmussen, E. (1997) Assessment of recovery from ocular irritancy using a human tissue equivalent model. *Toxicol. In vitro*, **11**, 81–88.

Fukuda, K., Chikama, T., Nakamura, M. and Nishida, T. (1999) Differential distribution of subchains of the basement membrane components type IV collagen and laminin among the amniotic membrane, cornea, and conjunctiva. *Cornea*, **18**, 73–79.

Geggel, H. S., Friend, J. and Thoft, R. A. (1985) Collagen gel for ocular surface. *Invest. Ophthalmol. Vis. Sci.*, **26**, 901–905.

Goskonda, V. R., Khan, M. A., Hutak, C. M. and Reddy, I. K. (1999) Permeability characteristics of novel mydriatic agents using an *in vitro* cell culture model that utilizes SIRC rabbit corneal cells. *J. Pharm. Sci.*, **88**, 180–184.

Grant, R. L., Yao, C., Gabaldon, D. and Acosta, D. (1992) Evaluation of surfactant cytotoxicity potential by primary cultures of ocular tissues: I. Characterization of rabbit corneal epithelial cells and initial injury and delayed toxicity studies. *Toxicology*, **76**, 153–176.

Griffith, M., Osborne, R., Munger, R., Xiong, X., Doillon, C. J., Laycock, N. L., *et al.* (1999) Functional human corneal equivalents constructed from cell lines. *Science*, **286**, 2169–2172.

Halbert, C. L., Demers, G. W. and Galloway, D. A. (1991) The E7 gene of human papillomavirus type 16 is sufficient for immortalization of human epithelial cells. *J. Virol.*, **65**, 473–478.

Hansen, M. B., Nielsen, S. E. and Berg, K. (1989) Re-examination and further development of a precise and rapid dye method for measuring cell growth/cell kill. *J. Immunol. Methods.*, **119**, 203–210.

He, Y. G. and McCulley, J. P. (1991) Growing human corneal epithelium on collagen shield and subsequent transfer to denuded cornea in vitro. *Curr. Eye Res.*, **10**, 851–863.

Herzinger, T., Korting, H. C. and Maibach, H. I. (1995) Assessment of cutaneous and ocular irritancy: a decade of research on alternatives to animal experimentation. *Fundam. Appl. Toxicol.*, **24**, 29–41.

Houweling, A., van den Elsen, P. J. and van der Eb, A. J. (1980) Partial transformation of primary rat cells by the leftmost 4.5% fragment of adenovirus 5 DNA. *Virology*, **105**, 537–550.

Hronis, T. S., Steinberg, M. L., Defendi, V. and Sun, T. T. (1984) Simple epithelial nature of some simian virus-40-transformed human epidermal keratinocytes. *Cancer Res.*, **44**, 5797–5804.

Huang, H. S., Schoenwald, R. D. and Lach, J. L. (1983) Corneal penetration behavior of beta-blocking agents II: Assessment of barrier contributions. *J. Pharm. Sci.*, **72**, 1272–1279.

Hutak, C. M., Kavanagh, M. E., Reddy, I. K. and Barletta, M. A. (1997) Growth pattern of SIRC rabbit corneal cells in microwell inserts. *J. Toxicol.-Cut. Ocul. Toxicol.*, **16**, 145–156.

Hämäläinen, K. M., Kontturi, K., Murtomäki, L., Auriola, S. and Urtti, A. (1997) Estimation of pore size and porosity of biomembranes from permeability measurements of polyethylene glycols using an effusion-like approach. *J. Control. Release.*, **49**, 97–104.

Igarashi, H. and Northover, A. M. (1987) Increases in opacity and thickness induced by surfactants and other chemicals in the bovine isolated cornea. *Toxicol. Lett.*, **39**, 249–254.

Jumblatt, M. M. and Neufeld, A. H. (1983) Beta-adrenergic and serotonergic responsiveness of rabbit corneal epithelial cells in culture. *Invest. Ophthalmol. Vis. Sci.*, **24**, 1139–1143.

Kahn, C. R., Young, E., Lee, I. H. and Rhim, J. S. (1993) Human corneal epithelial primary cultures and cell lines with extended life span: in vitro model for ocular studies. *Invest. Ophthalmol. Vis. Sci.*, **34**, 3429–3441.

Kawazu, K., Shiono, H., Tanioka, H., Ota, A., Ikuse, T., Takashina, H., *et al.* (1998) Beta adrenergic antagonist permeation across cultured rabbit corneal epithelial cells grown on permeable supports. *Curr. Eye Res.*, **17**, 125–131.

Kawazu, K., Yamada, K., Nakamura, M. and Ota, A. (1999) Characterization of cyclosporin A transport in cultured rabbit corneal epithelial cells: P-glycoprotein transport activity and binding to cyclophilin. *Invest. Ophthalmol. Vis. Sci.*, **40**, 1738–1744.

Klyce, S. D. and Crosson, C. E. (1985) Transport processes across the rabbit corneal epithelium: a review. *Curr. Eye Res.*, **4**, 323–331.

Krejci, N. C., Smith, L., Rudd, R., Langdon, R. and McGuire, J. (1991) Epithelial differentiation in the absence of extracellular matrix. *In Vitro Cell. Dev. Biol.*, **27A**, 933–938.

Kruse, F. E. and Tseng, S. C. (1993) Serum differentially modulates the clonal growth and differentiation of cultured limbal and corneal epithelium. *Invest. Ophthalmol. Vis. Sci.*, **34**, 2976–2989.

Kruszewski, F. H., Walker, T. L., Ward, S. L. and DiPasquale, L. C. (1995) Progress in the use of human ocular tissues for in vitro alternative methods. *Comments. Toxicol.*, **5**, 203–224.

Kruszewski, F. H., Walker, T. L. and DiPasquale, L. C. (1997) Evaluation of a human corneal epithelial cell line as an *in vitro* model for assessing ocular irritation. *Fundam. App. Toxicol.*, **36**, 130–140.

Kuppuswamy, M. and Chinnadurai, G. (1988) Cell type dependent transformation by adenovirus 5 E1a proteins. *Oncogene.*, **2**, 567–572.

Kyyrönen, K. and Urtti, A. (1990) Improved ocular: systemic absorption ratio of timolol by viscous vehicle and phenylephrine. *Invest. Ophthalmol. Vis. Sci.*, **31**, 1827–1833.

Lee, V. H. L. and Robinson, J. R. (1979) Mechanistic and quantitative evaluation of precorneal pilocarpine disposition in rabbits. *J. Pharm. Sci.*, **68**, 673–684.

Lee, V. H. and Robinson, J. R. (1986) Topical ocular drug delivery: recent developments and future challenges. *J. Ocul. Pharmacol.*, **2**, 67–108.

Liaw, J., Rojanasakul, Y. and Robinson, J. R. (1992) The effect of charge type and charge density on corneal transport. *Int. J. Pharm.*, **88**, 111–124.

Mak, V. H., Cumpstone, M. B., Kennedy, A. H., Harmon, C. S., Guy, R. H. and Potts, R. O. (1991) Barrier function of human keratinocyte cultures grown at the air-liquid interface. *J. Invest. Dermatol.*, **96**, 323–327.

Maldonado, B. A. and Furcht, L. T. (1995) Involvement of integrins with adhesion-promoting, heparin-binding peptides of type IV collagen in cultured human corneal epithelial cells. *Invest. Ophthalmol. Vis. Sci.*, **36**, 364–372.

Mathias, N. R., Kim, K. J., Robison, T. W. and Lee, V. H. (1995) Development and characterization of rabbit tracheal epithelial cell monolayer models for drug transport studies. *Pharm. Res.*, **12**, 1499–1505.

Maurice, D. M. and Mishima, S. (1984) Ocular pharmacokinetics. In M. L. Sears (ed), *Handbook of Experimental Pharmacology vol. 69. Pharmacology of the Eye*, pp. 16–119. Berlin-Heidelberg: Springer-Verlag.

McPherson, S. D., Draheim, J. W. J., Evans, V. J. and Earle, W. R. (1956) The viability of fresh and frozen corneas as determined in tissue culture. *Am. J. Ophthalmol.*, **41**, 513–521.

Minami, Y., Sugihara, H. and Oono, S. (1993) Reconstruction of cornea in three-dimensional collagen gel matrix culture. *Invest. Ophthalmol. Vis. Sci.*, **34**, 2316–2324.

Morimoto, K., Nakai, T. and Morisaka, K. (1987) Evaluation of permeability enhancement of hydrophilic compounds and macromolecular compounds by bile salts through rabbit corneas in-vitro. *J. Pharm. Pharmacol.*, **39**, 124–126.

Niederkorn, J. Y., Meyer, D. R., Ubelaker, J. E. and Martin, J. H. (1990) Ultrastructural and immunohistological characterization of the SIRC corneal cell line. *In Vitro Cell. Dev. Biol.*, **26**, 923–930.

Nieminen, A. L., Gores, G. J., Bond, J. M., Imberti, R., Herman, B. and Lemasters, J. J. (1992) A novel cytotoxicity screening assay using a multiwell fluorescence scanner. *Toxicol. Appl. Pharmacol.*, **115**, 147–155.

Ohji, M., SundarRaj, N., Hassell, J. R. and Thoft, R. A. (1994) Basement membrane synthesis by human corneal epithelial cells in vitro. *Invest. Ophthalmol. Vis. Sci.*, **35**, 479–485.

Ohno, T., Itagaki, H., Tanaka, N. and Ono, H. (1995) Validation study on five different cytotoxicity assays in Japan – an intermediate report. *Toxicol. In Vitro*, **9**, 571–576.

Pasternak, A. S. and Miller, W. M. (1995) First-order toxicity assays for eye irritation using cell lines: parameters that affect *in vitro* evaluation. *Fundam. Appl. Toxicol.*, **25**, 253–263.

Ponec, M., Weerheim, A., Kempenaar, J., Mommaas, A. M. and Nugteren, D. H. (1988) Lipid composition of cultured human keratinocytes in relation to their differentiation. *J. Lipid Res.*, **29**, 949–961.

Prinsen, M. K. (1996) The chicken enucleated eye test (CEET): a practical (pre) screen for the assessment of eye irritation/corrosion potential of test materials. *Food Chem. Toxicol.*, **34**, 291–296.

Prunieras, M., Regnier, M. and Woodley, D. (1983) Methods for cultivation of keratinocytes with an air–liquid interface. *J. Invest. Dermatol.*, **81**, 28S–33S.

Quaroni, A., Wands, J., Trelstad, R. L. and Isselbacher, K. J. (1979) Epithelioid cell cultures from rat small intestine. Characterization by morphologic and immunologic criteria. *J. Cell Biol.*, **80**, 248–265.

Reddel, R. R., Ke, Y., Gerwin, B. I., McMenamin, M. G., Lechner, J. F., Su, R. T., *et al.* (1988) Transformation of human bronchial epithelial cells by infection with SV40 or adenovirus-12 SV40 hybrid virus, or transfection via strontium phosphate coprecipitation with a plasmid containing SV40 early region genes. *Cancer Res.*, **48**, 1904–1909.

Robinson, T. W. and Kim, K. J. (1994) Air-interface cultures of guinea pig airway epithelial cells: effects of active sodium and chloride transport inhibitors on bioelectric properties. *Exp. Lung Res.*, **20**, 101–117.

Saarinen-Savolainen, P., Jarvinen, T., Araki-Sasaki, K., Watanabe, H. and Urtti, A. (1998) Evaluation of cytotoxicity of various ophthalmic drugs, eye drop excipients and cyclodextrins in an immortalized human corneal epithelial cell line. *Pharm. Res.*, **15**, 1275–1280.

Saha, P., Kim, K. J. and Lee, V. H. (1996a) A primary culture model of rabbit conjunctival epithelial cells exhibiting tight barrier properties. *Curr. Eye Res.*, **15**, 1163–1169.

Saha, P., Uchiyama, T., Kim, K. J. and Lee, V. H. (1996b) Permeability characteristics of primary cultured rabbit conjunctival epithelial cells to low molecular weight drugs. *Curr. Eye Res.*, **15**, 1170–1174.

Saha, P., Yang, J. J. and Lee, V. H. (1998) Existence of a p-glycoprotein drug efflux pump in cultured rabbit conjunctival epithelial cells. *Invest. Ophthalmol. Vis. Sci.*, **39**, 1221–1226.

Schoenwald, R. D. and Ward, R. L. (1978) Relationship between steroid permeability across excised rabbit cornea and octanol-water partition coefficients. *J. Pharm. Sci.*, **67**, 786–788.

Sieg, J. W. and Robinson, J. R. (1976) Mechanistic studies on transcorneal permeation of pilocarpine. *J. Pharm. Sci.*, **65**, 1816–1822.

Sina, J. F., Ward, G. J., Laszek, M. A. and Gautheron, P. D. (1992) Assessment of cytotoxicity assays as predictors of ocular irritation of pharmaceuticals. *Fundam. Appl. Toxicol.*, **18**, 515–521.

Sina, J. F., Galer, D. M., Sussman, R. G., Gautheron, P. D., Sargent, E. V., Leong, B., *et al.* (1995) A collaborative evaluation of seven alternatives to the Draize eye irritation test using pharmaceutical intermediates. *Fundam. Appl. Toxicol.*, **26**, 20–31.

Spielmann, H., Liebsch, M., Moldenhauer, F., Holzhütter, H.-G. and de Silva, O. (1995) Modern biostatistical methods for assessing in vitro/in vivo correlation of severely eye irritating chemicals in a validation study of in vitro alternatives to the Draize test. *Toxicol. In Vitro*, **9**, 549–556.

Steinberg, M. L. and Defendi, V. (1983) Transformation and immortalization of human keratinocytes by SV40. *J. Invest. Dermatol.*, **81**, 131S–136S.

Sun, T.-T. and Green, H. (1963) Cultured epithelial cells of cornea, conjunctiva and skin: absence of marked intrinsic divergence of their differentiated states. *Nature*, **269**, 489–493.

Todaro, G. and Green, H. (1963) Quantitative studies of the growth of mouse embryo cells in culture and their development into established lines. *J. Cell Biol.*, **17**, 299–313.

Toropainen, E., Ranta, V.-P., Talvitie, A., Yli-Jaskari, T., Suhonen, P. and Urtti, A. (2000) Corneal cell culture model. *GPEN 13. –15.9.2000 meeting abstract*, Uppsala, Sweden.

Urtti, A., Salminen, L. and Miinalainen, O. (1985) Systemic absorption of ocular pilocarpine is modified by polymer matrices. *Int. J. Pharm.*, **23**, 147–161.

Urtti, A., Pipkin, J. D., Rork, G. S., Sendo, T., Finne, U. and Repta, A. J. (1990) Controlled drug delivery devices for experimental ocular studies with timolol. 2. Ocular and systemic absorption in rabbits. *Int. J. Pharm.*, **61**, 241–249.

Van Doren, K. and Gluzman, Y. (1984) Efficient transformation of human fibroblasts by adenovirus-simian virus 40 recombinants. *Mol. Cell. Biol.*, **4**, 1653–1656.

Wang, W., Sasaki, H., Chien, D. S. and Lee, V. H. (1991) Lipophilicity influence on conjunctival drug penetration in the pigmented rabbit: a comparison with corneal penetration. *Curr. Eye Res.*, **10**, 571–579.

Yang, J. J., Ueda, H., Kim, K. and Lee, V. H. (2000) Meeting future challenges in topical ocular drug delivery: development of an air-interfaced primary culture of rabbit conjunctival epithelial cells on a permeable support for drug transport studies. *J. Control Release.*, **65**, 1–11.

Zieske, J. D., Mason, V. S., Wasson, M. E., Meunier, S. F., Nolte, C. J., Fukai, N., *et al.* (1994) Basement membrane assembly and differentiation of cultured corneal cells: importance of culture environment and endothelial cell interaction. *Exp. Cell Res.*, **214**, 621–633.

# Chapter 16

# Cell cultures of the retinal pigment epithelium to model the blood–retinal barrier for retinal drug and gene delivery

*Türkan Eldem, Yusuf Durlu, Bora Eldem and Meral Özgüç*

## INTRODUCTION

Diseases related to the posterior segment of the eye, especially due to the aging population in the world, the high incidence of intraocular inflammatory and auto-immune diseases and the advance in understanding the molecular basis of retinal degenerative and genetic diseases, urgently necessitate the development of either new drugs or drug and gene-based drug delivery systems suitable for intraocular applications. The blood–retinal barrier (BRB) that restricts the passage of most of the drugs to the retina through systemic administration makes the development of local assess to the diseased site by intravitreal or subretinal injections of drugs and gene-based drug delivery systems or by intravitreal implants essential for the treatment of vitreoretinal diseases. In this respect, the pharmaceutical basis of drug development, and concepts in drug delivery design, must create appropriate formulations by taking the properties of the target biological compartments or site into consideration. Moreover, knowledge about the posterior part of the eye, especially the BRB and the cells forming this barrier, is necessary for the design of ideal intraocular drug and gene-based delivery systems.

   Consequently, the aim of this chapter is to present some of the basic characteristics of the posterior segment of the eye. These characteristics relate to the BRB in healthy and diseased states, to the cells forming the BRB, to the characteristics of the retinal pigment epithelium (RPE) cells, to *in vitro* cell culture of the RPE and retinal vascular endothelial (RVE) cells, and to their potential use in pharmaceutical and biopharmaceutical research areas.

## THE BRB AND THE RPE

The retina is a thin layer of tissue that covers nearly 72 per cent of the inner surface of the posterior eye wall. The specialized part of the retina, which is called the macula, is responsible for central and color vision, whereas the retina outside this area accounts for peripheral vision in dim illumination (Michels *et al.*, 1990). Internal aspects of the retina are in contact with the vitreous body, and externally it is adjacent to the RPE but separated by a potential space called the intraretinal space (Mc Donnell, 1994). The retina has the highest oxygen uptake in the body and the

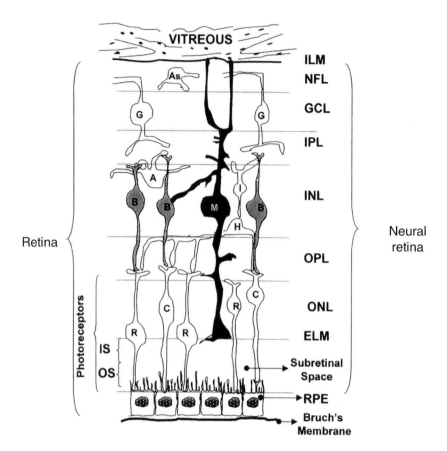

*Figure 16.1* Schematic diagram of cell types and histologic layers in the human retina. (ILM) inner limiting membrane; (NFL) nerve fiber layer; (GCL) ganglion cell layer; (IPL) inner plexiform layer; (INL) inner nuclear layer consisting of interneurons (amacrine [A], bipolar [B], horizontal cells) (H) and the major glial cell of the retina (Müller [M] cells); (OPL) outer plexiform layer; (ONL) outer nuclear layer (nuclei of photoreceptor cells); (ELM) external limiting membrane; (RPE) retinal pigment epithelium; (IS) inner segment of photoreceptors; (OS) outer segments of photoreceptors; (R) rod photoreceptors; (C) cone photoreceptors; (I) inner plexiform cell; (As) astrocytes (Redrawn from Blanks, 1994 with modifications).

blood supply to the inner two-thirds of it is provided by the retinal vasculature, whereas the outer third is provided by diffusion from choriocapillaris of the choroid (Michels *et al.*, 1990). The retina consists of ten layers, including nine layers within the neural retina and the RPE (Blanks, 1994; Michels *et al.*, 1990). The layers of neural retina with various cells, the RPE and their relationship are illustrated in Figure 16.1.

Although there is no anatomic adherence between the photoreceptors (PR) and the RPE, certain forces act to keep the neural retina firmly adherent to the RPE. The retina remains attached to the RPE under physiological conditions. However, when the retinal detachment occurs, the neural retina separates from the RPE and fluid accumulates in the subretinal space. Several physiological mechanisms tend to maintain retinal attachment and prevent fluid from accumulating in the subretinal space and extracellular space of the retina. Among them, the dynamics of fluid flow within the retina and the fluid movement from the vitreous across the retina into the choriocapillaris (vitreoretinal-choroidal outflow) have great importance. This fluid flow can be controlled by three anatomic barriers: the vitreous itself, the sensory retina, and the RPE (Pederson, 1994). The RPE is responsible for the net movement of ions and fluid in an apical-to-basal direction (retinal-to-choroidal) and removes fluids from the subretinal space not only for maintaining retinal adhesion but also keeping the neurosensory retina in a proper state of dehydration and providing optical clarity (Figure 16.2) (Anand and Tasman, 1994; Bok, 1993; Michels *et al.*, 1990). In addition, the BRB, which is localized to two anatomical sites, the tight junctions between the RVE cells and the tight junctions between adjacent RPE cells, prevents entry of water to the extracellular space of the retina. This prevention helps to keep the inner retina (neural retina) dehydrated. In this respect, the tight junctions of the RPE serve as a part of the BRB (outer barrier) that separates the choriocapillaris from the outer retina (Figure 16.2), and

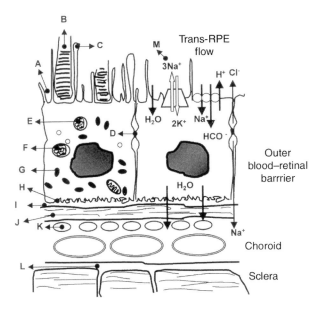

Figure 16.2 Diagrammatic cross-section of retinal pigment epithelium (RPE) cells showing morphologic, cytoplasmic and some functional characteristics, such as interdigitation of outer segments with apical microvilli, transepithelial fluid flow across RPE through choriocapillaris (retinal-choroidal) and the localization of blood–retinal barrier (BRB). (A) interphotoreceptor matrix; (B) photoreceptor outer segment; (C) microvilli; (D) tight junction; (E) phagosome; (F) phagolysosome; (G) melanin granule; (H) basal plasma membrane infoldings; (I) RPE basement membrane; (J) Bruch's membrane; (K) choriocapillaris; (L) suprachoroidal space, (M) sodium-potassium ATPase (sodium pump) (Redrawn from Hewitt and Adler, 1994 and Anand and Tasmann, 1994 with modifications).

the tight endothelial junctions of the RVE cells form the inner barrier as part of the BRB.

The RPE tight junctions retard diffusion between the RPE cells, while they also separate the apical and basolateral membrane domains and maintain the surface-specific distribution of the RPE membrane proteins, which is essential for trans-epithelial transport functions. Consequently, molecular exchanges, such as the transport of nutrients and the regulation of fluid flow between the neural retina and the fenestrated capillaris of the choroid, must occur across the RPE cells themselves. The presence of transport processes and the existence of tight junctions between RPE and the RVE cells prevent the entrance of toxic molecules and serum components into the retina. Moreover, these barriers restrict the paracellular passage of ions of metabolites that escape from the leaky choriocapillaris (Adler and Hewitt, 1994; Cunha-Vaz *et al.*, 1975; Do Carmo *et al.*, 1998; Jampol and Po, 1994). The BRB also has relevant functions for restricting the passage of systemically applied drugs to the intraocular sites and for drug elimination from vitreoretinal compartments (Lee *et al.*, 1994). Consequently, this barrier is functionally identical with that of the blood–brain barrier and its integrity is very important (Greenwood *et al.*, 1994). The disruption of either the inner or outer BRB results in the entry of plasma constituents and water, which in turn cause significant expansion in the extracellular space of the retina (Jampol and Po, 1994). This expansion is a well-established phenomenon that clinically leads to edema and loss of vision. The breakdown of the BRB occurs in several pathological conditions such as in metabolic diseases (diabetic retinopathy, diabetic maculopathy), ischemic diseases (retinal vascular occlusions), mechanical disruption (epiretinal membrane formation or proliferative vitreoretinopathy, PVR), hydrostatic factors (increased intravascular pressure and fall in tissue hydrostatic pressure in the eye), inflammation with release of chemical mediators (uveitis), hereditary diseases (retinitis pigmentosa), degenerative conditions (age-related macular degeneration, AMD), iatrogenic conditions (laser photocoagulation or cryotherapy) and toxic conditions (drug-induced toxicity) (Do Carmo *et al.*, 1998; Greenwood *et al.*, 1994; Jampol and Po, 1994).

Consequently, the RPE has a very strategic location within the retina. Apart from forming the BRB, it has several other functions that are essential for the health of the retina and these are directed by the distinct RPE cell characteristics. It is well-established that the RPE in the adult has a highly specialized internal and external organization, including both morphologic and functional polarity. Morphologically, the RPE cells have a polygonal shape, tight junctions, apical microvilli and basal membrane infoldings. The apical microvilli of RPE cells inter-digitates with the outer segments of the rod and cone PR. The interaction between the RPE and PR cells occurs through the interphotoreceptor matrix (IPM) which is a complex and highly ordered structure consisting of soluble and insoluble components. The basal surface of the RPE cells faces on Bruch's membrane (BM) which also has a complex five-layered extracellular matrix (ECM) structure (Figure 16.2). The innermost layer of BM is the basal lamina of the RPE which contains laminin, fibronectin, type-IV collagen and various proteoglycans (Adler and Hewitt, 1994; Mousa *et al.*, 1999). Several vital properties of the neural retina for vision are maintained by the physical, optical, metabolic, biochemical, developmental, trophic support, immunological and transport functions of the RPE

(Durlu and Tamai, 1997; Hewitt and Adler, 1994). These characteristics are summarized in Table 16.1, and understanding the rationale of RPE cell functions is important because their impairment results in several retinal diseases. In this respect, a significant amount of research has been done in the field of ophthalmology and the studies dealing with the culture of retinal cells, especially the RPE cell culture. This research has contributed to the relevant progress for identifying the molecular basis of these diseases.

Conversely, the amount of research dealing with the treatment of retinal diseases with intraocular drugs or drug delivery systems, including their pharmacokinetic properties, has not been sufficient enough to understand the several aspects of the drug effects in these sites. This lack may be due to the documented retinal toxicity of intravitreally applied free drugs, restricted administration of drugs to intraocular compartments, ethical reasons and constrains in the analysis of drugs in intraocular fluids. In spite of these facts, dedicated effort has been made to deal with the disposition and toxicity aspects of either free drugs or drug delivery systems (Lee *et al.*, 1994). Although a significant reduction in retinal toxicity was demonstrated after intravitreal application of encapsulated drugs (liposomes, microspheres), prolonged drug levels on the order of months were not achievable. Knowing these problems also exist in parenteral and intracellular drug delivery systems (Eldem and Speiser, 1989; Eldem, 1990), coupled with the great improvement in the biological stability of liposomes in the blood by steric stabilization (Klibanov *et al.*, 1990), we were interested in sterically stabilized liposomes (SSL) for intraocular applications. In this respect, intravitreal pharmacokinetics of SSL containing cyclosporine A (CsA) and the biodistribution of gallium deferoxamine ($^{67}$Ga-DF) labeled SSL were examined in healthy rabbits (Eldem *et al.*, 1997a,b; Eldem, 1997). The results indicated that SSL had profound effects in changing the pharmacokinetic properties of CsA by reducing its volume of distribution and vitreal clearance. These alterations increased the mean residence time of CsA in the vitreous and also lead a slow but constant passage of it to the aqueous humor. The accumulation pattern of SSL as compared with conventional liposomes (CL) in the vitreous, retina/uvea, and lens were also different. At the end of 72 hours the percentage of intact SSL remaining in the vitreous was four-fold higher than CL. After intravitreal application, there was a decrease with time in the percentage uptake in retina/uvea with SSL, whereas there was an increased uptake with time, reaching almost 80 per cent with CL at the end of 72 hours post injection. In addition, the accumulation of SSL in lens was found to be lower than CL, which could be an important finding for a possible reduction in drug-induced lens toxicity. Based on these results it was concluded that intraocular drug delivery with SSL was a feasible approach and could be optimized. The same rules regarding the fate of CL in the blood can be applied to their fate in vitreous. Thus, the structure and composition of the vitreous seems to play important roles for the elimination of liposomal drug delivery systems owing to the interactions between the liposome surface and the glycoaminoglycans present in the vitreous, in the vitreoretinal juncture and the internal limiting membrane. In addition, the vitreous was reported to contain a low number of cells such as hyalocytes (Green, 1994; Bishop, 1996) that were shown to phagocytose large latex and colloidal carbon particles (Grabner *et al.*, 1980; Uehara *et al.*, 1996). Consequently, these results indicate that

*Table 16.1* Characteristics and functions of the retinal pigment epithelium (RPE) cells

| Characteristics | Functions and/or requirements | References |
|---|---|---|
| *Morphologic* | | |
| Epithelial cell morphology | Polygonal (cuboidal) shape | Adler and Hewitt, 1994; Hewitt and Adler, 1994 |
| Apical microvilli | Interdigitation with PR and providing support for them; increase in surface area for exchange of nutrients and catabolites | |
| Basal membrane infoldings | Attachment of the cell to the Bruch's membrane | |
| *Physical/ultrastructural* | | |
| Tight junctions and associated actin filaments | Formation of BRB, structural polarity and cell shape | Adler and Hewitt, 1994; Hewitt and Adler, 1994; Ban and Rizzolo, 1997 |
| Uniform distribution of cytoskeletal elements and specific cytokeratin expression | Epithelial cell marker | |
| *Optical/ultrastructural* | | |
| Pigment granules (melanosomes) | Absorption of scattered light, prevention of light damage to the retina | Adler and Hewitt, 1994 |
| Pigmentation | Visible indicator for RPE development and change of phenotype *in vivo* and *in vitro* | |
| *Metabolic/biochemical* | | |
| Phagocytosis of PR outer segments | Internalization and degradation of shed outer segment fragments and rhythmic membrane turnover of PR | Bosch *et al.*, 1993; Bok, 1993 |
| Retinoid (visual) cycle | Vitamin A uptake, processing, transport and release by the aid of cytosolic retinoid-binding proteins (CRALBP and CRBP) and receptors for the plasma retinol-binding protein. | Bok, 1993; Hewitt and Adler, 1994 |
| Melanin granules/melanin | Detoxification of peroxides, drug and zinc binding, protection of the cell from oxidative stress | Schraermeyer and Heimann, 1999 |
| *Transport and secretion* | | |
| Sodium pump (Na, K⁺-ATPase $\alpha$ and $\beta$ subunits in apical membrane domain) and cytoskeletal proteins (ankyrin and fodrin) | Transepithelial ion transport, establishing membrane potentials, contributing to the attachment of retina and keeping it in dehydrated state as a result of transepithelial fluid flow | Gundersen *et al.*, 1993; Bok, 1993; Hewitt and Adler, 1994 |

**Development/trophic support**

| | | |
|---|---|---|
| Synthesis, vectorial secretion and degradation of ECM molecules | ECM molecules secreted (basal) form the part of Bruch's membrane and ECM molecules having different composition secreted (apical) contribute to the IPM | Hewitt and Adler, 1994 |
| IPM | Retinal attachment and trophic support for PR | Gundersen et al., 1993; Bok, 1993 |
| IRBP | Vitamin A visual cycle | Bok, 1993 |

**Immunologic**

| | | |
|---|---|---|
| Activated RPE cells express MHC class II molecules and ICAM-1 | Antigen presentation to lymphocytes, antigen specific immune response and regulation of leukocyte adhesion, | Konda et al., 1994; Akaishi et al., 1998 |
| IL-6, IL-8 and MCP-1 secretion after stimulation with IL-1$\beta$ and TNF-$\alpha$ and IL-7 | Leukocyte chemotactic factors for regulating ocular inflammatory response | Elner et al., 1996, 1997; Kociok et al., 1998 |
| TGF-$\beta$ | Immunomodulator, autocrine/paracrine cell growth regulator | Kociok et al., 1998 |
| Fas ligand, expression and production of soluble apoptotic factors | RPE cell and soluble factor-mediated apoptosis in activated T cells | Farrokh-Siar et al., 1999; Wenkel and Streilein, 2000 |

**Cell surface proteins**

| | | |
|---|---|---|
| Mannose 6-fosfat, CD36, Fc, C3bi and vitronectin receptor | Probable role in phagocytosis | Ryeom et al., 1996; Wilt et al., 1999 |
| Apical and basal transmembrane integral proteins (integrins) | Binding of PR outer segments, attachment to the ECM and IPM components | Mousa et al., 1999; Adler and Hewitt, 1994 |

**Other proteins**

| | | |
|---|---|---|
| aFGF, bFGF, PDGF-A, PDGF-B, TNF-$\alpha$, insulin, IGF, NGF, BDGF, Neurotrophin-3 | Autocrine survival factors, growth regulation | Kociok et al., 1998; Tombran-Tink et al., 1995 |
| VEGF | Survival promoting effect in basal levels and stimulate the growth of choriocapillaris and responsible for neovascularization in high levels | Mousa et al., 1999 |
| PEDF | Suggested as a neuroprotective factor | Cao et al., 1999; Jablonski et al., 2000 |
| Thrombospondin-1 | Natural inhibitor of angiogenesis and control factor in neovascularization | Miyajima-Uchida et al., 2000 |

Abbreviations: BDGF, brain-derived growth factor; CRALBP, cellular retinaldehyde-binding protein; CRBP, Cellular retinol-binding protein; FGF, fibroblast growth factor; ICAM, intracellular adhesion molecule; IL, interleukin; IGF, insulin-like growth factor; IPM, interphotoreceptor matrix; IRBP, interphotoreceptor retinoid-binding protein; MCP, monocyte chemoattractant protein; MHC, major histocompatibility complex; NGF, nerve growth factor; PDGF, platelet-derived growth factor; PEDF, pigment epithelial-derived growth factor; PR, photoreceptors; TGF, transforming growth factor; TNF, tumor necrotizing factor; VEGF, vascular endothelial growth factor.

the steric barrier created by the polymer may prevent the binding or adsorption of other molecules present in the vitreous and may inhibit early phagocytic removal of liposomes. Further studies are needed to understand the exact mechanism of interactions between SSL, the vitreous components and the retina. Elucidation of these interactions could be useful for an effective and safe intraocular drug delivery design with SSL suitable for intravitreal and subretinal applications.

## *IN VITRO* CELL CULTURE OF THE RPE CELLS

*In situ*, the RPE cells do not show any signs of proliferation. However, in pathological states (for instance in PVR, AMD and after laser photocoagulation), the RPE cells proliferate. *In vitro*, the induction of RPE cells entering into the mitosis phase may be influenced by several factors: the species, the age of the donor, process time for the isolation (death–culture period), the culture conditions (medium and the addition of growth factors, such as basic fibroblast growth factor (bFGF)) (Durlu and Tamai, 1995). Furthermore, these factors have influence for establishing the normal RPE cell phenotype, while the RPE cells can express a spectrum of cell shape from epithelioid to fibroblastic. The RPE cells grown on plastic culture dishes rapidly lose their cuboidal configuration, did not produce melanin and did not express mRNA for cellular retinaldehyde-binding proteins (CRALBP). The RPE cells grown on microporous filters or surfaces coated with ECM components have been shown to possess improved structural and functional properties when compared with cells grown on plastic (Defoe and Easterling, 1994; Durlu and Tamai, 1995; McKay and Burke, 1994; Song and Lui, 1990).

During primary culture, the characterization of the RPE cells is important for establishing the purity of the cell population. The immunostaining for cytoskeletal components (cytokeratin, an epithelial marker) can be used for distinguishing the other cell types, such as glia, fibroblast and melonocytes that could contaminate the RPE culture. In addition, the CRALBP, which is specifically localized in the RPE and Müller cells, is a useful marker to exclude contamination of choroidal cells (Durlu and Tamai, 1995; McKay and Burke, 1994; Song and Lui, 1990).

In brief, we have described an isolation, culture technique and cryopreservation of the RPE cells that were previously published elsewhere (Durlu and Tamai, 1995, 1997; Durlu *et al.*, 1999). Human RPE cell cultures were initiated using fetal eyes at an estimated gestational age, 16–21 weeks. The transfer of the fetal eyes was done in ice-cold culture medium (10 per cent fetal bovine serum (FBS) in minimum essential medium (MEM) containing antibiotics, 100 units penicillin and 0.1 mg streptomycin) (FBS/MEM) within two hours. The anterior segment and vitreous was removed and the eyecups were treated with trypsin (0.05 per cent)/ethylenediaminetetraacetic acid (EDTA) (0.53 mM) solution in calcium and magnesium-free Hank's balanced salt solution (HBSS) (trypsin/EDTA solution) for 15 minutes at $37\,^{\circ}C$ in a five per cent $CO_2$ gassed incubator. Then, trypsin/EDTA solution in the eyecups was aspirated and the neural retina was dissected from the RPE layer with the use of a spatula under microscope. The eyecups were washed with HBSS twice and retreated with trypsin/EDTA solution for ten minutes at $37\,^{\circ}C$. Following incubation, trypsin/EDTA solution was aspirated from the eyecups and they were filled with FBS/MEM. The micropipette was used for collecting sheets of RPE in FBS/MEM. The collection material; including RPE cell-sheets, was centrifuged at 1000 rpm for five minutes, the supernatant was discarded and the RPE pellet was resuspended and incubated with trypsin/EDTA solution in HBSS for five minutes at 37 °C. Thereafter, the total number of recovered RPE cells was calculated using Burker-Turk hemocytometer. The incubation was terminated by the addition of five times ice-cold FBS/MEM. The suspension was then centrifuged at 1000 rpm for five minutes at 4 °C, and finally the pellets were washed twice with FBS/MEM. The final pellet was resuspended in FBS/MEM, seeded at a density of 10,000 cells/cm$^2$ in modified polystyrene dishes and cultured at 37 °C in, five per cent CO$_2$ gassed incubator. The medium (FBS/MEM) was changed every three–four days. After proliferation of RPE cells in primary culture within one–three weeks, the passages were done.

The characterization of the RPE cells was made by several means, including morphology (light-phase and electron microscopy), immunocytochemistry (Figures 16.3 and 16.4), enzyme-linked immunosorbent assay (ELISA), immunoblotting, and polymerase chain reaction (PCR) analysis of RPE cell marker proteins. We have chosen the specific proteins which are cytokeratin (epithelial cell marker, Figures 16.3 and 16.4), CRALBP (a critical protein involved in vitamin A transport in RPE), tyrosinase, tyrosinase-related protein-I and -II (pigment cell marker proteins) (TRP-I and TRP-II), and Na,K-ATPase $\alpha$-1 and $\beta$-1 (polarity marker). The details of the assays were described elsewhere (Durlu et al., 1999).

Cryopreservation of the cultured RPE cells (but not uncultured) facilitates the usage of RPE cells for research and, potentially, for clinical transplantation studies. For this purpose, the culture medium of RPE cells was aspirated and the culture plate was washed twice with HBSS. The RPE cells were detached from the bottom of the plate by using trypsin/EDTA solution that was incubated at 37 °C for ten minutes. The RPE cells were collected from the plate by pipetting. Five times

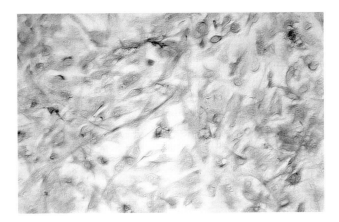

*Figure 16.3* Immunocytochemistry of cytokeratin in cultured human fetal retinal pigment epithelium (RPE) cells, showing diaminobenzidin staining (×120).

*Figure 16.4* Control by substitution of human immunoglobulin (IgG) for the monoclonal cyto-keratin antibody, no staining is seen (×120).

the volume of FBS/MEM was added to stop the trypsin/EDTA effect, then the solution was centrifuged at 1000 rpm for five minutes. The pellet was resuspended and washed twice in FBS/MEM and the final suspension was divided into two portions; the number of cells in the first portion was calculated by Burker-Turk hemocytometer and the viability was estimated by trypan blue dye exclusion test. The viability of cultured RPE cells by this method was found to be as 92 per cent at this step. The other portion of RPE cells was used for cryopreservation. The pellet of RPE cells was resuspended in ice-cold cryopreservation medium consisting of ten per cent dimethyl sulfoxide (DMSO)-containing FBS (10 per cent DMSO/90 per cent FBS) at a concentration of 1,000,000 cells/ml. One ml of RPE cell suspension in cryopreservation medium was filled in 2 ml tight-cap tubes which were placed in a biological freezing vessel kept at 4 °C and frozen at a cooling rate of 1 °C/min in −80 °C freezer overnight and moved to −196 °C under liquid nitrogen. By this method it was found that the viability was 70 per cent after a month at −80 °C.

Essentially, with the emphasis on culture properties, the characteristics of RPE cells from several mammalian species did not differ. However, the fetal RPE cells in all species showed rapid proliferation in culture and undifferentiated nature after several passages. The RPE cells are hexagonal *in situ* (Figure 16.5), but lost their epithelioid shape even one day after the initiation of cell culture, *in vitro*. The RPE cells rapidly undergo undifferentiation *in vitro* (Figure 16.6), but defined medium conditions, including the addition of retinal extract and bFGF to the culture media and coating conditions of the culture plates with ECM proteins (laminin and collagen), may enhance the differentiated properties of cultured RPE cells. RPE cells disclose plasticity that may lead to redifferentiation *in vitro* according to different culture conditions (Table 16.2) (Durlu *et al.*, 1999).

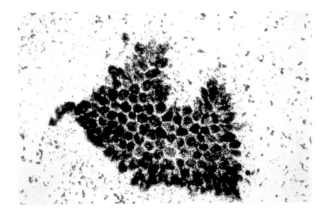

Figure 16.5 Human fetal primary retinal pigment epithelium (RPE) cell culture; pigmented RPE cells with epithelioid shape after isolation (×240).

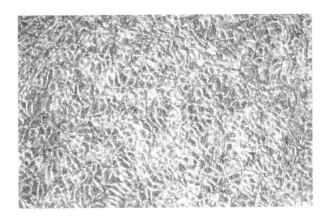

Figure 16.6 Human cultured fetal retinal pigment epithelium (RPE) cells at passage one showing that RPE cells lose their phenotypic characteristics with less pigmentation and epithelioid shape (×120).

Table 16.2 Retinal pigment epithelium (RPE) cell markers[a]

| RPE Cells | Tyrosinase | TRP-I | TRP-II | CRALBP | Na,K-ATPase-$\alpha$ | Na,K-ATPase-$\beta$ |
|---|---|---|---|---|---|---|
| In situ | + | + | + | + | + | + |
| In vitro | | | | | | |
| Passage 3 | + | + | + | + | + | + |
| Passage 9 | + | + | − | + | + | + |
| Cryopreserved | | | | | | |
| At − 80 °C (nine months) | + | + | − | + | + | + |
| At − 196 °C (four months) | + | + | − | + | + | + |
| Cultured at laminin-coated collagen gel | + | + | + | + | + | + |

Note
a  The results of PCR assay; modified from Durlu et al., 1999.

## *IN VITRO* CELL CULTURE OF RVE CELLS

Remarkable heterogeneities concerning the technique of isolation and culture of RVE cells exist among different species, organs or vascular beds. Yan *et al.* (1996) described an improved method of microvascular endothelial cell culture from Macaca monkey retina, which is a widely accepted model for the human retina. After enucleation of eyes from deeply anesthetized Macaca monkeys, the eyeballs were placed in ice-cold calcium-free and magnesium-free HBSS containing 10 mM *N*-(2-hydroxyethyl)piperazine-*N'*-(2-ethane sulphonic acid) (HEPES) buffer (pH 7.4), 0.1 per cent weight/volume bovine serum albumin (BSA), 100 U/ml penicillin G, and 100 μg/ml streptomycin sulfate that was described as preparation buffer. The anterior segment and vitreous were removed under a dissecting microscope. The retina was separated from the RPE, then washed in a preparation buffer and minced with microscissors. Minced retinal tissue was washed by centrifugation and dissociated into tissue suspension by flushing through a glass pipette. This suspension was washed twice with preparation buffer. The final pellet was resuspended in 0.3 per cent collagenase-dispase containing 20 U/ml DNAse-I at 37 °C with gentle agitation for two hours. The retinal tissue was washed twice with Dulbecco's modified Eagle's medium (DMEM) containing ten per cent FBS. The final pellet was resuspended in DMEM containg ten per cent FBS and was filtered through a 100 μm Nylon sieve. The filtrate was washed twice with this medium and resuspended in DMEM containing 20 per cent FBS, and cultured onto dishes precoated with 25 μg/ml bovine fibronectin. The RVE cells containing aggregates and capillary fragments adhered to the dishes within two hours. The medium was then aspirated and the dishes were washed four times with DMEM containing ten per cent FBS to remove other tissues and cells Culture medium (serum-free QB-58 medium containing ten per cent FBS, ten per cent adult monkey serum, 20 μg/ml retinal extract, 90 μg/ml heparin, 100 U/ml penicillin G, and 100 μg/ml streptomycin sulfate) was added and the plates were maintained at 37 °C in five per cent $CO_2$. The characterization of the cultured RVE cells was made with uptake of acetylated low-density lipoprotein, and immunocytochemistry of von Willebrand factor and cell adhesion protein CD31, markers for RVE cells, was made. The RVE cells cultured with this method typically secrete fibronectin, collagen types-I and -IV, laminin and SPARC (secreted protein, acidic and rich in cysteine) (Yan *et al.*, 1996).

## APPLICATIONS OF THE RPE AND RVE CELLS

*In vitro* culture studies with the RPE or the RVE cells should be performed while preserving their inherent functional properties. The *in vivo* biological functions of the RPE cells, such as pigmentation, phagocyte activity, fluid transport ability and electrophysiological barrier properties and their morphological characteristics are important tools for the studies of these functions. Insights into the drug effects on these functions, as well as many cell lines, were established by preserving the inherent functional properties of RPE and RVE cells. (Durlu *et al.*, 1999; Durlu and Tamai, 1995, 1997; Song and Lui, 1990).

While the RPE plays a pivotal role in the development of certain RPE-based retinal diseases (like AMD, PVR and some genetic retinal dystrophies caused by mutations of CRALBP, RPE-65 and others), the current interest in the cell culture studies of RPE is to find out new therapeutic approaches for pharmaceutical management of these diseases. Consequently, the functions, their impairment and the substances secreted or degraded by the RPE cells (Table 16.1) and the RVE cell-derived substances are important targets for developing new pharmaceutical opportunities. In this respect, increasing evidence suggests that vascular endothelial growth factor (VEGF) plays an important role for the development of retinopathy in diabetic patients, as well as complications in other retinal vascular diseases that may result with retinal edema (a sign of breakdown of BRB) and retinal neovascularization. As a result, the RVE cell culture studies for understanding the effects of several growth factors, such as VEGF, will not only help to identify the mode of treatment in retinal vascular diseases, but will also provide detailed information about the toxic effects of several systemically and intraocularly applied drugs (Chang *et al.*, 2000; Yan *et al.*, 1996).

The RPE cells grown on porous support can be used as excellent models for retinal drug transport, permeability, metabolism, cellular toxicity and *in vitro* pharmaco-dynamic studies, as well as phagocytic or endocytic uptake mechanisms of drugs and gene-based drug delivery systems. The RPE and RVE cells in culture are also suggested as systems for screening and testing the response to drugs prior to the use of animal models. As reported, mitomycin C (Ho *et al.*, 1997), 5-fluorourasil (Kon *et al.*, 1998), thiotepa (Kon *et al.*, 1998), vitamin E succinate (Sakamoto *et al.*, 1996), c-myc antisense oligonucleotide (Capeans *et al.*, 1998) and drug-induced apoptosis with genistein (Krott *et al.*, 2000) and transforming growth factor (TGF-*β*) (Esser *et al.*, 1997) were evaluated for their effects and toxic effects in PVR by using the RPE cells in culture.

In recent research, a systemically applied photosensitizer dye (Lu-Tex)-induced cytotoxicity was studied both in the RVE and RPE cells in culture. Lu-Tex is in Phase 3 clinical trials for photodynamic therapy (PDT) in AMD patients to treat choroidal neovascularization. The RPE and RVE cells in culture were used as a preliminary step for the design of treatments in PDT with Lu-Tex and angiostatin. The results of this study showed that Lutex/PDT and angiostatin had combined effects on the RVE cells. However, when angiostatin was administered after PDT, the combination did not potentiate the effects of PDT. Angiostatin was shown to exhibit a specific antiproliferative effect on the RVE cells and had no notable effect on the RPE cells, but angiostatin combined with Lu-Tex/PDT was demonstrated to potentiate cytotoxicity in the RVE cells. The results also indicated that Lu-Tex/PDT induced rapid caspase-dependent apoptosis in both of the cells (Renno *et al.*, 2000).

Another application of retinal cells in culture is their potential use in gene ther-apy and gene-based drug delivery approaches, although current studies dealing with retinal gene delivery mostly cover applications of viral delivery systems to experimental animals by intravitreal or subretinal injections for transfecting the cells within the retina. The existence of the BRB has been regarded as an advant-age for its ability to concentrate vectors in the target area and to reduce their spreading out of the eye. However, the major concern with several viral vectors examined up to date are their vector-induced toxicity, evoked immune response, poor transfection efficiency due to the non-dividing nature of the retinal cells, and co-transfection of the cells other than those targeted (Ali *et al.*, 1997; Nussenblatt and Csaky, 1997). In this respect, prior to *in vivo* studies, the evaluation of viral or non-viral gene delivery systems with different *in vitro* retinal cell culture systems, including the RPE or RVE cells, could be suggested as a necessary requirement for establishing the cellular safety aspects of delivery systems within the retina. The importance of this assessment has been demonstrated in a recent *in vitro* RPE cell culture study as the overproduction of the desired protein, after transduction of the RPE cells with adenovirus resulted in mitochondrial toxicity (Ali *et al.*, 1997; Sullivan *et al.*, 1996).

More recently another approach, which is a cell-based delivery of therapeutic genes as an *ex-vivo* gene therapy, has been applied for the transplantation of the RPE cell. In this study three novel applications (the transduction, the RPE transplantation and the intraocular drug delivery) were used together. The cultured human RPE cells were transduced with the gene for green fluorescent protein (GFP) using a lentiviral vector for noninvasive tracking by GFP fluorescence scanning laser ophthalmoscopy. After transplantation into the subretinal space, intravitreal CsA was administered either by weekly injections or by slow release from a capsule sutured into the vitreal cavity. As a result, the local administration of CsA was shown to prolong the survival of human RPE xenografts in the subretinal space of rabbits, which supported the hypothesis that classic rejection played a role in the survival of foreign transplants in the subretinal space (Lai *et al.*, 2000).

A non-viral gene delivery application was reported by Urtti *et al.* (2000). The fetal human primary RPE cells cultured with or without ECM were used for evaluating the transfection efficiency and toxicity of plasmid DNA (pRSVLuc, pCluc4, pSV2luc) in which AMD was chosen as the target disease. In this respect, DNA was complexed with several cationic lipids and dendrimers and their efficiency was examined in the presence of serum. Although the transfection efficiency was found to be low, the transgene expression per RPE cell was reported to be high. The results of this study demonstrated the transfection of the RPE cells with synthetic vectors.

Jääskeläinen *et al.* (2000) also examined the efficiency of cellular delivery of phosphorothioate oligonucleotides (PS-ODNs) to human RPE cells by using cationic carriers such as cationic polymers, polylysines, polyethyleneimines and fractured dendrimers with different charge ratios. A membrane active peptide was also added to the complexes performed and the effect of complexation medium was examined. As reported, the results indicated that the complexation medium, the size of the complex and the presence of membrane active peptides had a great influence in the efficient delivery of PS-ODNs. The lipid carrier with a membrane active component was also required in transfection.

In another non-viral gene delivery approach to the RPE cells, the efficacy of human RPE cell uptake and expression of GFP and neomycin resistance marker genes were evaluated by using complexes of plasmid DNA and commercially available cationic liposomes and a dendrimer. The study showed that polyplex-mediated DNA transfer into the RPE cells was effective, but the mechanism of DNA uptake was saturable. Both DNA and the dendrimer at the high concentration were found to be toxic (Chaum *et al.*, 1999).

## CONCLUDING REMARKS

The important characteristics and functions of the RPE and RVE cells make their well-characterized *in vitro* models ideal to investigate several aspects of retinal drug and gene-based drug delivery systems. The validity of the culture model depends on whether *in vivo* properties are retained during the cell culture, and there exists significant information and progress about this concept. Consequently, this information can be integrated into the pharmaceutical and biopharmaceutical aspects of ocular/intraocular drug research and development for predicting their effects and toxic effects or permeability *in vivo*, which can facilitate the design of ideal intraocular delivery systems. Well-characterized cell culture systems with the RPE and PVE cells can be established by having collaborations with other disciplines of interest for meeting the requirements and ensuring the purpose in pharmaceutical research.

## ACKNOWLEDGMENT

The RPE cell culture – liposome interactions study at our university has been supported by grants from the National Science and Technology Research Council of Turkey (TÜBİTAK) (Project Nr: SBAG-2119) and the authors are grateful to Genzyme Corp. (Cambridge, MA, USA) for kindly providing the lipids.

## REFERENCES

Adler, R. and Hewitt, A. T. (1994) The Outer Retina: Structural and Functional Considerations in Retina, S. J. Ryan (ed), 2nd edition, Mosby – Year Book Inc., St. Louis, pp. 1857–1868.

Akaishi, K., Ishiguro, S., Durlu, Y. K. and Tamai, T. (1998) Quantitative analysis of major histocompatibility complex class II-positive cells in posterior segment of rat eyes Royal College of Surgeons. *Jpn. J. Ophthalmol.*, **42**, 357–362.

Ali, R. R., Reichel, M. B., Hunt D. M. and Bhattacharya, S. S. (1997) Gene Therapy for inherited retinal degeneration. *Br. J. Ophthalmol.*, **81**, 795–801.

Anand, R. and Tasman, W. S. (1994) Nonrhegmatogenous Retinal Detachment in Retina, S. J. Ryan (ed), 2nd edition, Mosby – Year Book Inc., St. Louis, pp. 2463–2488.

Ban, Y. and Rizzola, L. J. (1997) A Culture Model of Development Reveals Multiple Properties of RPE Tight Junctions. *Mol. Vis.*, **3**, 18–26.

Bishop, P. (1996) The Biochemical Structure of Mammalian Vitreous. *Eye*, **10**, 664–670.

Blanks, J. C. (1994) Morphology of the Retina in Retina, S. J. Ryan (ed), 2nd edition, Mosby – Year Book Inc., St. Louis, pp. 37–53.

Bok, D. (1993) The Retinal Pigment Epithelium: A Versatile Partner in Vision. *J. Cell Sci.*, Suppl. **17**, 189–195.

Bosch, E., Horwitz, J. and Bok, D. (1993) Phagocytosis of Outer Segments by Retinal Pigment Epithelium: Phagosome-Lysosome Interaction. *J. Histochem. Cytochem.*, **41**, 253–263.

Cao, W., Tombran-Tink, J., Chen, W., Mrazek, D., Elias, R. and McGinnis, J. F. (1999) Pigment Epithelium-Derived Factor Protects Cultured Retinal Neurons Against Hydrogen Peroxide-induced Cell Death. *J. Neurosci. Res.*, **57**, 789–800.

Capeans, C., Pineiro, A., Dominguez, F., Loidi, L., Buceta, M., Carneiro, C. *et al.* (1998) A c-myc Antisense Oligonucleotide Inhibits Human Retinal Pigment Epithelial Cell Proliferation. *Exp. Eye Res.*, **66**, 581–589.

Chang, Y. S., Munn, L. L., Hillsley, M. V., Dull, R. O., Yuan, J., Lakshminarayanan, S. *et al.* (2000) Effect of Vascular Endothelial Growth Factor on Cultured Endothelial Cell Monolayer Transport Properties. *Microvasc. Res.*, **59**, 265–277.

Chaum, E., Hatton, M. P. and Stein, G. (1999) Polyplex-Mediated Gene Transfer into Human Retinal Pigment Epithelial Cells In Vitro. *J. Cell Biochem.*, **76**, 153–160.

Cunha-Vaz, J. G., Faria de Abreu, J. R., Campos, A. J. and Figo, G. M. (1975) Early Breakdown of the Blood-Retinal barrier in Diabetes. *Br. J. Ophthalmol.*, **59**, 649–656.

Defoe, D. M. and Easterling, K. C. (1994) Reattachment of Retinas of Cultured Pigment Epithelium Monolayers from Xenopus Laevis. *Invest. Ophthalmol. Vis. Sci.*, **35**, 2466–2476.

Do Carmo, A. D., Ramos, P., Reis, A., Proenca, R. and Cunha-Vaz, J. G. (1998) Breakdown of the Inner and Outer Blood Retinal Barrier in Streptozin-Induced Diabetes. *Exp. Eye Res.*, **67**, 569–575.

Durlu, K. and Tamai, M. (1995) In Vitro Expression of Epidermal Growth Factor Receptor by Human Retinal Pigment Epithelial Cells in Degenerative Diseases of the Retina. R. E. Anderson (ed), Plenum Press, New York, pp. 69–76.

Durlu, K. and Tamai, M. (1997) Transplantation of Retinal Pigment Epithelium Using Viable Cryopreserved Cells. *Cell Transplant.*, **6(2)**, 149–162.

Durlu, Y. K., Ishiguro, S.-I., Akaishi, K., Abe, T., Chida, Y., Shibahara, S. *et al.* (1999) The Retinal Pigment Epithelial Cell Differentiation and Cell Marker Expression Following Cryopreservation at $-80°C$ and Under Liquid Nitrogen at $-196°C$ in Retinal Degenerative Diseases and Experimental Therapy, J. Hollyfield (ed), Plenum Publishing Corporation, New York, pp. 559–567.

Eldem, T. (1990) Studies with a Carrier Glycerodiphosphate and Its Prodrug Estrogen Glycerodiphosphate, Diss. ETH. No. 9110, Zurich.

Eldem, T. (1997) Ocular and Parenteral Polymer-Coated Drug Delivery Systems, Turkish Patent Institute, Patent Application 97/01683.

Eldem, T. and Speiser, P. (1989) Endocytosis and Intracellular Drug Delivery. *Acta Pharm. Technol.*, **35**, 109–115.

Eldem, T., Eldem, B., Özerdem, U., Erdemli, I. and Merkle, H. P. (1997a) Intraocular Administration of PEG-Coated Liposomal Cyclosporine A, NATO-ASI, Targeting of Drugs: Strategies for Stealth Therapeutic Systems, 24 June–5 July 1997, Cape Sounion, Greece.

Eldem, T., Eldem, B., Özerdem, U., Erdemli, I., Ercan, M. and Merkle, H. P. (1997b) Bioavailability and Biodistribution of Sterically Stabilized and Conventional Cyclosporine A and $^{67}Ga$ Labeled Liposomes In Ocular Fluids/Tissues After Intravitreal and Subconjunctival Administration to Healthy Rabbits. 2nd International Symposium on Experimental and Clinical Ocular Pharmacology and Pharmaceutics, September 14, Munich, Germany.

Elner, V. M., Burnstine, M. A., Strieter, R. M., Kunkel, S. L. and Elner, S. G. (1997) Cell-Associated Human Retinal Pigment Epithelium Interleukin-8 and Monocyte Chemotactic Protein-1: Immunochemical and In-Situ Hybridization Analyses. *Exp. Eye Res.*, **65**, 781–789.

Elner, V. M., Elner, S. G., Standiford, T. J., Lukacs, N. W., Strieter, R. M. and Kunkel, S. L. (1996) Interleukin-7 (IL-7) Induces Retinal Pigment Epithelial Cell MCP-1 and IL-8. *Exp. Eye Res.*, **63**, 297–303.

Esser, P., Heimann, K., Bartz-Schmidt, K.-U., Fontana, A., Schraermeyer, U., Thumann, G. *et al.* (1997) Apoptosis in Proliferative Vitreoretinal Disorders: Possible Involvement of TGF-$\beta$-Induced RPE Cell apoptosis. *Exp. Eye Res.*, **65**, 365–378.

Farrokh-Siar, L., Rezai, K. A., Semnani, R. T., Patel, S. C., Ernest, J. T., Peterson, E. J. *et al.* (1999) Human Retinal Pigment Epithelial Cell Induce Apoptosis in the T-Cell Line Jurkat. *Invest. Ophthalmol. Vis. Sci.*, **40**, 1503–1511.

Grabner, G., Boltz, G. and Foerster, D. (1980) Macrophage Properties of Human Hyalocytes. *Invest. Ophthalmol. Vis. Sci.*, **19**, 333–340.

Green, W. R. (1994) Vitreoretinal Juncture in Retina. S. J. Ryan (ed), 2nd Edition, Mosby – Year Book Inc., St. Louis, pp. 1869–1930.

Greenwood, J., Howes, R. and Lightman, S. (1994) The Blood–Retinal Barrier in Experimental Autoimmune Uveoretinitis. *Lab. Invest.*, **70**, 39–52.

Gundersen, D., Powell, S. K. and Rodriguez-Boulan, E. (1993) Apical Polarization of N-CAM in Retinal Pigment Epithelium is Dependent on Contact with the Neural Retina. *J. Cell. Biol.*, **121**, 335–343.

Hewitt, A. T. and Adler, R. (1994) The Retinal Pigment Epithelium and Interphotoreceptor Matrix: Structure and Specialized Functions in Retina. S. J. Ryan (ed), 2nd Edition, Mosby – Year Book Inc., St. Louis, pp. 58–71.

Ho, T.-C., Del Priore, L. V. and Hornbeck, R. (1997) Effect of Mitomycin-C on Human Retinal Pigment Epithelium in Culture. *Curr. Eye Res.*, **16**, 572–576.

Jablonski, M. M., Tombran-Tink, J., Mrazek, D. A. and Iannaccone, A. (2000) Pigment Epithelium-Derived Factor Supports Normal Development of Photoreceptor Neurons and Opsin Expression After Retinal Pigment Epithelium Removal. *J. NeuroSci.*, **20**, 7149–7157.

Jääskeläinen, I., Peltola, S., Honkakoski, P., Mönkkönen, J. and Urtti, A. (2000) A Lipid Carrier with a Membrane Active Component and a Small Complex Size are Required for Efficient Cellular Delivery of Anti-Sense Phoshorothioate Oligonucleotides. *Eur. J. Pharm. Sci.*, **10**, 187–193.

Jampol, L. E. and Po, S. M. (1994) Macular Edema in Retina. S. J. Ryan (ed), 2nd edition, Mosby – Year Book Inc., St. Louis, pp. 999–1008.

Klibanov, A. L., Maruyama, K., Torchilin, V. P. and Huang, L. (1990) Amphipathic Polyethylene-Glycols effectively Prolong the Circulation Time of Liposomes. *FEBS Lett.*, **268**, 235–237.

Kociok, N., Heppekausen, H., Schraermeyer, U., Esser, P., Thuman, G., Grisanti, S. *et al.* (1998) The mRNA Expression of Cytokines and Their Receptors in Cultured Iris Pigment Epithelial Cells: A Comparison with Retinal Pigment Epithelial Cells. *Exp. Eye Res.*, **67**, 237–250.

Kon, C. H., Occleston, N. L., Foss, A., Sheridan, R., Aylward, G. W. and Khaw, P. T. (1998) Effects of Single, Short Term Exposures of Human Retinal Pigment Epithelial Cells to Thiotepa or 5-Fluorourasil: Implications for the Treatment of Proliferative Vitreoretinopathy. *Br. J. Ophthalmol.*, **82**, 554–560.

Konda, B. R., Pararajasegaram, G., Wu, G.-S., Stanforth, D. and Rao, N. A. (1994) Role of Retinal Pigment epithelium in the Development of Experimental Autoimmune Uveitis. *Invest. Ophthalmol. Vis. Sci.*, **35**, 40–47.

Krott, R., Leek, J., Grisanti, S., Esser, P. and Heimann, K. (2000) Antiproliferative Wirkung von Genistein auf kultivierte Pigmentzellen wom Schwein. *Ophthalmologica*, **214**, 296–300.

Lai, C.-C., Gouras, P., Doi, K., Tsang, S. H., Goff, S. P. and Ashton, P. (2000) Local Immunosuppression Prolongs Survival of RPE Xenografts Labelled by Retroviral Gene Transfer. *Invest. Ophthalmol. Vis. Sci.*, **41**, 3134–3141.

Lee, V. H. L., Pince, K. J., Frambach, D. A. and Martenhed, B. (1994) Dug Delivery to the Posterior Segment in Retina. S. J. Ryan (ed), 2nd edition, Mosby – Year Book Inc., St. Louis, pp. 533–551.

Mc Donnell, J. M. (1994) Ocular Embryology and Anatomy in Retina. S. J. Ryan (ed), 2nd Edition, Mosby – Year Book Inc., St. Louis, pp. 5–17.

McKay, B. S. and Burke, J. M. (1994) Separation of Phenotypically Distinct Subpopulations of Cultured Human Retinal Pigment Epithelial Cells. *Exp. Cell Res.*, **213**, 85–92.

Michels, R. G., Wilkinson, C. P. and Rice, T. A. (1990) Retinal Detachment. The C. V. Mosby Company, St. Louis.

Miyajima-Uchida, H., Hayashi, H., Beppu, R., Kuroki, M., Fukami, M., Arakawa, F. *et al.* (2000) Production and Accumulation of Thrombospondin-1 in Human Retinal Pigment Epithelial Cells. *Invest. Ophthalmol. Vis. Sci.*, **41**, 561–567.

Mousa, S. A., Lorelli, W. and Campochiaro, P. A. (1999) Role of Hypoxia and Extracellular Matrix-Integrin Binding in the Modulation of Angiogenic Growth Factors Secretion by Retinal Pigment Epithelial Cells. *J. Cell. Biochem.*, **74**, 133–143.

Nussenblatt, R. B. and Csaky, K. (1997) Perspectives on Gene Therapy in the Treatment of Ocular Inflammation. *Eye*, **11**, 217–221.

Pederson, J. E. (1994) Fluid Physiology of the Subretinal Space in Retina. S. J. Ryan (ed), 2nd edition, Mosby – Year Book Inc., St. Louis, pp. 1955–1967.

Renno, R. Z., Delori, F. C., Holzwer, R. A., Gragoudas, E. S. and Miller, J. W. (2000) Photodynamic Therapy Using Lu-Tex Induces Apoptosis In Vitro, and Its Effect is Potentiated by Angiostatin in Retinal Capillary Endothelial Cells. *Invest. Ophthalmol. Vis. Sci.*, **41**, 3963–3971.

Ryeom, S. W., Sparrow, J. R. and Silverstein, R. L. (1996) CD36 Participates in the Phagocytoses of Rod Outer Segments by Retinal Pigment Epithelium. *J. Cell Sci.*, **109**, 387–395.

Sakamoto, T., Hinton, D. R., Kimura, H., Spee, C., Gopalakrishna, R. and Ryan, S. J. (1996) Vitamin E Succinate Inhibits Proliferation and Migration of Retinal Pigment Epithelial Cells In Vitro: Therapeutic Implication for Proliferative Vitreoretinopathy. *Graefes. Arch. Clin. Exp. Ophthalmol.*, **234**, 186–192.

Schraermeyer, U. and Heimann, K. (1999) Current Understanding on the Role of Retinal Pigment Epithelium and Its Pigmentation. *Pigment Cell Res.*, **12**, 219–236.

Song, M.-K. and Lui, G. M. (1990) Propagation of Fetal Human RPE Cells: Preservation of Original Culture Morphology After Serial Passage. *J. Cell Physiol.*, **143**, 196–203.

Sullivan, D. M., Chung, D. C., Anglade, E., Nussenblatt, R. B. and Csaky, K. G. (1996) Adenovirus-Mediated Gene Transfer of Ornithine Aminotransferase in Cultured Human Retinal Pigment Epithelium. *Invest. Ophthalmol. Vis. Sci.*, **37**, 766–774.

Tombran-Tink, J., Shivaram, S. V., Chader, G. J., Johnson, L. V. and Bok, D. (1995) Expression, Secretion, and Age-Related Downregulation of pigment Epithelium-Derived Factor, a Serpin with Neurotrophic Activity. *J. Neurosci.*, **15**, 4992–5003.

Uehara, M., Imagawa, T. and Kitagawa, H. (1996) Morphological Studies of the Hyalocytes in the Chicken Eye: Scanning Electron Microscopy and Inflammatory Response After the Intravenous injection of Carbon Particles. *J. Anat.*, **188**, 661–669.

Urtti, A., Polansky, J., Lui, G. M. and Szoka, F. C. (2000) Gene Delivery and Expression in Human Retinal Pigment Epithelial Cells: Effects of Synthetic Carriers, Serum, Extracellular Matrix and Viral Promoters. *J. Drug Target.*, **7**, 413–421.

Wenkel, H. and Streilein, J. W. (2000) Evidence that Retinal Pigment Epithelium Functions as an Immune-Privileged Tissue. *Invest. Ophthalmol. Vis. Sci.*, **41**, 3467–3473.

Wilt, W., Greaton, C. J., Lutz, D. A. and McLaughlin, B. J. (1999) Mannose Receptor is Expressed in Normal and Dystrophic Retinal Pigment Epithelium. *Exp. Eye Res.*, **69**, 405–411.

Yan, Q., Vernon, R. B., Hendrickson, A. E. and Sage, E. H. (1996) Primary Culture and Characterization of Microvascular Endothelial Cells from Macaca Monkey Retina. *Invest. Ophthalmol. Vis. Sci.*, **37**, 2185–2184.

# Chapter 17

# Human skin and skin equivalents to study dermal penetration and permeation

*Heike Wagner, Nadial Zghoul, Claus-Michael Lehr and Ulrich F. Schäfer*

## INTRODUCTION

Transdermal or dermal drug delivery is gaining more and more interest in the pharmaceutical industry, owing to the high acceptance of these formulations by the patients. Moreover, in the field of cosmetic research, as well as for safety aspects of active substances like pesticides, there is also a need of knowledge of the invasion behavior of foreign substances on the skin. For ethical reasons this information normally cannot be obtained by *in vivo* studies. Additionally, for drugs where the effect is related to the concentration in the systemic circulation, the concentration of the drug or metabolite in blood or urine may be considered for safety and efficacy aspects. If the site of action of the drug is local, e.g. in the case of antimycotics, the determination of the concentration at this site *in vivo* is nearly impossible. Therefore, other techniques must be applied to this field of research to obtain the desired information. One of these possibilities is the use of *in vitro/ex vivo* penetration and permeation models.

## STRUCTURE AND FUNCTION OF HUMAN SKIN

### Functions of the skin

The skin is one of the largest organs of our body with a size of about $1.8\,m^2$. It performs many different functions. First of all the skin covers our body, keeping the body fluids and tissues inside and determines our appearance (mechanical function). Furthermore, the skin has several protective functions:

- A *chemical barrier* in two directions, controlling the loss of water, electrolytes, and other body constituents while barring the entry of harmful or unwanted molecules from the external environment.
- The *temperature regulation* of the body. The skin is responsible for maintaining the body temperature at $37\,°C$. This means that when the external temperature drops the skin shuts down the peripheral blood flow minimizing the heat loss at the surface, whilst in a warm environment the blood vessels dilate to maximize diffusive thermal loss.

- A *microbiological barrier* preventing the penetration of microorganisms through the intact stratum corneum.
- A *radiation barrier* as a result of producing melanin by the melanocytes in the basal layer upon ultraviolet light stimulation.

## Anatomical structure of the skin

All the different functions of the skin can only be realized by its specialized anatomical structure.

The human skin is made up of three tissue layers :

- the stratified, avascular, cellular epidermis;
- the underlying dermis of connective tissue, and;
- the subcutaneous fat layer.

Additionally, hairy skin contains hair follicles and sebaceous glands.

### The epidermis

The epidermis ranges in its thickness from 0.8 mm on the palms and soles to 0.06 mm on the eyelids. Normally the epidermis is divided into the viable epidermis and the stratum corneum, the outermost layer of the skin.

The *viable epidermis* is made up of several layers, the stratum germinativum (the basal layer), the stratum spinosum, the stratum granulosum and the stratum lucidum which is only present at the palm of the hand and the sole of the foot.

The non-viable almost impermeable *stratum corneum* consists of flattened, stacked, hexagonal and cornified cells which originally stems from the migrating dividing cells of the basal layer. These cells are mainly made up of insoluble bundled keratins accounting for about five per cent of the stratum corneum weight (Roberts and Walters, 1998). The intercellular region consists mainly of lipid bilayers (for composition see Wertz *et al.*, 1989) for corneocyte cohesion. This stratum corneum layer is crucially important in controlling the percutaneous absorption of topically applied substances. It is about 10–20 μm in thickness but swells severalfold when hydrated. A model of the stratum corneum is given in Figure 17.1, the so-called brick and mortar model.

### The dermis

Ranging from 3 to 5 mm in thickness the dermis consists of a matrix of connective tissue (produced by collagen, elastin, reticulin), pierced by skin appendages like sweat glands and pilosebaceous units. The dermis is also innervated by nerves, lymphatics and blood vessels. The blood supply reaches to within 0.2 mm of the skin, conveying nutrients and removing waste products. The blood acts as a sink for diffusing molecules, promoting percutaneous absorption by maximizing the epidermal gradient in keeping the concentration in the dermis low.

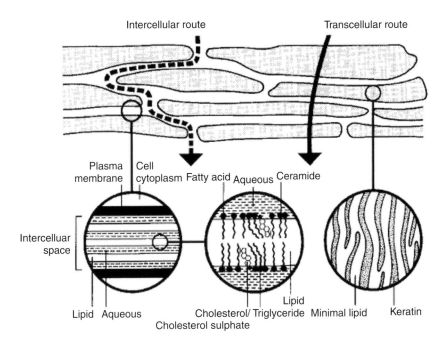

*Figure 17.1* Structure of the stratum corneum and possible penetration routes (according to Elias, 1983).

### The subcutaneous tissue

The subcutaneous fat layer acts as a heat insulator, mechanical cushion and stores readily available high-energy chemicals.

### The skin appendages

*Sweat glands* include the eccrine glands which produce sweat to aid heat control, and the apocrine sweat glands which are found at pilosebaceous follicles which secrete a milky or oily odorless liquid. When metabolized by surface bacteria, it produces the characteristic body smell.

*Hair follicles* with their associated sebaceous glands are developed over all the skin except on the lips, palms and soles. The sebum produced from cell integration by these glands is made up of glycerides, free fatty acids and cholesterol. It protects and lubricates the skin, and also provides a pH of 5 (Roberts and Walters, 1998).

The *nails*, like the hair, consist of hard keratin proteins with a high sulfur component. Unlike the stratum corneum, nails are quite permeable to hydrophilic substances.

### Skin metabolism

The skin contains the major enzymes that also found in other tissues of the body (Pannatier *et al.*, 1978). The effect of skin metabolism on the biological response to dermal applied substances is not exhaustingly investigated because there are many difficulties, e.g. independent measurement of the other systemic enzymes and relatively low enzyme activity when compared with the liver (cytochrome P-450 system). For more details see Bronaugh and Maibach (1999).

## Penetration pathways through the skin

The following paragraph is a brief summary of the penetration pathways through the skin. For more details see Bronaugh and Maibach (1999).

When a diffusant reaches intact skin two major routes through the stratum corneum to the viable tissue are feasible (Figure 17.1):

- through the skin appendages, like hair follicles and sweat ducts;
- across the intact stratum corneum in between these appendages

  i   the intercellular route
  ii  the transcellular route.

It would be a simplification to assume that one route prevails under all conditions, yet after steady-state is reached, transdermal diffusion through the intact stratum corneum is most likely.

### Appendages route of penetration

Skin appendages form about 0.1 per cent of the total skin area (Moghimi *et al.*, 1999) and usually this route does not contribute considerably to the steady-state flux. However, in the early stages of penetration, and in the case of large polar molecules and ions, diffusion through the appendages might play a significant role as they cross the intact stratum corneum only with difficulties. Recent publications report that the appendages route of penetration may play an important role for the absorption of liposomes, nanoparticles and cyclodextrin-inclusion complexes.

### Intercellular route

The intercellular lipids of the stratum corneum have been considered a major component of the skin barrier function. This consideration is not only valid for the normal skin barrier homeostasis but also for maintaining the normal levels of tissue hydration (Elias, 1983). The major route is via this tortuous but continuous intercellular lipid region. The penetration rate depends largely on the physicochemical characteristics of the penetrant which influences the following parameters relevant to the invasion rate: the concentration of the permeant applied; the partitioning of the permeant between stratum corneum and vehicle, and; the diffusivity of the component within the stratum corneum. Because the intercellular region of the stratum corneum is structured as bilayers with *lipophilic and hydrophilic* regions, it is stated that lipophilic substances penetrate predominantly in the lipophilic and hydrophilic in the more hydrophilic part of the bilayer.

### Transcellular route

Normally the transcellular route is not considered to be the major invasion way for dermally applied substances. The reason is the low permeability of the corneocytes and the need of partitioning several times between the more hydrophilic medium of the corneocytes and the lipophilic intercellular region of the stratum corneum. For some penetration enhancers, like e.g. urea, it is assumed that due to specific interaction with the matrix of the ceratinocytes this pathway might be opened.

The skin barrier function resides mainly in the stratum corneum layer and the rate-limiting step in percutaneous absorption is given by this layer. However, for very lipophilic drugs the rate determining step changes of the permeation through the stratum corneum to the clearance of the stratum corneum (Moghimi *et al.*, 1999) owing to the poor solubility of these compounds in the more aqueous phase of the viable epidermis.

## IN VITRO TEST SYSTEMS

*In vivo* experiments on humans are and will always be the 'gold standard' by which the behavior of drugs in the local dermal therapy is investigated. But considering ethical aspects as well as high costs, alternatives had to be found. Also, other problems had to be overcome, e.g. the difficulty of detecting the drug within the skin (Idson, 1975), or the lack of a drug to show a pharmacodynamic response on the skin's surface, which would allow quantitative statements about the effectiveness of the dermal therapy (e.g. vasoconstriction). So, different methods were developed: the detection of the drug residuals in topically applied products, the stratum corneum stripping using adhesive tape, and; the investigations of target tissue levels (e.g. by punch biopsies). For bioavailability studies blood and/or urine sampling was performed (Roberts and Walters, 1998). Nevertheless, the need of *in vitro* test systems to study *in vitro* pharmacokinetic characteristics became obvious. In a first step, these test systems should allow the evaluation of the drug's diffusion within the skin by carrying out *in vitro* test series' using different barriers which will combine the following advantages (ECETOC Monograph No. 20 (1993)):

- the conditions can be more easily controlled;
- normally they are easier to perform;
- radiolabeled as well as highly toxic chemicals can be investigated;
- a large number of experiments can be run simultaneously;
- special parameters (e.g. release of the drug from different vehicles, influence of temperature, moisture, pretreatment) can be studied more readily;
- the amount of penetrating substance can be measured directly in the skin or in an acceptor medium beneath the skin without dilution in tissue fluids or organs. In this manner the number of *in vivo* experiments may be reduced, resulting in a decrease of costs.

There are also disadvantages which have to be borne in mind when carrying out *in vitro* test series:

- the acceptor medium can influence the transfer of substance through the skin or the conditions of the skin itself;
- the absence of blood flow may lead to changes in the barrier properties of the different skin layers;
- when exploring excised skin specimens, the tissue can only be maintained unchanged for a relatively short period of time;
- *in vivo* conditions cannot be fully simulated *in vitro* (e.g. the shedding of the stratum corneum).

*In vitro* test series are not only carried out with excised human skin, but also with animal skin and artificial model membranes. As recommended by the European Commission on Validation of Alternative Methods (ECVAM) (Howes *et al.*, 1996), the following physiological hierarchy of methods for measuring percutaneous absorption is generally accepted (Figure 17.2).

Today, two different experimental designs are commonly established: permeation and penetration studies.

## Permeation versus penetration test systems

### Infinite versus finite dosing

Permeation and penetration studies can be carried out using a so-called infinite as well as a finite dose. Under infinite dose conditions the drug reservoir in the donor is very high and the drug and the vehicle concentration are more or less constant during the whole incubation time (Franz *et al.*, 1993). The advantage of this dosage regime is the possibility of achieving a quasi constant adsorption rate of the drug through the skin, which will end in quasi steady-state conditions in the stratum corneum. Infinite dose conditions can be maintained by applying more than $10 \, \text{mg/cm}^2$ of the drug preparation on the skin's surface (Diembeck *et al.*, 1999).

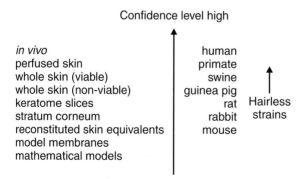

Confidence level high

| *in vivo* | human |
| perfused skin | primate |
| whole skin (viable) | swine |
| whole skin (non-viable) | guinea pig |
| keratome slices | rat   Hairless |
| stratum corneum | rabbit   strains |
| reconstituted skin equivalents | mouse |
| model membranes | |
| mathematical models | |

*Figure 17.2* Hierarchy of skin types for measuring percutaneous absorption, according to Howes *et al.*, 1996.

Finite dosing is often considered closer to the reality of drug application. With this dosing regime one is able to determine the fraction of absorbed dose and the influence of the changes of the formulation on the skin surface due to evaporation of vehicle compounds. With respect to safety aspects, one is able to check the potential of toxic effects by calculating the fraction of the absorbed dose. However, other kinetic calculations are difficult and, therefore, finite dosing is not described in the following paragraphs. For detailed information see Bronaugh and Maibach (1999).

## Permeation test system: Franz diffusion cell

Today, the most widely accepted permeation test system is the Franz diffusion cell (FD-C) (Akhter *et al.*, 1986; Brain *et al.*, 1998; Bronaugh, 1993; Franz, 1975). Permeation experiments with this model lead to data which provide information about the drug amount permeated through the skin after different time points, for example steady-state flux, permeability and diffusion constant and lag time.

The FD-C consists of two compartments (Figure 17.3): a donor compartment and an acceptor compartment. The donor compartment holds the drug preparation – an ointment as well as a solution or a patch can be investigated. The acceptor compartment contains an aqueous or alcoholic acceptor medium with or without solubilizers; the addition of preservatives to inhibit microbial growth in the acceptor medium may be necessary depending on the length of the incubation time and the barrier but is not unproblematic (Brain *et al.*, 1998; Sloan *et al.*, 1991). The composition of the acceptor medium is determined by the solubility of the drug (as a general rule, the concentration of the permeating drug in the acceptor medium should not be allowed to exceed approximately ten per cent of saturation solubility). It is important to consider the maintenance of sink conditions during the whole incubation time. The used barrier is sandwiched between the two compartments. Either artificial membranes, animal skin, human skin or reconstructed skin equivalents can be investigated. While carrying out the experiments, samples of the acceptor medium are drawn over the sampling port of the FD-C after different time points and are immediately replaced by fresh medium. The whole apparatus is kept at $32 \pm 1\,^{\circ}$C by a water jacket. The receptor fluid is mixed with a magnetic stirring bar. To achieve higher reproducibility, the barrier should be prehydrated with the basolateral receptor medium for a suitable time interval, e.g. for human full thickness skin, three hours.

The FD-C combines the following advantages: it is inert, robust and easy to handle; it allows the use of barriers of different thickness; it provides a thorough mixing of the acceptor medium and ensures intimate contact between the barrier and the acceptor phase; it is maintainable at a constant temperature and has precisely calibrated volumes and diffusion areas, and; it facilitates easy sampling and replenishment of the acceptor medium.

## Artificial membranes

### Liberation experiments

Different types of artificial membranes are usually utilized for FD-C experiments, which can be described as sheets of solid or semisolid material of fixed dimension,

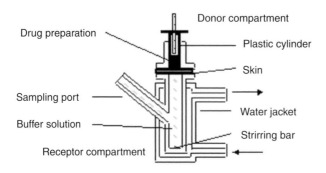

*Figure 17.3* Franz diffusion cell.

insoluble in the surrounding media, separating two different phases – one of which is usually fluid (Flynn *et al.*, 1974). The following materials are taken to obtain appropriate membranes:

- silastic (polydimethylsiloxane);
- cellulose acetate;
- polyurethane;
- supor (modified polysulfone);
- zeolite (aluminosilicates), or;
- miscellaneous types (collagen, egg shell) (Brain *et al.*, 1998; ECETOC Monograph No. 20 (1993); Shah *et al.*, 1989).

The artificial membranes are employed with or without wetting agents. The major task is to prevent mixing of the acceptor medium and the drug preparation (Nakano and Patel, 1970). They can be used alone or in combinations (multimembrane systems), in which the combination will facilitate the opportunity to investigate ointment bases with different polarities taking the same membrane system and, therefore, allowing a direct comparison of these ointment bases (Loth *et al.*, 1978, 1979). No matter which membrane is put in the FD-C experiments, it must be considered that liberation experiments not permeation experiments, are carried out. Normally, the membrane does not influence the drug diffusion. At the beginning of the experiments the diffusion of the drug through an artificial membrane is rate-limiting; this means the diffusion resistance of the membrane is much higher than the diffusion resistance in the drug preparation. As more drug diffuses out of the ointment base, the concentration profile penetrates deeper into the vehicle, slowing down the flux in the semisolid preparation. Normally, diffusion from the vehicle becomes slower than permeation through the membrane, and drug appearance in the acceptor medium becomes linear with the square root of time according to the equation of Higuchi (Guy and Hadgraft, 1990; Parks *et al.*, 1997), which can be illustrated for drugs as follows (Higuchi, 1961):

Drug in solution

$$Q = A \times 2C_0 \sqrt{\frac{Dt}{\pi}}$$ (17.1)

Drug in suspension

$$Q = A \sqrt{D \times t \times C_s (2 \times C_0 - C_s)}$$ (17.2)

where $Q$=amount released at time t; $A$=diffusion area (cm²); $D$=diffusion coefficient (cm/h); $t$=time (h); $C_s$=saturation concentration, and; $C_0$=concentration in the ointment at $t$=0.

More details for such liberation experiments are given in the Food and Drug Administration (FDA) Guidance SUPAC-SS.

### Artificial membranes as surrogate for the stratum corneum

Artificial membranes are also used to imitate the stratum corneum (Brain *et al.*, 1998). For this purpose, for example, liposome vesicles prepared from a mixture of stratum corneum lipids are fused onto a support membrane. Another alternative is the construction of structured lipid matrices achieved with cholesterol, water and fatty acids. These experimental approaches have nothing in common with the liberation experiments mentioned above and must be classified as permeation experiments.

## Animal skin

For permeation studies using animal skin, the skin of different animal species is sandwiched between the two compartments. The following animal species are often used as substitutes for human skin: mouse, rat and pig (mostly for the testing of cosmetics). Also used are the skin of guinea-pig, rabbit, hairless mouse, hairless guinea-pig, hairless dog, cat, dog, miniature pig, horse, cow udder, goat, snake, squirrel monkey, rhesus monkey and chimpanzee. Concerning this list, the question arises how far all these animal species can be considered to be suitable tools for permeation studies of drugs for the dermal therapy of humans. Important differences may not only be the number of hair follicles and sebaceous glands, but also the composition of the stratum corneum lipids and the thickness of the skin. Therefore, human skin is still the 'gold standard' for *in vitro* testing of drugs in the dermal field of research (Hadgraft, 1999).

## Human skin

Excised human skin is most favorably used in FD-C experiments (Figure 17.4). It can be differentiated between investigations with full thickness skin, dermatomed skin, separated epidermis, isolated stratum corneum and stripped skin. This range of barriers, which all originate from full thickness skin, was established because of the high thickness and high barrier properties of full thickness skin which lead to several hours of incubation until the drug reaches the acceptor medium.

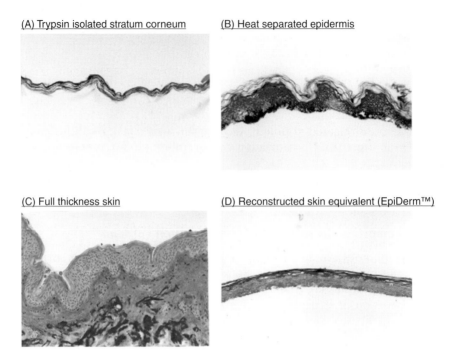

(A) Trypsin isolated stratum corneum

(B) Heat separated epidermis

(C) Full thickness skin

(D) Reconstructed skin equivalent (EpiDerm™)

*Figure 17.4* Cross-sections of different skin membranes (×104) (Wagner *et al.*, 2001).

Using only parts of the skin, a control concerning the integrity of the barriers, which should be used in FD-C experiments, is indispensable.

### Dermatomed skin

Dermatomed skin is often used in a thickness between 200 to 600 μm. Here, the stratum corneum, the epidermis and a small part of the dermis are present during the drug's permeation. It can be prepared as follows (Brain *et al.*, 1998): the skin is placed, dermal side down, onto a metal plate. A thin sheet of plastic is put on the stratum corneum to protect it before a second metal plate is applied. The two plates are clamped together and cooled at −20 °C until the skin is completely frozen. After removal from the freezer, the upper plate is gently warmed to ease removal from the stratum corneum. The plastic sheet is also removed and a dermatome is used to remove strips of skin of the desired thickness.

### Separated epidermis

To acquire epidermis different methods are available. The most current method is the heat separation (Parry *et al.*, 1990; Swarbrick *et al.*, 1982; Walker *et al.*, 1997). The epidermis is separated by putting full thickness skin in water of 60 °C for one minute, removing the skin from the water and placing it, dermal side down, on a filter paper where the SC-epidermis layer of the skin can be peeled off the dermis using forceps. The epidermis can be stored on teflon pieces (less than three days) in a desiccator until use. Other methods to separate the epidermis and the dermis are described as follows: incubation of the skin sample in 1.5 M solution of potassium bromide for 12 hours (Wiedmann, 1988), or in 0.22 M solution of ammonium chloride (pH 9.4) for two minutes at 4°C (Freinkel and Traczyk, 1983), or above concentrated ammonium hydroxide solution for 30 minutes (Kligmann and Christophers 1963). As mentioned before, the epidermis can be peeled off afterwards using all methods. A last alternative is the maximal stretching and mounting of a strip of skin and the direct separation of the epidermis using a razor blade (van Scott, 1952). For the handling of the isolated stratum corneum pieces, rehydration of the epidermis is also necessary before the experiments, and the use of a supporting membrane is helpful.

The drawback of all aforementioned methods is the influence of mechanical stress or chemical agents on the skin. This may lead to a change in the characteristics of the skin and, therefore, should be minimized, especially the chemical separation techniques.

### Isolated stratum corneum

Isolated stratum corneum is obtained using a trypsin solution in an appropriate concentration (trypsin concentration varies between 0.0001 per cent and 1.0 per cent). According to Kligmann and Christophers (1963) full thickness skin is transferred, dermal side down, into a Petri dish, which contains trypsin solution in phosphate-buffered saline (PBS) buffer, and incubated for 24 hours at 32°C. This procedure is repeated with fresh trypsin solution until the stratum corneum is fully isolated. The isolated stratum corneum pieces are washed three times with PBS buffer and distilled water and stored (less than three days) on teflon pieces in a desiccator until use. Drying prior to use will not affect the permeation results, as long as the stratum corneum pieces are rehydrated again before the experiments. Mostly, the stratum corneum pieces are sandwiched on a supporting membrane for better handling (Brain *et al.*, 1998).

### Skin with reduced barrier properties

To reduce the barrier function of the skin, stripped skin where only the epidermis and dermis are present can be used.

## Reconstituted skin equivalents

The technology behind the reconstruction of human skin equivalents (example presented in Figure 17.4) has its roots predominantly in the research into the treatment of burns (Brain *et al.*, 1998). Different techniques and culture conditions have been established and can be distinguished into methods based on raised collagen membranes and methods on the reconstitution of skin (Prunieras *et al.*, 1983). There is one fact that joins up all equivalents available: only the stratum corneum

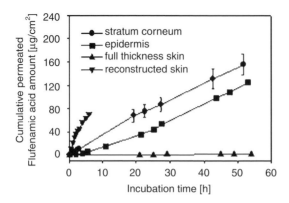

*Figure 17.5* Permeation of a model drug preparation (0.9 per cent flufenamic acid in wool alcohol ointment [WAO]) through different skin layers in dependence on the incubation time (Wagner *et al.*, 2001).

and the epidermis can be modeled. Until now, these skin equivalents were mostly used to investigate both cutaneous metabolic events and dermal irritation. Nowadays, permeation studies are also carried out, but an evaluation of the results obtained is difficult. In fact, this methodology is very expensive. In comparison with excised human skin it has the advantage of a smaller batch-to-batch variation, but reconstructed skin equivalents, in general, are more permeable than human skin (compare Figure 17.5). This might be due to the differences in the lipid composition compared with human skin (Asbill *et al.*, 2000). Nevertheless, an investigation of different drug preparations (solution, ointment, patch) is possible (Gysler *et al.*, 1999; Michel *et al.*, 1995; Regnier *et al.*, 1993; Specht *et al.*, 1998; Zghoul *et al.*, 2001).

## Results and treatment of the data

To calculate the parameters achieved using the FD-C, the measured content (C) of the acceptor medium must be corrected for the removed drug amounts using the following formula:

$$C_{n,corrected} = C_{n,measured} + volume_{sample}/volume_{acceptor} (C_{1,measured}$$
$$+ C_{2,measured} + \cdots + C_{n-1,measured}). \tag{17.3}$$

As an example, permeation parameters are obtained from the cumulative amounts of flufenamic acid permeated per cm$^2$ versus incubation time plots (Figure 17.5). The steady-state flux ($J$) – representing the absorption rate per unit area – is determined from the slope of the linear portion of the plots. Comparing several experiments, the same number of data points shall be taken to calculate the steady-state flux. The lag time ($t_{lag}$) – symbolizing the time of delay which describes the first contact of the drug with the skin's surface until a steady-state

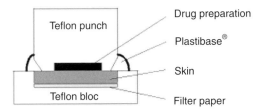

*Figure 17.6* Saarbrucken penetration model (Wagner *et al.*, 2000).

flux is established – is found as the intercept of the plots. The permeability constant (P) can be calculated according to Fick's first law of diffusion, based on the steady-state flux and the applied drug concentration ($C_i$) of the donor, or with the help of the diffusion constant (D) and the thickness of the skin sample (h). The D itself can be obtained with the square of the thickness of the skin divided by the product of the $t_{lag}$ and a constant (Foldvari *et al.*, 1998; Kubota *et al.*, 1993; Ritschel *et al.*, 1989):

$$P(cm/h) = \frac{J(\mu g/cm^2/h)}{C_i(\mu g/cm^3)} = \frac{D}{h(cm)} \tag{17.4}$$

$$D(\mu m^2/h) = \frac{h^2(\mu m^2)}{6t_{lag}(h)}. \tag{17.5}$$

## Penetration test system: Saarbrucken penetration model

Penetration studies with the Saarbrucken penetration model (SB-M) allow the determination of drug amounts within different skin layers (Blasius, 1985; Borchert, 1994; Hailer, 1981; Wagner *et al.*, 2000; Wild, 1988). Using segmentation techniques, the drug content can be separately obtained for the stratum corneum and the deeper skin layers. In contrast to the FD-C, non-physiological hydration of the tissue is avoided here because there is no liquid acceptor medium beneath the skin. But concerning sink conditions, the incubation time is limited to the time interval until the drug reaches the lowest layers of the dermis and starts to accumulate: this is caused by the fact that the skin is the only acceptor for the penetrating drug.

Using the SB-M (Figure 17.6), the skin is put on a filter paper which is soaked with Ringer solution to prevent water loss of the skin during the experimental time. Skin and filter paper together are placed into a cavity of a teflon bloc. The drug preparation is filled into a cavity of a teflon punch 2 mm in depth to maintain infinite dose conditions during the experiment. Ointments and patches can be investigated without further changes in the experimental design, while solutions can only be applied using pieces of cotton soaked with this solution. The teflon punch is applied on the surface of the skin, and a weight of 0.5 kg is placed on the top of the punch for two minutes to improve the contact between the skin and the drug preparation. After two minutes, the teflon punch is fixed in its position, and the gap between the two teflon parts is sealed with Plastibase® to avoid water loss from the skin. The whole apparatus is transferred into a plastic box and placed in a water bath at $32 \pm 1$ °C for thermostatization.

## Segmentation techniques

To obtain information about the drug content in the different layers of the skin, the stratum corneum, as well as the deeper skin layers (epidermis+dermis), is horizontally segmented at the end of each experiment using a standardized tape stripping technique and subsequent cryosectioning. Afterwards, the drug can be extracted from each sample using an appropriate extraction method, and the amount can be analyzed, for example, using a high-performance liquid chromatography (HPLC) method.

First, the remaining drug preparation is removed from the skin's surface by wiping the skin with cotton. Second, the skin is transferred into a special apparatus, where it is mounted on cork discs using small pins (Borchert, 1994). The incumbent stretching helps to overcome problems of furrows in the subsequent tape stripping procedure (van der Molen *et al.*, 1997). For this procedure, the surface of the skin is covered with a teflon mask with a central hole of 15 mm in diameter. The skin is successively stripped with 20 pieces of adhesive tape (size=15×19 mm) placed on the central hole. In a standardized procedure, each tape is charged with a weight of 2 kg for ten seconds, removed rapidly and combined in six pools, e.g. one, two, three, four, five and six strips for analytical purpose (Figure 17.7).

Due to this procedure, each of the removed cell layers has nearly the same thickness (Borchert, 1994; Pershing *et al.*, 1994; Theobald, 1998; Wagner *et al.*, 2000). The first tape strip is always discarded because of potential contamination by residual drug on the surface of the skin (Howes *et al.*, 1996). After the tape stripping, the skin is rapidly frozen in a stream of expanding carbon dioxide, and a specimen with a diameter of 13 mm is taken out of the stripped area and transferred into a cryomicrotome (Figure 17.8), (Zesch and Schaefer, 1974).

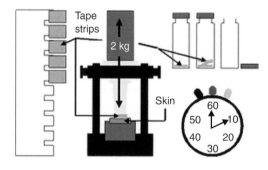

*Figure 17.7* Tape stripping procedure.

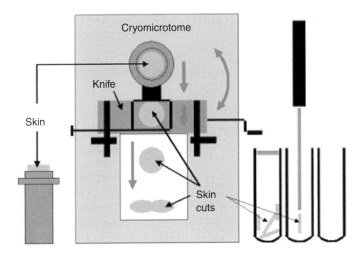

*Figure 17.8* Cryosectioning.

The skin is cut into surface-parallel sections and collected, e.g. according to the following scheme: # 1=incomplete cuts, # 2–5=2×25 μm sections, # 6–9=4× 25 μm sections, # 10–11=8×25 μm sections, # 12=rest of the residual tissue.

## Results and treatment of the data

The described segmentation provides the possibility to illustrate the data in two different ways. On the one hand, the totally detected drug amounts within the stratum corneum or the deeper skin layers can be recorded (Figure 17.9). On the

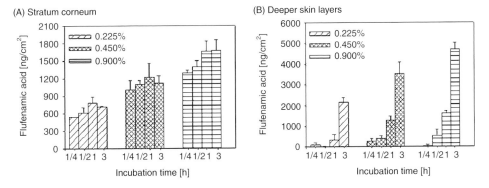

*Figure 17.9* Drug amounts detected in the skin after different incubation times. Model drug preparation: flufenamic acid in wool alcohols ointment (WAO) – varying concentrations (Wagner *et al.*, 2001).

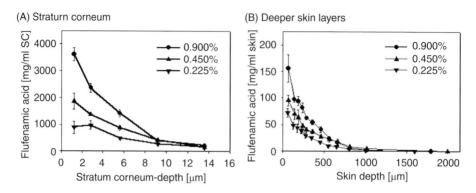

Figure 17.10 Drug concentration – skin depth – profiles. Model drug preparation: flufenamic acid in wool alcohols ointment (WAO) – varying concentrations (Wagner *et al.*, 2001).

other hand, drug concentration – skin depth – profiles can be generated (Figure 17.10). To calculate the respective data, the following parameters have to be known (Schaefer and Loth, 1996):

- size of the investigated skin area (given by the hole in the teflon mask and the punch biopsy);
- extraction volume;
- thickness of the stratum corneum and of all samples obtained by the cryosectioning of the deeper skin layers.

The parameters could be obtained using the following equations:

*Drug amount within the stratum corneum*

$$Drug(ng/cm^2) = \frac{(\text{extraction volume}(ml) \times \Sigma samples_{\text{stratum corneum}}(ng/ml))}{\text{investigated skin area } (cm^2)305}. \quad (17.6)$$

*Drug amount within the deeper skin layers*

$$Drug(ng/cm^2) = \frac{(\text{extraction volume}(ml) \times \Sigma samples_{\text{deeper skin layers}}(ng/ml))}{\text{investigated skin area } (cm^2)}. \quad (17.7)$$

*Drug concentration – stratum corneum depth – profiles*

$$\text{Thickness of each removed stratum stratum corneum layer } (\mu m) = \frac{\text{stratum corneum thickness}(\mu m)}{\text{numbers of the used tape strips}}. \quad (17.8)$$

$$
\begin{array}{l}
\text{Volume of the stratum} \\
\text{corneum strips (cm}^3)
\end{array}
=
\frac{\begin{array}{c}\text{thickness of each removed} \\ \text{stratum corneum layer }(\mu m)\end{array}}{\begin{array}{c}1000 \times \text{investigated (skin area (cm}^2)) \times \\ \text{number of the used tape strip.}\end{array}}. \quad (17.9)
$$

*Drug concentration – skin depth – profiles*: calculation of the thickness of the incomplete cuts and the skin rest over their weights relating to a standard cut with known weight and thickness.

$$
\text{Volume of the skin slices (cm}^3) = \frac{\text{thickness of the slices }(\mu m)}{1000 \times \text{investigated skin area(cm}^2)}. \quad (17.10)
$$

If the experiments are carried out as a function of time, the drug amount present in the stratum corneum under steady-state conditions can be registered (Figure 17.9). For this purpose, the following equation can be used (Wagner *et al.*, 2000):

$$
m_{act} = \frac{m_{ss} \times t_{inc}}{t_{ss/2} + t_{inc}} \quad (17.11)
$$

where $m_{ss}$ = quasi steady-state drug amount in the stratum corneum (ng/cm$^2$), $t_{ss/2}$ = incubation time after which 1/2 $m_{ss}$ is reached (minutes), $t_{inc}$ = incubation time [minutes] and $m_{act}$ = actual drug amounts at time $t_{inc}$ (ng/cm$^2$). The results according to equation 17.11 are given in Table 17.1 and Figure 17.11.

A certain advantage in using this equation is the calculation of the quasi steady-state drug amount in the stratum corneum $m_{ss}$ as a function of the drug formulation. The influence of the different formulation parameters can be estimated when comparing these values for different drug preparations.

Table 17.1 'Best fit' calculations of the detected drug amounts in the stratum corneum versus incubation time using Equation 17.11 $(n=2-4)$ (Wagner et al., 2001)

| | Flufenamic acid concentration | | |
|---|---|---|---|
| | 0.225 per cent | 0.450 per cent | 0.900 per cent |
| $m_{ss}$ [ng/cm$^2$] | 735.90 ± 7.31 | 1282.97 ± 19.34 | 1672.46 ± 48.16 |
| $t_{ss/2}$ [min] | 5.57 ± 0.34 | 4.29 ± 0.54 | 4.88 ± 1.06 |
| R | 0.999 | 0.999 | 0.998 |

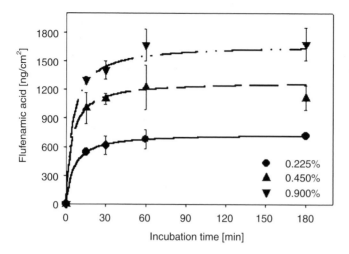

Figure 17.11 'Best fit' calculation of the means of the detected drug amounts in the stratum corneum against the incubation time and the corresponding parameters. Model drug preparation: Flufenamic acid in wool alcohols ointment (WAO) – varying concentrations.

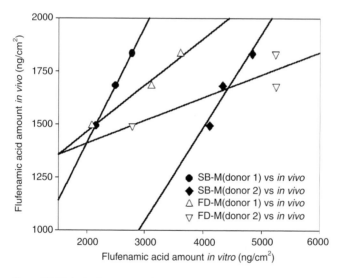

Figure 17.12 In vitro/in vivo correlation of experimental data shown as the means of the amount in the stratum corneum (ng/cm$^2$) ($n=2$–5 (in vitro) and $n=12$ (in vivo)) (Wagner et al., 2000).

## SUMMARY

*In vitro* permeation and penetration experiments with excised human skin are useful tools to obtain information on drug transport across or into the skin. Information on particular phenomena, such as, for example, drug distribution to the deeper skin layers or enrichment in a distinct skin layer is not readily available from *in vivo* experiments. Although *in vitro/in vivo* correlations for the skin are not often presented, experiments have shown that *in vitro* experiments are predictive. In this laboratory, e.g. Theobald (1998) has shown an *in vitro/in vivo* correlation for hydrocortisone and Wagner *et al.* (2000) for flufenamic acid (Figure 17.12). One prerequisite for all these experiments is a highly standardized experimental procedure. Therefore, it is desirable to define worldwide experimental designs and protocols.

# REFERENCES

Akhter, S. A. and Barry, B. W. (1986) Permeation of drugs through human skin: method and design of diffusion cells for *in vitro* use. In R. Marks and G. Plewig (eds), *Skin Models*. Springer Verlag, New York, pp. 358–370.

Asbill, C., Kim, N., El-Kattan, A., Creek, K., Wertz, P. and Michniak, B. (2000) Evaluation of a human bio-engineered skin equivalent for drug permeation studies. *Pharm. Res.*, **17**, 1092–1097.

Blasius, S. (1985) Sorptionsvermittler – Einfluß auf die Liberation von Indomethacin aus Salben und auf die Arzneistoffaufnahme durch exzidierte Haut. *Dissertation Saarbruecken*.

Borchert, D. (1994) Methoden zur Untersuchung der simultanen Penetration von Arzneistoffen und Vehikelbestandteilen aus Salben in exzidierter Humanhaut. *Dissertation Saarbruecken*.

Brain, K. R., Walters, K. A. and Watkinson, A. C. (1998) Investigation of skin permeation *in vitro*. In M. S. Roberts and K. A. Walters (eds), *Dermal Absorption and Toxicity Assessment*. Vol. 91. Marcel Dekker, New York, pp. 161–187.

Bronaugh, R. L. and Maibach, H. I. (1999) *Percutaneous Absorption*, Marcel Dekker, New York, Basel.

Bronaugh, R. L. (1993) Diffusion cell design. In V. P. Shah and H. I. Maibach (eds), *Topical Drug Bioavailability, Bioequivalence and Penetration*, Plenum Press, New York/London, pp. 117–125.

Diembeck, W., Beck, H., Benech-Kiefer, F., Courtellemont, P., Dupuis, J., Lovell, W. *et al.* (1999) Test guidance for *in vitro* assessment of dermal absorption and percutaneous penetration of cosmetic ingredients. *Food Chem. Toxicol.*, **37**, 191–205.

ECETOC Monograph No. 20 (1993) URL http://www.oecd.org/ehs/test/health.htm.

Elias, P. M. (1983) Epidermal lipids, barrier function, and desquamation. *J. Invest. Dermatol.*, **80**, 44S–49S.

Flynn, G. L., Yalkowsky, S. H. and Roseman, T. J. (1974) Mass transport phenomena and models: theoretical concepts. *J. Pharm. Sci.*, **63**, 479–510.

Foldvari, M., Oguejiofor, C. J. N., Wilson, T. W., Afridi, S. K. and Kudel, T. A. (1998) Transcutaneous delivery of prostaglandin $E_1$: *In vitro* and laser doppler flowmetry study. *J. Pharm. Sci.*, **87**, 721–725.

Franz, T. J. (1975) Percutaneous absorption – on the relevance of *in vitro* data. *J. Invest. Dermatol.*, **64**(3), 190–195.

Franz, T. J., Lehman, P. A., Franz, S. F., North-Root, H., Demetrulias, J. L., Kelling, C. K., *et al.* (1993) Percutaneous penetration of *N*-nitrosodiethanolamine through human skin (*in vitro*): comparison of finite and infinite dose applications from cosmetic vehicles. *Fundam. Appl. Toxicol.*, **21**, 213–221.

Freinkel, R. and Traczyk, T. N. (1983) Acid hydrolases of the epidermis: subcellular localisation and relationship to cornification. *J. Invest. Dermatol.*, **80**, 441–446.

Guy, R. H. and Hadgraft, J. (1990) On the determination of drug release rates from topical dosage forms. *Int. J. Pharm.*, **60**, R1–R3.

Gysler, A., Königsmann, U. and Schäfer-Korting, M. (1999) Dreidimensionable Hautmodelle zur Erfassung der perkutanen Resorption. *ALTEX*, **16**, 67–72.

Hadgraft, J. (1999) Advantages and limitations of the use of human skin. APV-Kurs 366 – *Bioaequivalenz dermaler Arzneizubereitungen*, D-Halle.

Hailer, M. (1981) Freisetzung von Salicylsaeure aus Suspensionssalben: Liberation im Membran-Modell im Vergleich zur Wirkstoffabgabe an exzidierter Haut. *Dissertation Saarbruecken*.

Higuchi, T. (1961) Rate of release of medicaments from ointment bases containing drugs in suspension. *J. Pharm. Sci.*, **50**, 874–875.

Howes, D., Guy, R., Hadgraft, J., Heylings, J., Hoeck, U. and Kemper, F. (1996) Methods for assessing percutaneous absorption. ECVAM workshop report 13. *Alternatives Lab. Anim.*, **24**, 81–106.

Idson, B. (1975) Percutaneous absorption. *J. Pharm. Sci.*, **64**, 901–924.

Kligman, A. M. and Christophers, E. (1963) Preparation of isolated sheets of human stratum corneum. *Arch. Dermatol.*, **88**, 702–705.

Kubota, K., Sznitowska, M. and Maibach, H. I. (1993) Percutaneous absorption. A single-layer model. *J. Pharm. Sci.*, **82**, 450–456.

Loth, H. and Holle-Benninger, A. (1978) Untersuchung der Arzneistoffliberation aus Salben. 1. Mitteilung: Entwicklung eines in-vitro Liberationsmodells. *Pharmazeutische Industrie*, **40**, 256–261.

Loth, H., Holle-Benninger, A. and Hailer, M. (1979) Untersuchungen der Arzneistoffliberation aus Salben. 2. Mitteilung: Einfluesse der Eigenschaften wasserfreier Salbengrundlagen auf die Wirkstofffreisetzung aus Suspensionssalben. *Pharmazeutische Industrie*, **41**, 789–796.

Michel, M., Germain, L., Belanger, P. M. and Auger, F. A. (1995) Functional evaluation of anchored skin equivalent cultured *in vitro*: percutaneous absorption studies and lipid analysis. *Pharm. Res.*, **12**, 455–458.

Moghimi, H. R., Barry, B. W. and Williams, A. C. (1999) Stratum Corneum and Barrier Performance: A Model Lamellar Structural Approach. In R. L. Bronaugh and H. I. Maibach (eds), *Drugs and the Pharmaceutical Sciences*, Vol. 97. *Percutaneous Absorption: Drugs – Cosmetics – Mechanisms – Methology*, Marcel Dekker, New York/Basel/Hong Kong, pp. 515–553.

Nakano, M. and Patel, N. K. (1970) Release, uptake and permeation behaviour of salicylic acid in ointment bases. *J. Pharm. Sci.*, **59**, 985–988.

Pannatier, A., Jenner, P., Testa, B. and Etter, J. C. (1978) The skin as a drug metabolizing organ. *Drug Metab. Rev.*, **8**, 319–343.

Parks, J. M., Cleek, R. L. and Bunge, A. L. (1997) Chemical release from topical formulations across synthetic membranes: infinite dose. *J. Pharm. Sci.*, **86**, 187–192.

Parry, G. E., Bunge, A. L., Silcox, G. D., Pershing, L. K. and Pershing, D. W. (1990) Percutaneous absorption of benzoic acid across human skin. I. In-vitro experiments and mathematical modeling. *Pharm. Res.*, **7**, 230–236.

Pershing, L. K., Corlett, J. and Jorgensen, C. (1994) *In vivo* pharmacokinetics and pharmacodynamics of topical ketoconazole and miconazole in human stratum corneum. *Antimicrob. Agents Chemother.*, **38**, 90–95.

Prunieras, M., Regnier, M. and Woodley, D. (1983) Methods for cultivation of keratinocytes with an air-liquid interface. *J. Invest. Dermatol.*, **81**, 28s–33s.

Regnier, M., Caron, D., Reichert, U. and Schäfer, H. (1993) Barrier function of human skin and human reconstructed epidermis. *J. Pharm. Sciences*, **82**, 404–407.

Ritschel, W. A., Sabouni, A. and Hussain, A. S. (1989) Percutaneous absorption of coumarin, griseofulvin and propranolol across human scalp and abdominal skin. *Methods Find. Exp. Clin. Pharmacol.*, **11**, 643–646.

Roberts, M. S. and Walters, K. A. (1998) The relationship between structure and barrier function of skin. In M. S. Roberts and K. A. Walters (eds), *Drugs and the Pharmaceutical Sciences*, Vol. 91. *Dermal Absorption and Toxicity Assessment*. Marcel Dekker, New York/Basel/Hong Kong, pp. 1–42.

Schaefer, U. and Loth, H. (1996) An ex-vivo model for the study of drug penetration into human skin. *Pharm. Res.*, **13** (Suppl.), 366.

Shah, V. P., Elkins, J., Lam, S.-Y. and Skelly, J. P. (1989) Determination of *in vitro* drug release from hydrocortisone creams. *Int. J. Pharm.*, **53**, 53–59.

Sloan, K. B., Beall, H. D., Weimar, W. R. and Villanueva, R. (1991) The effect of receptor phase composition on the permeability of hairless mouse skin in diffusion cell experiments. *Int. J. Pharm.*, **73**, 97–104.

Specht, C., Stoye, I. and Mueller-Goymann, C. C. (1998) Comparative investigations to evaluate the use of organotypic cultures of transformed and native dermal and epidermal cells for permeation studies. *Eur. J. Pharm. Biopharm.*, **46**, 273–278.

SUPAC-SS, FDA (1997) *Guidance for Industry. Nonsterile Semisolid Dosage Forms*, SUPAC-SS, CMC 7.

Swarbrick, J., Lee, G. and Brom, J. (1982) Drug permeation through human skin: I. Effect of storage conditions of skin. *J. Invest. Dermatol.*, **78**, 63–66.

Theobald, F. (1998) *In vitro* Methoden zur biopharmazeutischen Qualitätsprüfung von Dermatika unter Berücksichtigung der Lipidzusammensetzung des Stratum corneum. *Dissertation Saarbruecken*.

van der Molen, R. G., Spies, F., van't Noordende, J. M., Boelsma, E., Mommaas, A. M. and Koerten, H. K. (1997) Tape stripping of human stratum corneum yields cell layers that originate from various depths because of furrows in the skin. *Arch. Dermatol. Res.*, **289**, 514–518.

van Scott, E. J. (1952) Mechanical separation of the epidermis from the corium. *J. Invest. Dermatol.*, **18**, 377–379.

Wagner, H., Kostka, K. H., Lehr, C.-M. and Schaefer, U. F. (2000) Drug distribution in human skin using two different *in vitro* test systems: comparison with in-vivo data. *Pharm. Res.*, **17**, 1475–1481.

Wagner, H., Kostka, K. H., Lehr, C.-M. and Schaefer, U. F. (2001) Interrelation of permeation and penetration parameters obtained from *in vitro* experiments with human skin and skin equivalents. *J. Control. Release*, 75, 283–295.

Walker, M., Hulme, T. A., Rippon, M. G., Walmsley, R. S., Gunnigle, S. and Lewin, M. (1997) In-vitro model(s) for the percutaneous delivery of active tissue repair agents. *J. Pharm. Sci.*, **86**, 1379–1384.

Wertz, P. W., Downing, D. T. (1989) Stratum corneum: biological and biochemical considerations. In J. Hadgraft and R. H. Guy (eds), *Transdermal Drug Delivery*. New York: Marcel Dekker, New York 1–22.

Wiedmann, T. S. (1988) Influence of hydration on epidermal tissue. *J. Pharm. Sci.*, **77**, 1037–1041.

Wild, T. (1988) Einfluß der physikochemischen Eigenschaften von Arzneistoffen und Vehikeln auf die Permeabilitaet der menschlichen Hornschicht. *Dissertation Saarbruecken*.

Zesch, A. and Schaefer, H. (1974) Penetration kinetics of four drugs in the human skin. *Acta. Dermatovener (Stockholm)*, **54**, 91–98.

Zghoul, N., Fuchs, R., Lehr, C.-M., Schaefer, U.F. (2001) Reconstructed skin equivalants for assessing percutaneous drug absorption from pharmaceutical formulations. *ALTEX*, **18**(2), 103–106.

# Chapter 18

# *In vitro* models of the human buccal epithelium: the TR146 cell culture model and the porcine *in vitro* model

*Hanne Mørck Nielsen*

## INTRODUCTION

The buccal mucosa is attracting increasing interest as a site of administration for systemic delivery of drugs because the acidic and enzyme-rich gastrointestinal (GI) milieu is bypassed and hepatic first-pass metabolism is avoided (Ho *et al.*, 1992; Merkle and Wolany, 1992). However, an important limitation to delivery via the buccal route is that the epithelium constitutes a physical barrier to the permeability of hydrophilic, as well as some lipophilic drugs, and it also represents a biochemical barrier to enzyme labile drugs. To evaluate potential drug substances for buccal drug delivery it is necessary that both *in vivo* and *in vitro* studies are conducted. The permeability of the drug across the epithelium should be investigated *in vitro* along with studies of, e.g. metabolism and toxicity. For that purpose, valid *in vitro* models of the human buccal epithelium are required. In recent decades several different models have been used, most of which involves the use of experimental animals. The human TR146 cell culture model has been set up in our laboratory and is currently being used as an *in vitro* model for, e.g. permeability studies with human buccal epithelium. The TR146 cell culture model has been characterized and compared with excised buccal mucosa from man and pig by the use of Ussing chambers. In this chapter, an introduction to the barrier properties of buccal mucosa is given. In addition, the TR146 cell culture model and the porcine *in vitro* model are described and their applications are compared and discussed.

## Buccal morphology

The buccal mucosa is defined as the lining mucosa of the outer oral vestibule extending from the upper to the lower vestibular arch and proceeding into the lip mucosa at the level of the angles of the mouth (Schroeder, 1981). The total area of the adult human oral cavity is approximately $200\,cm^2$, of which the buccal area takes up about one-third (Collins and Dawes, 1987; Weatherell *et al.*, 1994). The average median turnover time for the buccal mucosa has been estimated to be two–three weeks, which is low compared with the turnover time for, e.g. skin (Squier and Hill, 1994).

The non-keratinized human buccal mucosa has a moist, pliable surface. The human buccal epithelium consists of 20–40 cell layers. It has a total thickness of

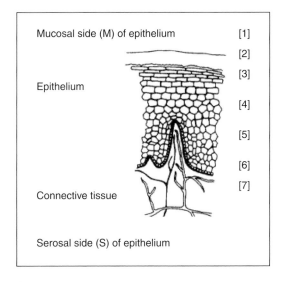

Mucosal side (M) of epithelium                    [1]

                                                  [2]

                                                  [3]

Epithelium

                                                  [4]

                                                  [5]

                                                  [6]

                                                  [7]

Connective tissue

Serosal side (S) of epithelium

*Figure 18.1*  Model of the human buccal mucosa (modified from Veillard and Anthony, 1990 – with permission).

450–600 μm (Meyer and Gerson, 1964; Schroeder, 1981) and is divided into different strata (Squier and Hill, 1994), which are outlined in Figure 18.1. The different layers are briefly described:

1   *Unstirred water layer* (UWL) exists adjacent to the mucus layer.
2   *Mucus layer* lines the oral mucosa as an aqueous, viscous layer of approximately 70 μm (Collins and Dawes, 1987).
3   *Superficial layers* representing the outer one-fourth–one-third of the buccal epithelium consisting of well-differentiated flat cells, as compared with the basal (BA) cells. Towards the mucosal (M) surface, the epithelial cells flatten increasingly owing to more extensive differentiation, and the surface cells are flat with microvilli-like extensions (Meyer and Gerson, 1964).
4   *Intermediate cell layers* consisting of slightly more flattened cells than BA cells with a thicker plasma membrane (Hashimoto *et al.*, 1966) and characterized by the presence of lipophilic intercellular substance excreted from the membrane coating granules (MCG) (Squier and Hill, 1994).
5   *Prickle cell layers* consisting of cells in which differentiation is initiated, resulting in larger, less columnar cells than the BA cells and with the occurrence of MCG (Squier and Hill, 1994).
6   *BA cell layers* consisting of mitotically active columnar cells anchored to the mechanical supportive BA lamina by hemidesmosomes (Squier, 1971).
7   *BA lamina*, which is approximately 1 μm thick, functions as a mechanical support for the epithelium (Meyer and Gerson, 1964; Schroeder, 1981). It separates the epithelium from the connective tissue, which is richly vascularized.

The morphology of the buccal mucosa has been described in more detail by other authors (e.g. Ganem-Quintanar *et al.*, 1997; Gandhi and Robinson, 1994; Squier, 1992).

## Barrier properties

The oral mucosa constitutes a physical and biochemical barrier between the organism and its environment.

### Physical barrier properties

In general, the UWL constitutes a barrier to the permeability of lipophilic substances (Jacobsen *et al.*, 1995; Karlsson and Artursson, 1991). The mucous layer also constitutes a permeability barrier mainly for lipophilic drugs (Kurosaki *et al.*, 1987) owing to viscosity, hydrophilicity and slight negative charge, but it also acts as a permeability barrier for protein binding substances (Bhat *et al.*, 1995).

The superficial cell layers constitute a tighter barrier to transcellular permeability than the BA cell layers. The explanation is that as the cells mature they accumulate lipids (mainly phospholipids, cholesterol, glycosylceramides, and moderate levels of triglycerides and cholesterol esters) resulting in a thickening of the plasma membrane (Squier, 1971). Within the oral cavity the high permeability of buccal mucosa has been shown to reflect the absence or low occurrence of ceramides and higher presence of phospholipids compared with the keratinized regions of the oral cavity (Squier *et al.*, 1991; Wertz *et al.*, 1986).

Tight intercellular barriers, tight junctions, are present throughout other parts of the oral cavity but are rarely seen in the buccal mucosa, where desmosomes are the main intercellular junctions (Squier, 1992). The important barrier for paracellular permeability is probably generated by the glycolipids and glycoproteins excreted from the MCG, rather than by the amount of desmosomes between the cells (Squier, 1984). By microscopic techniques it has been shown that the barrier to permeation of e.g. horseradish peroxidase occurs at the outer one-fourth–one-third of the non-keratinized buccal epithelium, coinciding with the area where MCG appears close to the plasma membrane (Hill *et al.*, 1982).

The BA lamina has been proposed to present a rate-limiting barrier for the permeability of hydrophilic macromolecules, such as inulin and large dextrans (molecular weight (MW) 70 kDa) (Alfano *et al.*, 1977).

In summary, the superficial (and to a minor extent also the intermediate) cell layers are believed to constitute the primary physical barrier to passive permeability across the buccal mucosa. This belief is with regard to both lipophilic substances believed to permeate via the transcellular route and hydrophilic substances, which mainly permeate via the paracellular route (Squier and Wertz, 1993).

### Biochemical barrier properties

Even though the buccal route of delivery lacks the pancreatic proteases, exopeptidases as well as endopeptidases are present in the buccal mucosa and drugs are susceptible to degradation on the surface of the mucosa, in the epithelium, in the connective tissue, as well as in the blood (Lee, 1988). Apart from peptidases and esterases (Nielsen and Rassing, 2000b), oxidases and reductases (Yamahara and Lee, 1993), lipooxygenases (Green, 1989) and glucosidases (Chang *et al.*, 1991) are represented in the buccal epithelium. In studies with whole tissue from hamster cheek pouch, the rate of metabolism of an aminopeptidase substrate was shown to be higher than the rate of permeation through the tissue. This indicates that buccal mucosa may present a significant enzyme barrier to permeability of intact peptides and proteins (Garren *et al.*, 1989). However, results concerning this issue are scarce.

In order to evaluate the buccal route as an alternative to other delivery routes, it is crucial that both the physical and the biochemical barriers of the human buccal mucosa are evaluated. The barrier properties of *in vitro* models used instead of human buccal mucosa should be characterized and compared with the barrier properties of human buccal mucosa.

## *IN VITRO* MODELS AND METHODS FOR STUDIES OF BUCCAL DRUG DELIVERY

Ideally, human buccal mucosa should be used for studies of buccal drug delivery, but for a number of reasons cell culture models and animal models are most often used as an alternative.

### TR146 cell line

The TR146 cells were isolated from a female lymph node metastasis of a buccal squamous cell carcinoma and the cell line has been characterized morpho-logically as a cell line of stratified squamous epithelium with a high degree of differentiation (Boukamp *et al.*, 1985; Rupniak *et al.*, 1985). As the cell line origin-ates from a carcinoma situated in the buccal area of the oral cavity, studies were initiated to establish an *in vitro* model resembling the normal human buccal epithelium.

### TR146 cell culture model

Culturing of the cells on permeable supports for approximately 30 days resulted in a multi-layered epithelium with structural (Jacobsen *et al.*, 1995) as well as differential (Jacobsen *et al.*, 1999) characteristics of simple stratified epithelium, similar to previous descriptions (Boukamp *et al.*, 1985). The model has four–eight cell layers with the BA cells being more columnar with larger nuclei than the apical (AP) flatter cells. The total thickness of the epithelium is 30–60μm (Jacobsen *et al.*, 1995; Nielsen, 2000). Furthermore, transmission electron micrographs reveal desmosome junctions, no tight junctions and an absence of complete keratinization. The cell layers exert polarity, and the integrity of the epithelial barrier is described by the transepithelial electrical resistance (TEER) and the permeability of model substances (Jacobsen *et al.*, 1995). Figure 18.2 schematizes the set-up of the TR146 cell culture model.

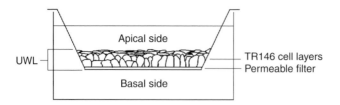

*Figure 18.2* The TR146 cell culture model set-up.

## Animal model

Canine, porcine and monkey buccal epithelium exhibit differential features similar to human non-keratinized buccal epithelium, whereas rodents, such as guinea-pig, rat, hamster and rabbit, show a keratin profile of keratinized epithelium (Wertz *et al.*, 1993). Comparison of the degree of keratinization with permeability of a hydrophilic drug in the hamster model showed that the degree of keratinization is an important factor with regard to the permeability across an epithelium (Kurosaki *et al.*, 1991).

Light microscopic examinations of hematoxylin and eosin stained buccal biopsies have shown that the thickness of canine, porcine and monkey buccal epithelium is approximately 700–850 μm (Squier and Hall, 1984), 600–800 μm (de Vries *et al.*, 1991; Nielsen, 2000; Squier and Hall, 1985) and 450–550 μm (Mehta *et al.*, 1991), respectively. Buccal tissue biopsies from rodents have been measured to be in the range of 100–200 μm (Dowty *et al.*, 1992; Kurosaki *et al.*, 1991; Nair and Chien, 1993). A potential gradient ($\Delta P_d$) across the human buccal mucosa is generated *in vivo*, and reported to be $-20$ mV to $-40$ mV, the M side being negative relative to the S side of the epithelium (Huston, 1978; Orlando *et al.*, 1988; Whittle *et al.*, 1981). The $\Delta P_d$ value measured *in vivo* on buccal mucosa from rabbit was similar to the values measured in human samples (Orlando *et al.*, 1988). The $\Delta P_d$ values measured *in vitro* on buccal mucosa from, for example, dog (Quadros *et al.*, 1991) and pig (Hansen *et al.*, 1992; Quadros *et al.*, 1991), seemed to be slightly lower, probably owing to decreased viability of the tissue. To the author's knowledge there are no reports on the transmucosal electrical resistance (TMER) of the human buccal mucosa. In general, *in vitro* measurements of TMER show that rodents (Orlando *et al.*, 1988; Quadros *et al.*, 1991), dog (Orlando *et al.*, 1988; Quadros *et al.*, 1991; Yang and Knutson, 1995) and pig (de Vries *et al.*, 1991; Hansen *et al.*, 1992; Nielsen, 2000) exhibit similar TMER values in the range of 1–1.4 kΩ cm$^2$.

In summary, porcine buccal mucosa is often preferred as an alternative to the human buccal mucosa *in vitro*, as the availability of tissue from dog or monkey is often limited, and as rodents have keratinized and often thinner buccal mucosa.

## Porcine *in vitro* model

The Ussing chamber technique has been used for studies with porcine buccal mucosa (Hansen *et al.*, 1992; Hoogstraate *et al.*, 1994; Le Brun *et al.*, 1989; Nielsen *et al.*, 1999;

*Table 18.1* The main differences of the two *in vitro* models

|  | *TR146 cell culture model* | *Porcine in vitro model* |
| --- | --- | --- |
| Epithelial orientation | Horizontal | Vertical |
| Stirring | Mechanical | Gas-lift |
| Exposed area | 0.9 cm$^2$ or 4.2 cm$^2$ | 0.5 cm$^2$ or 0.13 cm$^2$ |
| Epithelial thickness | 30–60 µm | 550–800 µm |
| Epithelial cell layers | 4–8 | 20–40 |
| Electrical resistance | ~200 Ω cm$^2$ | ~1000 Ωcm$^2$ |
| Basal lamina | Not present | Present |
| Connective tissue | Not present | Present to a minor extent |

Nielsen and Rassing, 1999, 2000a) and with dog buccal mucosa (Yang and Knutson, 1995). The human buccal mucosa has been studied to a limited extent in Ussing chambers (Nielsen and Rassing, 2000a), even though excised human buccal mucosa has been used in side-by-side diffusion chambers (Van der Bijl *et al.*, 1997, 1998) or flow-through cells (Lesch *et al.*, 1989; Squier *et al.*, 1997).

The Ussing chamber technique has the advantage that it enables determination of the $\Delta P_d$ across the mucosa, and by clamping of the mucosa the $I_{sc}$ (short circuit current) may be determined (Ussing and Zehran, 1951). Thus, the viability can be monitored during the experiment. Histological examination of the mucosa, changes in ATP levels (Hayashi *et al.*, 1999) or marker enzymes (Imbert and Cullander, 1999) may also contribute to viability estimates. The mucosa integrity can be estimated by the TMER along with the permeability profiles of selected substances (de Vries *et al.*, 1991).

## Comparison of the *in vitro* models

In summary, when interpreting data obtained with the TR146 cell culture model it is crucial to note that the thickness and number of cell layers, as well as the integrity, are reduced compared with the native human and porcine epithelium. Neither the BA lamina nor any connective tissue is present in the TR146 cell culture model. In Table 18.1 some of the differences of the models are summarized. Furthermore, the carcinogenic origin of the cell line should be considered.

To evaluate and to appraise the usefulness and the validity of the TR146 cell culture model for permeability studies, the barrier characteristics are currently compared with the barrier characteristics of excised porcine buccal mucosa, a more often used *in vitro* model of human buccal epithelium. Also, comparison with human buccal mucosa is carried out to a limited extent. The following sections summarize the results as well as the methods used for some of these studies.

## APPLICATIONS OF THE TR146 CELL CULTURE MODEL

### Physical barrier studies

So far, experiments with the TR146 cell culture model have been performed in order to characterize the physical barrier properties of the TR146 cell culture

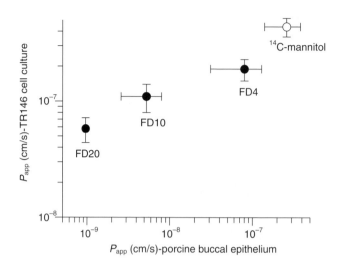

*Figure 18.3* Correlation of the apparent permeability rate coefficients ($P_{app}$) of $^{14}$C-mannitol, fluorescein isothiocyanate-labeled dextran FD4, FD10 and FD20 for permeability across the TR146 cell culture model correlated to the $P_{app}$ values for permeability of the same substances across the porcine buccal epithelium *in vitro* ($R^2$ 0.94, $P < 0.05$). Mean $\pm$ SD, $n \geq 2$.

model (Jacobsen *et al.*, 1995) and to compare the permeability across porcine buccal mucosa (Nielsen *et al.*, 1999). To some extent human buccal mucosa has also been included in these studies (Nielsen and Rassing, 2000a).

Figure 18.3 depicts the correlation of the apparent permeability coefficients ($P_{app}$) obtained in the two models for four hydrophilic substances with a neutral overall charge and MW between 182 and 20,000. In the same study, confocal laser scanning microscopy (CLSM) was used with the cell culture model, as some of the substances were labeled with a fluorophore. Figure 18.4 is a CLSM image of the permeability of one of the substances, fluorescein isothiocyanate-labeled dextran 10,000 (FD10). These studies exemplify that visualization of the permeability pathway (i.e. localization) of the substances may be investigated using the TR146 cell culture model.

A permeability study with $\beta$-adrenoceptive drugs of different lipophilicity, but with approximately the same MW (MW 248–336), revealed a correlation of the permeability across the two different epithelial models (Nielsen and Rassing, 2000a). Furthermore, a recent study with nicotine (MW 162) also showed a correlation between the permeability across the TR146 cell culture model and the porcine buccal mucosa at different degrees of ionization (unpublished results).

These results show that the TR146 cell culture model is as applicable as the porcine *in vitro* model for comparative (screening) studies of permeability. However, there seems to be a factor of difference in permeability when comparing the permeability across the TR146 cell layers and the excised human or porcine mucosa. Therefore, the physical barrier should be studied further and especially with regard to the presence of transporters (i.e. uptake and efflux mechanisms).

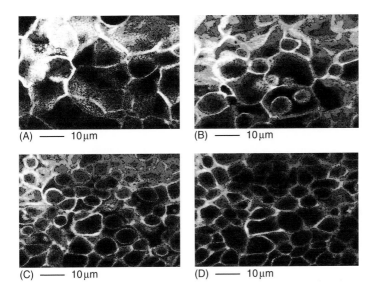

*Figure 18.4* Confocal laser scanning microscopy (CLSM) images of fluorescein isothiocyanate-labeled dextran (FD10) permeability across the TR146 cell culture model four hours after application. The cell layers are pictured horizontally in steps of 10 μm from the apical (AP) to the basal (BA) cell layer. (A) AP cell layer, (B) and (C) intermediate cell layers, (D) BA cell layer.

## Biochemical barrier studies

The enzyme activity of aminopeptidase, carboxypeptidase and esterase has been determined in homogenates and, as shown in Table 18.2, the activity was comparable with buccal epithelium homogenates from both man and pig (Nielsen and Rassing, 2000b).

These results indicate that some of the biochemical barrier properties of the TR146 cell culture model are comparable with human and porcine buccal mucosa. However, with this type of study, the epithelial distribution of the enzyme is not verified. To reveal this, an enzyme substrate (or an enzyme labile drug) may be applied to the intact epithelium and the metabolism determined relative to metabolism after application to homogenates. To visualize the localization of the

*Table 18.2* Homogenate enzyme activity of aminopeptidase, carboxypeptidase and esterase relative to protein contents in the TR146 cell culture model, human buccal epithelium and porcine buccal epithelium. The results are presented as nmol product formed per minutes per mg protein. Mean ± SD ($n = 5$–16).

|  | TR146 cell culture model (nmol/min/mg protein) | Human epithelium (nmol/min/mg protein) | Porcine epithelium (nmol/min/mg protein) |
|---|---|---|---|
| Aminopeptidase | 13.70 ± 2.10 | 8.82 ± 3.59 | 12.26 ± 3.24 |
| Carboxypeptidase | 3.73 ± 0.53 | 4.44 ± 0.87 | 4.34 ± 0.21 |
| Esterase | 223.39 ± 69.82 | 173.73 ± 90.35 | 221.72 ± 97.82 |

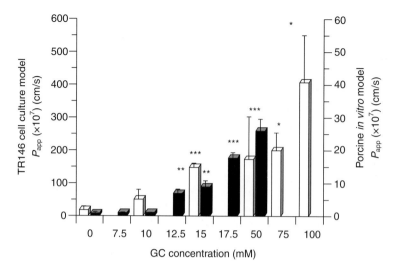

*Figure 18.5* The apparent permeability coefficients ($P_{app}$) for permeability of $^{14}$C-mannitol across the TR146 cell culture model (■) and the porcine buccal mucosa (□) exposed to glycocholate (GC). Mean ± SD, $n = 3$–6. * $P < 0.05$; ** $P < 0.01$; *** $P < 0.001$.

enzyme, microscopic techniques could be performed. A study carried out with an esterase substrate showed comparable degradation profiles in the TR146 cell culture model and the porcine *in vitro* model (Nielsen and Rassing, 2000b).

## Effect studies

The TR146 cell culture model might be a valuable tool to study the sensitivity of the buccal epithelium to, for example, chemical enhancers or other approaches taken to enhance the permeability of poorly absorbable drugs.

Mechanistic studies with a surface active bile salt (glycocholate, GC) showed that with regard to the integrity of the epithelium, the TR146 epithelium was as resilient as the porcine buccal mucosa. The integrity was measured as the permeability of the hydrophilic marker, $^{14}$C-mannitol. Figure 18.5 depicts the steady-state permeability of this paracellular marker during treatment of the epithelium with different concentrations of GC. Surprisingly, comparable relative increase in the permeability was observed.

Toxicity studies performed with the TR146 cell culture model might be very informative, both with regard to the degree and the mechanism by which the substance acts. Currently, assays for the determination of epithelial ATP content, the release of protein and enzymes, as well as the staining of dead cells are being developed and/or optimized for 24-hour-old TR146 cells, as well as for the cells cultured on filters. Some of these assays will also be performed with the excised porcine buccal mucosa. The aim is to compare the effect of drugs, chemical enhancers, toxins, etc. on the 24-hour-old cells in proliferation with the effect on the well-differentiated and multilayered TR146 cell culture model.

## MATERIALS AND METHODS

Description and sources for the materials and equipment necessary for the experiments follow below.

## TR146 cell culture model

The TR146 cell line was kindly provided by Imperial Cancer Research Technology (London, UK). Handling of the cell line is done aseptically and *everything* observed or done with the cell line is noted in the logbook. Along with routine visual inspection for contamination, the cell line is tested for mycoplasma every three months by the polymerase chain reaction (PCR) technique. The cell line is cultured at 37 °C, five per cent $CO_2$/95 per cent air in 98 per cent humidity. The culturing medium (DMEM-SUP) consists of DMEM (Dulbecco's modified Eagle medium) made from DMEM powder (Gibco BRL) supplemented with 0.225 per cent (weight/volume, w/v) $NaHCO_3$, ten per cent (volume/volume, v/v) FBS (fetal bovine serum) (HyClone Inc.), 100 IU penicillin and 100 µg/ml streptomycin (both from Gibco BRL). The pH is adjusted to 7.4 with five per cent $CO_2$, if necessary.

The TR146 cell stock ($10^6$ cells/ml/vial) is kept in cryo vials in liquid $N_2$ in a solution of 15 per cent (v/v) FBS, five per cent (v/v) glycerol. Immediately after adding glycerol to the vials, freezing is initialized with 1 °C/minute in a −80 °C freezer and, thereafter, moved to liquid $N_2$ storage. Thawing is done by immersing the cryo vial in 37 °C water, then transferring the cell suspension to a 75 cm$^2$ culturing flask (T75, Falcon®), adding 14 ml of 37 °C DMEM-SUP followed by incubation. The DMEM-SUP is changed the day after and then every other day until 80–90 per cent confluence. Subculturing is done by rinsing the cell layer with 10 ml of FBS-*free* DMEM-SUP and incubation with 0.7 ml of a solution of 0.5 per cent (w/v) trypsin and 0.2 per cent (w/v) ethylenediaminetetraacetic acid (EDTA) (Gibco BRL) until the cells detach from the flask. The cells are suspended in DMEM-SUP and seeded on Falcon® 0.9 cm$^2$ or 4.2 cm$^2$ polyethylene terephthalate inserts with a pore size of 0.4 µm (Becton Dickinson Labware) in a density of 24,000 cells/cm$^2$. The filter material in these inserts is translucent, which is advantageous in the case of microscopic inspection. The amount of DMEM-SUP is 0.5 ml on the apical (A) side and 1 ml on the basal (B) side of the filter-grown cells for the 0.9 cm$^2$ inserts and 2.5 ml over 2 ml for the 4.2 cm$^2$ inserts. For stock culturing, $10^5$ cells are transferred to T75 flasks and 14 ml DMEM-SUP added. At 80–90 per cent confluence, the cells are subcultured. For seeding in 96-well plates (Corning Inc.) the cell density is 20,000 cells/well, 100 µl well. Cell counting in the suspension is done using a hemocytometer.

### Permeability studies

Studies of permeability are often performed by application of the test substance to the A side with the receptor buffer on the B side, equivalent to the M and S side of the buccal mucosa. We find it important that the liquid level is equal in both chambers, i.e. 0.8 ml over 2 ml in the 0.9 cm$^2$ inserts and 2.5 ml over 2.5 ml in the 4.2 cm$^2$ inserts. Permeability experiments are performed on a temperated shaker

(Edmund Bühler) at a temperature of 37 °C and a horizontal stirring rate of 150 rpm to reduce the thickness of the UWL. The maximal stirring rate is represented by 150 rpm, which can reasonably be used without damage to the cell layers. The buffer used is most often Hank's balanced salt solution (HBSS, Gibco BRL) at pH 7.4. The experiment is started by rinsing the cell layers with HBSS at both sides, then applying the donor solution on the A side and transferring the insert to the well with receptor solution. Every 15–30 minutes for four–five hours samples of 100 µl are withdrawn from the receptor phase and replaced with the appropriate buffer. At the beginning and the end of the experiment samples are also withdrawn from the donor solution.

EVOM™ epithelial measurement chambers or chopstick electrodes with a volt-ohmmeter (World Presicion Instruments Inc.) are used for measuring the TEER before and after each experiment. The initial TEER is approximately $200 \, \Omega \, cm^2$ and the integrity of the TR146 cell culture model is maintained for up to nine hours, after which the TEER value is significantly lower than the initial TEER value ($P < 0.05$). The permeability of $^{14}C$-mannitol is at steady-state from $t = 10$ minutes to $t = 11$ hours, and as it does not increase significantly after nine hours, measuring the TEER might be a more sensitive method for monitoring the integrity of the epithelium (Nielsen, 2000).

### Studies using 24-hour-old cell monolayers

For these types of studies, the cells are used approximately 24 hours after seeding in 96-well plates. The MTS/PMS sensitivity assay, which is a modification of the 3-(4,5-dimethylthiazol-2-yl)-2,5-diphenyltetrazolium bromide (MTT) assay (Mosmann, 1983), has been optimized for the TR146 cell line (Jacobsen *et al.*, 1996). The assay is based on a color reaction and a Multiskan® plate reader is used for analysis (Labsystems). To estimate the sensitivity of the cells toward drugs or additives, our group routinely uses the assay.

### Porcine *in vitro* model

Porcine cheeks are obtained from domestic pigs immediately after sacrificing the animal at the local slaughterhouse. The cheeks are kept in ice-cold glucose-Ringer (GR) pH 7.4 until and during isolation. The buccal mucosa is discarded if there are visual wounds due to cheek biting. Mechanical isolation by scissors and a tissue slicer (Thomas Scientific®) is carried out. The thickness of the epithelium (including the BA lamina and a minimum of connective tissue) mounted between two microscope glass slides and measured with a sensitivity of $\pm 10 \, \mu m$ using a modified displacement transducer (Penny and Giles Position Sensors Ltd.) connected to a multimeter. The mucosa is mounted on the Ussing chamber within three hours of sacrificing the animal.

Traditional Ussing chambers (Ussing and Zehran, 1951) are used. Custom made Plexiglas chambers with a vertically exposed surface area of $0.5 \, cm^2$ and a volume of 1–1.5 ml in each half-chamber are used. Inserts for smaller tissue samples (area $0.13 \, cm^2$) as well as the multi-channel voltage and clamp equipment is also custom made. This type of equipment is commercially available from a number of companies, e.g. World Presicion Instruments Inc. and Warner Instrument Corp.

The $\Delta P_d$ and the $I_{sc}$ is monitored throughout the equilibration and the experimental period using a PicoLog® datalogger and software (Pico Technology Ltd.), and the TMER is calculated. $\Delta P_d$ is measured using two per cent (w/v) agar-GR bridges placed close to each surface of the mucosa and connected to calomel electrodes, and $I_{sc}$ is measured with Ag/AgCl electrodes. Compensation for offset between electrodes and fluid resistance is done without mucosa in the chamber.

### Permeability studies

Routinely, permeability experiments are performed with unclamped mucosa, except for brief clamping to measure the $I_{sc}$. The studies are performed at 37 °C and most often with GR adjusted to pH 7.4. A carbogen (95 per cent $O_2$/five per cent $CO_2$) gas-lift provides stirring (approximately one bubble per second). The experiment is started after 60 minutes of equilibration in GR by exchanging the GR with 1 ml of the appropriate donor and receptor solutions. If the liquid level in the half-chambers is not equal, more donor or receptor is added. For each chamber, the exact volume, as well as the precise concentration of test substance, is calculated from the dilution factor of the donor solution samples. Usually, the permeability is measured from the M to the S side. Every 30 minutes for four–five hours samples are withdrawn from the receptor phase and replaced with the appropriate buffer. At the beginning and at the end of the experiment samples are withdrawn from the donor.

Under these conditions, the initial TMER is approximately $1000 \, \Omega\text{cm}^2$ and the initial $\Delta P_d$ is approximately $-15 \, \text{mV}$. The integrity of the porcine buccal mucosa is maintained for at least eight hours after mounting on the Ussing chamber as measured by TMER and $^{14}\text{C}$-mannitol permeability. This indicates that studies of passive permeability processes can be carried out for this period of time. Yet, a decrease in viability is probably not reflected by a decrease in integrity. So with regard to studies of active processes the viability should also be monitored.

## PROBLEM SOURCES AND QUALITY CONTROL

### TR146 cell culture model

Standardization of cell culture and experimental conditions, of course, is of importance to be able to replicate experiments. The constituents of the culturing medium and the time for replenishment should be standardized over a range of passages. The passage numbers that we use are within a range of 15 passages routinely cultured for 27–30 days. It is important that the medium is replenished the day before the experiment. Due to inter-passage variation, studies to be compared are always performed using one passage and always include control experiments on each day of the study. For characterization of, for example, permeability mechanisms (transporters, etc.) the results should be repeatable in at least two–three different passages. The cell layers are inspected visually and the TEER measured before and after each experiment. If damage to the epithelium or very low or high TEER values are observed the inserts should be excluded from the study.

The TEER is dependent on the temperature and to some extent also on the medium in which the measurement is made. Even a few centigrade decrease results in an increase of TEER. Therefore, the inserts are equilibrated in the buffer used (most often HBSS adjusted to pH 7.4) to room temperature prior to measuring the TEER. For start and end point values the Endohm™ chamber is used as we have experienced less variation than with the Endohm™ chopstick electrode. However, the sticks have the advantage that the inserts do not need to be removed from the well and continuous TEER measurements can be carried out at time intervals throughout a study. Precaution should be taken that salts from buffers do not precipitate on the electrode.

## Porcine *in vitro* model

The viability of the tissues decreases as soon as the animal is sacrificed, so it is important to use the tissue as quickly as possible. Consequently, the cheek is cut out immediately after the death of the animal and kept in ice-cold GR of pH 7.4. For this purpose, this glucose-rich carbonate buffer seems to be better than, for example, PBS. Isolation of the mucosa should be done as quickly as possible and kept on ice until it is mounted in the Ussing chamber. It is important that five per cent $CO_2$ is in the gas-lift in order to maintain the pH of the carbonate-buffered GR. An easy method to visually ensure that the barrier properties are intact after the experimental period is to add methyl blue to one side of the mucosa and to inspect the receptor volume after a certain time. Of course, markers for paracellular and transcellular permeability might also be used to control the effect of the experimental conditions.

Maintenance of the Ag/AgCl electrodes and/or calomel electrodes as well as the agar bridges, if these are used in the experimental set-up, is important. The occurrence of airbubbles or dry ends in agar bridges will give unstable readings. To ensure accurate readings, in addition to preventing the contamination of experiments, the agar bridges should be freshly made or at least cut off at the end and kept in KCl solution until the next experiment. Also, the length of the agar bridges is of some importance for the readings.

General considerations, such as degradation or adsorption of test substance, accumulation in the tissue and evaporation of buffer should, of course, also be assessed.

## CONCLUDING REMARKS

The TR146 cell culture model seems to be a valid *in vitro* model of the human buccal epithelium, with regard to biochemical and physical barrier characteristics, compared with the human and porcine buccal epithelium/mucosa *in vitro*. The cell culture model has proved applicable to a range of different types of studies. Rank orders of human buccal permeability for selected drugs may be achieved and permeability enhancing strategies may be investigated in detail by the use of the model. As for other *in vitro* models of the human buccal epithelium, an increased knowledge of correlation to the human buccal epithelium, or at least to existing *in vitro* models of the buccal epithelium, is necessary to validate and fully show the potential of the TR146 cell culture model.

When comparing the TR146 cell culture model with the porcine *in vitro* model it can be seen that the cell culture model has several advantages. The TR146 cell culture model has a high capacity, it is simple, reproducible and the sensitivity of the epithelium is high. Furthermore, the model is relatively easy to handle and a lot of different studies and techniques can easily be applied to the model, e.g. visualization by CLSM and mechanistic studies of chemical enhancers. Under optimal conditions, the viability may be maintained for a long period of time allowing for long-term studies. Yet, the fact that the porcine *in vitro* model reflects the morphological features of human buccal epithelium with regard to thickness, degree of stratification, TMER and $\Delta P_d$ may give this model some advantages. On the other hand, the viability of the tissue is limited and the biological deviation is observed in the results from experiments with this model.

In general, it is crucial to be fully aware of the limitations of the model system used when interpreting the results. Furthermore, it is important to be aware whether the model system is confirmatory or comparative, i.e. whether a point-to-point correlation or a rank order correlation is obtained. An *in vitro* model should be validated with respect to its intended application, and characterization of the model should be implemented at regular intervals.

## ACKNOWLEDGMENT

The author's research was supported in part by PharmaBiotec Center, Center for Drug Design and Transport, Fertin Pharma A/S, Danish Animal Welfare Society, Erik Hørslev and wife Birgit Hørslev Foundation and Kong Christian den Tiendes Foundation.

## REFERENCES

Alfano, M. C., Chasens, A. I. and Masi, C. W. (1977) Autoradiographic study of the penetration of radiolabelled dextrans and inulin through non-keratinized oral mucosa *in vitro*. *J. Periodontal. Res.*, **12**, 368–377.

Bhat, P. G., Flanagan, D. R. and Donovan, M. D. (1995) Limiting role of mucus in drug absorption: drug permeation through mucus solution. *Int. J. Pharm.*, **126**, 179–187.

Boukamp, P., Rupniak, H. T. R. and Fusenig, N. E. (1985) Environmental modulation of the expression of differentiation and malignancy in six human squamous cell carcinoma cell lines. *Cancer Res.*, **45**, 5582–5592.

Chang, F., Wertz, P. W. and Squier, C. A. (1991) Comparison of glycosidase activities in epidermis, palatal epithelium and buccal epithelium. *Comp. Biochem. Physiol.*, **100B**, 137–139.

Collins, L. M. C. and Dawes, C. (1987) The surface area of the adult human mouth and thickness of the salivary film covering the teeth and oral mucosa. *J. Dent. Res.*, **66**, 1300–1302.

de Vries, M. E., Boddé, H. E., Verhoef, J. C., Ponec, M., Craane, W. I. H. M. and Junginger, H. E. (1991) Localization of the permeability barrier inside porcine buccal mucosa:

a combined *in vitro* study of drug permeability, electrical resistance and tissue morphology. *Int. J. Pharm.*, **76**, 25–35.

Dowty, M. E., Knuth, K. E. and Robinson, J. R. (1992) Enzyme characterization studies on the rate-limiting barrier in rabbit buccal mucosa. *Int. J. Pharm.*, **88**, 293–302.

Gandhi, R. B. and Robinson, J. R. (1994) Oral cavity as a site for bioadhesive drug delivery. *Adv. Drug Del. Rev.*, **13**, 43–74.

Ganem-Quintanar, A., Kalia, Y. N., Falson-Rieg, F. and Buri, P. (1997) Mechanisms of oral permeation enhancement. *Int. J. Pharm.*, **156**, 127–142.

Garren, K. W., Topp, E. M. and Repta, A. J. (1989) Buccal absorption. III. Simultaneous diffusion and metabolism of an aminopeptidase substrate in the hamster cheek pouch. *Pharm. Res.*, **6**, 966–970.

Green, F. A. (1989) Lipoxygenase activities of the epithelial cells of the human buccal cavity. *Biochem. Biophys. Res. Commun.*, **160**, 545–551.

Hansen, L. B., Christrup, L. L. and Bundgaard, H. (1992) Enhanced delivery of ketobemidone through porcine buccal mucosa *in vitro* via more lipophilic ester prodrugs. *Int. J. Pharm.*, **88**, 237–242.

Hashimoto, K., Dibella, R. J. and Shklar, G. (1966) Electron microscopic studies of the normal human buccal mucosa. *J. Invest. Dermatol.*, **47**, 512–525.

Hayashi, M., Sakai, T., Hasegawa, Y., Nishikawahara, T., Tomioka, H., Iida, A. *et al.* (1999) Physiological mechanism for enhancement of paracellular drug transport. *J. Control Release*, **62**, 141–148.

Hill, M. W., Squier, C. A. and Linder, J. E. (1982) A histological method for the visualization of the intercellular permeability barrier in mammalian stratified squamous epithelia. *Histochem. J.*, **14**, 641–648.

Ho, N. F. H., Barsuhn, C. L., Burton, P. S. and Merkle, H. P. (1992) Routes of delivery: case studies. (3) Mechanistic insights to buccal delivery of proteinaceous substances. *Adv. Drug Del. Rev.*, **8**, 197–235.

Hoogstraate, A. J., Cullander, C., Nagelkerke, J. F., Senel, S., Verhoef, J. C., Junginger, H. E. *et al.* (1994) Diffusion rates and transport pathways of fluorescein isothiocyanate (FITC)-labeled model compounds through buccal epithelium. *Pharm. Res.*, **11**, 83–89.

Huston, G. J. (1978) The effects of aspirin, ethanol, indomethacin and 9α-fludrocortisone on buccal mucosal potential difference. *Br. J. Clin. Pharmacol.*, **5**, 155–160.

Imbert, D. and Cullander, C. (1999) Buccal mucosa *in vitro* experiments. I. Confocal imaging of vital staining and MTT assays for the determination of tissue viability. *J. Control Release*, **58**, 39–50.

Jacobsen, J., van Deurs, B., Pedersen, M. and Rassing, M. R. (1995) TR146 cells grown on filters as a model for human buccal epithelium: I. Morphology, growth, barrier properties, and permeability. *Int. J. Pharm.*, **125**, 165–184.

Jacobsen, J., Pedersen, M. and Rassing, M. R. (1996) TR146 cells as a model for human buccal epithelium: II. Optimisation and use of a cellular sensitivity MTS/PMS assay. *Int. J. Pharm.*, **141**, 217–225.

Jacobsen, J., Nielsen, E. B., Brøndum-Nielsen, K., Christensen, M. E., Olin, H.-B. D. and Tommerup, N. (1999) Filter-grown TR146 cells as an *in vitro* model of human buccal epithelial permeability. *Eur. J. Oral Sci.*, **107**, 138–146.

Karlsson, J. and Artursson, P. (1991) A method for the determination of cellular permeability coefficients and aqueous boundary layer thickness in monolayers of intestinal epithelial (Caco-2) cells grown in permeable filter chambers. *Int. J. Pharm.*, **71**, 55–64.

Kurosaki, Y., Hisaichi, S.-I., Hamada, C., Nakayama, T. and Kimura, T. (1987) Studies on drug absorption from oral cavity. II. Influence of the unstirred water layer on absorption from hamster cheek pouch *in vitro* and *in vivo*. *J. Pharmacobio-Dyn.*, **10**, 180–187.

Kurosaki, Y., Takatori, T., Nishimura, H., Nakayama, T. and Kimura, T. (1991) Regional variation in oral mucosal drug absorption: permeability and degree of keratinization in hamster oral cavity. *Pharm. Res.*, **8**, 1297–1301.

Le Brun, P. P. H., Fox, P. L. A., de Vries, M. E. and Boddé, H. E. (1989) *In vitro* penetration of some β-adrenoreceptor blocking drugs through porcine buccal mucosa. *Int. J. Pharm.*, **49**, 141–145.

Lee, V. H. L. (1988) Enzymatic barriers to peptide and protein absorption. *CRC Crit. Rev. Ther. Drug Carrier Syst.*, **5**, 69–97.

Lesch, C. A., Squier, C. A., Cruchley, A., Williams, D. M. and Speight, P. (1989) The permeability of human oral mucosa and skin to water. *J. Dent. Res.*, **68**, 1345–1349.

Mehta, M., Kemppainen, B. W. and Stafford, R. G. (1991) *In vitro* penetration of tritium-labelled water (THO) and [$^3$H]PbTx-3 (a red tide toxin) through monkey buccal mucosa and skin. *Toxicol. Lett.*, **55**, 185–194.

Merkle, H. P. and Wolany, G. (1992) Buccal delivery for peptide drugs. *J. Control Release*, **21**, 155–164.

Meyer, J. and Gerson, S. J. (1964) A comparison of human palatal and buccal mucosa. *Periodontics*, **2**, 284–291.

Mosmann, T. (1983) Rapid colorimetric assay for cellular growth and survival: application to proliferation and cytotoxicity assays. *J. Immunol. Methods*, **65**, 55–63.

Nair, M. and Chien, Y. W. (1993) Buccal delivery of progestational steroids. I. Characterization of barrier properties and effect of penetrant hydrophilicity. *Int. J. Pharm.*, **89**, 41–49.

Nielsen, H. M. (2000) Characterstics of the TR146 Cell Culture Model – an *in vitro* model for Studies of Buccal Drug Delivery. *Ph.D. Thesis*, HCØ Tryk, Copenhagen.

Nielsen, H. M. and Rassing, M. R. (1999) TR146 cells grown on filters as a model of human buccal epithelium: III. Permeability enhancement by different pH values, different osmolalities, and bile salts. *Int. J. Pharm.*, **185**, 215–225.

Nielsen, H. M., Verhoef, J. C., Ponec, M. and Rassing, M. R. (1999) TR146 cells grown on filters as a model of human buccal epithelium: permeability of FITC-labelled dextrans in the presence of sodium glycocholate. *J. Control Release*, **60**, 223–233.

Nielsen, H. M. and Rassing, M. R. (2000a) TR146 cells grown on filters as a model of human buccal epithelium: IV. Permeability to water, mannitol, testosterone and ß-adrenoceptor antagonists. Comparison to human, monkey and porcine buccal mucosa. *Int. J. Pharm.*, **194**, 155–167.

Nielsen, H. M. and Rassing, M. R. (2000b) TR146 cells grown on filters as a model of human buccal epithelium: V. Enzyme activities of the TR146 cell culture model, human buccal epithelium and porcine buccal epithelium, and permeability of enkephalin. *Int. J. Pharm.*, **200**, 261–270.

Orlando, R. C., Tobey, N. A., Schreiner, V. J. and Readling, R. D. (1988) Active electrolyte transport in mammalian buccal mucosa. *Am. J. Physiol.*, **255**, G286–G291.

Quadros, E., Cassidy, J., Gniecko, K. and LeRoy, S. (1991) Buccal and colonic absorption of CGS 16617, a novel ACE inhibitor. *J. Control Release*, **19**, 77–86.

Rupniak, T. H., Rowlatt, C., Lane, E. B., Steele, J. G., Trejdosiewicz, L. K., Laskiewicz, B. *et al.* (1985) Characteristics of four new human cell lines derived from squamous cell carcinomas of the head and neck. *J. Natl. Cancer Inst.*, **75**, 621–635.

Schroeder, H. E. (1981) Differentiation of Human Oral Stratified Epithelia. 1st edition, S. Karger AG., Basel, Switzerland, pp. 1–156.

Squier, C. A. (1971) Ultrastructural features of oral epithelium. In C. A. Squier and J. Meyer (eds), Current Concepts of the Histology of Oral Mucosa, 1st edition, Charles C. Thomas Publisher, Springfield, IL, USA, pp. 5–33.

Squier, C. A. (1984) Effect of enzyme digestion on the permeability barrier in keratinizing and non-keratinizing epithelia. *Br. J. Dermatol.*, **3**, 253–264.

Squier, C. A. (1992) Structure and barrier function of the epithelium of gastrointestinal and oral mucosa. In H. E. Junginger (ed), Drug Targeting and Delivery. Concepts in Dosage Form Design, 1st edition, Ellis Horwood Ltd., Chichester, West Sussex, England, pp. 45–56.

Squier, C. A. and Hall, B. K. (1984) The permeability of mammalian nonkeratinized oral epithelia to horseradish peroxidase applied *in vivo* and *in vitro*. *Arch. Oral Biol.*, **29**, 45–50.

Squier, C. A. and Hall, B. K. (1985) The permeability of skin and oral mucosa to water and horseradish peroxidase as related to the thickness of the permeability barrier. *J. Invest. Dermatol.*, **84**, 176–179.

Squier, C. A. and Hill, M. W. (1994) Oral mucosa. In A. R. Ten Cate (ed), Oral Histology. Development, Structure, and Function. 4th edition, Mosby – Year Book, Inc., St. Louis, MO, USA, pp. 389–431.

Squier, C. A. and Wertz, P. W. (1993) Permeability and the pathophysiology of oral mucosa. *Adv. Drug. Del. Rev.*, **12**, 13–24.

Squier, C. A., Cox, P. and Wertz, P. W. (1991) Lipid content and water permeability of skin and oral mucosa. *J. Invest. Dermatol.*, **96**, 123–126.

Squier, C. A., Kremer, M. and Wertz, P. W. (1997) Continuous flow mucosal cells for measuring the in-vitro permeability of small tissue samples. *J. Pharm. Sci.*, **86**, 82–84.

Ussing, H. H. and Zerahn, K. (1951) Active transport of sodium as the source of electric current in the short-circuited isolated frog skin. *Acta Physiol. Scand.*, **23**, 110–127.

van der Bijl, P., Thompson, I. O. C. and Squier, C. A. (1997) Comparative permeability of human vaginal and buccal mucosa to water. *Eur. J. Oral Sci.*, **105**, 571–575.

van der Bijl, P., van Eyk, A. D., Thompson, I. O. C. and Stander, I. A. (1998) Diffusion rates of vasopressin through human vaginal and buccal mucosa. *Eur. J. Oral. Sci.*, **106**, 958–962.

Veillard, M. and Antony, F. (1990) Buccal and gastrointestinal drug delivery systems. In R. Gurny and H. E. Junginger (eds), Bioadhesion – Possibilities and Future Trends, 1st edition, Wissenschaftliche Verlagsgesellschaft mbH, Stuttgart, Germany, pp. 124–139.

Weatherell, J. A., Robinson, C. and Rathbone, M. J. (1994) Site-specific differences in the salivary concentrations of substances in the oral cavity – implications for the aetiology of oral disease and local drug delivery. *Adv. Drug Del. Rev.*, **13**, 23–42.

Wertz, P. W., Cox, P. S., Squier, C. A. and Downing, D. T. (1986) Lipids of epidermis and keratinized and non-keratinized oral epithelia. *Comp. Biochem. Physiol.*, **83B**, 529–531.

Wertz, P. W., Swartzendruber, D. C. and Squier, C. A. (1993) Regional variation in the structure and permeability of oral mucosa and skin. *Adv. Drug Del. Rev.*, **12**, 1–12.

Whittle, B. J. R., Makki, K. A. and O'Grady, J. (1981) Changes in potential difference across human buccal mucosa with buffered or unbuffered aspirin and salicylate. *Gut*, **22**, 798–803.

Yamahara, H. and Lee, V. H. L. (1993) Drug metabolism in the oral cavity. *Adv. Drug Del. Rev.*, **12**, 25–40.

Yang, B. and Knutson, K. (1995) Molecular size and ionization effects on transbuccal permeability. *J. Control. Release Bioact. Mater.*, **22**, 13–14.

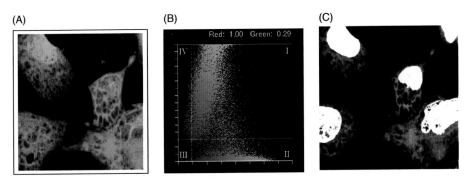

Color plate 1 Co-localization analysis. Cos-1 cells treated with an inductor of the transcription factor NFκB were stained for NFκB using a primary anti-NFκB antibody and a secondary fluorescein isothiocyanate (FITC)-labeled anti-mouse immunoglobulin (IgG1) F(Ab)2 fragment. The nucleus was counter-stained with propidium iodide (PI) (red). (A) Confocal cut through the stained cells on the nuclear level. (B) 2D dot-plot of the pixel intensities of Figure A. Each pixel of the image is assigned an intensity value for the green and the red signal. After definition of threshold values, the whole pixel population can be divided into four sub-populations: (I) positive for red and green, (II) positive for green, negative for red, (III) positive for red, negative for green; and (IV) negative for both. (C) Co-localization analysis. (See Figure 5.2)

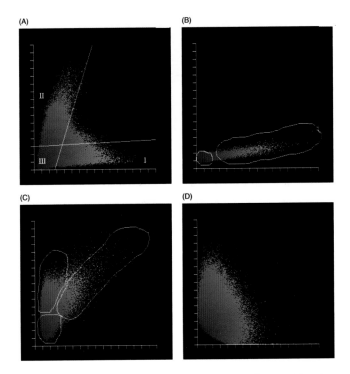

Color plate 2 Typical problems in co-localization analysis. (A) A cell stained for the nucleus (red) and caveolae (green) that do not co-localize was analyzed. Population (I) represents green pixels, (II) represents red pixels and (III) represents negative pixels. There are very few positive pixles for the red and green signal. However, the relatively intense red signal caused bleed-through into the green detector channel, resulting in higher values for the green signal. (B–D) A sample with intense green fluorescence, but weak red fluorescence is examined. At low signal amplification, only green and negative pixels are recorded (B). If the signal from the red channel is amplified red pixels also appear, but all the pixels with high green values also have high red values. These values could lead to interpretation as co-localization. However, after careful examination, two sub-populations can be distinguished (C). By compensation of bleed-through from the green signal into the red channel, these two sub-populations are separated. The positive pixels for green fluorescence no longer appear red (D). (See Figure 5.3)

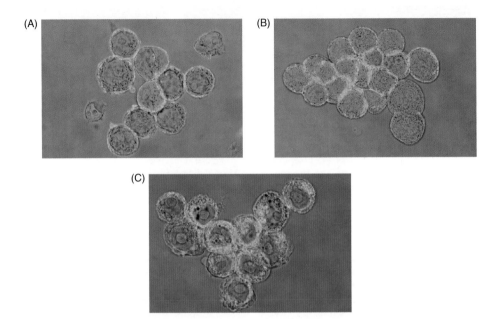

*Color plate 3* (A) Confocal images of paraformaldehyde-fixed Caco-2 cells stained with fluorescein-labeled wheat germ agglutinin for one hour at 4°C, (B) for 30 minutes at 37°C, and (C) for two hours at 37°C. The cell diameter refers to about 20 mm. Reproduced with permission (Wirth et al., 1998). (See Figure 5.14)

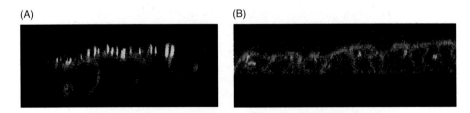

*Color plate 4* Discrimination between surface binding and internalization. Caco-2 cells are incubated with green fluorescent fluorescein-labeled wheat germ agglutinin (10 μg/ml) for five minutes at 4°C (A) or five minutes at 4°C followed by washing and further incubation at 37°C for 30 minutes (B). Cytoplasmic actin filaments including microvilli are stained with actin-reactive TRITC-phalloidine. After fixation with methanol at −20°C, the cells are incubated with 0.2 ng/ml of the dye (incubation volume 200 μl) for 30 minutes and washed three times in phosphate buffered saline. Images are obtained by performing a vertical line scan on a confocal scanning microscope. The lectin is observed to be in close contact with the red stained microvilli, but it is not taken up into the cell body at 4°C (A). At 37°C, internalization of the lectin occurs (B). (See Figure 5.15)

(A) (B)

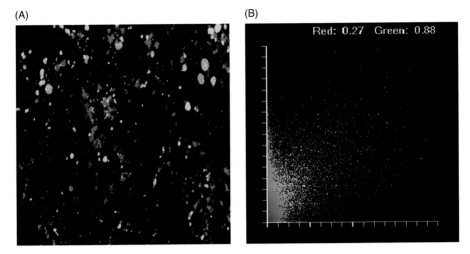

Red: 0.27    Green: 0.88

*Color plate 5* Co-localization analysis to assess processing of wheat germ agglutinin (WGA) via the endo-/lysosomal pathway. (A) Confocal image of Caco-2 cells incubated with red fluorescent Rh-WGA (10 μg/ml) for five minutes at 4°C followed by washing and further incubation at 37°C for 30 minutes. The lysosomes are counterstained with an anti-cathepsin D antibody and a fluorescein isothiocyanate (FITC)-anti mouse F(Ab)2 fragment. (B) 2D dot-plot of pixel intensities in image (A). Co-localization analysis shows that 88 percent of the volume representing green fluorescent lysosomes are filled with red fluorescent lectin, while only 27 percent of the lectin are found within cathepsin B positive lysosomes. (*See Figure 5.16*)

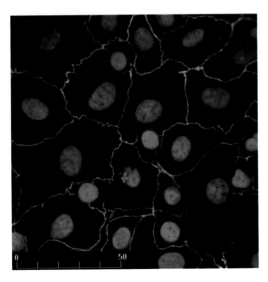

*Color plate 6* Zonula occluden (ZO)-1 staining on human alveolar cell culture, cell nuclei are stained by propidium iodide depicted in red, the tight junction protein ZO-1 is depicted in green, bar 50 μm. (*See Figure 12.7*)

(A) (B)

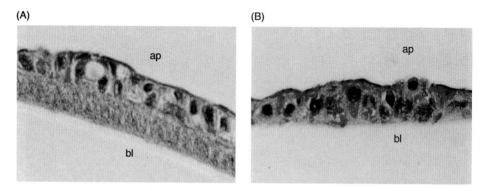

*Color plate 7* Staining of mucus produced by Calu-3 cells cultured under submerged conditions (A) and at an air interface (B). Staining was done with periodic Schiff's reagent and counterstained with Alcian blue after fixation in paraformaldehyde and embedding in paraffin. Magnification 40×, ap: apical membrane, bl: basolateral membrane. (*See Figure 13.3*)

(A) (B)

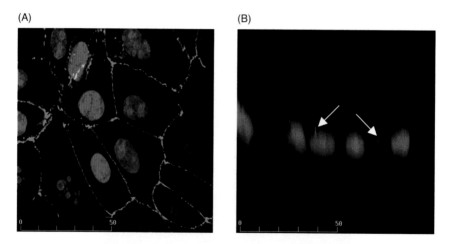

*Color plate 8* Expression of zonula occludens (ZO)-1 in 16HBE14o- cell cultures. Cells were seeded at a density of $10^5$ cells/cm$^2$ on Transwell® inserts and cultured under submerged conditions. Cells were fixed one week after seeding and stained for the tight junctional protein ZO-1 (green) as described. Top view (A) and cross section (B). Nuclei were counter stained with propidium iodide (red). Bars represent μm. (*See Figure 13.4*)

*Color plate 9* Specific staining of mucin produced in percine tracheal epithelial cell culture (PTC) grown at an air interface by periodic acid–Schiff's (PAS) reagent and counterstained with Alcian blue. Note the presence of cilia at the apical cell membranes. Microporous membrane, (×100). (*See Figure 13.9*)

(A)                                    (B)

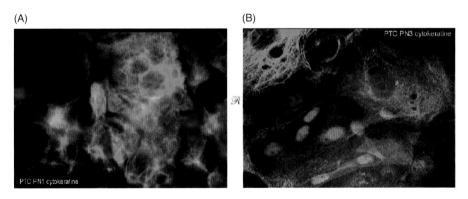

*Color plate 10* Staining of cytokeratin in percine tracheal epithelial cell culture (PTC) of passage numbers 1 (A) and 3 (B). Cytokeratin was stained with an fluorescein isothiocyanate (FITC)-labeled monoclonal anti-cytokeratin 18 antibody (green fluorescence), nuclei were counterstained with Hoechst 33258 (red). (*See Figure 13.10*)

  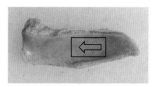

*Color plate 11* Excision of bovine nasal mucosa. Left panel: bovine snout. Middle panel: bovine snauze after cutting out specimen. Right panel: nasal specimen, indicating area and direction for subsequent stripping of mucosa. (*See Figure 14.2*)

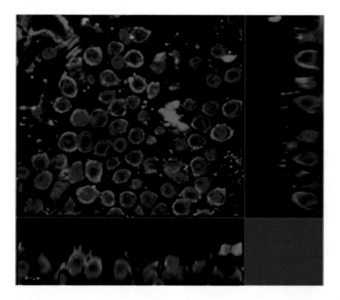

*Color Plate 12* Confocal laser scanning microscopy (CLSM) image for the intracellular uptake of human calcitonin (hCT). Mucosa was fixed (paraformaldehyde) and permeabilized (Triton X-100) for staining after hCT uptake experiment. Cell nuclei were stained by DAPI (blue), hCT in vesicular structures was labeled by immunofluorescence (green) (Schmidt *et al.*, 1998b). (*See Figure 14.5*)

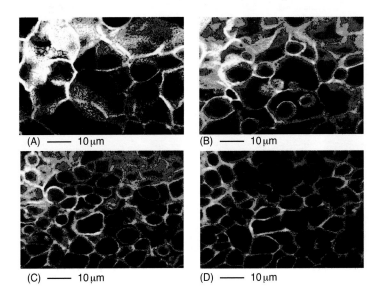

Color plate 13 Confocal laser scanning microscopy (CLSM) images of fluorescein isothiocyanate-labeled dextran (FD10) permeability across the TR146 cell culture model four hours after application. The cell layers are pictured horizontally in steps of 10 μm from the apical (AP) to the basal (BA) cell layer. (A) AP cell layer, (B) and (C) intermediate cell layers, (D) BA cell layer. (See Figure 18.4)

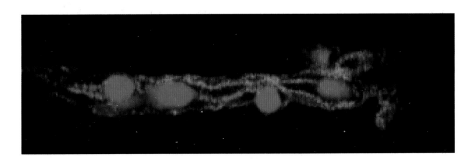

Color plate 14 Immunohistochemical localization of p-glycoprotein (p-gp) in freshly isolated brain capillaries from porcine brain. Green lines show detection of p-gp with highest density at the luminal surface, red areas are cell nuclei stained with propidium iodide (with permission from Nobmann et al., 2001). (See Figure 19.3)

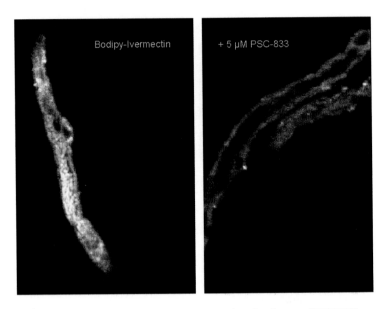

*Color plate 15* Extrusion of substrates of a *p*-glycoprotein (*p*-gp) substrate (BODIPY-Ivermectin) into the lumen of freshly isolated brain capillaries from porcine brain in the absence and in the presence of the *p*-gp blocker PSC-833. The *p*-gp inhibitor totally abolishes BODIPY-Ivermectin excretion (with permission from Nobmann *et al.*, 2001). (*See Figure 19.4*)

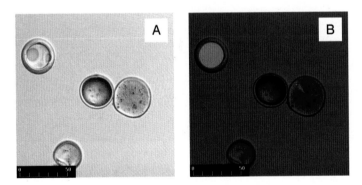

*Color plate 16* Visualization of microcapsules containing a nile red stained oil phase by a light microscopy image (A) and by confocal laser scanning microscopy (CLSM) using the red fluorescence channel (B). The fluorescence signal allows the oil-containing and air-containing microcapsules to be distinguished unambiguously. (*See Figure 22.4*)

# Chapter 19

# Drug transport across the blood–brain barrier – a molecular and functional perspective

*Gert Fricker, Helmut Franke and Hans-Joachim Galla*

## INTRODUCTION

Disorders of the brain, including neurodegenerative diseases, epilepsy, Morbus Parkinson, brain tumors or human immunodeficiency virus (HIV) related encephalopathy, belong to the principal causes of morbidity and death. For example, Alzheimer's disease, which afflicts more than four million people at an annual cost of over 50 billion US$, is currently the third largest medical problem in the United States (Friden, 1996). A major problem in the management of these diseases is the limited access of drugs to the central nervous system (CNS) across the endothelial cells of brain capillaries, the so-called blood–brain barrier. Studying the cellular and molecular structure, as well as the function of this barrier is hampered because it is not directly accessible *in vivo*. At present, immense efforts are ongoing to develop representative cellular *in vitro* models that mimic the structural and functional characteristics of the blood–brain barrier. This article provides a short overview about a general scheme to isolate brain endothelial cells, to maintain their major properties and to study drug transport *in vitro*.

## ANATOMIC PRINCIPLES OF THE BLOOD–BRAIN BARRIER

The first documented observation of a barrier preventing the entry of a xenobiotic into the CNS was made by the German pharmacologist Paul Ehrlich, who found that the peripherally administered inorganic dye Evans blue was unable to stain the brain tissue (Ehrlich, 1885). In contrast, when the dye was injected into the cerebrospinal fluid, staining of the brain tissue occurred, suggesting some kind of barrier on the level of the cerebral capillaries. These findings led to the term 'blood–brain barrier' some years later (Goldmann, 1913). In the 1960s, it was established by the utilization of electron microscopy and the application of horseradish peroxidase that the endothelium is indeed the principal anatomical site of the blood–brain barrier (Brightman and Reese, 1969; Reese and Karnovsky, 1967). The cerebral capillary endothelium (Figure 19.1) exhibits several structural differences compared with peripheral endothelial capillaries. Whereas peripheral capillaries are fenestrated (gaps to 50 nm wide), the endothelial cells of brain capillaries are closely connected to each other by tight junctions and zonulae

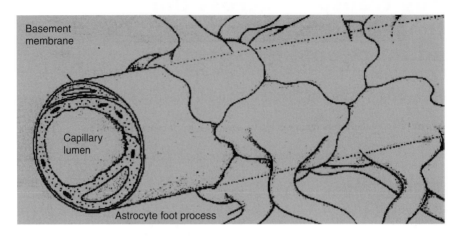

*Figure 19.1* Cross-section of a brain capillary. The endothelial cells are sealed by tight junctions and covered by a basement membrane including the pericytes. The capillary is unsheated by astrocytic foot processes.

occludentes. As a consequence, extremely high electrical resistances of approximately 1500–2000 ($\Omega\,cm^2$) have been measured at the blood–brain barrier *in vivo* (Crone and Olesen, 1982). In addition to these tightly connected cells, the capillaries are surrounded by a continuous basal membrane enclosing pericytes, an intermittent cell layer, which has been postulated to be involved in defense mechanisms. The outer surface of the basement membrane is covered by astrocytic or glial foot processes (Bradbury, 1993; Goldstein and Betz, 1986). The function of these cells has not yet been completely elucidated, but there is evidence that secretion of soluble growth factor(s) by astrocytes may play a role in endothelial cell differentiation (Schlosshauer, 1993).

## TRANSPORT PROCESSES AT THE CEREBRAL ENDOTHELIUM

For a long time brain capillaries have been considered a passive anatomical lipid barrier against hydrophilic compounds, except for some nutrients from the brain. Furthermore, brain capillaries have also been considered permeable for uncharged lipophilic substances, thereby determining the brain entry of molecules by lipophilicity and molecular weight. This traditional point of view has been considerably changed within the last ten years and the blood–brain barrier is now regarded as a dynamic interface with all possibilities of physiologic transport systems (Figure 19.2).

Diffusion through paracellular spaces is almost negligible at the blood–brain barrier owing to the restriction by tight junctions. The junctions effectively close off diffusion through intercellular pores; as a result, most solutes cross the blood–brain barrier either by diffusing across the lipoid endothelial cell membranes or by

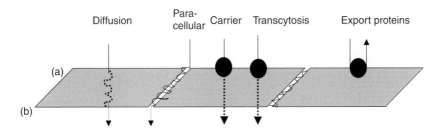

*Figure 19.2* Potential transport routes across the blood–brain barrier. (a) vascular surface of the cells, (b) cerebral surface of the cells.

carrier-mediated specific transport. However, with regard to passive diffusion, several studies indicate that a correlation between lipophilicity and permeation can only be made for compounds with a molecular weight between 400 and 600 Da (Schinkel *et al.*, 1995, 1996). For more hydrophilic nutrients, active transport is essential to satisfy the metabolic needs of the CNS. For example, the transport of amino acids through the endothelial wall is an important control point for the overall regulation of cerebral metabolism, including protein synthesis and neurotransmitter production (Pardridge, 1998; Smith, 2000). Glucose uptake is predominantly mediated by the glucose transporter Glut-1, which is expressed in high levels at the blood–brain barrier (Maher *et al.*, 1994). The vesicular pathway, which includes absorptive endocytosis and specific receptor-mediated endocytosis, mediates protein uptake into the brain. Here, special attention is given to the transferrin receptor which may be used for drug delivery by transcytosis across the capillary wall (Huwyler *et al.*, 1996, 1997).

In addition to the various transport mechanisms controlling the permeation of molecules from the blood into the brain, active efflux transport proteins, like the multi-drug resistance (MDR1)-gene product, *p*-glycoprotein (*p*-gp), and several proteins belonging to the MRP (multi-drug resistance associated proteins) family, are present at the blood–brain barrier (Miller *et al.*, 2000; Nobmann *et al.*, 2001; Zhang *et al.*, 2000). Both types of carriers belong to the large class of ABC (ATP binding cassette) transport proteins. Although discovered originally in tumor cells, *p*-gp and MRP2 are present at high levels in a variety of normal tissues with various physiological functions, which include the adrenal cortex, the brush-border membrane of renal proximal tubules, the apical membrane of enterocytes in the gut, testis, or endometrium of pregnant uterus. At the blood–brain barrier, expression of *p*-gp has been shown to be a critical element in preventing the entry of many drugs into the CNS (Schinkel *et al.*, 1994, 1995, 1996). For example, this prevention has been impressively demonstrated for HIV-protease inhibitors. A number of studies have shown that indinavir, ritonavir and saquinavir are substrates of *p*-gp in brain capillary endothelial cells (Choo *et al.*, 2000; Drewe *et al.*, 1999; Kim *et al.*, 1998; ). Other clinically relevant drugs, which have been shown to be actively transported by *p*-gp, include anticancer drugs, such as vinca alkaloids or doxorubicin. Similarly to the HIV protease inhibitors, the presence of *p*-gp results in a reduced drug permeation and hence a diminished therapeutic efficacy in the chemotherapy of brain tumors (Tsuji, 1998). Considering the potential clinical significance of *p*-gp,

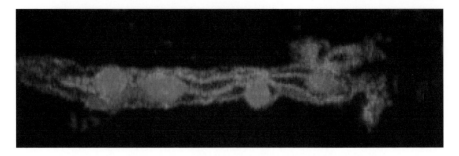

*Figure 19.3* Immunohistochemical localization of *p*-glycoprotein (*p*-gp) in freshly isolated brain capillaries from porcine brain. Green lines show detection of *p*-gp with highest density at the luminal surface, red areas are cell nuclei stained with propidium iodide (with permission from Nobmann *et al.*, 2001). (*See Color plate 14*)

inhibitors like PSC-833, a cyclosporin A analog, or verapamil are currently being used in clinical trials.

Although there is convincing functional evidence of the protective action of *p*-gp, the exact localization of this export pump at the blood–brain barrier is still being discussed. Whereas some authors claim substantial expression of the protein in astrocytic foot processes (Golden and Pardridge, 1999; Pardridge *et al.*, 1997), others (Nobmann *et al.*, 2001) unequivocally show its localization at the luminal surface of the capillaries (Figure 19.3).

The extent of expression and the localization of other export pumps, the MRP proteins, are currently under discussion. MRP1 has been shown to be expressed at a very low level in intact brain capillaries, but seems to be upregulated in cells kept in culture (Regina *et al.*, 1998). Results concerning the presence of other MRPs at the blood–brain barrier are also contradictory: reverse transcription-polymerase chain reaction (RT-PCR) analysis demonstrated the presence of MRP1, MRP4, MRP5 and MRP6 in bovine brain endothelium. Low levels of MRP3 were detected in the cultured cells, but not in a capillary-enriched fraction (Zhang *et al.*, 2000). Another study gives functional and molecular evidence for the expression of MRP2 in porcine brain endothelial capillaries (Miller *et al.*, 2000).

## *IN VITRO* MODELS TO STUDY DRUG PERMEATION ACROSS THE BLOOD–BRAIN BARRIER

Cell culture models mimicking the real characteristics of the blood–brain barrier *in vitro* would have a broad field of application in experimental, pharmaceutical and clinical studies. In contrast to *in vivo* studies, they offer direct access to brain capillary endothelial cells without interference with other structures of the brain. At present, two models gain broad attention: isolated, functionally intact brain capillaries and isolated brain capillary endothelial cells, kept in monolayer cultures on permeable filter supports.

## Isolated cerebral capillaries

Isolated capillaries have been used for a long time to study transport and metabolic functions of the blood–brain barrier (Goldstein *et al.*, 1975; Mrsulja *et al.*, 1976; Williams *et al.*, 1980). Capillaries can easily be isolated from the whole brain by dissecting little pieces of tissue with forceps. To obtain larger quantities brain homogenates may be subjected to dextran centrifugation. The resulting pellet is resupended in an appropriate buffer and filtered through a 200 µm nylon mesh. Then, the filtrate is passed over a glass bead column and washed with buffer. The beads are carefully removed from the column and adhering capillaries are removed by gentle agitation. For a detailed protocol the reader is referred to Nobmann *et al.* (2001). However, one disadvantage in using capillaries for the study of transport processes is that tested substrates approach the capillaries from the abluminal side, which is exactly opposite to the *in vivo* situation. Therefore, isolated capillaries are of special value when the influence of ABC transporters has to be investigated. Recent experiments applying confocal laser scanning microscopy and fluorescein-labeled substrates show the active extrusion of both *p*-gp and MRP substrates into the lumen (Figure 19.4) of freshly isolated porcine and rat brain capillaries (Miller *et al.*, 2000; Nobmann *et al.*, 2001). In contrast to monolayer cell cultures, which are susceptible to induction or inhibition of carrier protein expression, these experiments directly reflect the expression of ABC transporters at the luminal side of brain capillaries.

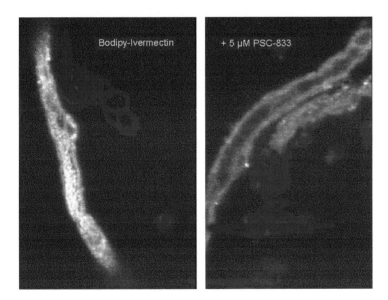

*Figure 19.4* Extrusion of substrates of a *p*-glycoprotein (*p*-gp) substrate (BODIPY-Ivermectin) into the lumen of freshly isolated brain capillaries from porcine brain in the absence and in the presence of the *p*-gp blocker PSC-833. The *p*-gp inhibitor totally abolishes BODIPY-Ivermectin excretion (with permission from Nobmann *et al.*, 2001). (*See Color plate 15*)

## Brain capillary endothelial cell culture

To study permeation of drugs into the brain *in vitro*, freshly isolated or passaged brain microvascular cells can be grown as monolayers on culture plates or permeable membrane supports. The cells retain the major characteristics of brain endothelial cells *in vivo*, such as the morphology, specific enzyme markers of the blood–brain barrier (e.g. γ-glutamyl transpeptidase, alkaline phosphatase or the von Willebrand factor-related antigen) and the intercellular tight junctional network. The methods of isolation are similar for most mammalian species. Here, the isolation procedure of porcine brain capillary endothelial cells is exemplarily described.

## Preparation of porcine brain microvessel endothelial cells

Porcine brain microvessel endothelial cells (PBEC) are prepared following a method of Bowman *et al.* (1983) modified by Franke *et al.* (2000). Refer to the latter reference for a detailed protocol, troubleshooting and further discussion. In brief, meninges and secretory areas of the brains of freshly slaughtered animals are removed and cerebral matter is homogenized mechanically. The homogenate undergoes a digestion in one per cent dispase followed by a dextran density gradient centrifugation to obtain capillary fragments as a pellet. These are further treated in a trituration step and a second digestion in 0.1 per cent collagenase/dispase solution. Released endothelial cells are collected from the interface of a density gradient centrifugation. Cells are plated on collagen-coated culture flasks and cultivated in M199 medium containing ten per cent oxen serum, 0.7 mM L-glutamine, 10,000 U/mL penicillin/streptomycin, and 10 mg/mL gentamicin. Twenty four hours after initial plating, cells are washed with phosphate buffered saline containing $Ca^{2+}$ and $Mg^{2+}$ and supplied with fresh culture medium.

For further purification, primary cultures of PBEC are passaged on the third day of culture by gentle trypsination at room temperature. This enzymatic treatment selectively releases endothelial cells, leaving behind contaminating cells such as pericytes and smooth muscle cells. Purified endothelial cells are then seeded at 30,000 cells/$cm^2$ on rat-tail collagen (Bornstein, 1958) coated cell culture inserts (Figure 19.5).

## Uptake studies

Uptake assays are usually performed at 37 °C using confluent monolayers of endothelial cells approximately seven–ten days after cell isolation. Studies can be performed in single or multiwell plates, up to 96-well dimensions. Cells should be incubated with serum-free Krebs–Ringer solution, unless binding studies with serum proteins are to be carried out. The substance of interest is added: in case of a water-unsoluble compound it should be dissolved in dimethyl sulfoxide (DMSO) with a final concentration not exceeding 1 per cent in the incubation buffer. After designated time intervals the supernatant is carefully removed. After two washings with ice-cold (4 °C) Krebs–Ringer solution the cell monolayers can be solubilized with one per cent Triton X-100 and subjected to further analysis by high-performance liquid chromatography (HPLC), fluorescence quantification or others.

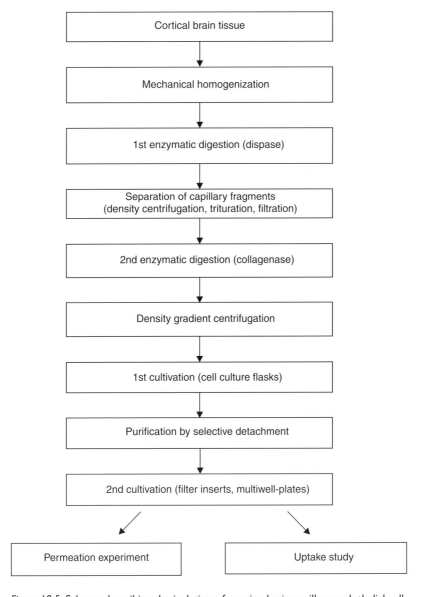

*Figure 19.5* Scheme describing the isolation of porcine brain capillary endothelial cells.

## Permeation experiments

Brain endothelial cells can be grown on permeable filter supports, like polycarbonate or polyester filters. Commonly used filter pore size is approximately 0.4 µm. Larger pore sizes can also be used, however, cells may then grow through filter pores. For permeation experiments, 6-well or 12-well filter dishes are recommended. Cells normally grow for seven–ten days before forming confluent monolayers. Prior to permeation experiments the monolayer integrity should be assessed either by determining the transendothelial electrical resistance or by running control experiments with paracellular markers such as radiolabeled mannitol or sucrose or fluorescein-labeled compounds such as fluorescein isothiocyanate (FITC)-dextrans or others.

## Problems during cell culturing and assaying

Several problems may occur during the cultivation of brain capillary cells on filter membranes. For example, the cell monolayers may not grow to sufficient confluence. The integrity of the cell monolayer can easily be damaged while replacing the cell culture media. Care has to be taken not to harm the cell monolayer when aspirating medium from the filter insert. Futhermore, pipetting fresh medium into the culture well has to be carried out whilst applying minimal force to the cells. Another reason for non-confluent monolayers may be an unequal coating of the well or filter support. Coating conditions have to be optimized for filter inserts of different manufacturers. As long as primary cell cultures are involved, the contamination of cultures by pericytes, smooth muscle, and other cell-types remains a serious problem. Brain microvessel cultures have to be carefully characterized to ensure their purity. A sufficient plating density has to be chosen in order to obtain confluent monolayers within the limited time of viability of primary cell cultures.

Another frequently occurring problem may be the low solubility of test compounds in aqueous solvents. Organic solvents such as DMSO or ethanol can be used, however, due to limited cell viability final concentrations above one per cent have to be avoided.

## CONCLUSION AND PERSPECTIVES

In recent years our understanding of the complexity of the blood–brain barrier has greatly increased. Beside pharmacokinetic and pharmacodynamic *in vivo* studies, *in vitro* techniques such as the isolation of functionally intact capillary fragments or the development of cell culture models have allowed a more profound insight into the mechanisms underlying the barrier function on a cellular and a molecular level. The improvement of culture conditions of isolated primary cells, as well as the development of immortal cell lines, is continuously ongoing. Thus, in the future, *in vitro* models resembling the characteristics of the intact blood–brain barrier will become available. Clearly, to improve access of new and current therapeutics to the brain, and to be able to better predict potentially toxic drug interactions, we need a deeper understanding of the mechanisms limiting permeation across the blood–brain barrier. Therefore, from applying the appropriate models, we may be able to identify specific probes to distinguish transporter subtypes, as well as tools, to transiently modify barriers to drug permeation. Finally, a deeper understanding of the molecular mechanisms involved may lead to simple tests that will allow us to better tailor drug dose to patient physiology.

# REFERENCES

Bradbury, M. W. B. (1993) The blood brain barrier. *Exp. Physiol.*, **78**, 453–462.

Bornstein, M. B. (1958) Reconstituted rat-tail collagen used as substrate for tissue cultures on coverslips in maximow slides and roller tubes. *Lab. Invest.*, **7**, 134–137.

Bowman, P. D., Ennis, S. R., Rarey, K. E., Betz, A. L., Goldstein, G. W. (1983) Brain microvessel endothelial cells is tissue culture: a model for study blood brain barrier permeability. *Annals of Neurology*, **14**, 396–402.

Brightman, M. W. and Reese, T. S. (1969) Junctions between intimately apposed cell membranes in the vertebrate brain. *J. Cell Biol.*, **40**, 648–677.

Choo, E. F., Leake, B., Wandel, C., Imamura, H., Wood, A. J., Wilkinson, G. R., *et al.* (2000) Pharmacological inhibition of P-glycoprotein transport enhances the distribution of HIV-1 protease inhibitors into brain and testes. *Drug Metab. Disp.*, **28**, 655–660.

Crone, C. and Olesen, S. P. (1982) Electrical resistance of brain microvascular endothelium. *Brain Res.*, **241**, 49–55.

Drewe, J., Gutmann, H., Fricker, G., Török, M., Beglinger, C. and Huwyler, J. (1999) HIV protease inhibitor ritonavir: a more potent inhibitor of p-glycopotein than the cyclosporine analogue SDZ PSC-833. *Biochem. Pharmacol.*, **57**, 1147–1152.

Ehrlich, P. (1885) Das Sauerstoffbedürfniss des Organismus. Eine farbenanalytische Studie, edited by Hirschwald, A. , Berlin.

Franke, H., Galla, H.-J. and Beuckmann, C. T. (2000) Primary cultures of brain microvessel endothelial cells: a valid and flexible model to study drug transport through the blood–brain barrier *in vitro*. *Brain Res. Protoc.*, **5**, 248–256.

Friden, P. M. (1996) Utilisation of an endogeneous cellular transport system for the delivery of therapeutics across the blood–brain barrier. *J. Control. Release*, **46**, 117–128.

Golden, P. L. and Pardridge, W. M. (1999) P-Glycoprotein on astrocyte foot processes of unfixed isolated human brain capillaries. *Brain Res.*, **819**, 143–146.

Goldmann, E. E. (1913) Vitalfärbung am Zentralnervensystem, Berlin.

Goldstein, G. W. and Betz, A. L. (1986) The blood brain barrier. *Sci. Am.*, **255**, 70–79.

Goldstein, G. W., Wolinsky, J. S., Csejtey, J. and Diamond, I. (1975) Isolation of metabolically active capillaries from rat brain. *J. Neurochem.*, **25**, 715–717.

Huwyler, J., Wu, D. and Pardridge, W. M. (1996) Brain drug delivery of small molecules using immunoliposomes. *Proc. Nat. Acad. Sci. USA*, **93**, 14164–14169.

Huwyler, J., Yang, J. and Pardridge, W. M. (1997) Receptor mediated delivery of daunomycin using immunoliposomes: pharmacokinetics and tissue distribution in the rat. *J. Pharmacol. Exp. Ther.*, **282**, 1541–1546.

Kim, R. B., Fromm, M. F., Wandel, C., Leake, B., Wood, A. J., Roden, D. M., *et al.* (1998) The drug transporter P-glycoprotein limits oral absorption and brain entry of HIV-1 protease inhibitors. *J. Clin. Invest.*, **101**, 289–294.

Lee, C. G. L. and Gottesman, M. M. (1998) HIV-1 protease inhibitors and the MDR-1 multidrug transporter. *J. Clin. Invest.*, **101**, 287–288.

Maher, F., Vannucci, S. J. and Simpson, I. A. (1994) Glucose transporter proteins in brain. *FASEB J.*, **8**, 1003–1011.

Miller, D. S., Nobmann, S., Gutmann, H., Török, M., Drewe, J. and Fricker, G. (2000) Xenobiotic transport across isolated brain microvessels studied by confocal microscopy. *Mol. Pharmacol.*, **58**, 1357–1363.

Mrsulja, B. B., Mrsulja, B. J., Fujimoto, T., Klatzo, I. and Spatz, M. (1976) Isolation of brain capillaries: a simplified technique. *Brain Res.*, **110**, 361–365.

Nobmann, S., Bauer, B. and Fricker, G. (2001) Ivermectin excretion by isolated functionally intact brain endothelial capillaries *Br. J. Pharmacol.*, **132**, 722–728.

Pardridge, W. M. (1998) Blood brain barrier carrier-mediated transport and metabolism of amino acids. *Neurochem. Res.*, **23**, 635–644.

Pardridge, W. M., Golden, P. L., Kang, Y. S. and Bickel, U. (1997) Brain microvascular and astrocyte localization of P-glycoprotein. *J. Neurochem.*, **68**, 1278–1285.

Reese, T. S. and Karnovsky, M. J. (1967) Fine structural localisation of a blood–brain barrier to exogeneous peroxidase. *J. Cell Biol.*, **34**, 207–217.

Regina, A., Koman, A., Piciotti, M., El Hafny, B., Center, M. S., Bergmann, R., *et al.* (1998) Mrp1 multidrug resistance-associated protein and P-glycoprotein expression in rat brain microvessel endothelial cells. *J. Neurochem.*, **71**, 705–715.

Schinkel, A. H., Smit, J. J., van Tellingen, O., Beijnen, J. H., Wagenaar, E., van Deemter, L., *et al.* (1994) Disruption of the mouse mdr-1 p-glycoprotein gene leads to a deficiency in the blood brain barrier and to increased sensitivity to drugs. *Cell*, **77**, 491–502.

Schinkel, A. H., Wagenaar, E., van Deemter, L., Mol, C. A. and Borst, P. (1995) Absence of the mdr-1a p-glycoprotein in mice affects tissue distribution and pharmacokinetics of dexamethason, digoxin, and cyclosporin A. *J. Clin. Invest.*, **96**, 1698–1705.

Schinkel, A. H., Wagenaar, E., Mol, C. A. and van Deemter, L. (1996) P-glycoprotein in the blood brain barrier of mice influences the brain penetration and pharmacological activity of many drugs. *J. Clin. Invest.*, **97**, 2517–2524.

Schlosshauer, B. (1993) The blood–brain barrier: morphology, molecules, and neurothelin. *Bioessays*, **15**, 341–346.

Smith, Q. R. (2000) Transport of glutamate and other amino acids at the blood–brain barrier. *J. Nutr.*, **130**, 1016S–1022S.

Tsuji, A. (1998) P-glycoprotein-mediated efflux of anticancer drugs at the blood–brain barrier. *Ther. Drug Monit.*, **20**, 588–590.

Williams, S. K., Gillis, J. F., Matthews, M. A., Wagner, R. C. and Bitensky, M. W. (1980) Isolation and characterization of brain endothelial cells: morphology and enzyme activity. *J. Neurochem.*, **35**, 374–381.

Zhang, Y., Han, H., Elmquist, W. F. and Miller, D. W. (2000) Expression of various multidrug resistance-associated protein (mrp) homologues in brain microvessel endothelial cells. *Brain Res.*, **87**, 148–153.

# Chapter 20

# BeWo cells: an *in vitro* system representing the blood–placental barrier

*Amber Young, Akira Fukuhara and Kenneth L. Audus*

## INTRODUCTION

Current medical and pharmaceutical opinion strongly advises women to avoid drugs during pregnancy. However, numerous women with conditions such as asthma, diabetes, and human immunodeficiency virus (HIV) must often make difficult choices affecting their own health as well as that of the child's (Yaffe, 1998).

For obvious reasons, drug trials are not performed on pregnant women to determine the risks. Instead, pregnant women are told to avoid any drugs, alcohol, and drugs of abuse. Most information has been gathered from negative experiences, such as thalidomide and drug-addicted mothers (Yaffe, 1998).

In the past, the placenta has been imagined to be a sort of 'leaky filter' that allowed unrestricted exchange between maternal and fetal blood supplies. However, recent research involving placental tissue has shown that the transport processes of the maternal/fetal interface are much more active in regulating the exchange between maternal and fetal blood supplies than previously believed. Efflux proteins, such as *p*-glycoprotein (*p*-gp), appear to have a protective effect on the fetus by transporting xenobiotic compounds back into the maternal blood supply (Lankas *et al.*, 1998).

If a system could somehow be devised to screen compounds for safety, it would be a possibility that drugs could be designed and developed for the administration to mother and/or fetus with minimal or reduced risk. The following section discusses some of the available models for such studies and, specifically, the cell line BeWo.

## MODELS OF THE HUMAN PLACENTA

### Overview of the blood–placental barrier

One of the major challenges facing researchers in this area is that each species has a different placental structure. Applying animal data to humans can be difficult, if not impossible. The human placenta is described as hemachorial (Enders and Blankenship, 1999), meaning that the maternal blood comes into direct contact with the chorionic villi containing fetal capillaries.

In fact, the human 'blood–placental barrier' is composed of one layer of cells called trophoblasts. As pregnancy progresses, the mononucleated cytotrophoblasts synctialize to form large multinucleate cells, termed synctiotrophoblasts. In a term placenta, the blood–placental barrier is largely formed of synctiotrophoblasts, with a few precursor cytotrophoblasts.

Unlike most body organs, placentas can be easily retrieved from living humans. Most placental tissue used for research is donated by women who have given birth. Harvested placental tissue can be used as perfused organ systems, slices, isolated cells or cell membranes, or can be processed to isolated trophoblasts in cell culture systems. Placental tumors, called choriocarcinomas, have also been used as sources of trophoblasts, isolated and immortalized for cell culture.

The following paragraphs will describe three basic models: perfused tissues, primary cell culture, and immortalized cell lines.

## Perfused lobes, slices, and vesicles

Many different methods have been developed over the years to isolate placental tissue for use in transport configurations (Sastry, 1999). Each preparation has its benefits and drawbacks, which will be discussed in the following paragraphs.

One of the earliest and most characterized techniques is a dually perfused placental cotyledon. Human term placenta is perfused with buffer to remove the blood and examined to find a single, intact cotyledon (the functional unit of placental structure). The cotyledon is placed in a chamber where two buffer reservoirs are attached to simulate maternal and fetal blood supplies. The drug in question is loaded into one of the two reservoirs and the other is monitored for the drug's appearance. The apparatus can be designed to have open or recirculating perfusion (Sastry, 1999).

The perfused cotyledon model has several applications. It can be used to assess the placental transfer of substances, as well as the effects of endogenous and exogenous substances on transport (Sastry, 1999). The applications of this method are listed in several references at the end of this chapter (Brandes *et al.*, 1983; Contractor and Stannard, 1983; Schneider *et al.*, 1972; Zakowski *et al.*, 1993).

Data obtained from perfused cotyledon cannot be directly applied to 'real life' scenarios for several reasons. The placenta may have been damaged during the birth process, the tissue may be no longer metabolically active, and results from term placenta do not apply to early pregnancy (Sastry, 1999). All of these factors can significantly impact experimental data.

Another major method of studying placental uptake and/or transport is the use of placental slices or villus preparations. In this case, the fetal capillaries are no longer intact, so the assumption is that materials taken up by the trophoblasts diffuse into the fetal blood supply. These tissues are suspended in a physiological medium, which is loaded with drugs, nutrients, etc. The major application of these preparations is to measure the uptake of substances by a trophoblast, as well as by the release of endogenous and exogenous compounds (Gusseck *et al.*, 1975; Miller *et al.*, 1974; Sastry *et al.*, 1977, 1989; Schneider and Dancis, 1974).

Creating vesicles from surface microvilli is another commonly used method. These vesicles are primarily used for drug metabolism studies, such as the effect of phospholipids-N-methyltransferase 1 (PMT1) on amino acid uptake (Barnwell and

Sastry, 1987), and for confirming the presence of muscarinic receptors with radio-ligands (Fant and Harbison, 1981).

All of these methods provide insight into the function and characteristics of the blood–placental barrier, but they can be quite time-consuming and inconvenient. Advances in cell culture have allowed researchers to use techniques more amenable to high-throughput and industrial applications.

## Cytotrophoblasts in primary culture

Using the method of Kliman *et al.* (1986), cytotrophoblasts can be isolated from human term placenta using a series of Percoll gradient separations. These isolated cells can be grown in tissue culture dishes and used in uptake studies (Utoguchi *et al.*, 2000). The cells continue to secrete hormones when growing in culture. While convenient for some types of experiments, such as uptake, this method also has drawbacks.

Primary cultures of cytotrophoblasts cannot be used for transport studies because they synctialize in culture and leave large spaces among the synctia. Their inability to form a monolayer makes it impossible to assess transtrophoblast transport. Also, these cultures cannot be passaged. A more convenient cell culture model would involve cells which form monolayers and can maintain morphology and function through multiple passages.

## BeWo and other immortalized cell lines

Several cell lines have emerged that are able to form monolayers and remain stable with passage. Some are derived from animal tissue (e.g. HRP-1), while others are from human sources. The human trophoblast cultures are choriocarcinomas.

HRP-1 is a rat-derived cell line obtained from midgestation placenta. It is a trophoendodermal stem cell population that resembles the labyrinthine layer, the transporting layer, of the rat placenta. Although the rat placenta is composed of three layers, HRP-1 cells form monolayers in culture. They also exhibit enzymatic activity similar to human trophoblasts, so HRP-1 can be used to assess metabolic and transport properties of the trophoblast (Shi *et al.*, 1997).

Human choriocarcinoma-derived lines include JAr, JEG, and BeWo. These cell lines have been extensively studied to compare their morphology and biochemistry with normal trophoblasts. For example, JAr has an identical serotonin transporter compared with placental brush-border membrane (Cool *et al.*, 1991). All of them are commercially available from American type culture collection (ATCC) (Sastry, 1999). However, there are multiple BeWo clones. The BeWo cell culture described below does not apply to the ATCC clone.

BeWo is of particular interest because it is one of the few trophoblast cultures to form confluent monolayers and is of human origin. BeWo is easy to maintain by passage, is stable, and grows to a confluent monolayer in a relatively short period of time (Pattillo and Gey, 1968). This cell line displays morphological properties and biochemical marker enzymes common to normal trophoblasts (Kenagy *et al.*, 1998; Liu *et al.*, 1997; Pattillo and Gey, 1968) and has been shown applicable to the characterization of asymmetric transcellular transport of serotonin and monoamine uptake

(Prasad et al., 1996), asymmetric amino acid transport (Eaton and Sooranna, 2000; Furesz et al., 1993; Moe et al., 1994; Utoguchi et al., 1999; Utoguchi and Audus, 2000; Way et al., 1998), asymmetric transferrin transport (van der Ende et al., 1990; Cerneus et al., 1993), asymmetric fatty acid transport (Liu et al., 1997), asymmetric vincristine and vinblastine transport (Ushigome et al., 2000), choline uptake (Eaton and Sooranna, 1998a), glucose modulation of arginine transport (Eaton and Süranna, 1998b), and asymmetric immunoglobulin (IgG) transport (Ellinger et al., 1999).

Among the choices for a trophoblast model, BeWo is one of the most convenient and promising candidates. The remaining portion of this chapter deals with the BeWo cell culture system and the future of blood–placental barrier research.

## METHODS

### Materials

The BeWo clone (b30) was provided by Dr. Alan Schwartz (Washington University, St. Louis, MO, USA). Fetal bovine serum (FBS) is purchased from Atlanta Biologicals. Penicillin-streptomycin, glutamine, and minimum essential medium (MEM) non-essential amino acids are from Gibco/BRL. Dulbecco's modified Eagle's medium (DMEM), trypsin-ethylenediaminetetraacetic acid (EDTA) 10x, Hank's balanced salt solution (HBSS) and all other chemicals are purchased from Sigma Chemical Company. Tissue culture plates and flasks are obtained from Fisher.

### Solutions and reagents

All of the solutions should be made in advance, sterile filtered, and stored under refrigeration unless specifically indicated in the text.

*DMEM, pH 7.4*: Prepare by adding one bottle of DMEM (with L-glutamine and 4500 mg glucose/l; without sodium bicarbonate) powdered medium (Sigma Cat #D-5648) to approximately 900 ml of deionized water. The volume is brought to one liter with the addition of the FBS in later steps. Add 3.5 g of sodium bicarbonate and allow to stir for 20 minutes. Adjust the pH to 7.4, then sterile filter. Add the following ingredients in a sterile manner: 10 ml of L-glutamine, 200 mM (Cat #25030-081); 10 ml of penicillin-streptomycin, 10,000 units/ml (Cat #15140-122); and 10 ml of MEM non-essential amino acids, 10 mM, 10x (Cat # 11140-050).

*Phosphate-buffered saline (PBS), pH 7.4*: Add the following salts in the given concentrations to deionized water. Adjust the pH to 7.4 and sterile filter. The ingredients are: NaCl (129 mM), KCl (2.5 mM), $Na_2HPO_4$ (7.4 mM), and $KH_2PO_4$ (1.3 mM).

*Plating/culture medium*: Prepare a 10 per cent (v/v) solution of FBS (Atlanta Biologicals Cat #S11550) with DMEM, pH 7.4.

*Freezing medium*: Prepare a 5–10 per cent (v/v) dimethyl sulfoxide (DMSO) solution with the plating/culture medium.

*HBSS*: Modified with sodium bicarbonate and without phenol red. Order ready-made from Sigma (Cat #H-8264).

# BeWo cell culture

## *Revival of BeWo*

BeWo cells may be preserved in liquid nitrogen or stored in a −70 °C freezer for prolonged periods. It takes approximately two weeks to have cells ready for experiments when starting from frozen vials. BeWos must be passaged in gradually larger flasks to get them acclimated after revival. Begin by thawing a 1.5 ml vial of approximately $10^6$ cells/ml in a 37 °C water-bath for about five minutes. As soon as the suspension is melted, remove from the water-bath and centrifuge for two minutes at 1000 rpm. Gently remove the vial, being very careful not to disturb the pellet. Under a tissue culture hood, add 8 ml of prewarmed, fresh culture/plating medium to a 25 cm$^2$ tissue culture flask (Falcon #353108).

Carefully remove the medium from the vial with a sterile Pasteur pipet. It is not necessary to remove all of the supernatant, but getting rid of the DMSO is important. Draw up some of the medium from the small culture flask with a sterile Pasteur pipet and use it to resuspend the pellet. When the pellet has been properly disrupted (no chunks are visible), transfer the contents of the vial to the small flask and cap tightly. Label and incubate at 37 °C, 95 per cent H$_2$O and 5 per cent CO$_2$.

After two days, check to see if any cells have adhered to the bottom of the flask. This can be done by gently rocking the medium over the cell surface and observing under a light microscope. Many of the cells will be free in suspension, while a number should appear to be stuck to the bottom. If a number of cells have adhered, remove the medium under sterile conditions and replace it with 8 ml of fresh prewarmed culture medium.

Observe the growth of the cells every day and replace the medium every other day. When the cells cover roughly 70 per cent of the flask bottom, passage on to the next size flask (75 cm$^2$, Falcon #353111).

Begin passaging by aspirating the medium away and rinsing the cell surface twice with about 8 ml of warm PBS. This step removes the serum which can inhibit the detachment abilities of the trypsin. Dilute the stock Trypsin-EDTA 10x (Cat #T-4174) to 1/10 with warm PBS. Add 5 ml of the dilute trypsin solution and place the vial horizontally so that the trypsin covers the monolayer. Leave for 30 seconds at room temperature and promptly remove. Quickly aspirate off the trypsin before the cells detach.

Place the flasks in the incubator for approximately two minutes. Locate and label 75 cm$^2$ tissue cultures flasks (one for each 25 cm$^2$ flask) and pipet 12 ml of culture medium into each one. Remove the small flasks from the incubator and examine them under the microscope to ensure that the cells have detached. Resuspend the cells in the small vials with a portion of the medium from the 75 cm$^2$ flask. After cells are suspended, transfer this medium back to the 75 cm$^2$ flask in order to seed the flask.

Again, incubate and observe cell growth every day and replace medium (12 ml) every other day until cells are 70 per cent confluent. This time, passage to flasks used for maintaining cell culture (150 cm$^2$, Corning #430825). The only differences are as follows: use 10 ml of the dilute trypsin solution, resuspend detached cells with 10 ml of medium, and split the suspension between two regular-sized flasks, each filled with 25 ml medium. Replace medium every other day until cells are roughly 70 per cent confluent.

### Maintenance of BeWo culture

From this point on, cells can be passaged and plated in the typical way. Passaged cells can be used for experiments until about passage 50. Cells past this age begin to change morphologically and biochemically; they also have problems with adhesion. Old cells must be discarded and new ones revived as described previously.

Cells should be passaged at about 70–90 per cent confluency. Experience shows that passaging cells with >90 per cent coverage yields poor growth. Start by aspirating the medium from the flasks and add 9 ml of warm PBS and 2 ml of undiluted trypsin-EDTA 10x. Cap the flask and place in the incubator (same conditions as above) for three–five minutes. Check the flask after three minutes for detachment. Shaking the solution over the cell surface and holding the flask up to the light should indicate the amount of detachment. If the cells are not yet detached, place the flask back in the incubator and recheck in a couple of minutes. It should be visually apparent when cells have detached from the flask wall.

Under a culture hood, draw up the suspension from the flask and pipette it into a 50 ml centrifuge tube. There should be approximately 11 ml of suspension for every flask being passaged. Tightly cap and centrifuge at 1500 rpm for eight minutes. Gently remove the tube and check for a pellet. Aspirate off supernatant in tissue culture hood.

Resuspend the pellet in 10 ml of fresh culture medium by gently pipetting repeatedly until there are no visible chunks remaining (about 20 times). Add enough medium to make the total volume 10 ml for every flask that was passaged (i.e. two flasks = 20 ml total, three flasks = 30ml total). Mix gently and resuspend thoroughly.

Obtain 150 cm$^2$ culture flasks and label. Be sure to add one to the previous passage number. Add 25 ml of fresh culture medium to each of the flasks. The amount of cell suspension added is somewhat arbitrary. Usually, 1.5–2 ml of cell suspension is sufficient to achieve confluency in five days. Check the cell density under the microscope after seeding to get a general feel for how much needs to be added. Experiment with the amount added until the appropriate volume is determined.

Cap tightly and incubate using the conditions stated above. Replace medium (25 ml) every other day until 70–90 per cent confluency (about four to five days) and repeat this procedure.

### Freezing BeWo cells

BeWo cells can be stored in liquid nitrogen or at −70 °C for later use. To prepare vials for freezing, follow the protocol above, up to and including the centrifugation step. Instead of using culture/plating medium to resuspend the pellet, use the freezing medium (DMEM + 10 per cent FBS + 10 per cent DMSO). Take a small sample and count the cells using a hemacytometer. Dilute the suspension to a density of approximately 10$^6$ cells/ml. Aseptically add 1 ml of this suspension to a 2.0 ml Nalgene Cryo-Vial (Cat #5000-0020). Properly label the tubes with cell line, passage number, date, etc. Place in −20 °C freezer overnight and transfer to liquid nitrogen or −70 °C freezer shortly thereafter. Nalgene also manufactures a freezing container which slowly freezes the cells at the recommended 1 °C/minute (Cat #5100-0001).

## Uptake and transport configurations

### Preparing plates for uptake

BeWos have been successfully grown in 12-, 24-, and 96-well plates (Corning Costar #3513, 3524, and 3596, respectively). Since adhesion can be a problem with this cell line, it is advisable to coat tissue culture treated plates with some type of collagen before seeding. Human placental collagen, HPC, (Cat #C-7521) seems to work best for this purpose. Dissolve the HPC in 0.1 per cent glacial acetic acid solution to make a 3 mg/ml stock. This stock can be stored in the refrigerator for later use. Just prior to use, dilute the stock one part HPC to three parts 70 per cent ethanol. Mix thoroughly.

Using a micropipettor, apply roughly 17.5 µl/cm$^2$ of well area (i.e. 70 µl for 12-well, 35 µl for 24-well, etc.). Application does not need to be under a cell culture hood, as the plates will be sterilized later. Try to spread the solution evenly over the bottom of the well. Place in a cell culture hood and allow to dry for about three hours under incandescent light. Prolonged exposure, especially for Transwells®, causes damage to the plate and/or membranes. After time has elapsed, expose the plates to ultraviolet (UV) light for roughly one hour.

Place the lids on the plates and tape them so that the inside remains sterile. The plates can then be removed from the hood and stored in the refrigerator for up to one week before being used.

### Plating BeWo for uptake

Use the protocol above to passage the cells. When preparing flasks, collect a small sample (<0.5 ml) to count. Using a hemacytometer, count the number of viable cells/ml in the suspension. Typically, a 2:1 mixture of cell suspension to trypan blue is used to obtain the count.

After the cell density is determined, calculate the dilution necessary to seed 10$^5$ cells/ml (or 25,000 cells/cm$^2$). Next, calculate the volume of a 10$^5$ cells/ml suspension needed to seed the plates. At least 12 ml are needed to seed one 12-well (1 ml per well) or 24-well (0.5 ml per well) plate. It is usually desirable to prepare more volume than necessary in case of error.

Determine the volumes of original suspension and culture medium needed to make the proper dilution. Prepare the new suspension in a clean tube. Make certain that the suspension is mixed well.

Using a repipettor with sterile tips, seed each well with the appropriate volume of cell suspension. Shake gently to distribute cells throughout the well, and place in incubator. Feed cells the next day with the same volume of medium used to seed the well. Repeat every other day until 100 per cent confluency is reached (about four–six days). At this point, uptake studies can be performed.

### Uptake protocol

Before the cells can be prepared, HBSS must be warmed to 37 °C in a water-bath, and a hot box/shaking platform apparatus must be warmed to 37 °C. At this point, cells can be removed from the incubator. This experiment may be performed on the benchtop, because sterile conditions are no longer necessary.

Aspirate medium from the cells and rinse three times with warm HBSS. This removes any medium components and serum that may interfere with the experiment. Wash and dose cells with the same volume used to feed cells in culture. Leave the last HBSS wash in the plates and equilibrate the cells in the hot box for at least 30 minutes to one hour. Since BeWos have a tendency to peel (see PROBLEM SOURCES), it is not advisable or necessary to shake at this stage.

While cells are equilibrating, prepare pre-incubation solutions (drug or inhibitor in HBSS), if necessary. When the time is up, aspirate away the buffer and add pre-incubation solutions. Place in hot box for relevant length of time (usually 30 minutes to one hour) and shake gently to disturb the unstirred aqueous layer.

Remove the cells from the hot box and aspirate away the solutions. Add the dosing solutions to each well (drug, drug+inhibitor, etc.) and place back in the hot box, making sure to turn on the shaker. Incubate for the desired time period.

Immediately remove plate after time is up and stop the reaction by aspirating away dosing solution and washing three times with ice-cold HBSS or appropriate stop solution. Add lysing solution (recipe depends on drugs used) and shake in hot box for at least one–two hours. Cells can also be lysed overnight in the refrigerator (wrap in aluminum foil if using fluorescent compounds to prevent quenching).

When cells are lysed, mix each well thoroughly with a micropipettor. Lysing solution typically contains detergent and will form bubbles very easily, which could interfere with spectroscopic readings. Be sure to use separate pipet tips for each well, or at least for each set of replicates.

Evaluate amount of drug taken up by measuring fluorescence, counts, or by using another desired technique. These values can be normalized by performing an appropriate assay to determine the protein content of each well.

### Preparing plates for transport

Follow the directions above for preparing uptake plates (Costar 12 mm diameter Transwell® #3401, or Transwell® Clear #3460 – 12-well format), except apply the HPC mixture (70 μl for 12-well inserts) directly to Transwell® inserts. Dry for four hours and UV for one, because the inserts take quite a while to dry.

### Plating BeWo for transport

Follow the steps above for 'Plating BeWo for uptake.' Under sterile conditions, wet the Transwell® membrane by pipetting 1 ml of warm PBS into the upper compartment, and 2 ml into the lower compartment. Let the plate sit in the culture hood for at least 30 minutes while preparing the cell suspension. Transwell® compartments are also referred to as apical (upper) and basolateral (lower). These terms apply to the polarization of monolayers. In BeWo cells, apical refers to the maternal blood side and basolateral refers to the fetal compartment inside the chorionic villi. BeWos orient in Transwells® such that the apical side faces the upper compartment and the basolateral side faces the lower compartment.

Prepare a dilution of $10^5$ cells/ml. Aspirate away the PBS from both compartments. Add 1.5 ml of culture/plating medium to the lower compartment, and 0.5 ml (50,000 cells/cm$^2$) of the cell suspension to the upper compartment. Allow cells to attach over one day's time, and replace medium in both compartments (the same volume used to seed).

Regular Transwell® plates do not have transparent membranes, so confluency cannot be assessed with a microscope. Several methods are discussed in the next section. The polyester Transwell® Clear membranes allow the researcher to view the monolayer and get a general idea when cells are ready for experiments. Continue replacing the medium every other day until cells reach 100 per cent confluency (about three–five days).

### Transport protocol

Begin by warming up the hot box/shaker apparatus to 37 °C, whilst warming the HBSS buffer in a 37 °C water-bath. Once buffer is warmed, aspirate away medium and wash twice with 1 ml in the apical compartment and 2 ml in the basolateral compartment. Equilibrate in the hot box (without shaking) for at least 30 minutes to one hour.

At this point, confluency and presence of tight junctions can be assessed by measuring the TEER (transepithelial electrical resistance). Since BeWos do not form tight junctions, the TEER values for complete and incomplete monolayers do not vary significantly. Using a paracellular marker such as $^{14}$C-mannitol is advised. If working with non-radioactive materials, perform this assay after the experiment to prevent contamination.

Aspirate away the HBSS and add pre-incubation solutions if needed (0.5 ml to apical compartment, and 1.5 ml to basolateral compartment). Incubate for the desired time (usually 30 minutes to one hour) with shaking. When pre-incubation is complete, remove the solutions and dose the cells.

If performing an apical (upper) to basolateral (lower) transport experiment, pipette 0.5 ml of drug±inhibitor to the apical compartment and 1.5 ml of HBSS±inhibitor to the basolateral compartment. Basolateral-to-apical (B-to-A) experiments begin by dosing the basal compartment with 1.5 ml of drug±inhibitor and adding 0.5 ml HBSS±inhibitor to the apical compartment.

Remove samples at the desired time points (i.e. 5, 15, 30, 45, 60, 90 minutes). For a 12-well Transwell®, 100 μl is sufficient to assess the appearance of drug (fluorescence, radioactivity, etc.). Replace the volume removed with warm HBSS±inhibitor. For an apical-to-basolateral (A-to-B) experiment, the basolateral compartment is sampled. The reverse is true for a B-to-A experiment.

Put the plate back in the hot box, remove, and sample at each of the desired time points. Samples can be loaded into an assay plate or scintillation vials for easy analysis. When the full time has elapsed, wash the cells (the apical surface) with 1 ml ice-cold HBSS or appropriate stop solution. This process limits any transporter-mediated uptake/transport processes.

If desired, perform $^{14}$C-mannitol flux for one hour. This performance can be used to assess whether the drug is passing primarily by the transcellular or paracellular route. A high mannitol flux suggests a 'leaky' cell layer, and can compromise the validity of experimental results.

Use preferred calculation methods to determine the permeability of the drug A-to-B and B-to-A, in the absence and presence of inhibitors.

## PROBLEM SOURCES

Adherence to uptake plates is the major problem when dealing with BeWo cells. The cells appear to grow in tissue culture flasks and uptake plates perfectly well. During the washing steps of the uptake protocol, cells begin to 'peel off' of their plastic support. If shaken or washed too vigorously, the cells will be aspirated away with the buffer. Loss of some of monolayer is typical, but a high percentage must be retained to get valid results. Although spectrophotometric and radioactivity values can be normalized for protein content, experience shows that large errors occur when significant peeling off occurs.

This problem can be reduced or eliminated by the use of adhesion factors such as poly-L-lysine, fibronectin, and various types of collagen. Instructions for preparing plates with human placental collagen, one of the more successful treatments attempted, are listed above. For this particular cell line, purchasing plates pre-treated with collagen has not remedied the problem.

Another common problem is slow cell growth, which can be attributed to several factors. First, make sure that the medium and all medium components are fresh. The DMEM and culture/plating medium recipes listed above are fresh for one month if refrigerated. Also, make sure that FBS has been added to the culture medium.

Cells that have been recently thawed often take some time to divide at the normal rate. Allow extra time to reach confluency in the first few passages after thawing. If cells continue to grow too slowly and other factors have been ruled out, discard cells and start with a new frozen vial.

Seeding too few cells is also a common mistake when new to cell culture. Make sure that cell counts are accurate and dilutions are properly calculated. BeWo cells will not grow to confluency properly if the seeding density is too low.

All of the problems described above can be circumvented with experience and a little experimentation. Different labs, and even different individuals within a lab, use varying techniques to achieve optimal experimental results.

## CONCLUSION

Understanding the contributions of the single layer of trophoblasts separating maternal and fetal compartments to transplacental transport processes is a vitally important area of pharmaceutical research. The extreme limitations placed on drug exposure during pregnancy can be hazardous to the health of both mother and child. Rather than learning of the dangers through such tragedies as thalidomide, researchers should strive to find a screening process that could warn clinicians in advance. Characterization of the placenta may also allow rational drug design that could significantly reduce fetal exposure (Audus, 1999).

Although the BeWo cell line has some limitations, it provides a tool to evaluate uptake, transport, and efflux of drugs across a layer of trophoblasts. At this point, BeWo cannot begin to simulate the human placental barrier in entirety, however, it does allow researchers to gather information about the trophoblasts themselves.

One major issue is that BeWo is primarily composed of cytotrophoblasts. As pregnancy progresses, the cytotrophoblasts differentiate to synctiotrophoblasts. In fact, very few cytotrophoblasts are present in a term placenta. Does this mean that BeWo is not representative of the blood–placental barrier? Not necessarily. In Liu *et al.*, 1997, demonstrated that BeWo cells could be stimulated to synctialize with the addition of such compounds as forskolin. A comparison of the behavior and expression of the synctial and cytotrophoblastic states could shed some light on the overall picture. The differences between single and multinucleate trophoblasts are not well established. There is some argument regarding multidrug resistance (*p*-gp) expression in different stages of gestation and cell type (Mylona *et al.*, 1996; Nakamura *et al.*, 1997). BeWo may also prove to be an accurate model for early pregnancy, when the cell line more closely resembles placenta morphology. The human placentas available for research are nearly always term, so it is quite difficult to make comparisons.

At this stage, research shows that the BeWo cell line is currently the most reliable, convenient human cell line available. One of its major advantages is that, aside from HRP-1, it is the only known trophoblast model that forms monolayers in culture. Therefore, BeWo appears to be a valuable tool for the future of blood–placental research.

## ACKNOWLEDGMENT

The authors wish to acknowledge past support of the National Institute on Drug Abuse in funding studies on the development of trophoblast cell culture systems and Corning Costar for the Cellular and Molecular Biopharmaceutics Handling Laboratory.

## REFERENCES

Audus, K. L. (1999) Controlling drug delivery across the placenta. *Eur. J. Pharm. Sci.*, **8**, 161–165.

Barnwell, S. and Sastry, B. V. R. (1987) S-adenosyl-L-methionine mediated enzymative methylations in the plasma membranes of the human trophoblast. *Trophoblast Res.*, **2**, 95–120.

Borchardt, R. T., Smith, P. L. and Wilson, G. (1996) General principles in the characterization and use of model systems for biopharmaceutical studies. In R. T. Borchardt *et al.* (eds), *Models for Assessing Drug Absorption and Metabolism*, pp. 1–11, New York, Plenum Press.

Brandes, J. M., Tavolini, N., Potter, B. J., Sarkozi, L., Shepard, M. D. and Berk, P. D. (1983) A new recycling technique for the human placental cotyledon perfusion: application to the studies of the fetomaternal transfer of glucose, insulin, and antipyrine. *Am. J. Obstet. Gynecol.*, **146**, 800–806.

Cerneus, D. P., Strous, G. J. and van der Ende, A. (1993) Bidirectional transcytosis determines the steady state distribution of the transferrin receptor at opposite plasma domains of BeWo cells. *J. Cell Biol.*, **122**, 1223–1230.

Contractor, S. F. and Stannard, P. J. (1983) The use of AIB transport to assess the suitability of a system of human placental perfusion for drug transport studies. *Placenta*, **4**, 19–29.

Cool, D. R., Leibach, F. H., Balla, V. K., Mahesh, V. B. and Ganapathy, V. (1991) Expression and cyclic AMP-dependent regulation of a high affinity serotonin transporter in the human placental choriocarcinoma cell line ( JAr). *J. Biol. Chem.*, **255**, 15750–15757.

Eaton, B. M. and Sooranna, S. R. (1998a) Regulation of the choline transport system in superfused microcarrier cultures of BeWo cells. *Placenta*, **19**, 663–669.

Eaton, B. M. and Sooranna, S. R. (1998b) *In Vitro* modulation of L-arginine transport in trophoblast cells by glucose. *Eur. J. Clin. Invest.*, **28**, 1006–1010.

Eaton, B. M. and Sooranna, S. R. (2000) Transport of large neutral amino acids into BeWo cells. *Placenta*, **5–6**, 558–564.

Ellinger, I., Schwab, M., Stefanescu, A., Hunziker, W. and Fuchs, R. (1999) IgG transport across trophoblast-derived BeWo cells: a model system to study IgG transport in the placenta. *Eur. J. Immunol.*, **29**, 733–744.

Enders, A. C. and Blankenship, T. A. (1999) Comparative placental structure. *Adv. Drug Deliv. Rev.*, **38**, 3–15.

Fant, M. E. and Harbison, R. D. (1981) Synctiotrophoblast membrane vesicles: a model for examining the human placental cholinergic system. *Teratology*, **24**, 187–199.

Furesz, T. C., Smith, C. H. and Moe A. J. (1993) ASC system activity is altered by development of cell polarity in trophoblast from human placenta. *Am. J. Physiol.*, **265**, C212–C217.

Gusseck, D. J., Yuen, P. and Longo, D. (1975) Amino acid transport in placental slices: mechanisms of increased accumulation by prolonged incubation. *Biochim. Biophys. Acta.*, **401**, 278–284.

Kenagy, J., Avery, M., Liu, F., Soares, M. J. and Audus, K. L. (1998) Gestational and smoking effects on peptidase activity in the placenta. *Peptides*, **19**, 1659–1666.

Kliman, H. J., Nestler, J. E., Sermasi, E., Sanger, J. M. and Strauss, J. F. III. (1986) Purification, characterization and *in vitro* differentiation of cytotrophoblasts from human term placentae. *Endocrinology*, **118**, 1567–1582.

Lankas, G. R., Wise, L. D., Cartwright, M. E., Pippert, T. and Umbenhauer, D. R. (1998) Placental p-glycoprotein deficiency enhances susceptibility to chemically induced birth defects in mice. *Reprod. Toxicol.*, **12**, 457–463.

Liu, F., Soares, M. J. and Audus, K. L. (1997) Permeability properties of monolayers of the human trophoblast cell line BeWo. *Am. J. Physiol.*, **273**, C1596–C1604.

Miller, R. K. and Berndt, W. O. (1974) Characterization of neutral amino acid accumulation by human term placenta slices. *Am. Physiol.*, **277**, 1236–1242.

Moe, A. J., Furesz, T. C. and Smith, C. H. (1994) Functional characterization of L-alanine transport in a placental choriocarcinoma cell line (BeWo). *Placenta*, **15**, 797–802.

Mylona, P., Glazier, J. D., Greenwood, S. L., Slides, M. K. and Sibley, C. P. (1996) Expression of the cystic fibrosis (CF) and multidrug resistance (MDR1) genes during development and differentiation in the human placenta. *Mol. Hum. Reprod.*, **2**, 693–698.

Nakamura, Y., Ikeda, S., Furkawa, T., Sumizawa, T., Tani, A., Akiyama, S. *et al.* (1997) Function of P-glycoprotein expressed in placenta and mole. *Biochem. Biophys. Res. Commun.*, **235**, 849–853.

Pattillo, R. A. and Gey, G. O. (1968) The establishment of a cell line of human hormone-synthesizing trophoblastic cell *in vitro*. *Cancer Res.*, **28**, 1231–1236.

Prasad, P. D., Hoffmans, B. J., Moe, A. J., Smith, C. H., Leibach, F. H. and Ganapathy, V. (1996) Functional expression of the plasma membrane serotonin transporter but not the vesicular monoamine transporter in human placental trophoblasts and choriocarcinoma cells. *Placenta*, **17**, 201–207.

Sastry, B. V. R. (1999) Techniques to study human placental transport. *Adv. Drug Deliv. Rev.*, **38**, 17–39.

Sastry, B. V. R., Olubadewo, J. O. and Boehm, F. H. (1977) Effects of nicotine and cocaine on the release of acetylcholine from isolated human placental villi. *Arch. Int. Pharmacodyn. Ther.*, **229**, 23–36.

Sastry, B. V. R., Horst, M. A. and Naukam, R. J. (1989) Maternal tobacco smoking and changes in amino acid uptake by human placental villi: induction of uptake systems, gamma-glutamyl-transpeptidase, and membrane fluidity. *Placenta*, **10**, 345–358.

Schneider, H. and Dancis, J. (1974) Amino acid transport in human placental slices. *Am. J. Obstet. Gynecol.*, **120**, 1092–1098.

Schneider, H., Panigel, M. and Dancis, J. (1972) Transfer across the perfused human placenta of antipyrine, sodium, and leucine. *Am. J. Obstet. Gynecol.*, **114**, 822–828.

Shi, F., Soares, M. J., Avery, M., Liu, F., Zhang, X. and Audus, K. L. (1997) Permeability and metabolic properties of a trophoblast cell line (HRP-1) derived from normal rat placenta. *Exp. Cell. Res.*, **234**, 147–155.

Ushigome, F., Takanaga, H., Matsuo, H., Yanai, S., Tsukimori, K., Nakano, H. *et al.* (2000) Human placental transport of vinblastine, vincristine, digoxin, and progesterone: contribution of P-glycoprotein. *Eur. J. Pharmacol.*, **408(1)**, 1–10.

Utoguchi, N. and Audus, K. L. (2000) Carrier-mediated transport of valproic acid in BeWo cells, a human trophoblast cell line. *Int. J. Pharm.*, **195**, 115–124.

Utoguchi, N., Magnusson, M. and Audus, K. L. (1999) Carrier-mediated transport of monocarboxylic acids in BeWo cell monolayers as a model of the human trophoblast *J. Pharm. Sci.*, **88**, 1288–1292.

Utoguchi, N., Chandorkar, G., Avery, M. and Audus, K. L. (2000) Functional expression of P-glycoprotein in primary cultures of human cytotrophoblasts and BeWo cells. *Reprod. Toxicol.*, **14**, 217–224.

van der Ende, A., du Maine, A., Schwartz, A. L. and Strous, G. J. (1990) Modulation of transferrin-receptor activity and recycling after induced differentiation of BeWo choriocarcinoma cells. *Biochem. J.*, **270**, 451–457.

Way, B. A., Furesz, T. C., Schwarz, J. K., Moe, A. J. and Smith, C. K. (1998) Sodium-independent lysine uptake by the BeWo choriocarcinoma cell line. *Placenta*, **19**, 323–328.

Yaffe, S. J. (1998) Introduction. In G. Briggs, R. K. Freeman and S. J. Yaffe (eds), *Drugs in Pregnancy and Lactation*, pp. xiii–xix, Baltimore, Williams & Wilkins.

Zakowski, M., Schlessinger, J., Dumberoff, S., Grant, G. and Turndorf, H. (1993) *In vitro* human placental uptake and transfer of fentanyl. *Anesthesiology*, **79**, A1007, (abstract).

# Part 3

# Emerging tools for studying biological barriers and drug transport

# Chapter 21

# Predicting drug absorption by computational methods

*Michael B. Bolger, Thomas M. Gilman, Robert Fraczkiewicz, Boyd Steere and Walter S. Woltosz*

## INTRODUCTION

Drug discovery and development are expensive undertakings and any technology that can decrease the length of time prior to New Drug Application (NDA) submission, or restrict the number of experimental procedures required for compound selection and development, will save money. The philosophy of 'kill early, kill often' during preclinical drug development is a mantra frequently chanted by pharmaceutical industry managers today. The observed oral bioavailability of a particular therapeutic agent can be broken down into components that reflect delivery to the intestine (gastric emptying, intestinal transit, pH, food), absorption from the lumen (dissolution, lipophilicity, particle size, active transport), intestinal metabolism, active efflux, and subsequent first-pass hepatic extraction (Hall *et al.*, 1999). Experimental *in vitro* and estimated *in silico* biopharmaceutical properties can be used to predict drug absorption, distribution, metabolism, excretion, and toxicity (ADMET). This chapter will focus on *in silico* approaches that have the potential to save valuable resources in the drug discovery and development process. The discussion of property estimation will begin by defining some of the statistical terms and pitfalls commonly observed in theoretical model building. A discussion of the literature associated with the futile attempts to build statistical models of drug absorption directly from structure will follow. Next, will be the presentation of some of the more successful recent attempts to build models of individual biopharmaceutical properties, followed by a review of the theoretical methods employed to simulate gastrointestinal (GI) absorption. Finally, we will discuss our results in estimating biopharmaceutical properties and simulating GI absorption using Biopharmaceutical property Estimation and the Advanced Compartmental Absorption and Transit (BEstACAT) model.

### Structure–activity relationship (SAR) model building methods and pitfalls

Several characteristics of a statistical model that estimates a biological property are common to all methods of model development. First, all of the model parameters are determined using a regression or machine-learning method to minimize the error between a 'training set' of data and the estimated values. When the number of total data points is small (<50) the training set is usually the same as the test set and a procedure called cross-validation (CV) is used to determine the 'quality and generalizability' of the model. In this technique, all of the data is used to build the complete model. A series of validation models are also built by leaving some fraction of the complete data set out of the training process. Then the performances of these validation models are assessed with the cross-validated correlation coefficient ($Q^2$) by comparing the performance of the complete model. When the total data set includes hundreds of molecules, then the training set is typically selected to be 50–75 per cent of the total. The rest of the molecules are 'held out' for use as an external test set.

The goal of a statistical model of a biological property is two-fold. First, it should 'accurately' estimate the desired biological property for a given external test set. Second, it should be 'generalizable' to the degree that any diverse set of molecules used in an external test set should give similar results. Calculating the root mean square error (RMSE), the mean absolute error (MAE), and the coefficient of determination ($R^2$) can unambiguously determine the accuracy for a given model. However, the definition of whether a given estimate has been determined 'accurately' is a relative question that changes with the nature of each model. In general, the RMSE of model estimates for an external test set should never be less than the experimental error associated with the training set. When the RMSE of an external test set is within ten per cent of the experimental error, a model can be considered to 'accurately' estimate the biopharmaceutical property.

The question of 'generalizability' is more difficult. The chemical diversity used to develop the model will limit its ability to accurately estimate the true value of a biopharmaceutical property for molecules that are novel. If the training set of a model is based on a single molecular scaffold, and the molecules represented in the training set are congeners, the resulting model may not be very 'generalizable' but may be accurate for estimating that congeneric series. If the molecules of the training set are selected to have a wide degree of chemical diversity (Zheng *et al.*, 1999), then the resulting model is expected to be more 'generalizable'. However, accuracy may suffer from the difficulty of finding a set of molecular descriptors that capture the diversity of the training and test data sets. The classical approach taken in quantitative structure property relationships (QSPR) studies has been to limit the training and test sets to molecules that belong to a congeneric series and are associated with biopharmaceutical properties through a similar mechanism. For example, models that estimate permeability for molecules that are only absorbed by carrier-mediated transport, or are only passively absorbed, have been known for many years (Lien, 1981; Plakogiannis *et al.*, 1970).

More recently, the application of partial least squares (PLS) and artificial neural networks (ANN) to model development, combined with dramatic improvements in low-cost computing power, has enabled the building of models that involve larger data sets with greater diversity of structure and mechanism of activity. ANN grew out of research in artificial intelligence; specifically, out of the attempts to mimic the fault-tolerance and capacity to learn of biological neural systems by modeling the low-level structure of the brain (Patterson, 1996). An artificial neuron is defined as follows: it receives a number of inputs (either from original data, or from the output of other neurons in the neural network). Each input comes via a connection that has a strength (or weight); these weights correspond to synaptic efficacy in a biological

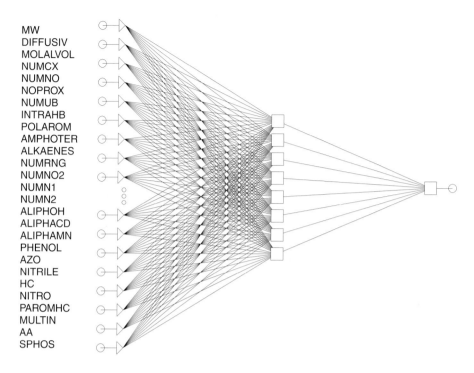

MW
DIFFUSIV
MOLALVOL
NUMCX
NUMNO
NOPROX
NUMUB
INTRAHB
POLAROM
AMPHOTER
ALKAENES
NUMRNG
NUMNO2
NUMN1
NUMN2
ALIPHOH
ALIPHACD
ALIPHAMN
PHENOL
AZO
NITRILE
HC
NITRO
PAROMHC
MULTIN
AA
SPHOS

*Figure 21.1* Schematic diagram of a typical artificial neural network (ANN) with X-input nodes, eight hidden layer nodes, and one output node.

neuron. Each neuron also has a single threshold value. The weighted sum of the inputs is formed, and the threshold subtracted, to compose the activation of the neuron (also known as the post-synaptic potential [PSP] of the neuron). The activation signal is passed through an activation function (also known as a transfer function) to produce the output of the neuron (see Figure 21.1) (Hunter, 1996).

Another important requirement for the use of a neural network is knowing (or at least strongly suspecting) a relationship between the proposed inputs and outputs. This requirement means that a judicious choice of molecular descriptors, based on an understanding of the physicochemical properties associated with the desired biopharmaceutical property, is essential to the development of a predictive model. Training the ANN involves the application of a regression method or back propagation method to adjust the large number of weights connecting the input molecular descriptors, the hidden nodes, and the output property. In principle, training the ANN results in a decrease in the training set error, which is determined by comparing the estimated outputs and the actual experimental values. For this reason, an important concept used with ANNs is the inclusion of a 'verification set' of molecules. During training, the verification set is not used to adjust the weights connecting the input values of the training set and the output properties. Rather, during each iteration of training, the verification set error is

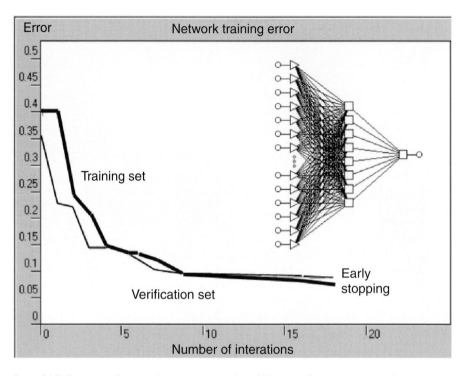

*Figure 21.2* Progress of network training procedure. When verification set error begins to rise the network training stops to maximize the generalizability of the subsequent model.

calculated based on the current set of model weights and is compared with the error of the training set. At the start of training, all weights are randomized and the training and verification sets have their highest error. Assuming that a similar distribution of chemical diversity is represented in both the training and verification sets, the error of both sets will decrease as training begins. However, as training continues the training set error will continue to decrease and the verification set error will eventually begin to increase as the generalizability of the model deteriorates and the model becomes 'over-trained'. ANN training must stop at this point, in order to preserve the 'generalizability' of the final model. In evaluation of the accuracy and generalizability of ANNs, it is always important to achieve comparable estimates of error for the training, verification, and external test sets (see Figure 21.2). This brief overview of the methods and pitfalls of computational methods in predicting drug absorption should be kept in mind when evaluating any models of biopharmaceutical properties.

## SAR models of absorption

There have been several attempts to predict human intestinal absorption (%HIA) into the portal vein based on estimated properties and molecular descriptors

(Andrews *et al.*, 2000; Clark, 1999; Lien, 1981; Oprea and Gottfries, 1999; Palm *et al.*, 1997; Wessel *et al.*, 1998). In 1902, Overton realized that cellular membranes are more readily penetrated by lipid-soluble substances than by lipid-insoluble substances. Thus, one of the most important physicochemical properties of drug molecules was discovered (Overton, 1902). Many subsequent studies have reported correlations of intestinal absorption for small congeneric series of molecules (Lien, 1981). More recently, the development of fast computers and powerful algorithms have given rise to more 'generalizable' models of intestinal absorption utilizing data for chemically diverse series of drug molecules. Using a database of 86 measured %HIA values, Wessel *et al.* (1998) developed a non-linear computational neural network model. The calculated per cent HIA (cHIA) model had a RMSE of 9.4 %HIA units for the training set, 19.7 %HIA units for the CV set, and 16.0 %HIA units for the external prediction set. It is useful to examine the six descriptors selected by the genetic algorithm for their model to get some idea of the kinds of physicochemical properties that help to predict %HIA. The descriptors used include: number of single bonds (a measure of flexibility), two descriptors that relate to size, shape, and hydrophobicity, and three descriptors that relate to the drug's potential to form H-bonds. The nature of a neural network model does not allow us to determine if these factors contribute an increase or a decrease in absorption to the model. However, the rough correlation of single descriptors with fraction absorbed indicates that hydrophobicity increases absorption and hydrogen bonding decreases absorption.

One measure of H-bond potential is the fraction of total van der Waals surface area that is contributed by 'polar' atoms (N, O, or H attached to N and O), the so-called 'polar surface area' (PSA). Dynamic polar surface area is a Boltzmann-weighted average value of PSA computed from an ensemble of low-energy conformers obtained by conformational searching. Dynamic polar surface area has been found to be inversely correlated to passive intestinal absorption (Krarup *et al.*, 1998; Palm *et al.*, 1996, 1997). The calculation of dynamic polar surface area is a computationally intensive process, as 1,000–25,000 conformations of each molecule must be generated with geometry optimization using a Monte Carlo search method. For 20 drugs of various classes, an excellent correlation between the polar surface area and the fraction absorbed was obtained. However, Palm *et al.* (1997) qualified their result as being applicable only to passively absorbed drugs with good solubility and negligible presystemic metabolism. Clark tested *static* polar surface area (a much faster calculation) and found it to be highly correlated with dynamic polar surface area (Kelder *et al.*, 1999). The calculation was also found to be equally capable of correlating with fraction absorbed (Clark, 1999). However, Clark's correlation was greatly improved by removing compounds from the data set that were actively transported. Furthermore, a compound outlier (etoposide), whose erratic oral bioavailability has been attributed to low aqueous solubility, slow intrinsic dissolution rate, and chemical instability at pH = 1.3, was also removed (Shah *et al.*, 1989). Thus, one can assume that formulation properties of the poorly soluble drug molecules will also play an important role in the %HIA.

Recently, a large database ($N = 591$) of human oral bioavailability was used to develop a regression model using 85 2D molecular descriptors. The model was derived from the simple molecular input line entry system (SMILES) representation of each molecule (Andrews *et al.*, 2000; Weininger, 1988). Bioavailability data were divided into two groups, good and bad, based on a 20 per cent cut-off. This model was able to predict oral bioavailability with only three per cent false negatives, 53 per cent false positives, and RMSE of 18 per cent. Oprea and Gottfries (1999) applied the parsimony principle (*less is better*) to the development of a partial least squares (PLS) model of %HIA. Their minimalistic PLS model was based on one principle component derived from a set of molecular descriptors that included: static polar surface area, simple 1D (atom counts) and 2D (connectivity relationships) estimates of hydrogen bonding, and estimates of hydrophobicity. Although the performance of the minimalistic model was acceptable, the authors concluded that the variability of their model was due to the fact that %HIA depends on other factors besides epithelial permeability. They pointed out that the area of the absorbing surface, the residence time of the drug at the absorbing surface, metabolic reactions in the intestines, the regional dependence of small intestinal (SI) absorption, and drug transport and efflux considerations were all complicating factors.

If it were simple to estimate %HIA from structure, then one of the methods described above would be useful in building a 'generalizable' model of drug absorption. Absorption is much too complex to be directly estimated from structure alone, and for these reasons it has been proposed that methods for predicting drug absorption directly from structure are doomed to failure.

## SAR models of biopharmaceutical properties

In order to overcome the problems associated with estimating a complex biological process, such as absorption directly from structure, it has been proposed that the estimation of individual biopharmaceutical properties coupled with a physiologically based simulation of dissolution, intestinal transit, and absorption can be quite successful.

### Permeability

Many biological models of permeability have been developed and are discussed in Chapters 9–13, 15 and 18–20 in this book (Artursson and Borchardt, 1997; Audus *et al.*, 1990; Borchardt *et al.*, 1996; Hillgren *et al.*, 1995; Lennernas, 1998). In order to develop *in silico* models of permeability, we have focused our attention on *in vivo* human jejunum effective permeability, *in situ* and *in vitro* rat intestinal permeability, and *in vitro* Madin–Darby canine kidney (MDCK) cell culture permeability. Lennernas *et al.* (1992) have developed a method for measuring human effective permeability (H-$P_{eff}$) using a regional intestinal perfusion technique. In this method a perfusion apparatus, consisting of a multichannel tube with two inflatable balloons (10 cm apart), is swallowed by the patient and eventually located in the proximal jejunum. Diluted solutions of the test drug are introduced at the inlet located at the center of the 10 cm section, and the loss of drug is determined from the concentration in the outlet intestinal perfusate. It has been determined that the intestinal segment that is isolated by the two inflated balloons behaves as a well-mixed compartment. The recovery of a non-absorbable marker polyethylene glycol, (PEG 4000) in the outlet intestinal perfusate is used to correct the outlet concentration of drug for the effect of net water flux into or out of the isolated intestinal segment. In such a fashion, the H-$P_{eff}$ for 22 carefully selected drug molecules has been determined and a theoretical model of H-$P_{eff}$ has been developed (Winiwarter *et al.*, 1998). The small size of the published H-$P_{eff}$ database is most likely due to the expense of the human measurement. Fortunately, a rat model of *in situ* intestinal permeability (R-$P_{eff}$) has been developed and used to determine the permeability of many more drugs (Amidon *et al.*, 1980; Friedman and Amidon, 1989; Sinko and Amidon, 1988). When R-$P_{eff}$ is compared with H-$P_{eff}$, a good correlation is demonstrated where the human permeability for passively absorbed compounds is on average 3.6 times higher than in rats (Fagerholm *et al.*, 1996).

In order to overcome the expense and time-consuming nature of *in vivo* and *in situ* determination of permeability, human and canine epithelial cell culture models have been developed as a surrogate for the *in vivo* estimates. In general, epithelial cells from human colon cancer (Caco-2) and canine kidney cells (MDCK) can be cultured as monolayers on semi-permeable membranes and allowed to differentiate, become polarized with apical and basolateral surfaces, and establish tight junctions between cells in a manner very similar to the *in vivo* situation. These cellular monolayers are grown in Transwell™ culture dishes that allow for drugs in the donor chamber to be exposed to the apical or basolateral surfaces of the cell monolayer. Timed samples are collected from the receiver side of the monolayer and permeability is estimated by the kinetics of the flux of drug across the monolayer. This measure of cellular permeability has the potential to be exploited in a high-throughput fashion for determining the permeability of many diverse chemical molecules. Many models of cell culture permeability have been described both in the literature and in Chapters 9–13, 15 and 18–20. We have developed a novel model of MDCK cell permeability that will be discussed briefly in the experimental sections of this chapter.

### Solubility

Next to permeability, aqueous solubility is the most important biopharmaceutical property associated with oral drug absorption. These two properties have established the basis for a Biopharmaceutical Classification System (BCS) and have become the subject of a Food and Drug Administration (FDA) guidance entitled 'Waiver of *In vivo* Bioavailability and Bioequivalence Studies for Immediate-Release Solid Oral Dosage Forms Based on a Biopharmaceutics Classification System' (Amidon *et al.*, 1995). The fundamental parameters that define oral drug absorption in humans are used as a basis for this classification scheme. These four Biopharmaceutic Drug Classes are defined as follows:

- Class 1: High solubility–high permeability drugs;
- Class 2: Low solubility–high permeability drugs;
- Class 3: High solubility–low permeability drugs and;
- Class 4: Low solubility–low permeability drugs.

If one is able to accurately estimate permeability and solubility from molecular structure, then it becomes feasible to estimate the BCS classification and to begin to develop *in silico* methods for the simulation of GI absorption.

Theoretical models of aqueous solubility are very difficult to develop and even more difficult for the user of such a model to evaluate. Quite often the statistician developing the model does not know the source and exact experimental conditions for solubility measurements. Consequently, the resulting models give highly specific results that relate primarily to the data sets used in the development. This does not mean that solubility models are not 'generalizable' to diverse structural types, but it does mean that statistical estimates of solubility may be quite different depending on the nature of the experimental conditions used to collect the solubility data for a given training set.

Yalkowsky and Banerjee (1992) have published an extensive review of methods for estimating aqueous water solubility of organic compounds. Many methods have been developed based on measured properties such as partition coefficients, chromatographic parameters, and activity coefficients (Amidon et al., 1974; Meylan et al., 1996; Yalkowsky et al., 1988). Purely in silico methods are based on linear free energy relationships, and a variety of geometric, electronic, and topological molecular descriptors (Brooke and Jurs, 1998; Huuskonen et al., 1997, 1998; Klopman et al., 1992).

The earliest models of aqueous solubility were based on the activity coefficient of the hydrophobic portion of a molecule, its surface area, and its interfacial tension (Amidon et al., 1974). The solvatochromic method of estimating water solubility has been compared with the relationship between solubility and the octanol-water partition coefficient (log $P_{ow}$), and it has been demonstrated that the latter provides a better estimation (Yalkowsky et al., 1988). When applied to a diverse series of over 100 small non-electrolyte organic molecules, the aqueous solubility and log $P_{ow}$ were found to be related by the simple equation log $S_w = -1.016$ log $P_{ow} + 0.515$, where $S_w$ is the molar solubility of liquid solutes in water and PC is the experimental partition coefficient of the solutes in the octanol-water system (Valvani et al., 1981). Other models using this approach have been developed for various small sets of drug molecules (Pinal and Yalkowsky, 1987). Meylan et al. (1996) compiled a large database of 1450 aqueous solubilities, including over 100 drug molecules, and developed a very predictive model using log $P_{ow}$, melting point, molecular weight, and 12 correction factors based on structure. This model was able to estimate the solubility of an external test set of 85 molecules with standard deviation (SD) of 0.96 log units. If a computed value of log $P_{ow}$ is available, this method of calculating solubility can be used without measured values.

We have been very interested in models of aqueous solubility solely based on in silico parameters. Klopman et al. (1992) developed a method for estimating water solubility based on the group contribution approach. Two models were developed, based on 21 organic compounds, with SD of 0.58 log units for the first test set of 13 compounds and SD of 1.25 log units for the second test set of all 21 compounds. Using topological descriptors (connectivity indices, kappa indices, and electrotopological state indices) and two small databases of steroids and barbituric acids, an ANN model of aqueous solubility was developed (Huuskonen et al., 1997). Separate models were built for each class of drugs using back-propagation neural networks with one hidden layer and five topological indices as input parameters. The results indicate that neural networks can produce useful models of the aqueous solubility of a congeneric set of compounds, even with simple structural parameters.

When applied to a much larger data set ($N=210$) of primarily drug molecules, the same method produced a model that was able to estimate an external test set ($N=51$) with SD of 0.58 log units (Huuskonen *et al.*, 1998). Genetic algorithm and simulated annealing routines, in conjunction with multiple linear regression and ANNs, were used to develop a model of aqueous solubility from 332 small organic molecules (Brooke and Jurs, 1998). This model was able to predict an external test set of 32 molecules with RMSE of 0.34 log units. Thus, it would appear that purely *in silico* methods might be successfully used to estimate aqueous solubility. Likewise, molecular diffusivity can be estimated from the molal volume using computational approaches (Schroeder, 1949; Hayduk and Laudie, 1974).

## Simulation of oral drug absorption

Absorption of drugs from the GI tract is very complex and can be influenced by many factors that fall into three classes (Yu *et al.*, 1996b). The first class represents physicochemical factors, including $pK_a$, solubility, stability, diffusivity, lipophilicity, salt forms, surface area, drug particle size, and crystal form. The second class comprises physiological factors, including GI pH, gastric emptying, small and large bowel transit times, active transport and efflux, and gut wall metabolism. The third class comprises formulation factors, such as solution, tablet, capsule, suspension, emulsion, gel, and modified release.

One of the original concepts governing oral absorption of organic molecules is called the 'pH-partition' hypothesis. Under this hypothesis, only the unionized form of ionizable molecules will partition into the membranes of epithelial cells lining the GI tract (Hogben *et al.*, 1959; Jacobs, 1940). Testing this hypothesis has resulted in the recognition of the important contribution of GI pH to permeability and to the dissolution rate of solid dosage forms, but it has not been found to be universally applicable. Ho *et al.* (1972) developed one of the most sophisticated theoretical approaches to simulating drug absorption based on the diffusional transport of drugs across a compartmental membrane (Suzuki *et al.*, 1970a,b). Their physical model consisted of a well-stirred bulk aqueous phase, an aqueous diffusion layer, and a heterogeneous lipid barrier composed of several compartments ending in a perfect sink. Their model represented the first example of the rigorous application of a physical model to the quantitative and mechanistic interpretation of *in vivo* absorption (Ho and Higuchi, 1971). The simultaneous chemical equilibria and mass transfer of basic and acidic drugs were modeled and compared favorably with *in situ* measurements of intestinal, gastric, and rectal absorption in animals. The pH-partition theory was shown to be a limiting case of the more general model they developed. Given its complexity, the diffusional mass transit model has not been widely utilized. In the 1980s, a simple and intuitive alternative approach based on a series of mixing tank compartments was developed (Dressman *et al.*, 1984). Pharmacokinetic (PK) models incorporating discontinuous GI absorption from at least two absorption sites, separated by N non-absorbing sites, have been used to explain the occurrence of double peaks in plasma concentration versus time (Cp-time) profiles for ranitidine and cimetidine (Suttle *et al.*, 1992). A similar discontinuous oral absorption model based on two absorption compartments and two transit compartments was developed to explain the bioavailability

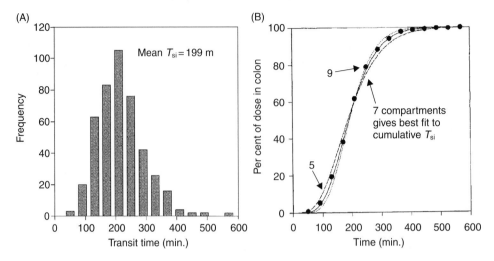

*Figure 21.3* (A) Mean transit time distribution for 400 human subjects. (B) Cumulative transit time expressed as percentage of dose reaching the colon. Data was modeled with an assumption of five, seven, or nine compartments. Best fit was obtained with seven compartments (Yu *et al.*, 1996a).

of nucleoside analogs (Wright *et al.*, 1996). Finally, using a poorly described GI simulation model, Norris *et al.* (2000) were able to estimate Cp-time profiles for ganciclovir.

Correct representation of the GI tract as a series of compartments requires an analysis of SI transit time frequency distribution in order to determine the correct number of compartments. Analysis of more than 400 human SI transit times revealed a log normal distribution with mean SI transit time of 3.3 hours (Yu *et al.*, 1996a) (see Figure 21.3A). In the case of no absorption or degradation, the cumulative distribution of the transit time data set can be viewed as the percentage of a dose entering the colon as a function of time (Figure 21.3B). By integrating the transit of material through the SI, as represented by various numbers of compartments and plotting the cumulative amount that reaches the colon, Yu determined that seven compartments give the best fit to the observed cumulative frequency distribution. The seven-compartment transit model may be visualized as having the first half of the first compartment representing the duodenum, the second half of the first compartment, along with the second and third representing the jejunum, and the rest representing the ileum. The corresponding transit times in the duodenum, jejunum, and ileum are 14, 71, and 114 minutes, respectively.

Based on this transit time distribution for seven SI compartments, Yu *et al.* (1996b) developed a compartmental absorption and transit (CAT) model to simulate the rate and extent of drug absorption. The CAT model is described by a set of differential equations that considers simultaneous movement of a drug solution through the GI tract and absorption of the dissolved material from each compartment into the portal vein. The constant rate for absorption from each compartment is based on measured values of H-$P_{eff}$. Both analytical and numerical methods have been used to solve the model equations. It was found that the fraction of dose absorbed can be estimated by %HIA$= 1 - (1 + 0.54 \text{ H-}P_{eff})$ (−7). A good correlation was found between the fraction of dose absorbed and the effective permeability for ten drugs covering a wide range of absorption characteristics when the effects of drug dissolution and dosage form could be neglected. The model was also able to explain the oral plasma concentration profiles of atenolol (Yu and Amidon, 1999). The CAT model has proven to be more versatile than earlier aggregate models, such as the single-tank mixing model and the macroscopic or microscopic mass balance models (Oh *et al.*, 1993; Sinko *et al.*, 1991).

Subsequently, a similar GI transit and segmental absorption model has been developed. It is based on five SI segments, plus the cecum and the large intestine coupled to a two compartment (PK) model. It is used to simulate the plasma concentration versus time (Cp-time) profiles for ampicillin, theophylline, propranolol, and cephalexin (Sawamoto *et al.*, 1997). The propranolol Cp-time profile was overestimated because first pass metabolism was not considered in the PK model. Another method based on the CAT model utilizes *in vitro* measurements of biopharmaceutical properties. It has been used to simulate the absorption of ganciclovir. The authors concluded from their simulation results that the low bioavailability of ganciclovir is limited by compound solubility, rather than by permeability owing to partitioning as previously speculated (Norris *et al.*, 2000).

Macheras has developed the heterogeneous tube model of the SI based on a stochastic simulation of drug molecules moving through a cylinder of fixed radius with random geometric placement of dendritic-type virtual 'villi'. Drug molecules are assigned a probability of forward movement using a Monte–Carlo blind ant random walk simulation. Contact between the drug molecule and virtual villi impede the movement, causing dispersion of the mass of drug as it passes through the cylinder. This model has been calibrated to accurately account for the observed human SI transit time distribution (Kalampokis *et al.*, 1999a,b). Finally, a theory of molecular absorption from the SI based on macrotransport analysis has resulted in a model that includes complex inter-relationships between lumen and membrane diffusion, convection, degradation and absorption mechanisms (Stoll *et al.*, 2000).

## SPECIFIC APPLICATIONS

### BEstACAT

We have developed statistical models for the estimation of biopharmaceutical properties (implemented in the software program QMPRPlus™) and have extended the CAT model of Yu and Amidon (1999) in the software program GastroPlus™. Extensions of the original CAT model include drug dissolution and precipitation, pH dependence of solubility and permeability, controlled release, luminal degradation, gut metabolism, and absorption from stomach and colon, in addition to the SI. A three-compartment PK model that includes saturable liver metabolism is used to simulate Cp-time profiles.

## Biopharmaceutical property estimation (QMPRPlus™)

We have now developed statistical models of the following biopharmaceutical properties:

- Log P ($\log_{10}$ of octanol-water partition coefficient for unionized molecules)
- Effective permeability (human jejunum) at pH 6.5 ($P_{eff}$, cm/s×$10^4$)
- MDCK cell monolayer permeability at pH 7.4 ($P_{app}$, nm/s)
- Blood–brain barrier permeation (high, low, undecided)
- Saturated aqueous solubility (mg/ml)
- Molal volume ($cm^3$/gmol)
- Diffusivity (diffusion coefficient, $cm^2$/s).

Our method is based on the calculation of approximately 170 molecular descriptors obtained by parsing the 2D or 3D molecular structure of a drug molecule as represented by either the SMILES string format or as ISIS- .RDF, .SDF, or .MOL file format (MDL Information Systems Inc., www.mdli.com). Molecular descriptor values are input to independent mathematical models in order to generate estimates for each of the biopharmaceutical properties listed above. Using either these property estimates or experimentally determined properties as inputs to the advanced compartmental absorption and transit (ACAT) model, drug molecules may be classified according to their absorption, distribution, metabolism, and excretion (ADME) qualities. While no computer program is able to estimate exact experimental values for these properties, we have demonstrated that the accuracy of estimated values generated by our method is often on the order of experimental error and is sufficiently accurate to allow rank ordering of a large number of compounds for 'overall ADME quality'. Our concept is to identify as many 'losers' as early as possible to avoid spending valuable resources on them. This concept, sometimes referred to as a 'quick kill' approach, provides guidance to research teams who are faced with the decision to promote selected potential drug molecules to development status. This approach can also be used to select compounds for synthesis from a large combinatorial library.

## Log P model

Our log P model is based on an ANN constructed from 9,651 sample compounds selected from the 'Starlist' of ion-corrected experimental log P values (Hansch *et al.*, 1995). The final model contains over 100 molecular descriptors that can be divided into 1D, 2D, 3D, physicochemical, and topological classifications. 1D descriptors are based on the molecular formula and represent the molecular weight and counts of various atom types (N, O, halogens, etc.). 2D descriptors are based on the molecular connectivity table and represent counts of various types of bonds and functional groups. In addition, 2D descriptors are used to calculate many of the physicochemical properties, such as fraction ionized, formal charges, number of donor and acceptor H-bonds and partial charges (by the Gasteiger method; Gasteiger and Marsili, 1980) and topological descriptors. Our 3D descriptors were calculated using 3D structures generated by the computer program CORINA

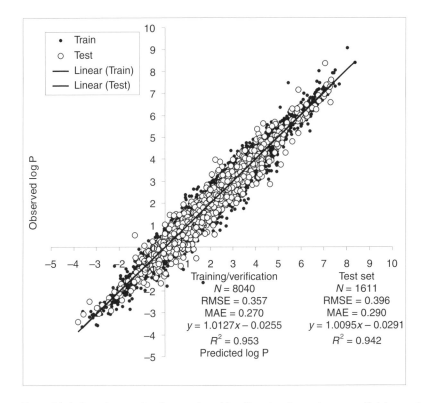

*Figure 21.4* Correlation plot for results of log P estimation using an artificial neural network (ANN). Dots represent the training and verification data sets. Open circles represent the external test set.

(Gasteiger *et al.*, 1990). 3D descriptors used in the log P model include polar and non-polar solvent accessible surface area, dipole moment, moments of inertia and solvation energy moment vector. The correlation between experimental and predicted log P values obtained from QMPRPlus™ is shown in Figure 21.4 for 3D inputs. For an external test set of 1,611 samples (not used for model development), input to both the 2D and 3D versions of the model, the RMSEs are 0.43 and 0.40 log units, the mean absolute errors (MAE) are 0.32 and 0.29 log units, and the explained variance $R^2$ values are 0.93 and 0.94, respectively.

## H-$P_{eff}$

Our model of H-$P_{eff}$ is based on *in vivo* permeability values measured in human subjects (Lennernas, 1998) and *in situ* rat wall permeability (R-$P_{app}$) measurements (Amidon *et al.*, 1980). The *in situ* R-$P_{app}$ values were converted to H-$P_{eff}$ values using a linear relationship (Fagerholm *et al.*, 1996). Data from 44 samples (38 drugs and six primary alpha amino acids) were used for model building. Four examples for which data were available were excluded from model building. These were

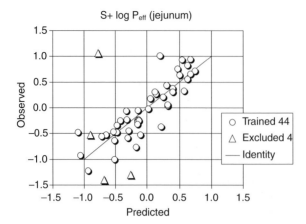

*Figure 21.5* Correlation plot for results of human jejunum permeability (log P$_{eff}$). 44 compounds were used in training. Four compounds were excluded from training as discussed in the text.

glucose, which is a hexose transporter substrate, carnosine which may be a dipeptide transporter substrate (Ganapathy and Leibach, 1982), and hydrochlorothiazide and furosemide which had low and variable *in vivo* permeability values that seemed inconsistent with reports of their intestinal absorption and bioavailability.

The predictive model was constructed using partial least squares regression (University of North Carolina QSAR Server). We found that molecular descriptors relating to lipophilicity, H-bond donors and acceptors, and H-bond topology were selected using a genetic algorithm based on a CV criterion. The performance of the model is shown in Figure 21.5. The MAE was 0.24 log units, the RMSE was 0.27 log units, and the explained variance ($R^2$) was 76 per cent.

## Solubility models

Our water solubility models are based on composites of ANNs and were constructed from 1,337 sample compounds with reliable water solubility measurements (Meylan *et al.*, 1996). Of the 1,337 substances used in the training and verification of the S+solubility models, 925 were solids and 412 were liquids at room temperature. All liquids were assigned the melting point value of 298 K. All solids had experimentally measured melting points. When melting point values of unknown drugs are available, a solubility model trained using melting point as one of the descriptors is selected. In the absence of melting point, another model trained without melting point is selected. Little difference in performance was observed between models using melting point and models without melting point.

The correlation between observed and predicted log S$_w$ values obtained from the QMPRPlus™ 3D S+S$_w$ model with melting point is shown in Figure 21.6. For an external test set of 133 sample compounds (not used for model development), the RMSEs for both models are 0.45 log units, the MAEs are 0.35 log units, and

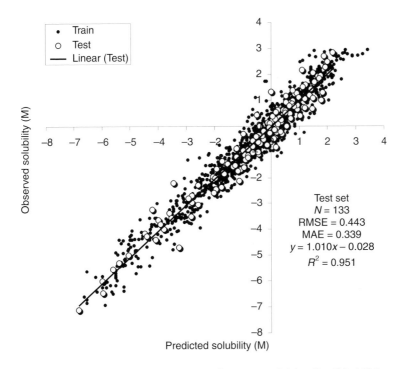

*Figure 21.6* Correlation plot for results of aqueous solubility (log $S_w$). 1204 compounds were used in training and verification sets. 133 compounds were used in the external test set.

the explained $R^2$ values are 0.95. There is no statistical difference between the performances of the 2D and the 3D S+MP models.

## ACAT Model

The published version of the original compartmental absorption and transit (CAT) model does not account for dissolution rate, the pH dependence of the solubility of drugs, controlled release, absorption in the stomach or colon, metabolism in gut or liver, degradation in the lumen, or the changes in such factors as surface area, transporter densities, efflux protein densities, and other regional factors within the intestinal tract. For drugs with low permeability or solubility, absorption may not be complete in the SI and the CAT model can be made more accurate by treating the colon as an additional absorbing compartment. For drugs with high permeability and high solubility, colonic absorption is a negligible fraction of the total for immediate release formulations. However, for many immediate release drugs with moderate-to-low permeability, and for most controlled release formulations, colonic absorption can be significant. We have now extended the CAT model to include these effects. GastroPlus™ is a widely used simulation software product based on a new model, which is called the ACAT model (Figure 21.7). As with the original

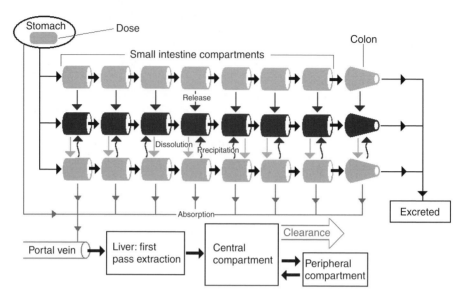

*Figure 21.7* Schematic diagram of the advanced compartmental absorption and transit (ACAT) model. Each arrow represents a differential equation.

CAT model, a basic assumption of the ACAT model is that drugs passing through the SI will have equal transit time in each compartment. Since the volume of fluid entering the small intestine (8–9 l/day) in the duodenum and jejunum is more than the fluid exiting the SI (0.5–1 l/day), volumes and flow rates of the upper SI compartments are considered to be larger than the lower compartments to satisfy the transit time constraint.

## Numerical integration of ACAT

The form of the ACAT model implemented in GastroPlus™ describes the release, dissolution, luminal degradation, metabolism, and absorption of a drug as it transits through successive compartments to represent the digestive tract. The kinetics associated with this process are modeled by a system of $6 \times N + 6$ (60 total) coupled non-linear rate equations. The equations include the consideration of six states (unreleased, undissolved, dissolved, degraded, metabolized, and absorbed), eighteen compartments (9 GI (stomach, 7SI, and colon) and 9 enterocyte), three states of excreted material (unreleased, undissolved, and dissolved), and the amount of drugs in up to three pharmacokinetic compartments (when pharmacokinetic parameters are available). The total amount of absorbed material is summed over the integrated amounts being absorbed from each absorption/transit compartment.

In general, the rate of change of drug concentration in a compartment depends on six different processes: (1) transit of drug into a compartment, (2) transit of drug out of a compartment, (3) release of drug from the formulation in the compartment, (4) dissolution of the drug particles, (5) degradation of the drug, and (6) absorption of the drug. The time scale associated with the transit process is set by a transfer rate constant, $k_t$, that is determined from the mean transit time within each compartment. The time scale of the dissolution process is set by a rate constant, $k_d$, that can be computed from a drug's solubility (as a function of pH), its effective particle size, its molecular density, its lumen concentration, its diffusion coefficient, and the diffusion layer thickness (Equation 21.1). The time scale associated with the absorption process is set by a rate constant, $k_a$, that depends on the effective permeability of the drug multiplied by an absorption scale factor (ASF with units of $cm^{-1}$) for each compartment (Equation 21.2). The time scale of the chemical degradation process is set by a rate constant, $k_{degrad}$, that is determined by interpolation from an input table of degradation rate (or half-life) versus pH and the pH in the compartment.

The system of differential equations is integrated using a fourth/fifth order Runge-Kutta numerical integration package with adaptive step size (Press, 1992). The fraction of dose absorbed is calculated as the sum of all drug amounts entering the gut wall as a function of time, divided by the dose, or by the sum of all doses if multiple dosing is used.

$$k_{(i)d} = 3\gamma \frac{C_S - C_L}{\rho r T} \tag{21.1}$$

$$k_{(i)a} = \alpha(i) P_{eff} \tag{21.2}$$

Where $k_{(i)d}$=dissolution rate constant for the $i$th compartment ($s^{-1}$), $k_{(i)a}$=absorption rate constant for the ith compartment ($s^{-1}$), $C_S$=aqueous solubility (g/l), $C_L$=intestinal lumen concentration (g/l), $\gamma$=molecular diffusion coefficient ($cm^2$/s), $\rho$=drug particle density ($g/cm^3$), $r$=effective drug particle radius (cm), $T$=diffusion layer thickness (cm), $\alpha$=compartmental absorption scale factor ($cm^{-1}$), and $P_{eff}$=H-$P_{eff}$ (cm/s).

## Automatic scaling of $k_a$ as a function of $P_{eff}$, pH, and log D

The size and shape of a drug molecule, its acid and base dissociation constants, and the pH of the GI tract all influence the absorption rate constant for specific regions of the GI. Pade and Stavchansky (1997) measured the Caco-2 cellular permeability for a diverse set of acidic and basic drug molecules at two pH values. They concluded that the permeability coefficients of the acidic drugs were greater at pH 5.4, whereas that of the basic drugs was greater at pH 7.2. The transcellular pathway was found to be the favored pathway for most drugs, probably owing to its larger accessible surface area. The paracellular permeability of the drugs was size- and charge-dependent. The permeability of the drugs through the tight junctions decreased with increasing molecular size. Further, the pathway also appeared to be cation-selective, with the positively charged cations of weak bases permeating the aqueous pores of the paracellular pathway at a faster rate than the negatively charged anions of weak acids. Thus, the extent to which the paracellular and transcellular routes are utilized in drug transport is influenced by the fraction of ionized and unionized species (which in turn depends upon the pK$_a$ of the drug

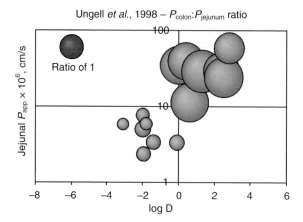

Figure 21.8 Regional dependence of rat small intestine permeability as a function of log D. Data from Ungell et al. (1998) are replotted.

and the pH of the solution), the intrinsic partition coefficient of the drug, the size of the molecule and its charge.

Figure 21.8 is a representation of regional permeability coefficients of 19 drugs with different physicochemical properties determined by Ungell et al. (1998) using excised segments from three regions of rat intestine: jejunum, ileum, and colon.

They observed a significant decrease in permeability to hydrophilic drugs and a significant increase in permeability for hydrophobic drugs aborally to the SI (P 0.0001). Figure 21.8 illustrates that for hydrophilic drugs (low permeability and low log D) the ratio of colon : jejunal permeability was less than one. While for hydrophobic drugs (higher permeability and higher log D) the ratio of colon : jejunal permeability is observed to be greater than one. At certain pH values, the permeability of small (molecular weight < 200) hydrophilic drugs may have a large paracellular component (Adson et al., 1995), and it is well-known that the transepithelial electrical resistance (TEER) of the colon is much higher than the SI. TEER increases as the width of tight junctions decreases and the tight junction width has been determined to be 0.75–0.8 nm in jejunum, 0.3–0.35 nm in ileum, and 0.2–0.25 nm in colon. The narrower tight junctions in the colon suggest that paracellular transport will be much less significant in the colon, which help to explain the lower ratio of colon : jejunal permeability for hydrophilic drugs. We have used the ACAT model with experimental biopharmaceutical properties for a series of hydrophilic and hydrophobic drug molecules to calibrate a 'log D model' that explains the observed rate and extent of absorption.

## BEstACAT example

An experimental data set composed of nine drug molecules with varying degrees of solubility, permeability, and fraction absorbed was collected and is shown in Table 21.1. QMPRPlus™ was applied to 3D representations of each of the nine

*Table 21.1* Experimental data used to illustrate the process of biopharmaceutical property estimation and the advanced compartmental absorption and transit (ACAT) simulation

| Name/ID | %HIA | $P_{eff}$ | log D | pH (log D) | $S_w$ | pH ($S_w$) | Dose | DiffCoef |
|---|---|---|---|---|---|---|---|---|
| Lisinopril | 25 | 0.12 | −1.00 | 2.0 | 57.00 | 7.0 | 20 | 0.750 |
| Atenolol | 56 | 0.30 | −1.84 | 4.0 | 30.90 | 9.0 | 100 | 0.771 |
| Ranitidine | 58 | 0.43 | 0.27 | 10.5 | 660.00 | 4.0 | 300 | 0.741 |
| Furosemide | 61 | 0.30 | −0.90 | 7.4 | 2.25 | 7.2 | 100 | 0.810 |
| Propranolol | 90 | 2.72 | −0.23 | 4.0 | 81.50 | 9.0 | 140 | 0.761 |
| Metoprolol | 95 | 1.34 | −1.72 | 4.0 | 534.00 | 3.0 | 156 | 0.741 |
| Piroxicam | 95 | 10.40 | 0.72 | 6.6 | 0.11 | 6.6 | 20 | 0.520 |
| Carbamazepine | 97 | 4.30 | 1.65 | 6.0 | 0.23 | 7.0 | 200 | 0.804 |
| Ketoprofen | 100 | 8.70 | 1.22 | 6.0 | 0.21 | 2.0 | 50 | 0.842 |

Notes

%HIA = fraction of dose absorbed into the portal vein; $P_{eff}$ = human effective permeability (cm/s × $10^4$); log D = octanol/water distribution coefficient at given pH; pH (log D) = pH at which the log D was measured; $S_w$ = water solubility (mg/ml) at given pH; pH ($S_w$) = pH at which the solubility was measured; Dose = dose of drug used in the simulation; and DiffCoef = diffusion coefficient (cm²/s × $10^5$).

molecules and the resulting biopharmaceutical properties are listed in Table 21.2. ASFs for each of the small intestinal compartments were automatically determined using the ACAT 'log D model'.

This automatic determination results in ASFs that account for the predicted effect of pH and GI location on the absorption rate constant. Figures 21.9A and 21.9B illustrate a comparison of the ASF for ketoprofen (a hydrophobic acidic drug) compared with lisinopril (a hydrophilic basic drug). As predicted from the pH-partition theory, it can be seen that the ASFs in the SI increase with increasing pH for lisinopril and decrease with decreasing pH for ketoprofen. In addition, the ratio of colon : jejunal ASF for each drug follows the expected result based on the experimental observations of Ungell *et al.* (1998).

*Table 21.2* In silico data from QMPRPlus™ used to illustrate the process of biopharmaceutical property estimation and the advanced compartmental absorption and transit (ACAT) simulation

| Name/ID | %HIA | $S + P_{eff}$ | $S + \log P$ | $S + S_w MP$ | pH ($S_w$) | Dose | DiffCoef |
|---|---|---|---|---|---|---|---|
| Lisinopril | 25 | 0.06 | −0.99 | 12.30 | 5.6 | 20 | 0.750 |
| Atenolol | 56 | 0.46 | 0.64 | 12.30 | 11.0 | 100 | 0.771 |
| Ranitidine | 58 | 0.75 | 1.15 | 0.48 | 9.7 | 300 | 0.741 |
| Furosemide | 61 | 0.17 | 3.03 | 0.02 | 4.3 | 100 | 0.810 |
| Propranolol | 90 | 1.88 | 2.99 | 0.52 | 10.3 | 140 | 0.761 |
| Metoprolol | 95 | 1.00 | 2.11 | 2.03 | 10.7 | 156 | 0.741 |
| Piroxicam | 95 | 1.27 | 0.90 | 0.12 | 4.3 | 20 | 0.520 |
| Carbamazepine | 97 | 3.61 | 2.47 | 0.02 | 7.0 | 200 | 0.804 |
| Ketoprofen | 100 | 5.17 | 2.99 | 0.09 | 3.9 | 50 | 0.842 |

Notes

%HIA = fraction of dose absorbed into the portal vein; $P_{eff}$ = human effective permeability (cm/s × $10^4$); $S_w$ = water solubility (mg/ml) at given pH; pH ($S_w$) = pH at which the solubility was measured; Dose = dose of drug used in the simulation; and DiffCoef = diffusion coefficient (cm²/s × $10^5$).

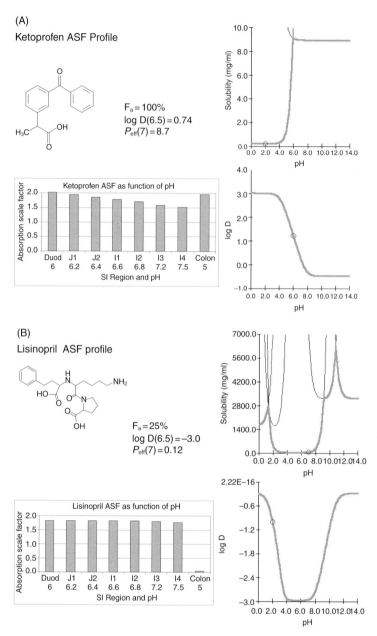

*Figure 21.9* (A) Compartmental absorption scale factor profile for Ketoprofen. Ketoprofen is highly permeable and highly hydrophobic with log D(6.5) = 0.74 and log P = 2.92. The aliphatic carboxylic acid has a pK$_a$ = 4.3 resulting in higher absorption scale factors at the lower pH conditions found in the duodenum. In addition, the absorption scale factor (ASF) for colon is large. (B) Compartmental absorption scale factor profile for Lisinopril. Lisinopril has relatively low jejunal permeability and is very polar with log D(6.5) = −3.0 and log P = −0.3. The (ASF) for colon is much smaller than for the small intestine, resulting in approximately 25 per cent of the total dose absorbed.

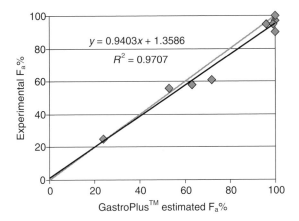

*Figure 21.10* Correlation plot for simulated fraction absorbed using the advanced compartmental absorption and transit (ACAT) model as implemented in the software package GastroPlus™. Nine drugs were simulated using experimental values for log P, solubility, pK$_a$, and permeability as shown in Table 21.1.

Using this calibrated ACAT model within GastroPlus™, with experimental values for permeability, solubility, pKa, and log P as shown in Table 21.1, the predicted %HIA is in very good agreement with experimental results (Figure 21.10). When a similar calculation was completed using purely *in silico* biopharmaceutical properties calculated by QMPRPlus™ as shown in Table 21.2, the correlation was only slightly worse (Figure 21.11).

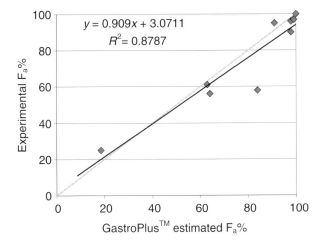

*Figure 21.11* Correlation plot for simulated fraction absorbed using the advanced compartmental absorption and transit (ACAT) model as implemented in the software package GastroPlus™. Nine drugs were simulated using purely *in silico* estimates for log P, solubility, pK$_a$, and permeability determined by QMPRPlus™ as shown in Table 21.2. pK$_a$ values were estimated using the Physchem Batch program from ACD laboratories (Advanced Chemistry Development Inc., Toronto, Canada).

## CONCLUDING REMARKS

The current methods for the prediction of drug absorption using computerized methods have been summarized. The application of biopharmaceutical property estimation and simulation using an ACAT model to predict the rate and extent of oral absorption for a diverse selection of drug molecules has also been demonstrated. We are currently extending the ACAT model to include the simulation of pharmacological response. The future potential of *in silico* methods in pharmaceutical research and development is limitless. Recently, a spokesperson from the US FDA speculated that we are rapidly moving towards a future situation in drug discovery research and development that is not just supplemented with electronic technology but is rather driven by such technology. Lawrence Yu, suggested a term for this future would be *eR&D* (Yu, 2001).

## REFERENCES

Adson, A., Burton, P. S., Raub, T. J., Barsuhn, C. L., Audus, K. L. and Ho, N. F. (1995) Passive diffusion of weak organic electrolytes across Caco-2 cell monolayers: uncoupling the contributions of hydrodynamic, transcellular, and paracellular barriers. *J. Pharm. Sci.*, **84**, 1197–1204.

Amidon, G. L., Yalkowsky, S. H. and Leung, S. (1974) Solubility of nonelectrolytes in polar solvents II: solubility of aliphatic alcohols in water. *J. Pharm. Sci.*, **63**, 1858–1866.

Amidon, G. L., Kou, J., Elliott, R. L. and Lightfoot, E. N. (1980) Analysis of models for determining intestinal wall permeabilities. *J. Pharm. Sci.*, **69**, 1369–1373.

Amidon, G. L., Lennernas, H., Shah, V. P. and Crison, J. R. (1995) A theoretical basis for a biopharmaceutic drug classification: the correlation of *in vitro* drug product dissolution and *in vivo* bioavailability. *Pharm. Res.*, **12**, 413–420.

Andrews, C. W., Bennett, L. and Yu, L. X. (2000) Predicting human oral bioavailability of a compound: development of a novel quantitative structure–bioavailability relationship [In Process Citation]. *Pharm. Res.*, **17**, 639–644.

Artursson, P. and Borchardt, R. T. (1997) Intestinal drug absorption and metabolism in cell cultures: Caco-2 and beyond. *Pharm. Res.*, **14**, 1655–1658.

Audus, K. L., Bartel, R. L., Hidalgo, I. J. and Borchardt, R. T. (1990) The use of cultured epithelial and endothelial cells for drug transport and metabolism studies. *Pharm. Res.*, **7**, 435–451.

Borchardt, R. T., Smith, P. L. and Wilson, G. (1996) General principles in the characterization and use of model systems for biopharmaceutical studies. *Pharm. Biotechnol.*, **8**, 1–11.

Brooke, M. E. and Jurs, P. C. (1998) Prediction of aqueous solubility of organic compounds from molecular structure. *J. Chem. Inf. Comput. Sci.*, **38**, 489–496.

Clark, D. E. (1999) Rapid calculation of polar molecular surface area and its application to the prediction of transport phenomena. 1. Prediction of intestinal absorption. *J. Pharm. Sci.*, **88**, 807–814.

Dressman, J. B., Fleisher, D. and Amidon, G. L. (1984) Physicochemical model for dose-dependent drug absorption. *J. Pharm. Sci.*, **73**, 1274–1279.

Fagerholm, U., Johansson, M. and Lennernas, H. (1996) Comparison between permeability coefficients in rat and human jejunum. *Pharm. Res.*, **13**, 1336–1342.

Friedman, D. I. and Amidon, G. L. (1989) Passive and carrier-mediated intestinal absorption components of two angiotensin converting enzyme (ACE) inhibitor prodrugs in rats: enalapril and fosinopril. *Pharm. Res.*, **6**, 1043–1047.

Ganapathy, V. and Leibach, F. H. (1982) Peptide transport in rabbit kidney. Studies with L-carnosine. *Biochim. Biophys. Acta.*, **691**, 362–366.

Gasteiger, J. and Marsili, M. (1980) Iterative partial equalization of orbital electronegativity – A rapid access to atomic charges. *Tetrahedron*, **36**, 3219–3228.

Gasteiger, J., Rudolph, C. and Sadowski, J. (1990) Automatic generation of 3D-atomic coordinates for organic molecules. *Tetrahedron Comp. Method*, **3**, 537–547.

Hall, S. D., Thummel, K. E., Watkins, P. B., Lown, K. S., Benet, L. Z., Paine, M. F. *et al.* (1999) Molecular and physical mechanisms of first-pass extraction. *Drug Metab. Dispos.*, **27**, 161–166.

Hansch, C., Leo, A. and Hoekman, D. H. (1995) *Exploring QSAR*, American Chemical Society, Washington, DC.

Hayduk, W. and Laudie, H. (1974) Prediction of diffusion coefficients for nonelectrolytes in dilute aqueous solutions. *Am. Inst. of Chem. Eng. J.*, **20**, 611–615.

Hillgren, K. M., Kato, A. and Borchardt, R. T. (1995) In vitro systems for studying intestinal drug absorption. *Med. Res. Rev.*, **15**, 83–109.

Ho, N. F. and Higuchi, W. I. (1971) Quantitative interpretation of in vivo buccal absorption of n-alkanoic acids by the physical model approach. *J. Pharm. Sci.*, **60**, 537–541.

Ho, N. F., Higuchi, W. I. and Turi, J. (1972) Theoretical model studies of drug absorption and transport in the GI tract. 3. *J. Pharm. Sci.*, **61**, 192–197.

Hogben, C. A. M., Tocco, D. J., Brodie, B. B. and Schanker, L. S. (1959) On the mechanism of intestinal absorption of drugs. *J. Pharmacol. Exp. Ther.*, **125**, 275–282.

Hunter, A. (1996) Trajan Neural Networks, Trajan, United Kingdom.

Huuskonen, J., Salo, M. and Taskinen, J. (1997) Neural network modeling for estimation of the aqueous solubility of structurally related drugs. *J. Pharm. Sci.*, **86**, 450–454.

Huuskonen, J., Salo, M. and Taskinen, J. (1998) Aqueous solubility prediction of drugs based on molecular topology and neural network modeling. *J. Chem. Inf. Comput. Sci.*, **38**, 450–456.

Jacobs, M. H. (1940) Some aspects of cell permeability to weak electrolytes. *Cold Spring Harb. Symp. Quant. Biol.*, **8**, 30–39.

Kalampokis, A., Argyrakis, P. and Macheras, P. (1999a) Heterogeneous tube model for the study of small intestinal transit flow. *Pharm. Res.*, **16**, 87–91.

Kalampokis, A., Argyrakis, P. and Macheras, P. (1999b) A heterogeneous tube model of intestinal drug absorption based on probabilistic concepts. *Pharm. Res.*, **16**, 1764–1769.

Kelder, J., Grootenhuis, P. D., Bayada, D. M., Delbressine, L. P. and Ploemen, J. P. (1999) Polar molecular surface as a dominating determinant for oral absorption and brain penetration of drugs. *Pharm. Res.*, **16**, 1514–1519.

Klopman, G., Wang, S. and Balthasar, D. M. (1992) Estimation of aqueous solubility of organic molecules by the group contribution approach. Application to the study of biodegradation. *J. Chem. Inf. Comput. Sci.*, **32**, 474–482.

Krarup, L. H., Christensen, I. T., Hovgaard, L. and Frokjaer, S. (1998) Predicting drug absorption from molecular surface properties based on molecular dynamics simulations. *Pharm. Res.*, **15**, 972–978.

Lennernas, H. (1998) Human intestinal permeability. *J. Pharm. Sci.*, **87**, 403–410.

Lennernas, H., Ahrenstedt, O., Hallgren, R., Knutson, L., Ryde, M. and Paalzow, L. K. (1992) Regional jejunal perfusion, a new in vivo approach to study oral drug absorption in man. *Pharm. Res.*, **9**, 1243–1251.

Lien, E. J. (1981) Structure-activity relationships and drug disposition. *Annu. Rev. Pharmacol. Toxicol.*, **21**, 31–61.

Meylan, W. M., Howard, P. H. and Boethling, R. S. (1996) Improved method for estimating water solubility from octanol-water partition coefficient. *Environ. Toxicol. Chem.*, **15**, 100–106.

Norris, D. A., Leesman, G. D., Sinko, P. J. and Grass, G. M. (2000) Development of predictive pharmacokinetic simulation models for drug discovery. *J. Control Release.*, **65**, 55–62.

Oh, D. M., Curl, R. L. and Amidon, G. L. (1993) Estimating the fraction dose absorbed from suspensions of poorly soluble compounds in humans: a mathematical model. *Pharm. Res.*, **10**, 264–270.

Oprea, T. I. and Gottfries, J. (1999) Toward minimalistic modeling of oral drug absorption. *J. Mol. Graph. Model.*, **17**, 261–274, 329.

Overton, E. (1902) Beitrage zur allgemeinen Muskel- und Nervenphysiologie. *Pfleugers Arch. Gesamte Physiol. Menschen Tiere*, **92**, 115–280.

Pade, V. and Stavchansky, S. (1997) Estimation of the relative contribution of the transcellular and paracellular pathway to the transport of passively absorbed drugs in the Caco-2 cell culture model. *Pharm. Res.*, **14**, 1210–1215.

Palm, K., Luthman, K., Ungell, A. L., Strandlund, G. and Artursson, P. (1996) Correlation of drug absorption with molecular surface properties. *J. Pharm. Sci.*, **85**, 32–39.

Palm, K., Stenberg, P., Luthman, K. and Artursson, P. (1997) Polar molecular surface properties predict the intestinal absorption of drugs in humans. *Pharm. Res.*, **14**, 568–571.

Patterson, D. (1996) *Artificial Neural Networks*, Prentice Hall, Singapore.

Pinal, R. and Yalkowsky, S. H. (1987) Solubility and partitioning. VII: Solubility of barbiturates in water. *J. Pharm. Sci.*, **76**, 75–85.

Plakogiannis, F. M., Lien, E. J., Harris, C. and Biles, J. A. (1970) Partition of alkylsulfates of quaternary ammonium compounds: structure dependence and transport study. *J. Pharm. Sci.*, **59**, 197–200.

Press, W. H. (1992) *Numerical recipes in C: the art of scientific computing*, Cambridge University Press, Cambridge, New York.

Sawamoto, T., Haruta, S., Kurosaki, Y., Higaki, K. and Kimura, T. (1997) Prediction of the plasma concentration profiles of orally administered drugs in rats on the basis of gastrointestinal transit kinetics and absorbability. *J. Pharm. Pharmacol.*, **49**, 450–457.

Schroeder, X. (1949) In J. Partington (ed), *Fundamental principles: the properties of gases*, Vol. 1 Longmans, Green, New York.

Shah, J. C., Chen, J. R. and Chow, D. (1989) Preformulation study of etoposide: identification of physicochemical characteristics responsible for the low and erratic oral bioavailability of etoposide. *Pharm. Res.*, **6**, 408–412.

Sinko, P. J. and Amidon, G. L. (1988) Characterization of the oral absorption of beta-lactam antibiotics. I. Cephalosporins: determination of intrinsic membrane absorption parameters in the rat intestine in situ. *Pharm. Res.*, **5**, 645–650.

Sinko, P. J., Leesman, G. D. and Amidon, G. L. (1991) Predicting fraction dose absorbed in humans using a macroscopic mass balance approach. *Pharm. Res.*, **8**, 979–988.

Stoll, B. R., Batycky, R. P., Leipold, H. R., Milstein, S. and Edwards, D. A. (2000) A theory of molecular absorption from the small intestine. *Chem. Eng. Sci.*, **55**, 473–489.

Suttle, A. B., Pollack, G. M. and Brouwer, K. L. (1992) Use of a pharmacokinetic model incorporating discontinuous gastrointestinal absorption to examine the occurrence of double peaks in oral concentration-time profiles. *Pharm. Res.*, **9**, 350–356.

Suzuki, A., Higuchi, W. I. and Ho, N. F. (1970a) Theoretical model studies of drug absorption and transport in the gastrointestinal tract. II. *J. Pharm. Sci.*, **59**, 651–659.

Suzuki, A., Higuchi, W. I. and Ho, N. F. (1970b) Theoretical model studies of drug absorption and transport in the gastrointestinal tract. I. *J. Pharm. Sci.*, **59**, 644–651.

Ungell, A. L., Nylander, S., Bergstrand, S., Sjoberg, A. and Lennernas, H. (1998) Membrane transport of drugs in different regions of the intestinal tract of the rat. *J. Pharm. Sci.*, **87**, 360–366.

Valvani, S. C., Yalkowsky, S. H. and Roseman, T. J. (1981) Solubility and partitioning IV: aqueous solubility and octanol-water partition coefficients of liquid nonelectrolytes. *J. Pharm. Sci.*, **70**, 502–507.

Weininger, D. (1988) SMILES, a chemical language and information system. 1. Introduction to methodology and encoding rules. *J. Chem. Inf. Comput. Sci.*, **28**, 31–36.

Wessel, M. D., Jurs, P. C., Tolan, J. W. and Muskal, S. M. (1998) Prediction of human intestinal absorption of drug compounds from molecular structure. *J. Chem. Inf. Comput. Sci.*, **38**, 726–735.

Winiwarter, S., Bonham, N. M., Ax, F., Hallberg, A., Lennernas, H. and Karlen, A. (1998) Correlation of human jejunal permeability (*in vivo*) of drugs with experimentally and theoretically derived parameters. A multivariate data analysis approach. *J. Med. Chem.*, **41**, 4939–4949.

Wright, J. D., Ma, T., Chu, C. K. and Boudinot, F. D. (1996) Discontinuous oral absorption pharmacokinetic model and bioavailability of 1-(2-fluoro-5-methyl-beta-L-arabinofurano-syl)uracil (L-FMAU) in rats. *Biopharm. Drug Dispos.*, **17**, 197–207.

Yalkowsky, S. H. and Banerjee, S. (1992) *Aqueous solubility. Methods of estimation for organic compounds*, Marcel Dekker, New York, USA.

Yalkowsky, S. H., Pinal, R. and Banerjee, S. (1988) Water solubility: a critique of the solvatochromic approach. *J. Pharm. Sci.*, **77**, 74–77.

Yu, L. (2001) eR&D the future of drug discovery and development. In G. L. Amidon (ed), *Oral Drug Discovery Workshop*, Garmish, Germany.

Yu, L. X. and Amidon, G. L. (1999) A compartmental absorption and transit model for estimating oral drug absorption. *Int. J. Pharm.*, **186**, 119–125.

Yu, L. X., Crison, J. R. and Amidon, G. L. (1996a) Compartmental transit and dispersion model analysis of small intestinal transit flow in humans. *Int. J. Pharm.*, **140**, 111–118.

Yu, L. X., Lipka, E., Crison, J. R. and Amidon, G. L. (1996b) Transport approaches to the biopharmaceutical design of oral drug delivery system: prediction of intestinal absorption. *Adv. Drug Deliv. Rev.*, **19**, 359–376.

Zheng, W., Cho, S. J., Waller, C. L. and Tropsha, A. (1999) Rational combinatorial library design. 3. Simulated annealing guided evaluation (SAGE) of molecular diversity: a novel computational tool for universal library design and database mining. *J. Chem. Inf. Comput. Sci.*, **39**, 738–746.

Chapter 22

# Confocal and two-photon fluorescence microscopy

*Alf Lamprecht and Claus-Michael Lehr*

## INTRODUCTION

The main advantage of a confocal laser scanning microscope, when compared with a conventional fluorescence microscope, is the former's ability to slice very clean thin optical sections out of thick fluorescent specimen, optically dissecting a specimen without any physical manipulation. Those sections can be assembled for 3D views at high resolution. The important technical difference between a conventional microscope and the confocal laser scanning microscope approach is the existence of two pinholes in the axis of light. In this way, light scattered from parts other than the illuminated point on the specimen is rejected from the optical system. The specimen is scanned through a moving point of light, and the variation of emitted light, modulated by the specimen, is captured by a photoelectric cell.

In general, the advantages of confocal imaging are a reduced blurring of the imaging from light scattering and an increased effective resolution. The method permits clear examination of thick light scattering objects if the sample material is sufficiently translucent. Moreover, xy-scans, as well as z-scans of the specimen are possible (Figure 22.1).

Using a technical trick, confocal microscopes allow high resolution images of a desired plane of the sample. First, a lens focuses the light which will be used for the excitation of the fluorescence (usually a laser beam); (1) into the opening of a pinhole (2) A mirror (3) directs the beam towards the objective (4) of the microscope, which bundles the beam to the desired depth of the sample (5) The emitted or rejected light is seized by the objective and, through the mirror which acts as beam separator, it is focused on the opening of a second pinhole (6) behind which the detector (7) is located. This crucial second pinhole shields the undesired scattered light, especially light from higher (red) or lower (orange) optical planes in the sample. Finally, the laser beam is moved very fast point by point and line by line over the sample, until the plane of interest is completely scanned. Images from confocal laser scanning microscopy (CLSM) are normally much sharper than non-confocal ones.

As already mentioned, the main principle of the confocal microscope is to focus the excitation light on a certain point in the sample. From this point the light is reflected through both the objective and the second pinhole to the detection system. 'Point' does not mean, in this case, that it is a spherical and homogenous

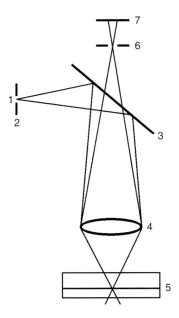

*Figure 22.1* Schematical figure of the principle of a confocal laser scanning microscope (CLSM) (1 = focusing lens and laser beam source; 2 = first pinhole; 3 = beam separating mirror; 4 = objective; 5 = sample; 6 = second pinhole; 7 = detector unit).

bright spot. Actually, it is a vertical ellipsoid depending on the wave nature of the light and is surrounded by bipolar zones of strongly reduced light intensity.

With a view to quantum mechanics, the distribution of intensity represents the probability for a photon at a certain distance from the focal point to be available for the excitation of a fluorescent molecule. This probability is quickly reduced from the nominal value of one if the distance from the focal point is increased. A photon which is emitted from the mentioned point also possesses a probability to be recorded from the detection system. If both the probability of elucidating the point in the sample and its probability to be detected is 0.1, then the total probability of this point being to be detected by the whole system is 0.01. Essentially, the distribution of illumination is squared. Subsequently, a narrow area restriction can be reached and nearly all light from outside of the focal point is suppressed.

Using the same wavelength of light and the same numerical aperture of the objective, the image resolution is about 40 per cent better than with a normal microscope. However, more interesting is the resolution in the vertical section. The restriction in depth of the elliptic light spot determines the diameter of the detected plane. As a value of resolution power, in confocal microscopy the distance from one point to another where the light intensity is diminished to 50% is used. Two points with a distance double this can be detected separately. With the best objectives optical plane thickness can be reduced down to 0.5 μm.

The most widely used visualization procedures are conventional light microscopy (LM) and scanning electron microscopy (SEM). Both techniques are very useful, although they both have certain limitations. LM is impeded by the scattering or emission of light from structures outside the optical focal plane, which reduces the image quality. On the other hand, SEM usually requires a relatively intensive sample pre-treatment and does not permit the visualization of internal structures of objects at all. Mechanical sections of embedded samples allow their inspection, but this risks building artifacts during the sample preparation.

CLSM allows the minimization of scattered light from out-of-focus structures, and permits the identification of several compounds by using different fluorescent labels. Moreover, CLSM allows visualization and characterization of structures not only on the surface, but also inside the sample, if the material is sufficiently transparent and can be labeled fluorescently.

The applications of CLSM for the characterization of cell culture systems vary and are mainly used for the detection and visualization of characteristic structures of the cells. Some cell structures can be stained with fluorescent dyes by their intracellular accumulation at these regions. Cellular structures that cannot be reached by simple incubation with those dyes mentioned above can be detected by fluorescently marked antibodies or lectins which specifically bind thereon and are imaged by CLSM afterwards.

In addition to biological characterization CLSM is also used for the visualization of many kinds of drug delivery systems. The distribution of fluorescently marked carriers, e.g. liposomes, micro- or nanoparticles, in or on the cell layer can be qualitatively characterized by CLSM. The visualization technique is also applied to the characterization of internal structures of solid dosage forms, tablets, pellets and microparticles. Moreover, its use is broadened to the field of drug delivery for the visualization and quantification of drug release kinetics of several drug delivery devices.

## SPECIFIC APPLICATIONS

### Characterization of cell culture or *ex vivo* tissue

Isolated cells are often used as *in vitro* test systems to study the effect of compounds on cell biology. It has become possible to study biochemical parameters, for example, ion concentrations, pH, membrane potentials, etc., in living cells. Such study is due to the development of sensitive cameras, efficient computer systems, and commercially available fluorescent probes which can translate the intensity of a specific cellular parameter into a fluorescent signal. Moreover, it is possible to determine and correlate multiple effects in one cell. Therefore, many different fluorescent probes are available for several applications (Table 22.1).

CLSM, in combination with data restoration, has been successfully applied to distinguish between mono- and multilayer formation in (Caco-2) human intestinal epithelial cell cultures under various culture conditions (Rothen-Rutishauser *et al.*, 2000). The method of staining is mainly used for the detection of characteristic structures of the cells. Staining of certain cell structures with fluorescent dyes reveals the specific intracellular distribution, based on their solubility properties.

*Table 22.1* Examples of fluorescent probes as indicators used in studies with living cells

| Structure of interest | | Fluorescent marker |
|---|---|---|
| $Ca^{2+}$, $Na^+$, $K^+$, $Mg^{2+}$, $Cl^-$, $Zn^{2+}$ | | Quin-2, Fura-2, Indo-2, Fluo-3 |
| Thiolgroups | | Bimanes |
| | | 5-chloromethylfluorescein-diacetate |
| Membrane potentials | | Rhodamine 123, tetramethylrhodamine |
| Lipid peroxidation | | Dihydrofluoresceins and dihydrorhodamines |
| Viability | | Propidium iodide, calceine |
| Localization of organelles | Acidic organelles | Neutral red |
| | E.R. | Carbocyanines |
| | Golgi | Ceramides |
| | Mitochondria | Rhodamines |
| | Nucleus | Acridines, propidium iodide |
| | Membranes | Fluorescent phospholipids, Dil |

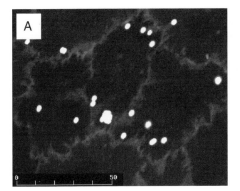 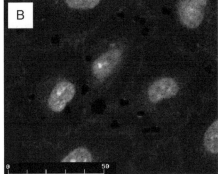

*Figure 22.2* Confocal laser scanning microscope (CLSM) images of human vascular endothelial cells in primary cell culture for the evaluation of nanoparticle bioadhesion. The green channel shown on image (A) represents the fluorescein-marked microparticles and the cell membranes after labeling, while image (B) represents the red channel with particles and the cell nucleus stained with propidium iodide.

Other cellular structures are reached by incubation with fluorescently marked antibodies or lectins which specifically bind to them (Figure 22.2).

The 3D analysis of confocal data sets, in particular the reconstruction of the z-axis, gives important clues on the characteristics of the cell layer. Although phase contrast microscopy, is in general, an important tool in cell culture, it is of limited use as soon as confluence is reached and thicker layers have to be investigated. In the past, electron microscopy has often been applied to study cell layers, but this approach can be insufficient if specific labeling or detection inside of the cells is required. CLSM, in combination with immunofluorescence methods, provides an important tool in the effort to optimize cell culture models, as several fluorescent markers can be used simultaneously.

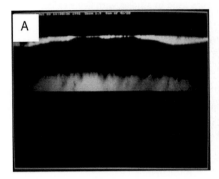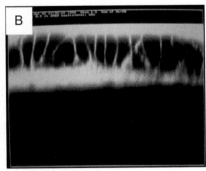

*Figure 22.3* Transport studies on Caco-2 cell culture in order to visualize the intact barrier against the model drug fluorescein isothiocyanate (FITC)-dextran (A) and the enhanced paracellular transport of FITC-dextran after adding a mucoadhesive polymer (B). Huruni *et al.*, 1993)

## Localization of drugs or drug delivery systems

CLSM can be used to visualize the transport pathway across the cell monolayers. Optical cross-sections of cell monolayers or tissues allow observations about drug transport (requiring the fluorescence labeling of the drug) across the cell layer. By this method a potential transport of the drug crossing the cell barrier can be visualized and, moreover, paracellular transport can be distinguished from the transcellular (Borchard *et al.*, 1996). Figure 22.3 demonstrates the opening of usually tight intercellular junctions between Caco-2 under the influence of some mucoadhesive polymer.

CLSM imaging was applied to the determination of pH value in the skin by using a pH-sensitive fluorescent dye (Turner *et al.*, 1998). By incubating the skin with the water-soluble carboxy seminaphthorodafluor (SNARF)-1 the fluorescent marker penetrates the skin and can be detected after cross-sectioning the tissue. This assay is rather unsusceptible, because the pK of the fluorescent marker makes it insufficiently sensitive to the required normal pH gradient. However, for a qualitative characterization it seems to be useful. If the drug carrier system is in a certain size range (0.5–100 µm) then there exists the option to depit and localize it inside the surrounding cell culture system or tissue section.

## Characterization of drug delivery systems

Recently, CLSM was applied as a non-invasive technique producing 3D images of the internal structure of tablets. CLSM can be used in characterizing the behavior and deformation of drug particles and excipients under compression forces. The compression behavior of both fluorescent drug and non-fluorescent excipient can be visualized simultaneously by using fluorescence and reflection modes (Guo *et al.*, 1999).

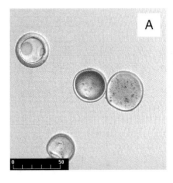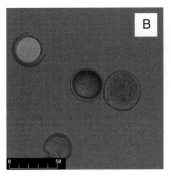

Figure 22.4 Visualization of microcapsules containing a nile red stained oil phase by a light microscopy image (A) and by confocal laser scanning microscopy (CLSM) using the red
fluorescence channel (B). The fluorescence signal allows the oil-containing and air-
containing microcapsules to be distinguished unambiguously. (See Color plate 16)

The characterization methods for microparticles are numerous and they reveal
many characteristics of the particulate system. Since CLSM allows the minimization of scattered light from out-of-focus structures and permits the identification
of several compounds by using different fluorescence labels, its use seems to be of
interest for the characterization of microparticles as well. Moreover, CLSM allows
visualization and characterization of structures not only on the surface, but also
inside the particles, provided the material is sufficiently transparent and can be
fluorescently labeled.

Due to these advantages CLSM has already been applied for the evaluation/
characterization of solid pharmaceutical formulations; this includes the determination
of the release kinetics of the entrapped drugs (Cutts *et al.*, 1996) and the examination
of the swelling of microparticles by using fluorescein-containing aqueous media
(Adler *et al.*, 1999; Bouillot *et al.*, 1999). Furthermore, microparticles were visualized
in order to depict both it the dispersion of the entrapped phase (Rojas *et al.*, 1999)
and polymer structures on the surface (Caponetti *et al.*, 1999).

For example, structural analysis has been performed for the complex coacervation process (Lamprecht *et al.*, 2000a). Staining the oil phase with nile red
distinguishes between encapsulated oil and other droplet-like structures, which by
their shape could be mistaken as oil phase (Figure 22.4).

Using the confocal approach it is possible to optically dissect the particle in any
desired number of coplanar cross-sections. This optical dissection shows the internal
structure of the fluorescently labeled internal polymeric wall component. Such data
can also be used to reconstruct the complete structure of an entire microparticle
(Figure 22.5).

## Applications for kinetic measurements

The CLSM method has been used widely in cell biology for its ability to produce
images of high resolution that are free from out-of-focus light. This ability to

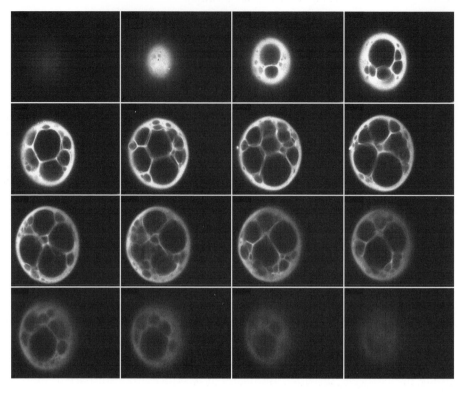

*Figure 22.5* 3D cross-sectioning through a microparticle prepared by complex coacervation after covalent labeling of gelatin with rhodamine-B-isothiocyanate.

produce non-invasive optical sections through translucent materials gives CLSM considerable potential for investigating the changing structure and following dynamic processes within pharmaceutical drug delivery systems (Cutts *et al.*, 1996). In this work, a new application for the use of CLSM was given. An experimental technique using CLSM and a mathematical model have been developed to describe drug release from non-swelling, non-eroding spherical matrices. The technique enabled the direct visualization of changes occurring within the drug carrier during hydration.

Fluorescence recovery after photobleaching (FRAP) offers the possibility of microscopically examining a sample and getting information on molecular motion in a special part of this sample (De Smedt *et al.*, 1999; Meyvis *et al.*, 1999). Three main advantages of this technique have been reported: speed of experiment, high resolution and the possibility of measuring intact samples both *in vitro* and *in vivo*. However, in contrast to its great versatility, the use of FRAP is not widespread. Some possible future applications may include the analysis of mobility of drugs in pharmaceutical dosage forms, mobility and binding of drugs to tissues, etc. The most promising evolution may be FRAP in combination with two-photon microscopy and the appropriate model for 3D diffusion.

## Two-photon fluorescence microscope

Molecular excitation by two-photon absorption holds great promise for vital imaging of biological systems using laser scanning microscopy (Denk *et al.*, 1995). Fluorescence microscopy, which can provide submicron spatial resolution of chemical dynamics within living cells, is frequently limited in its sensitivity and spatial resolution by out-of-focus background fluorescence. Two-photon excitation avoids this background in laser scanning microscopy by virtue of its non-linear optical absorption character, which almost completely limits the excitation to the high-intensity region at the focal point of the strongly focused excitation laser. Since excitation of background fluorescence is avoided, no confocal spatial filter is required, and all of the advantages of a linear (one-photon) confocal microscope are obtained, including the absence of out-of-focus photobleaching and photo-damage.

Two-photon excitation arises from the simultaneous absorption of two photons that combine their energies to cause the transition to the excited state of the chromophore. For example, two photons of red light can excite an ultraviolet (UV)-absorbing fluorophore, which then emits blue fluorescence. Because two-photon absorption requires two photons for each excitation, its rate depends on the square of the instantaneous intensity, just as the rate of a chemical reaction, $2A + B = C$, varies with the square of the concentration of A.

The set-up and operation of a two-photon laser scanning microscope (LSM) is very similar to that of a one-photon LSM. The main differences are in the type of laser that is used and in the increased number of options for detection. The two most commonly used lasers are the Ti:Sapphire laser (700 to above 1050 nm) and the Hybrid mode-locked laser (550 to above 700 nm).

The application of a two-photon LSM to answer biological questions has been demonstrated for several cases. Free calcium concentrations were measured as a function of space and time (4D imaging) using the indicator Indo-1 in motile cells and cardiac myocytes (Piston and Webb, 1991; Piston *et al.*, 1994). Moreover, concentrations of NAD(P)H, an important indicator of cellular energetics, were imaged using its intrinsic fluorescence (Eng *et al.*, 1990; Piston *et al.*, 1994). It appears, however, that considerable photodamage occurs even for two-photon excitation.

Two-photon excitation microscopy has been found to work very well in laser scanning microscopy for dynamic imaging of molecular distributions in living cells. It has been particularly successful in systems where imaging of ion or other chemical indicators requires UV excitation energies.

Phototoxicity in cells is poorly understood, in general, and for ultrashort pulse illumination, in particular. There is always some cell damage associated with fluorescence microscopy. However, due to its lower quantum energy and due to its often lower intrinsic absorption, red and infrared light is ordinarily found to be less biologically deleterious. But two-photon absorption of red light does occur for UV-absorbing molecules, including many intrinsic chromophores. In addition to fluorophore-induced effects, damage can then arise from chemical interactions between the sample and optically excited intrinsic chromophores, whose existence at significant concentrations is illustrated by two-photon excitation of NADH.

*Table 22.2* Excitation wavelengths of the different lasers

| Laser type | Excitation wavelengths (nm) |
| --- | --- |
| Argon Ion | 488 514 |
| Green HeNe | 543 |
| Krypton/argon mixed gas | 488 568 647 |
| HeCd RGB | 442 534 538 636 |

## MATERIALS AND METHODS

There are two different types of microscopes normally used, the upright and the inverted microscope. An inverted system is more advisable for applications in cell culturing since those systems allow a narrowing of the objective to the cell culture flask from downside. The usual magnification of the objectives is in between ×2.5 and ×100; higher than ×40 objective is usually performed with oil immersion.

There are also several types of laser systems available, each with a different specificity for the imaging procedure (Table 22.2).

Normally the laser is adjusted in a certain fluorescence mode which yields one, two, or more excitation wavelengths. Green, red, and sometimes blue fluorescence images can be obtained from separate channels, and another picture from the transmitted light detector might be optional if necessary. The final pictures were composed from the different channels to visualize the marked structures and also the signal from the ordinary transmission light microscope within the same image.

The next step is to find adequate fluorophores, the applicability of which essentially depends on the type of laser. One prerequisite is that the product's extinction time quantum yield is high. For a double or triple labeling a possible interference of the fluorescent probes has to be kept in mind. Moreover, the high extinction coefficients of probes (e.g. phycocyanines) are due to both their high molecular weight and the large number of fluorophores per molecule; their size may make them difficult to be introduced into a cell. Here, some frequently used dyes are given in Table 22.3 as an example (Haugland, 1996; Tsien and Waggoner, 1995).

If cell compartments, drugs, or polymer and fluorescence markers have similar solubility properties, a simple addition to the sample, or prior mixing of the marker and drug or polymer, may be sufficient, as long as any process will not be altered by the marker.

This is the favorable method if the fluorescent marker is simply accumulating exclusively in the right region of the cell or covalent labeling is not possible. For example, the covalent labeling of polymers can be inefficient owing to a lack of functional groups as observed for polyesters.

In case of different solubilities of marker and drug, polymer, or any cell structure, covalent coupling is necessary to ensure that the fluorescence marker represents the desired material. In fluorescent marking of cell structures there is one link needed between the fluorescent dye and the cell structure to be detected. Usually this is a specifically binding antibody or lectin. Most of these 'targeters' are nowadays commercially available. If this is not the case, fluorescent labeling has to be done 'in house'.

*Table 22.3* Different fluorophores with their excitation and emission wavelengths

| Fluorophore | Excitation maximum (nm) | Emission maximum (nm) |
|---|---|---|
| Fura-2 low $Ca^{2+}$ | 335 | 512–518 |
| high $Ca^{2+}$ | 360 | 505–510 |
| Indo-1 low $Ca^{2+}$ | 330 | 390–410 |
| high $Ca^{2+}$ | 350 | 482–485 |
| Fluorescein | 496 | 518 |
| Bodipy | 503 | 511 |
| Lucifer yellow | 428 | 533 |
| SNARF-1 | pH 5.5: 518, 548 | pH: 5.5: 587 |
| | pH 10: 574 | pH 10: 636 |
| Propidium iodide | 536 | 623 |
| Tetramethylrhodamine | 554 | 576 |
| CY 3.18 | 554 | 568 |
| Rhodamine B | 572 | 590 |
| Texas red | 592 | 610 |
| Allophycocyanine | 621, 650 | 661 |
| CY 5.18 | 649 | 666 |

Based on a method of protein labeling by Schreiber and Haimovich (1983), the covalent labeling protocol has to be adopted to the individual requirements. Here, we give an example for a protein labeling by fluorescein isothiocyanate (FITC): An aqueous protein or polymer solution is adjusted at pH 8.5 by sodium carbonate solution. The fluorescent dye should be dissolved in dimethyl sulfoxide (DMSO) at a concentration of 1 mg/ml. Ten–100 µl of the dye solution will be added to the polymer solution and stirred for one–four hours at 4 °C. The reaction will be stopped by adding ethanolamine, and free FITC can be removed by dialysis (dialysis tube with a 10,000 Da pore size) against distilled water or by any type of preparative chromatography.

In cell culture experiments the fixation of the cells is required if further staining of certain cell structures is planned. Therefore, cells are usually washed with buffer and fixed with formaldehyde at room temperature. The fixation time and the formaldehyde concentration vary according to the literature (Rothen-Rutishauser *et al.*, 1998; Zuidam *et al.*, 2000). If a fluorescent probe is to be detected from the cells, it is transferred onto a glass slide, preferably with some viscosity-increasing agent such as polyvinyl alcohol or equivalent. In the case where, immunofluorescent staining is desired, cells have to be permeabilized with agents like Triton X-100 in phosphate buffer in concentrations varying between 0.1 and 0.5 per cent. Thereafter, the cell sample may be incubated with the fluorescently marked antibodies.

The standard set-up for a transport study requires a fluorescently marked model compound. Since the transport of macromolecules is of high interest, a majority of these model compounds consist of fluorescein-labeled dextrans of different molecular weight, which are commercially available. Thus, the transport pathway of these fluorescent markers across the cell layer can be visualized with CLSM. The cell layer may be incubated with some compound of interest, e.g. absorption enhancer, which may influence the absorption behavior of the cells. The transport filters on which the cells are seeded are prepared by varying techniques, but generally the filters are carefully removed from the plastic inserts of the Transwell system, placed between glass coverslips and mounted on a heatable microscopic stage (37 °C). In order to ascertain the viability of the cells, the images should be taken within the first 15 minutes after removal from the transwell system (Kotzé *et al.*, 1998).

*Ex vivo* samples are usually prepared after their experimental treatment with the fluorescent probe, either by embedding the tissue in, for example, paraffin and afterwards sectioning with a microtome or by directly cryosectioning followed by specific sample treatment (Torché *et al.*, 2000). All these sections are normally performed vertically across the tissue sample. An x-y optical section of the tissue below the plane of the cut is advisable to avoid interference by fluorescence signals from damaged cells (Hoogstraate *et al.*, 1994). This procedure gives a true estimation of the distribution pattern of the fluorescent probe, without the attenuation that was evident in the optical sections which were made parallel to the surface in the intact specimen.

The imaging of solid dosage forms, such as microspheres, can usually be performed very easily with dry particles. Since aggregates may disturb the image analysis, particle dispersions are sometimes preferable. In this case, a dispersion in neutral oil is advised in order to avoid particle swelling during sample imaging.

## PROBLEM SOURCES

A limiting factor of CLSM is the sample translucency. Since tissue samples or cell layer density is not very high, sample translucency is not generally problematic in cell culturing. However, if larger tissue samples are analyzed, sectioning by a microtome prior to the imaging procedure is advisable. For example, in deeper cross-sections (>50 μm) the images still show the general structure through microparticles but the intensity of the signal is lowered and the images become blurred.

It must be kept in mind, that non-fluorescent material cannot be detected by CLSM without prior fluorescence labeling. If drugs or polymers and fluorescence markers have similar solubility properties, the mixing of markers and drugs or polymers may be sufficient, as long as any following processes will not be altered by the markers. This is the favored method if covalent labeling of macromolecules is inefficient, as observed for polyesters (Lamprecht *et al.*, 2000b). In the case of different solubilities of markers and drugs, covalent coupling is unavoidable to ensure that the fluorescence marker represents the desired material.

But if some covalent coupling of markers and drugs or polymers is required, it has to be done with caution, as the physicochemical properties may be considerably changed (especially in the case of lower molecular weight drugs). First, the concentrations of the fluorescent markers have to be adopted to their stability in the laser light (photobleaching). Moreover, in some cases, there is a chance that functional groups of the proteins or polymers could be blocked after the fluorescence labeling step, thereby changing the properties of the proteins or polymers. However, isoelectric focusing may analyze any change of the isoelectric point for these compounds.

It must be noticed that an interference of the two fluorescence marker signals cannot be excluded. For example, the emitted signal from fluorescein-labeled compounds may contribute to the excitation of rhodamine and its derivatives. This double labeling system cannot be used for quantitative determinations of the fluorescently labeled polymers. Therefore, simultaneous visualization of the two fluorescein- and rhodamine-labeled structures has to be proven by a single labeling of each polymer and separate visualization. Changing the type of laser light, e.g. argon-krypton laser and using a more suitable set of non-interfering fluorescent probes, should facilitate both imaging and a quantitative determination of different structures simultaneously.

Among the most commonly applied fluorochromes are fluorescein, rhodamine and coumarin derivatives. They provide both the three primary colors of the visible part of the electromagnetic spectrum and ample possibilities for multicolor studies (Lengauer *et al.*, 1993; Nederlof *et al.*, 1990). The minimal amount of target-bound fluorochrome that can be microscopically detected is dependent on the microscope set-up, as well as the absorption coefficient and quantum efficiency of the fluorophore. In addition, the fading properties of the fluorochrome have major impact on the applicability because extended periods of excitation time are necessary for visual evaluation and photography or image acquisition. Fading of fluorochromes upon excitation is a photochemical process. Light-induced damage to the fluorochrome is prominent in the presence of oxygen (Giloh and Sedat, 1982). Also, non-oxygen-mediated radical generation has been indicated as a source of photo chemical fluorochrome destruction (Johnson *et al.*, 1982). Consequently, the fading characteristics can be influenced by adding compounds to the embedding media, such as antioxidants and radical scavengers, which interfere with the photochemical reaction in such a way that the excited fluorochrome will not be (irreversibly) damaged. *p*-phenylenediamine (Pd) and 1,4-diaza-bicyclo-(2,2,2)-octane (DABCO), both antioxidents, are two well-known additives that considerably retard fading (Johnson *et al.*, 1982; Krenik *et al.*, 1989).

## CONCLUSIONS

Nowadays, CLSM is a very helpful characterization tool for both cell culture or drug delivery systems. Using this relatively simple method, structures can be visualized and unambiguously identified within the sample. Additional information on sample composition and distribution of any structure of interest throughout the particle wall can be obtained by CLSM. Furthermore, compounds can be localized three-dimensionally inside or on the surface of the sample after fluorescence labeling. In general, CLSM appears to be a very powerful technique which is easy to use and provides additional information when compared with other microscopical inspection methods.

## REFERENCES

Adler, J., Jayan, A. and Melia, C. D. (1999) A method for quantifying differential expansion within hydrating hydrophilic matrixes by tracking embedded fluorescent microspheres. *J. Pharm. Sci.*, **88**, 371–377.

Borchard, G., Lueßen, H. L., de Boer, A. G., Verhoef, J. C., Lehr, C. M. and Junginger, H. E. (1996) The potential of mucoadhesive polymers in enhancing intestinal peptide drug absorption. III: Effects of chitosan-glutamate and carbomer on epithelial tight junctions *in vitro*. *J. Control. Release*, **39**, 131–138.

Bouillot, P., Babak, V. and Dellacherie, E. (1999) Novel bioresorbable and bioeliminable surfactants for microsphere preparation. *Pharm. Res.*, **16**, 148–154.

Caponetti, G., Hrkach, J. S., Kriwet, B., Poh, M., Lotan, N., Colombo, P. *et al.* (1999) Microparticles of novel branched copolymers of lactic acid and amino acids: preparation and characterization. *J. Pharm. Sci.*, **88**, 136–141.

Cutts, L. S., Hibberd, S., Adler, J., Davies, M. C. and Melia, C. D. (1996) Characterising drug release processes within controlled release dosage forms using the confocal laser scanning microscope. *J. Control. Release*, **42**, 115–124.

De Smedt, S. C., Meyvis, T. K. L., Van Oostveldt, P. and Demeester, J. (1999) A new microphotolysis based approach for mapping the mobility of drugs in microscopic drug delivery devices. *Pharm. Res.*, **16**, 1639–1642.

Denk, D., Piston, D. W. and Webb, W. W. (1995) Two-photon molecular excitation in laser-scanning microscopy. In J. B. Pawley (ed), *Handbook of Biological Confocal Microscopy*, New York, Plenum Publishing Corp.

Eng, J., Lynch, R. M. and Balban, R. S. (1990) Two-photon laser scanning fluorescence microscopy, *Science*, **248**, 73–76.

Giloh, H. and Sedat, J. W. (1982) Fluorescence microscopy: reduced photobleaching of rhodamine and fluorescein protein conjugates by n-propyl gallate. *Science*, **217**, 1252–1255.

Guo, H. X., Heinamaki, J. and Yliruusi, J. (1999) Characterization of particle deformation during compression measured by confocal laser scanning microscopy. *Int. J. Pharm.*, **186**, 99–108.

Haugland, R. P. (1996) *Handbook of Fluorescent Probes and Research Chemicals*, Eugene, USA, Molecular Probes Inc.

Hoogstraate, A. J., Cullander, C., Nagelkerke, J. F., Senel, S., Verhoef, J. C., Junginger, H. E. *et al.* (1994) Diffusion rates and transport pathways of fluorescein isothiocyanate (FITC)-labeled model compounds through buccal epithelium. *Pharm. Res.*, **11**, 83–89.

Hurni, M. A., Noach, A. B., Blom-Roosemalen, M. C., de Boer, A. G., Nagelkerke, J. F. and Breimer, D. D. (1993) Permeability enhancement in Caco-2 cell monolayers by sodium salicylate and sodium taurodihydrofusidate: assessment of effect-reversibility and imaging of transepithelial transport routes by confocal laser scanning microscopy. *J. Pharm. Exp. Ther.*, **267**, 942–950.

Johnson, G. D., Davidson, R. S., McNamee, K. C., Russell, G., Goodwin, D. and Holborrow, E. L. (1982) Fading of immunofluorescence during microscopy: a study of its phenomenon and its remedy. *J. Immunol. Methods*, **55**, 231–242.

Kotzé, A. F., Luessen, H. L., de Leeuw, B. J., de Boer, A. G., Verhoef, J. C. and Junginger, H. E. (1998) Comparison of the effect of different chitosan salts and N-trimethyl chitosan chloride on the permeability of intestinal epithelial cells (Caco-2). *J. Control. Release*, **51**, 35–46.

Krenik, K. D., Kephart, G. M., Offord, K. P., Dunette, S. L. and Gleich, G. J. (1989) Comparison of antifading agents used in immunofluorescence. *J. Immunol. Methods*, **117**, 91–97.

Lamprecht, A., Schäfer, U. and Lehr, C. M. (2000a) Characterisation of microcapsules by confocal laser scanning microscopy: analysis of structure, capsule wall composition and of encapsulation rate. *Eur. J. Pharm. Biopharm.*, **49**, 1–9.

Lamprecht, A., Schäfer, U. and Lehr, C. M. (2000b) Structural analysis of microparticles by confocal laser scanning microscopy. *AAPS PharmSciTech.*, **1**, 3.

Lengauer, C., Speicher, M. R., Popp, S., Jauch, A., Taniwaki, M., Nagaraja, R. *et al.* (1993) Chromosomal bar codes produced by multicolor fluorescence in situ hybridization with multiple YAC clones and whole chromosome painting probes. *Human Mol. Genetics*, **2**, 505–512.

Meyvis, T. K., De Smedt, S. C., Van Oostveldt, P. and Demeester, J. (1999) Fluorescence recovery after photobleaching: a versatile tool for mobility and interaction measurements in pharmaceutical research. *Pharm. Res.*, **16**, 1153–1162.

Nederlof, P. M., van der Flier, S., Wiegant, J., Raap, A. K., Tanke, H. J., Ploem, J. S. and Ploeg, M. (1990) Multiple fluorescence in situ hybridization. *Cytometry*, **11**, 126–131.

Piston, D. W. and Webb W. W. (1991) Three dimensional imaging of intracellular calcium activity using two-photon excitation of the fluorescent indicator dye Indo-1. *Biophys. Journal*, **59**, 156–163.

Piston, D. W., Kirby, M. S., Cheng, H. and Lederer W. J. (1994) Two-photon excitation fluorescence imaging of three-dimensional calcium-ion activity. *Appl. Opt.*, **33**, 662–669.

Rojas, J., Pinto-Alphandary, H., Leo, E., Pecquet, S., Couvreur, P., Gulik, A. *et al.* (1999) A polysorbate-based non-ionic surfactant can modulate loading and release of beta-lactoglobulin entrapped in multiphase poly(DL-lactide-co-glycolide) microspheres. *Pharm. Res.*, **16**, 255–260.

Rothen-Rutishauser, B., Kramer, S. D., Braun, A., Gunthert, M. and Wunderli-Allenspach, H. (1998) MDCK cell cultures as an epithelial *in vitro* model: cytoskeleton and tight junctions as indicators for the definition of age-related stages by confocal microscopy. *Pharm. Res.*, **15**, 964–971.

Rothen-Rutishauser, B., Braun, A., Gunthert, M. and Wunderli-Allenspach, H. (2000) Formation of multilayers in the caco-2 cell culture model: a confocal laser scanning microscopy study. *Pharm. Res.*, **17**, 460–465.

Schreiber, A. B. and Haimovich, J. (1983) Quantitative fluorometric assay for detection and characterization of Fc receptors. *Methods Enzymol.*, **93**, 147–155.

Torché, A. M., Jouan, H., Le Corre, P., Albina, E., Primault, R., Jestin, A. *et al.* (2000) Ex-vivo and in situ PLGA microspheres uptake by pig ileal Peyer's patch segment. *Int. J. Pharm.*, **201**, 15–27.

Tsien, R. Y. and Waggoner, A. (1995) Fluorophores for confocal microscopy. In J. B. Pawley (ed), *Handbook of Biological Confocal Microscopy*, New York, Plenum Publishing Corp.

Turner, N. G., Cullander, C. and Guy, R. H. (1998) Determination of the pH gradient across the stratum corneum. *J. Invest. Dermatol. Symp. Proc.*, **3**, 110–113.

Zuidam, N. J., Posthuma, G., de Vries, E. T., Crommelin, D. J., Hennink, W. E. and Storm, G. (2000) Effects of physicochemical characteristics of poly(2-(dimethylamino)ethyl methacrylate)-based polyplexes on cellular association and internalization. *J. Drug Target.*, **8**, 51–66.

# Chapter 23

# On the application of scanning force microscopy in (cell) biology

*U. Bakowsky, C. Kneuer, V. Oberle, H. Bakowsky,*
*U. Rothe, I. Zuhorn and D. Hoekstra*

## INTRODUCTION

In many scientific fields, the visualization of surface morphologies is used to obtain insight that will allow an understanding of physical, chemical and biological phenomena within the area of interest. For many years, optical microscopy has been used as a tool to produce images of surfaces in the micrometer dimension. By light microscopy, it is possible to measure the size of features in the x and y plane of a sample surface on a micrometer scale. However, in general, measurements in the z-direction will not be possible. The resolution that can be achieved by traditional light microscopes is limited to approximately one micrometer by the Nyquist relation.

Developed in the 1940's, the next most widely used instrument for investigating surface morphology has been the scanning electron microscope (SEM). By this technique only the near surface of samples can be visualized, while it does not allow conclusions about the optical properties of an object. Similar to an analysis by optical microscopy, SEM only measures in the x and y dimension of a sample, and insight into the z-direction is not obtained. With today's general purpose SEMs, resolution is limited to about five nanometers owing to the properties of the electromagnetic lenses. However, this resolution can only be achieved under vacuum conditions. Furthermore, a rather laborious sample preparation is often required for SEM, as it may include steps such as freeze drying, staining or metal coating. Both sample preparation and the vacuum can produce a number of artifacts, and the structure that is imaged may be far from its native state. This is especially the case for normally hydrated biological samples. However, novel technology has become available, like environmental scanning electron microscopy (ESEM), that allows imaging under ambient conditions, though at the expense of a lower resolution far from the molecular level.

Scanning probe microscopes (SPM) are instruments used for imaging and measuring the properties of surfaces. In contrast to optical microscopy and SEM, in this case it is possible to characterize the morphology of surfaces in all three dimensions (x, y and z). The result of scanning probe microscopy will be a topographic image of the sample surface. As for SEM, the x- and y-resolution is in the order of some nanometers. With the optimal choice of equipment, resolutions in the range between 0.1 nm and 2.0 nm for the x-y dimension, and less than 0.1 nm for the z-dimension can routinely be obtained.

## Scanning tunneling microscopy

The first SPM technique, as developed by Binnig, Rohrer and coworkers in the early 1980s was the so-called scanning tunneling microscope (STM). This technique relies on the mechanical scanning of a sample with a sensor tip. Both tip and sample should display electrical conductivity. Accordingly, when the tip approaches the sample surface, a tunneling current will start to flow. This current is proportional to the distance between tip and sample surface and a topographic image can be reconstructed from the collected data. The STM technique allows imaging of solid conductive samples with atomic resolution. All other SPM methods are based on the same principle: a sensor tip is moved over a sample surface and the resulting interactions are monitored.

## SCANNING FORCE MICROSCOPY

## Introduction

Scanning force microscopy (SFM), also known as atomic force microscopy (AFM), was developed by Binnig and coworkers in 1986. The AFM allows the imaging of conducting as well as non-conducting samples. High lateral and vertical resolutions can be achieved in vacuum, in air, and even on liquid covered surfaces. In addition, with this microscopical approach, it is not only possible to analyze the topography of a sample, but also other physical properties, including friction forces, softness and viscoelasticity, and the charge density on a nanometer scale. All these options have made this instrument a very useful tool in many biological applications. Thus, large objects, such as whole cells can be imaged, as well as smaller structures, such as chromosomes, membranes, proteins and nucleic acids. Consequently, in this rapidly growing area of research, a vast amount of literature has been published in recent years, and some excellent reviews can be found elsewhere (Hansma *et al.*, 1997, 2000; Meyer and Heinzelmann, 1992; Morris, 1994).

## The principle of SFM

The main element of every scanning force microscope is the tip-cantilever system (Figure 23.1). The cantilever can be made of different materials, usually silicon or siliconnitride. The choice of material depends on the specific application. A sharp tip is mounted on one end of a 100–500 µm long lever. The geometry of this tip is crucial, as it represents one of the major parameters that determine the resolution of a measurement. The highest resolution can be achieved with a tip that mounts on a single atom. Imaging of such an AFM tip reveals it to be spherically shaped with a diameter approximately 5 nm.

   In order to be moved over the sample surface, the tip-cantilever system is integrated into a stylus profilometer. Such equipment has been used in the optics industry for many years. In essence, a stylus profilometer is a piezoelectric scanner that can generate movements with the accuracy and magnitude required to generate precise topographic images. The movement of the piezoelectric elements can be controlled by the voltage that is applied across its electrodes. Depending on the

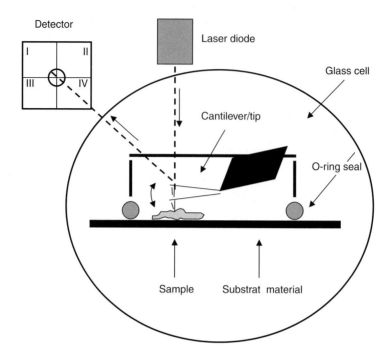

*Figure 23.1* Schematic diagram of a scanning force microscope (SFM) equipped with a liquid cell. This type of instrumentation is used in the MultiMode and Bioscope technique from Digital Instruments (Mannheim, Germany).

design of the SPM, the scanner is used either to move the sample underneath the cantilever or to move the cantilever over the sample. This feature of the equipment sets a limiting factor for the maximum scanning field. Hence, distances of about 125 microns in the *x*- and *y*-direction and ten microns in the *z*-direction can be covered. This range is sufficient for many but not all biological purposes.

When the cantilever is moved over a structured surface, the attractive and repulsive forces between tip and surface will change. These forces are measured by sensing the deflections of the cantilever. Deflections of the cantilever can be detected by a variety of methods (Meyer and Heinzelmann, 1992). In both past and present SFM applications, the cantilever deflection is sensed by electron interferometry, optical deflection or capacity methods.

## Techniques in SFM

In SFM, there are a number of different scanning modes. They primarily differ in the type of interaction between tip and sample that is measured, but also, at least in some cases, in the way the tip is moved across the surface. Commercially available microscopes can perform contact AFM, TappingMode™ AFM, non-contact AFM, lateral force microscopy (LFM), magnetic force microscopy (MFM), electric force microscopy (EFM), phase imaging, nanoindenting/scratching, scanning capacitance microscopy (SCM), scanning thermal microscopy (SThM) and force spectroscopy. The most relevant techniques for biological applications are contact AFM, TappingMode™ AFM, LFM and force spectroscopy.

### Contact AFM

With this technique, the tip is always in contact with the surface and the cantilever scans line by line across the sample. Generally, it is possible to scan in either the constant force modus or in the constant height modus. With the second, more widely distributed mode, the probe is positioned close to the surface during the scanning process and the cantilever is held in a fixed position during the scan cycle. Accordingly, the topography of the sample results in deflections of the cantilever which are detected and amplified in order to produce an image. This image shows a map of tip–sample interactions resulting from the interatomic repulsive forces between these two surfaces. To prevent damages of the sample, the force between the probe and the surface has to be kept at a minimum. Furthermore, this method is limited by additional interaction forces, such as friction, adhesion, electrostatic forces and other difficulties like liquid layers. The surface of the sample can be damaged, and the possibility of monitoring artifacts has to be seriously taken into account, especially when scanning soft material.

### TappingMode™ AFM

TappingMode™ AFM imaging is a key advance in the SFM of soft, adhesive or fragile samples, i.e. biological materials. This scanning mode overcomes many of the problems encountered in the contact mode and allows the imaging of soft samples with high resolution without damaging the sample. This technique can be used on air interfaces and also on liquid covered surfaces.

For TappingMode™ the cantilever has to be oscillated close to its own resonance frequency. This oscillation can usually be achieved with a piezoelectric element. The piezo motion causes the cantilever to oscillate with a high amplitude, typically more than 20 nm, when the tip is not in contact with the surface. The oscillating tip is then moved toward the surface until it senses or 'taps' the surface. This contact of the tip with the sample causes a change in the oscillation amplitude which is proportional to the amount of energy lost during the contact. This reduction in oscillation amplitude is used to identify and measure surface features.

### Force techniques

The SFM can be used to measure forces between the tip and the sample surface and to generate force–distance curves. A variety of forces can be detected. These include coulombic forces, van der Waals forces and electrostatic forces. These types of operations can be divided into scanning techniques, LFM, and force spectroscopy. Furthermore, these types of operations rely on the measurement of forces at different points of the sample, which allows one to construct force versus distance curves. By sensing the twisting of the cantilever, it is also possible to monitor frictional interactions between the tip and the sample. This form of imaging is called lateral force imaging or frictional force imaging. Lateral force microscopes are available commercially.

The contrast of SFM images generally depends on the mechanical properties of the surface and the probe, like adhesiveness and elasticity. To produce a good image of the surface topography, the sample material has to be relatively rigid compared with that of the probe. As the material becomes softer, the image produced will be influenced by the elastic properties of the surface. In extreme cases, the tip–surface interaction can cause damage or displacement of the sample.

Studies on the interactive forces between single molecules have contributed significantly to our understanding of major processes in nature. In this regard, SFM may further improve an in-depth understanding of a variety of such processes. For example, the binding force of complementary molecules (ligand/receptor pairs, drug/substrate pairs) can be characterized by the interpretation of force–distance curves in the piconewton range. It is possible to perform such measurements with SFM under conditions that are close to the natural environment. In such studies, one of the interacting molecules can be covalently bound to the tip, while its corresponding partner is adsorbed (or bound) to the substrate surface. The interaction of these recognition pairs will change the shape of the force distance curve. This effect can then be used for the characterization of interaction parameters under various conditions, such as pH, salt, temperature or blocking of the binding event.

Several interaction features along a typical force curve are shown schematically in Figure 23.2. Point A represents forces sensed by the cantilever in its free position, i.e. without being in direct contact with the surface. In this region, if the cantilever

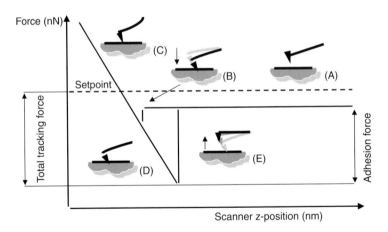

*Figure 23.2* Anatomy of a force–distance curve. (A) Tip approach: The cantilever, not touching the surface, is moved towards the sample. (B) Contact: The cantilever is close enough to the sample to be influenced by attractive forces. It will 'jump' into contact. (C) Contact and further approach: As the fixed end of the cantilever is moved further towards the sample, repulsive forces are generated. (D) Return: The cantilever is withdrawn from the sample and repulsive forces are reduced, while attractive forces are still active. (E) Loss of contact: Adhesive interactions are broken and the cantilever comes free from the surface.

feels a long-range attractive (or repulsive) force it will deflect downwards (or upwards) before making contact with the surface. In the case shown here, there is only a small long-range force. Hence, no deflection is measured in this 'non-contact' part of the force curve. As the probe tip is brought very close to the surface, it may jump into contact (B) if sufficient attractive forces are generated. Once the tip is in contact with the surface, the cantilever deflection will increase (C) as the fixed end of the cantilever is further moved towards the sample. If the cantilever is sufficiently stiff, the probe tip may indent into the surface at this point. In the later case, the slope and the shape of the 'contact' part of the force curve (C) can provide information about the elasticity of the sample surface.

After loading the cantilever with the desired force value, the tip approach is stopped and the process is reversed. As the cantilever is withdrawn, adhesion or bonds formed during the contact with the surface may cause the cantilever to stick to the sample (section D). These adhesive forces may be active beyond the initial point of contact during the tip approach (B). An important characteristic of every AFM force curve is point (E), where the adhesive interactions are disrupted and the cantilever detaches from the surface. This point can be used to measure either the rupture force required to break the bond or the adhesion between tip and sample.

## Substrates, sample preparation and probes for SFM

### Substrates

A proper sample preparation and the selection of a suitable substrate for the specific sample is crucial for high resolution and artifact-free SFM. The substrate must be flat and smooth and the surface roughness has to be substantially below the size of objects under investigation. Substrate materials that have been successfully used in SFM include: freshly cleaned surfaces of highly orientated pyrolytic graphite (HOPG), mica, evaporated or single crystal gold, glass slides and silicon wafers as known from the production of electronic chips. These surfaces can be further modified by chemical reactions in order to change their hydrophobicity, charge and charge density, the surface ion concentration and other parameters (Gould *et al.*, 1989; Hansma *et al.*, 1991; Radmacher, 1997; Radmacher *et al.*, 1994, 1996; Thundat *et al.*, 1992a,b,c; Wong *et al.*, 1998).

It is important that the interaction between the sample and the substrate surface is not too strong. Otherwise, the morphology of the sample might be changed dramatically during sample preparation. On the other hand, the interaction has to be strong enough to prevent delocation of the sample during the scanning process. A hydrophilic surface should be chosen for the examination of biological samples, as the hydrophilicity will reduce the contact angle of the wetting fluid that covers the sample and the surface. Such materials include mica and a chemically modified mica, which have been applied for the characterization of DNA (Hansma *et al.*, 1992a,b, 1993; Heckl *et al.*, 1991; Thundat *et al.*, 1992a,b,c,d), or lipids (Ohlsson *et al.*, 1995; Stephens and Dluhy, 1996; Tillmann *et al.*, 1994). In addition, the use of silicon wafers as substrate for DNA, DNA gene transfer complexes or lipids has been described (Bakowsky *et al.*, 2000; Kneuer *et al.*, 2000; Oberle *et al.*, 2000).

For additional, equally suitable substrate materials, the reader is referred to Morris (1994), Meyer and Heinzelmann (1992) and Marti *et al.* (1988).

### Sample preparation

Various methods are available for sample preparation. A very convenient procedure involves the evaporation of a droplet of the sample that has been dissolved or suspended in buffer. A few microliters are simply pipetted onto the substrate which is then dried in the air. This method sometimes generates artifacts, especially when the organization and distribution of the molecules within the sample is dependent on their concentration, the salt concentration, time and spreading of the droplet. Simple rehydration of a dried sample before the measurement is, therefore, not helpful. Instead, further artifacts may be induced owing to structural changes during the rehydration process. These problems can be overcome by using a self-assembly technique for preparation: the substrate material is placed into the sample solution and the molecules are allowed to adsorb under equilibrium conditions. These adsorbed layers are very stable and the sample can be handled under aqueous conditions to avoid dehydration during preparation and measurement. A special fluid chamber is commercially available for some scanning force microscopes. Another advantage of this type of equipment is its potential to change the milieu that surrounds the sample during or in between measurements without the need to remove the sample. Other preparation methods have been described by Thundat *et al.* (1992a–d) and Muller *et al.* (1997).

### Tips

A key factor for high resolution imaging is also the quality of the tip, especially its geometry, radius and chemical composition. Hence, an appropriate choice of the tip in conjunction with the sample, and the purpose of the experiment in terms of resolution requirements, is essential. For example, for very high, atomic resolution, special probes – the so-called supertips – are (commercially) available. These are also well-suited for lateral force measurements or point force measurements. The shape of such tips is shown in Figure 23.3. The typical tip material is silicon or silicon nitrate. Diamond, carbon or single crystals of minerals have been employed for some special applications (Albrecht and Quate, 1988; Binnig *et al.*, 1986). The most frequently used method for preparation of these probes is the microfabrication etching technique. With this technique, the cantilever/probe systems are etched from oxidized silicon wafers using photographic masks to define the shape of the cantilever. This process results in tips with a radius of less than 30 nm. Sharper 'supertips' have been obtained by the controlled growth of carbon filaments on the end of pyramidal $Si_3N_4$ tips (Keller and Chung, 1992). The surface of the tips produced by these methods can be further modified to yield a well-defined sensor system for the characterization of mechanical, elastic and chemical properties of samples at a resolution of several nanometers. Nanosensors can thus be created by the covalent binding of interesting molecules, such as lipids, DNA, proteins or antibodies, to the probe surface. These types of modifications can be produced based upon a great variety of chemical reactions (Micic *et al.*, 1999).

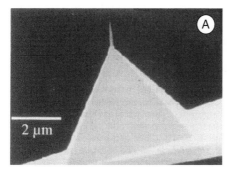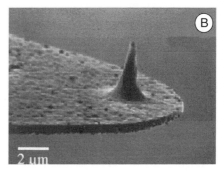

*Figure 23.3* Typical probe geometry for scanning force microscopy. (A) Etched silicon probe, (B) 'supertip' made from $Si_3N_4$ (reproduced from Fritzsche *et al.*, 1996).

## APPLICATIONS OF SFM

### Model membranes

The first instance of the SFM of an organized biological system involved the characterization of model membranes. Phospholipid bilayers are the major constituents of the basic barrier of cellular systems against the surrounding natural environment – the cell membrane. A detailed knowledge of the complex interplay between the lipids and other components of the membrane is necessary to understand such fundamental phenomena as cell–cell interaction, formation of lateral structures within a membrane (raft formation), specific recognition or transmembrane transport. For this reason, simple model membranes were developed to serve as a tool for the *in vitro* study of such processes. Membranes that are supported by a substrate can be obtained by a variety of techniques, including the Langmuir-Blodgett technique (Gains, 1965), self-assembly (Cyr *et al.*, 1996) and vesicle fusion (Ohlsson *et al.*, 1995). These have proven very useful in numerous studies.

In early studies, such systems included simple fatty acids and their salts, which were preorganized into monolayers on the air/water interface, and then transferred to a substrate by the Langmuir-Blodgett technique. Such samples are very thin (between 2 nm and 5 nm), very hard and smooth. A sharp probe and a high scanning speed is needed to produce images of these layers with a high resolution. The structures that have been observed with SFM were interpreted as revealing the lateral chain to chain distance in the fatty acid lattice. The first to report about the molecular resolution of alkyl chains in a liquid environment were Weisenhorn and colleagues (1991). Meine *et al.* (1997) used SFM to study the stability of lipoid films and the time-dependent formation of thermodynamically stable 3D systems. SFM did also allow investigation into the lateral-phase separation in membrane films containing fatty acids with hydrocarbon and fluorocarbon chains (Kato *et al.*, 1996). There is a vast number of studies in which SFM has been employed to describe the morphology and the phase behavior of lipids and mixtures with other membrane components (Day *et al.*, 1993; Hansma *et al.*, 1991). With this technology, it has also been possible to monitor the temperature-dependent phase transition of DMPC, DSPC and DPPE and the conversion of lateral structures, including ripple-phase formation (Hui *et al.*, 1995; Woodward and Zasadzinski, 1996). Some examples of imaged model membranes are shown in Figure 23.4.

Not only does SFM allow for the characterization of the lateral morphology of lipid systems, but complex protein/lipid assemblies can also be analyzed in this manner. The first investigations on such model systems were performed on 2D crystal forming membrane proteins (Butt *et al.*, 1990a,b; Guckenberger *et al.*, 1988; Hoh *et al.*, 1994; Müller *et al.*, 1995; Ohneorge and Binnig, 1993; Schabert and Engel, 1994; Wiegräbe *et al.*, 1991). Thus, a high resolution image (*z*-direction 0.1 nm and *x*-*y* direction 0.5 nm) of the purple membrane of *Halobacterium halobium* and the trimeric structure of the bacteriorhodopsin has been obtained (Müller *et al.*, 1995). This experiment was performed on self-assembled crystals supported by mica operating the AFM in the contact mode with the sample and the tip submerged in liquid. In this manner, it could also be demonstrated that the structure of a single protein can be modulated by the loading force of the cantilever. An AFM image of bacteriorhodopsin and its organization in a Langmuir-Blodgett film is shown in Figure 23.5. In recent years, lipid–protein interactions occuring between pulmonary surfactants and surfactant proteins have been a focus of intense research (Amrein *et al.*, 1997; Kramer *et al.*, 2000; Krol *et al.*, 2000; von Nahmen *et al.*, 1997; Yang *et al.*, 1993).

## Vesicular systems

Artificial membrane vesicles, prepared from lipids and also known as liposomes, have been used as model membranes for the study of both cell adhesion phenomena and dynamic processes within biological membranes. Another application of liposomes is their use as drug carriers, in which bioactive substances are encapsulated. In order to understand the biological activity of such liposomes and to relate these activities to the liposomal structure, it is very important to define the morphology of such lipid particles and their dynamic properties in different environments (Kalb and Tamm, 1992; Kalb *et al.*, 1992; Ohlssen *et al.*, 1995; Ottenbacher *et al.*, 1993; Plant *et al.*, 1994; Vikholm *et al.*, 1995). In such studies, vesicles could be visualized as spherical and sometimes water-filled objects. The transfer of proteoliposomes onto surfaces to generate planar supported lipid bilayers was visualized in a similar manner, but the proteins on the surface were not resolved. However, in general, the quality of vesicle imaging is often limited by tip–sample interactions. Therefore, the best resolutions can be achieved with the TappingMode™ and under fluids.

McMaster *et al.* (1996) reported very high resolution images down to the molecular level of proteins which form the gas-vesicles in cyanobacterium *Anabena flos-aquae*. Other natural vesicles, such as secretory vesicles or synaptic vesicles, have been studied in a similar manner (Jena *et al.*, 1997; Laney *et al.*, 1997; Parpura and Fernandez, 1996). It was also possible to obtain molecular resolution of antibodies that were covalently bound to the surface of vesicles (Bendas *et al.*, 1999). Thus, using AFM, it could be demonstrated that the lateral structure depends on the coupling strategy and the density of the antibody. In Figure 23.6 a vesicle that has been covalently modified with immunoglobulin (IgG) is shown. Finally, the resolution of the subdomains of a single vesicle-bound IgG molecule was demonstrated by Bakowsky

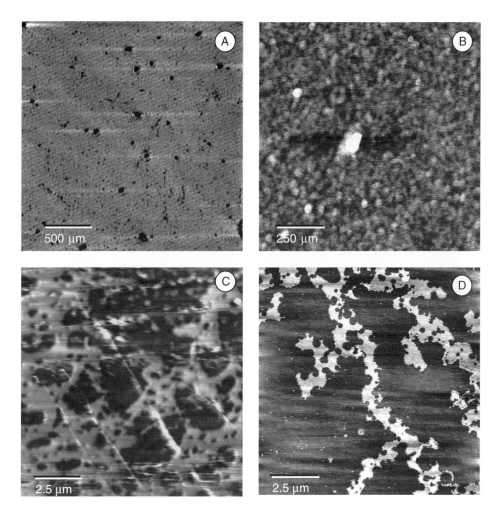

*Figure 23.4* Morphology of Langmuir-Blodgett films of pure skin lipids and their mixtures. (A) Ceramides IV: The lipid forms a very smooth crystalline surface film. The roughness of the film is about 0.3 nm and the total thickness is 3.6 nm. Within the film, homogeneously distributed small crystal defects with a diameter of 20–100 nm, so called 'pinholes', can be visualized. (B) Cholesterol: The cholesterol film consists of small crystallites with a diameter of about 20 nm. This amorphous thin layer has a total thickness of 2.3 nm and appears flat without defects. (C) Mixture of ceramides IV/cholesterol/trioleine. The structure of this mixed film is characterized by large-phase separated areas containing the different lipid components. The dark areas are softer and deeper (trioleine) than the brighter crystalline domains (ceramide IV/cholesterol). (D) Natural skin lipid mixture from sole of the foot: In this film, a number of different morphological structures can be visualized. One example is given in this image. Crystalline bright domains are surrounded by a darker, more fluid film. These lateral structures indicate that the natural skin lipids are organized in separated areas. All Images were obtained with a Nanoscope IIIa (Digital Instruments) in contact mode using $Si_3N_4$ I-type cantilevers with a nominal force constant of 0.1 N/m.

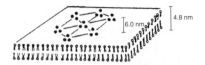

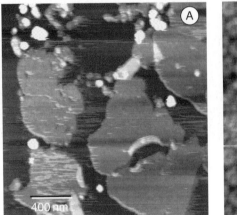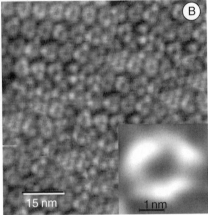

*Figure 23.5* Molecular organization of bacteriorhodopsin in lipid bilayers. The diagram shows the model for the molecular organization of bacteriorhodopsin as a hexagonal lattice of protein trimers within the purple membrane of *Halobacterium halobium*. (A) and (B) are atomic force microscopy (AFM) images of Langmuir-Blodgett films of purple membrane prepared according to Lu *et al.* (1999). The thickness of the purple membrane is 5.2 nm. The hexagonal organization of bacteriorhodopsin can be seen. Some defects within the hexagonal lattice, depending on the preparation method used, could be observed. Imaging was performed with a Nanoscope II (Digital Instruments) under liquid in contact mode with V-type $Si_3N_4$ cantilevers (nominal force constant 0.64 N/m).

*et al.* (2001a,b,c). The structure of the protein was identical to the one previously described by Hansma (1999).

In addition to the analysis of vesicle structures, SFM allows for the monitoring of the dynamic change of vesicle characteristics during interaction processes. For example, Pignataro *et al.* (2000) analyzed the specific adhesion of unilamellar vesicles to functionalized surfaces. Changes in size and shape during the adhesion process could be detected. Another example is the specific adhesion of immunoliposomes to cells. Figure 23.7 shows an AFM image that visualizes this event and shows the organization and the shape of cell-bound vesicles. Due to the complexity of such studies, the use of AFM in this field has not been well-explored. On the other hand, this method has the clear potential to follow all steps of vesicle–cell interaction, from recognition and binding to fusion or uptake.

## Cells and cell surfaces

One advantage of the SFM technique is that it allows for the measurement of biological objects in 'real time' under conditions that are close to the physiological

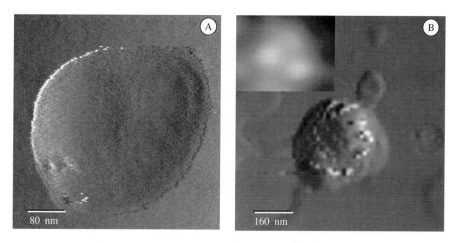

*Figure 23.6* Typical morphology of phosphaditylcholine PC/cholesterol/PE liposomes. (A) unmodified, (B) after covalent binding of immunoglobulin (IgG) molecules to the liposome surface according to Bendas *et al.* (1999). The vesicles were adsorbed to silicon wafers via self-assembly and imaged in buffer in TappingMode™ (Digital Instruments, Nanoscope IIIa) using a $Si_3N_4$ cantilever with a resonance frequency of 220 kHz and a nominal force constant of 36 N/m. The insert in (B) shows a single IgG molecule (dimension $50 \times 25$ nm) (from Bakowsky *et al.*, 2001c).

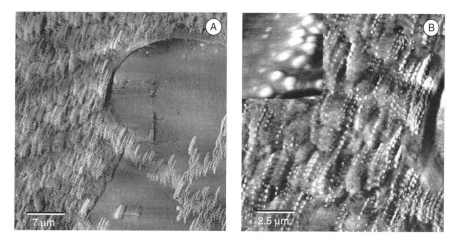

*Figure 23.7* Specific adhesion of anti-E-selectin immunoglobulin (IgG)-modified vesicles to CHO-I cells. IgG mediates specific binding of the modified liposomes to the E-selectin-containing cell membrane. It was possible to visualize the intact liposomes in a natural environment with good resolution. Cells were grown on silicon wafers close to confluency, incubated with the liposome suspension and washed three times as described by Bendas *et al.* (1999). Imaging was performed in buffer in the TappingMode™ (Digital Instruments, Nanoscope IIIa) using a $Si_3N_4$ cantilever with a resonance frequency of 220 kHz and a nominal force constant of 36 N/m (from Bakowsky *et al.*, 2001b).

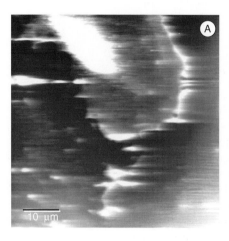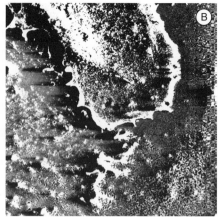

*Figure 23.8* Morphology of primary endothelial cells one hour after seeding. (A) Height mode, (B) amplitude mode. The height of the cell is about 1.5 μm. Cellular structures such as the nucleus and the ruffled membrane could be visualized. Imaging was performed in cell culture medium in TappingMode™ (Digital Instruments, Nanoscope IIIa) using a $Si_3N_4$ cantilever with a resonance frequency of 220 kHz and a nominal force constant of 36 N/m.

situation (Figure 23.8). Dynamic processes can be imaged from the microscopic (optical) scale down to single molecules. The first studies to be published were focused on bacterial cells (Butt *et al.*, 1990b; Hoh *et al.*, 1993). In fact, it has been possible to visualize objects as small as 10 nm on intact but dried cells (Butt *et al.*, 1990a,b). Different preparation methods were elaborated to produce stable samples, including fixation, surface coating and surface decoration (Kasas and Ikai, 1995).

Ohnesorge *et al.* (1992) succeeded to image single proteins from the S-layer of *Bacillus coagulans*. In addition to reproducing and confirming structures already known from earlier studies (often obtained by electron microscopy), force microscopy has provided new information about cellular structures. Hence, entities like the nucleus, actin fibers and secretory vesicles could be visualized (Chang *et al.*, 1993; Fritz *et al.*, 1994a,b; Henderson *et al.*, 1992; Hoh and Schoenenberger, 1994; Jena *et al.*, 1997). In addition to imaging, SFM may also serve as a tool to study the elastic and viscoelastic properties (Radmacher *et al.*, 1994, 1996). Mechanical properties (Fritz *et al.*, 1994a; Radmacher *et al.*, 1994) and the dynamics (Schoenenberger and Hoh, 1994; Schneider *et al.*, 1999, 2000; Shroff *et al.*, 1995; Thomson *et al.*, 1996) can also be characterized. Moreover, Schneider *et al.* (1999) have continuously monitored the extracellular ATP concentration on living cells by force microscopy. This was one of the first biological parameters to be detected on a molecular level with a very high lateral resolution. More information about force microscopy in cellular systems is given in Hansma *et al.* (1997) and Ohnesorge *et al.* (1997).

In conclusion, the visualization of processes involved in transport phenomena, molecular recognition and cellular signaling is a field in which the use of SFM is the most promising, adding the possibility of revealing structural details at a level that will lead to a better understanding of overall cellular functioning.

## Molecular interaction phenomena

Molecular interaction phenomena are important aspects in cellular physiology. In general, there is a recognition of complementary surfaces or structures, and the interactions are non-covalent. Examples of such phenomena are ligand/receptor or antibody/antigen interactions and complementary base pairing in nucleic acids. The smallest biological samples which can be characterized in their morphology and dynamic function by SFM are proteins. One can distinguish between the sole visualization of a binding event and the measurement of the forces which are involved in that binding process. For the visualization of a binding event, the lateral resolution has to be extremely high, usually below 1 nm. However, the identification of proteins or antibodies is very difficult, because all these biological molecules are more or less globular in shape when in their native state. Still, it is well-known that proteins or antibodies have different molecular weights, resulting in different sizes. Indeed, as measured by force microscopy, a correlation between the molecular weight and the molecular volume can be demonstrated (Quist *et al.*, 1995; Schneider *et al.*, 1995).

Another interesting aspect in the field of molecular recognition is the lateral organization of the binding partners. In Figure 23.9 it is illustrated how the lateral distribution of a ligand can influence the adhesion of the protein receptor. This aspect is of relevance in both model membranes and cell surface membranes, as changes in the lateral organization may frequently occur. So far, only a few authors have recognized this problem (Mazzola and Fodor, 1995; Muscatello *et al.*, 2000).

In addition to the topographic measurements described above, SFM can also provide information about the strength of the force acting between the cantilever and the sample. One of the interaction partners is usually bound to the probe (Figure 23.10). This nanosensor is then scanned over the ligand-presenting sample. When a ligand and receptor interact, an increase in adhesion force between tip and sample can be monitored. This monitoring can be used to visualize binding processes over large areas (Chen and Moy, 2000; Radmacher *et al.*, 1994; Wong *et al.*, 1999). With the same equipment, it is also possible to measure forces between single ligand/receptor pairs. This measurement requires the recording and analysis of force–distance curves (see paragraph TappingMode™ AFM). Florin *et al.* (1994) and Moy *et al.* (1994) were among the first to use this method to characterize the binding force between biotin and streptavidin, which is in the piconewton range. An example for the type of information that can be obtained with these measurements is given in Figure 23.11. Further improvement of the available equipment and the methods used has culminated in the measurement of a large number of molecular interactions. Examples include DNA–DNA pairing (Lee *et al.*, 1994), DNA–fibrin interaction (Guthold *et al.*, 1999), adhesion of an anti-ICAM-1 antibody to its antigen (Willemsen *et al.*, 1999), binding of fibronectin-3 to IgG (Rief *et al.*, 1998), the interaction of P-selectin with its glycoprotein ligand-1 (Fritz *et al.*, 1998)

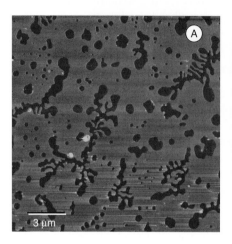
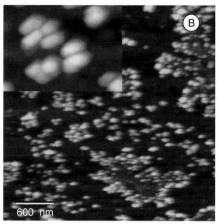

*Figure 23.9* Recognition of sialyl-Lewis[X]-containing glycolipid ligands by E-selectin in a model membrane system. (A) The lipid model membrane before protein adhesion. The dark areas represent the fluid glycolipid domains which are surrounded by the crystalline matrix lipid DSPC. (B) Lipid model membrane after specific adhesion of P-selectin. The protein has adhered on well-defined areas which are identical with the glycolipid rich domains. The insert (250 × 250 nm) shows single molecules of P-selectin. More information about this study is given in Bakowsky *et al.* (2001a). The images were obtained on a Nanoscope Dimension 5000 operated in Tapping-Mode™ (Digital Instruments, $Si_3N_4$ cantilever, 220 kHz resonance frequency, 36 N/m nominal force constant) in air under fully-hydrated conditions.

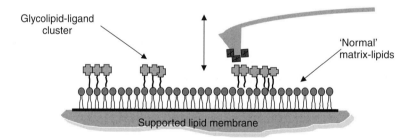

*Figure 23.10* Membrane model. The diagram shows a typical experimental set-up (Bakowsky *et al.*, 2000). On the tip, the lectin concanavalin A was covalently fixed. Its glycolipid ligand was preorganized at the air–water interface in a mixture with the matrix lipid DSPC and subsequently transferred by the Langmuir-Blodgett method onto a silicon wafer.

and the interaction between photolyase and DNA (van Noort *et al.*, 1998). These studies, which have been performed on isolated molecules and in biomimetic models, constituted the basis for more sophisticated experiments on living cells (Razatos *et al.*, 1998).

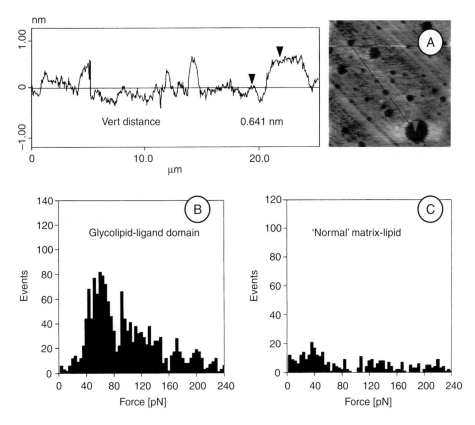

*Figure 23.11* Force measurement of molecular interactions. The binding events between the modified probe and the glycolipid domains or the matrix lipid rich areas can be detected by evaluation of the force curves. The binding force can also be calculated. (A) Height profile of the supported lipid model membrane. Higher values correspond to the glycolipid domains, while the surrounding DSPC matrix lipid films appears lower. (B) Binding events and forces sampled by scanning a glycolipid domain. Total number of measurements: 1805; binding events: 1369; calculated forces: between 20 and 300 pN mean value 60 pN. (C) Binding events and forces between tip and the DSPC matrix. Total number of measurements: 1637; binding events: 319.

## DNA and DNA complexes

Probably the most important biopolymer is DNA, which is also reflected by the fact that nucleic acid is the most frequently investigated polymer by SFM. Most of the studies deal with the conformation of nucleic acid under different conditions. Evidently, the extensive knowledge of both its molecular composition and structure, as clarified in previous studies, as well as its size, explains the interest in this particular molecule for SFM studies. In addition, the state and size of DNA strands

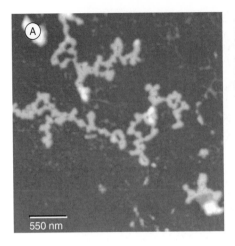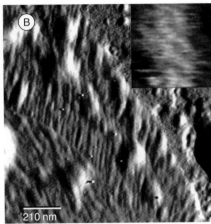

*Figure 23.12* Visualization of DNA. (A) Circular supercoiled structure as known for plasmid DNA from electron microscopy. (B) Organization into parallel strands. The insert shows a high resolution image of the helical structure of the DNA (dimension $25 \times 25$ nm). The typical periodicity of the helix is between 3.4 nm and 3.7 nm, which is in agreement with X-ray data. The thickness of the DNA is 2.2 nm. The images were obtained by Tapping™ atomic force microscopy (AFM) (Digital Instruments, Nanoscope Dimension 5000, $Si_3N_4$ cantilever, 220 kHz resonance frequency, 36 N/m nominal force constant) in air under fully-hydrated conditions.

can be easily manipulated between circular and supercoiled-to-linear double- and single-stranded restriction fragments.

The first high resolution image of nucleic acids was reported by Hansma *et al.* (1992b). They adsorbed a short-synthetic DNA onto a positively charged mica substrate. With this method, the visualization of single base pair could be achieved. An example of such an SFM image, showing a resolution at the molecular level, and some other examples of DNA conformations that have been observed by SFM, are shown in Figure 23.12.

The most common method for the preparation of DNA samples for SFM is the evaporation of a droplet (Henderson *et al.*, 1992; Thundat *et al.*, 1992a,d). Mica or modified mica has been especially popular as a substrate for DNA (Bustamate *et al.*, 1992; Hansma *et al.* 1992b; Henderson 1992a,b; Thundat *et al.*, 1992a–c). However, it has been reported that the history of sample preparation has a strong influence on the organization of the DNA on the substrate. Humidity and the stress applied during the evaporation procedure seem to be especially important. The influence of hydration and dehydration on the molecular structure of DNA has been systematically investigated by Hansma *et al.* (1993), using different mixtures of water and propanol. It has further been reported that the use of pure mica can lead to imaging artifacts. In some cases, DNA might even be removed from the substrate surface by the cantilever's motion. Lack of sample stability and substrate-induced changes to DNA conformation are general concerns in DNA imaging. Consequently, different methods were evaluated to overcome these problems. Hansma *et al.*, (1991) and

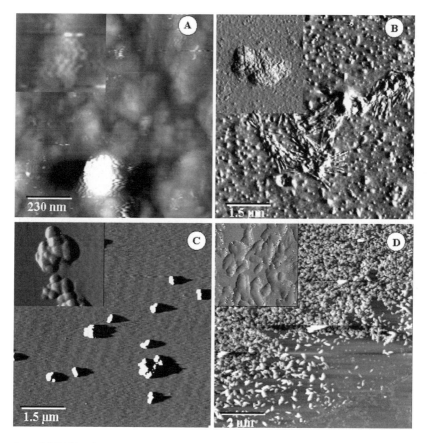

*Figure 23.13* Visualization of the structure of various gene transfection systems. (A) Vesicle-like spherical complexes of SAINT2 and plasmid DNA. The samples were prepared according to Oberle *et al.* (2000) on silicon wafers. The complexes have a diameter of approximately 250 nm. The DNA is largely surrounded by the lipid. Some DNA strains can be found on the surface of the vesicles. A higher magnification is given in the insert (dimension 50×50 nm). (B) Typical morphology of a poly-L-lysin/DNA complex. The mean diameter of the complexes is in the range of 100–200 nm. Some uncomplexed DNA can also be seen. The insert (dimension 125×125 nm) shows the ultrastructure of a 100 nm DNA/polymer complex. The polymer material was removed before imaging by a strong tip–sample interaction. By this procedure, the ordered and condensed structure of the DNA inside the complex can be visualized. (C) Nanoplexes formed between positively charged silica nanoparticles and plasmid DNA (pCMV?). The size of the complexes is between 200 nm and 300 nm. The DNA is protected by surrounding particles. The insert (dimension 250×250 nm) shows the structure of a single complex. More details about these complexes are given in Kneuer *et al.* (2000). (D) The complexes shown here are the results of a self-assembly of SAINT2 and DNA on a Langmuir film balance. The DNA was covered by the lipid under equilibrium conditions. The shape (egg-like) and the dimension (about 250 nm) of the complexes are both very uniform. This new method of complex formation is described in Oberle *et al.* (2000). Insert size: 1.2×1.2 μm. The images were obtained by Tapping™ atomic force microscopy (AFM) (Digital Instruments, Nanoscope Dimension 5000, Si₃N₄ cantilever, 220 kHz resonance frequency, 36 N/m nominal force constant) in air under full-hydrated conditions.

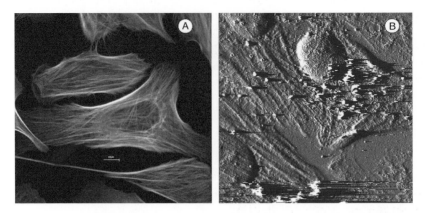

*Figure 23.14* Combination of fluorescence and scanning force microscopy (SFM) for biological samples. (A) Shows the organization of the cytoskeleton of an endothelial cell, labeled with a fluorescent dye. (B) Surface morphology of a similar cell, imaged by SFM. Image (A) is a courtesy of H. Sann (MPI Magdeburg, Germany). This combination technique is commercially available from Digital Instruments (Bioscope).

Weisenhorn *et al.* (1990), for instance, adsorbed DNA onto mica supported lipid membranes. Thundat *et al.* (1992c,d) investigated a number of different surfaces, such as quartz, nitrocellulose, clay, lead phosphate and others. On the basis of this background knowledge, SFM can be successfully applied to analyze more fundamental phenomena like the influence of different cations (Mn, Mg, Co) and ionic strengths on the DNA structure (Allison *et al.*, 1992; Thundat *et al.*, 1992a–d, 1994).

In recent years, most SFM studies of DNA have been performed in the context of artificial gene delivery systems. This context includes naturally occurring complexes of DNA with other biomolecules as well as fully synthetic gene transfection systems. The latter include DNA complexes based on positively charged lipids, liposomes, proteins, polymers and nanoparticles (Figure 23.13).

Thus, AFM has been used to study the influence of polymer chain length on the size and shape of poly-L-lysine/DNA complexes (Wolfert and Seymour, 1996), and to study the change in the structure of such complexes from globular to toroidal after the addition of hydrophilic polyethylene glycol (PEG) blocks to the polymer (Wolfert *et al.*, 1996). Indeed, scanning probe microscopy has proven very useful for the characterization of DNA complexes, including those that were assembled with lipids, organic and inorganic nanoparticles or artificial virus capsids (Oberle *et al.*, 2000; Kneuer *et al.*, 2000).

## SUMMARY

Scanning probe microscopy provides new methods for the imaging of biological objects. These methods have a number of advantages over established imaging techniques in so far as they (i) are non-destructive, (ii) allow examination of samples under 'natural' conditions, (iii) involve fast and convenient sample preparation procedures and (iv) provide a direct on-line image generation. In addition, not only can the topography of samples be obtained. It is also possible to collect data on the mechanical and elastic properties and on specific surface properties, such as adhesiveness. In the future, the development of new techniques in scanning probe microscopy and their combination with existing methods, such as fluorescence microscopy (Figure 23.14), will further extend the potential of this exciting technology.

## ACKNOWLEDGMENT

This work was supported by the Stiftung Deutscher Naturforscher Leopoldina BMBF 9901/8-6 and BMBF 03C0301D1.

## REFERENCES

Albrecht, T. R. and Quate, C. F. (1988) *J. Vac. Sci. Technol.*, **A6**, 271–274.

Allison, D. P., Bottomley, L. A., Thundat, T., Brown, G. M., Woychik, R. P., Schrick, J. J., *et al.* (1992) Immobilization of DNA for scanning probe microscopy. *Proc. Natl. Acad. Sci. USA.*, **89**(21), 10129–10133.

Amrein, M., von Nahmen, A. and Sieber, M. (1997) A scanning force- and fluorescence light microscopy study of the structure and function of a model pulmonary surfactant. *Eur. Biophys. J.*, **26**(5), 349–357.

Bakowsky, U., Rettig, W., Bendas, G., Vogel, J., Bakowsky, H. and Rothe, U. (2000) Characterization of the interactions between various alkylmannosid/phospholipid model membranes with the mannose-specific binding lectin Concanavalin A. *Phys. Chem. Chem. Phys.*, **2**, 4609–4614.

Bakowsky, U., Bakowsky, H., Kneuer, C., Bendas, H., Rothe, U. and Hoekstra, D. (2001a) Specific interaction of molecular P-Selectin and lectin modified liposomes with supported lipidbilayer containing glycolipids with Sialyl Lewis[x] epitop. *Biophys. J.*, submitted.

Bakowsky, U., Hoekstra, D., Krause, A., Bendas, G., Vogel, J., Kneuer, C., *et al.* (2001b) Vesicular target systems in interaction with biomembrane surfaces characterized by atomic force microscopy and quartz crystal microbalance. *Biochem. Biophys. Acta*, submitted.

Bakowsky, U., Oberle, V., Kneuer, C. and Hoekstra, D. (2001c) Non-invasive ultrastructural analysis of modern nanoscaled drug-delivery systems by atomic force microscopy. *Eur. J. Pharm. Biopharm.*, submitted.

Bendas, G., Krause, A., Bakowsky, U., Vogel, J. and Rothe, U. (1999) Targetiliby of novel immunoliposomes prepared by a new antibody conjugation technique. *Int. J. Pharm.*, **181**, 79–93.

Binnig, G., Rohrer, H., Gerber, C. and Weibel, E. (1982) Surface studies by scanning tunneling microscopy. *Phys. Rev. Lett.*, **49**, 57–60.

Binnig, G., Quate, C. F. and Gerber, C. (1986) Atomic force microscopy. *Phys. Rev. Lett.*, **56**, 930–933.

Bustamante, C., Vesenka, J., Tang, C. L., Rees, W., Guthold, M. and Keller, R. (1992) Circular DNA molecules images in air by scanning force microscopy. *Biochem.*, **31**, 22–26.

Butt, H. J., Wolff, E. K., Gould, S. A., Dixon Northern, B., Peterson, C. M. and Hansma, P. K. (1990a) Imaging cells with the atomic force microscope. *J. Struct. Biol.*, **105**(1–3), 54–61.

Butt, H.-J., Downing, K. H. and Hansma, P. K. (1990b) Imaging the membrane protein bacteriorhodopsin with the atomic force microscope. *Biophys. J.*, **58**, 1473–1480.

Chang, L., Kious, T., Yorgancioglu, M., Keller, D. and Pfeiffer, J. (1993) Cytoskeleton of living, unstained cells imaged by scanning force microscopy. *Biophys. J.*, **64**, 1282–1286.

Chen, A. and Moy, V. T. (2000) Cross-linking of cell surface receptors enhances cooperativity of molecular adhesion. *Biophys. J.*, **78**(6), 2814–2820.

Cyr, D. M., Venkataraman, B., Flynn, G. W., Black, A. and Whitesides, G. M. (1996) Functional group identification in scanning tunneling microscopy of molecular adsorbates. *J. Phys. Chem.*, **100**, 13747–13759.

Day, H. C., Allee, D. R., George, R. and Burrows, V. A. (1993) *Appl. Phys. Lett.*, **62**, 1629–1631.

Florin, E. L., Moy, V. T. and Gaub, H. E. (1994) Adhesion forces between individual ligand-receptor pairs. *Science*, **264**(5157), 415–417.

Fritz, M., Radmacher, M. and Gaub, H. E. (1994a) Granula motion and membrane spreading during activation of human platelets imaged by atomic force microscopy. *Biophys. J.*, **66**(5), 1328–1334.

Fritz, M., Radmacher, M., Peterson, N. and Gaub, H. E. (1994b) Visualization of intracellular structures via force modulation and chemical degradation. *J. Vac. Sci. Technol.*, B, in press.

Fritz, J., Katopodis, A. G., Kolbinger, F. and Anselmetti, D. (1998) Force-mediated kinetics of single P-selectin/ligand complexes observed by atomic force microscopy. *Proc. Natl. Acad. Sci. USA*, **95**, 12283–12288.

Fritzsche, W., Schaper, A. and Jovin, T. M. (1996) *Eur. Microsc. Anal.*, **41**, 5–7.

Gaines, G. L. (1966) Insoluble Monolayers. In I. Prigogine (ed), Wiley & Sons, New York.

Gould, S. A., Burke, K. and Hansma, P. K. (1989) Simple theory for the atomic-force microscope with a comparison of theoretical and experimental images of graphite. *Phys. Rev. B Condens Matter.*, **40**(8), 5363–5366.

Guckenberger, R., Wiegräbe, W., Hillebrand, A., Hartmann, T., Wang, Z. and Baumeister, W. (1989) Scanning tunneling microscopy of a hydrated bacterial surface protein. *Ultramicroscopy*, **31**, 327–332.

Guthold, M., Zhu, X., Rivetti, C., Yang, G., Thomson, N. H., Kasas, S. *et al.* (1999) Direct observation of one-dimensional diffusion and transcription by Escherichia coli RNA polymerase. *Biophys. J.*, **77**(4), 2284–2294.

Hansma, H. G. (1999) Varieties of imaging with scanning probe microscopes. *Proc. Natl. Acad. Sci. USA*, **96**(26), 14678–14680.

Hansma, H. G., Weisenhorn, A. L., Edmundson, A. B., Gaub, H. E. and Hansma, P. K. (1991) Atomic force microscopy: seeing molecules of lipid and immunoglobulin. *Clin. Chem.*, **37**(9), 1497–1501.

Hansma, H. G., Vesenka, J., Siegerist, C., Keldermann, G., Morrett, H. and Sinsheimer, R. L. (1992a) Reproducible imaging and dissection of plasmid DNA under liquid with the atomic force microscopy. *Science*, **256**, 1180–1184.

Hansma, H. G., Sinsheimer, R. L., Li, M. Q. and Hansma, P. K. (1992b) Atomic force microscopy of single- and double-stranded DNA. *Nucleic Acids Res.*, **20**(14), 3585–3590.

Hansma, H. G., Bezanilla, M., Zenhausern, F., Adrian, M. and Sinsheimer, R. L. (1993) Atomic force microscopy of DNA in aqueous solutions. *Nucleic Acids Res.*, **21**(3), 505–512.

Hansma, H. G., Kim, K. J., Laney, D. E., Garcia, R. A., Argaman, M., Allen, M. J. *et al.* (1997) Properties of biomolecules measured from atomic force microscope images: a review. *J. Struct. Biol.*, **119**, 99–108.

Hansma, H. G., Pietrasanta, L. I., Auerbach, I. D., Sorenson, C., Golan, R. and Holden, P. A. (2000) Probing biopolymers with the atomic force microscope: a review. *J. Biomater. Sci. Polym. Ed.*, **11**(7), 675–683.

Heckl, W. M., Smith, D. P. E., Binnig, G., Klagges, H., Hansch, T. W. and Maddocks, J. (1991) Two dimensional ordering of the DNA base guanine observed with scanning tunneling microscopy. *Proc. Natl. Acad. Sci. USA*, **88**, 8003–8005.

Henderson, E. (1992a) Atomic force microscopy of conventional and unconventional nucleic acid structures. *J. Microsc.*, **167**(1), 77–84.

Henderson, E. (1992b) Imaging and nanodissection of individual supercoiled plasmids by atomic force microscopy. *Nucleic Acids Res.*, **20**(3), 445–447.

Henderson, E., Haydon, P. G. and Sakaguchi, D. S. (1992) Actin filament dynamics in living glial cells imaged by atomic force microscopy. *Science*, **257**(5078), 1944–1946.

Hoh, J. H. and Schoenenberger, C.-A. (1994) Surface morphology and mechanical properties of MDCK monolayers by atomic force microscopy. *J. Cell Sci.*, **107**, 1105–1114.

Hoh, J. H., Sosinsky, G. E., Revel, J.-P. and Hansma, P. K. (1993) Structure of the extracellular surface of the gap junction by atomic force microscopy. *Biophys. J.*, **65**, 149–163.

Hui, S. W., Viswanathan, R., Zasadzinski, J. A. and Israelachvili, J. N. (1995) The structure and stability of phospholipid bilayers by atomic force microscopy. *Biophys. J.*, **68**, 171–178.

Jena, B. P., Schneider, S. W., Geibel, J. P., Webster, P., Oberleithner, H. and Sritharan, K. C. (1997) $G_i$ regulation of secretory vesicle swelling examined by atomic force microscopy. *Proc. Natl. Acad. Sci. USA*, **94**, 13317–13322.

Kalb, E. and Tamm, L. (1992) *Thin Solid Films*, **210–211**, 763–765.

Kalb, E., Frey, S. and Tamm, L. (1992) Formation of supported planar bilayers by fusion of vesicles to supported phospholipid monolayers. *BBA*, **1103**, 307–316.

Kasas, S. and Ikai, A. (1995) A method for anchoring round shaped cells for atomic force microscope imaging. *Biophys. J.*, **68**(5), 1678–1680.

Kato, T., Kameyama, M. and Kawano, M. (1996) Two-dimensional micronodule structure in monolayers of a partially fluorinated long-chain acid observed by atomic force microscopy. *Thin Solid Films*, **273**, 232–235.

Keller, D. and Chung, C. H. (1992) Imaging steep, high structures by scanning force microscopy with electron beam deposited tips. *Surf. Sci.*, **268**, 333–339.

Kneuer, C., Sameti, M., Bakowsky, U., Schiestel, T., Schirra, H., Schmidt, H., *et al.* (2000) Surface modified silica-nanoparticles can enhance transfection *in vitro*: A novel class of non-viral DNA vectors. *Bioconj. Chem.*, **11**, 926–932.

Kramer, A., Wintergalen, A., Sieber, M., Galla, H. J., Amrein, M. and Guckenberger, R. (2000) Distribution of the surfactant-associated protein C within a lung surfactant model film investigated by near-field optical microscopy. *Biophys. J.*, **78**, 458–465.

Krol, S., Ross, M., Sieber, M., Künneke, S., Galla, H. J. and Janshoff, A. (2000) Formation of three-dimensional protein-lipid aggregates in monolayer films induced by surfactant protein B. *Biophys. J.*, **79**, 904–918.

Laney, D. E., Garcia, R. A., Parsons, S. M. and Hansma, H. G. (1997) Changes in the elastic properties of cholinergic synaptic vesicles as measured by atomic force microscopy. *Biophys. J.*, **72**, 806–813.

Lee, G. U., Chrisey, L. A. and Colton, R. J. (1994) Direct measurement of the forces between complementary strands of DNA. *Science* (USA), **266**, 771–773.

Lu, T., Rothe, U., Liebau, M., Schaffer, H., Hufnagel, P. and Bakowsky, U. (1999) Investigation of the orientation of Purple Membrane sheets in Langmuir-Blodgett films by a quartz-crystal microbalance. *Life Sci.*, **48**(5), 549–556.

Marti, O., Elings, V., Haugan, M., Bracker, C. E., Schneir, J., Drake, B. *et al.* (1988) Scanning probe microscopy of biological samples and other surfaces. *J. Microsc.*, **152**, 803–809.

Mazzola, L. T. and Fodor, S. P. (1995) Imaging biomolecule arrays by atomic force microscopy. *Biophys. J.* **68**(5), 1653–1660.

McMaster, T. J., Miles, M. J. and Walsby, A. E. (1996) Direct observation of protein secondary structure in gas vesicles by atomic force microscopy. *Biophys. J.*, **70**, 2432–2436.

Meine, K., Weidemann, G., Vollhardt, D., Brezesinski, G. and Kondrashkina, E. A. (1997) Atomic force microscopy and X-ray studies of three-dimensional islands in monolayers. *Langmuir*, **13**, 6577–6581.

Meyer, E. and Heinzelmann, H. (1992) In R. Wiesendanger and H.-J. Guntherodt, *Scanning Tunnelling Spectroscopy II*, Springer Ser. In *Sur. Sci.* **28**, pp. 100–149, Springer-Verlag, Berlin, Heidelberg.

Micic, M., Chen, A., Leblanc, R. M. and Moy, V. T. (1999) Scanning electron microscopy studies of protein-functionalized atomic force microscopy cantilever tips. *Scanning*, **21**(6), 394–397.

Morris, V. J. (1994) Biological applications of scanning probe microscopies. *Prog. Biophys. Mol. Biol.*, **61**, 131–185

Moy, V. T., Florin, E. L. and Gaub, H. E. (1994) Intermolecular forces and energies between ligands and receptors. *Science*, **266**(5183), 257–259.

Müller, D. J., Schabert, F. A., Büldt, G. and Engel, A. (1995) Imaging purple membrane in aqueous solutions at sub-nanometer resolution by atomic force microscopy. *Biophys. J.*, **68**, 1681–1686.

Muller, D. J., Amrein, M. and Engel, A. (1997) Adsorption of biological molecules to a solid support for scanning probe microscopy. *J. Struct. Biol.*, **119**(2),172–188.

Muscatello, U., Alessandrini A., Valdre G., Vannini V. and Valdre U. (2000) Lipid oxidation deletes the nanodomain organization of artificial membranes. *Biochem. Biophys. Res. Commun.*, 13, **270**(2), 448–452.

von Nahmen, A., Post, A., Galla, H. J. and Sieber, M. (1997) The phase behavior of lipid monolayers containing pulmonary surfactant protein C studied by fluorescence light microscopy. *Eur. Biophys. J.*, **26**(5), 359–369.

van Noort, S. J. T., van der Werf, K. O., Eker, A. M., Wyman, C., de Grooth, B. G., van Hulst, N. F. *et al.* (1998) Direct visualization of dynamic protein–DNA interactions with a dedicated atomic force microscope. *Biophys. J.*, **74**, 2840–2849.

Oberle, V., Bakowsky, U., Zuhorn, I. and Hoekstra, D. (2000) Lipoplex formation under equilibrium conditions reveals a three step mechanism. *Biophys. J.*, **79**, 1447–1454.

Ohlsson, P.-A., Tjärnhage, T., Herbai, E., Löfas, S. and Puu, G. (1995) Liposome and proteoliposome fusion onto solid substrates, studied using atomic force microscopy, quartz crystal microbalance and surface plasmon resonance. Biological activities of incorporated components. *Bioelectrochem. Bioenerg.*, **38**, 137–148.

Ohnesorge, F. and Binnig, G. (1993) True atomic resolution by atomic force microscopy through repulsive and attractive forces. *Science*, **260**, 1451–1456.

Ohnesorge, F., Heckl, W. M., Haberle, W., Pum, D., Sara, M., Schindler, H. *et al.* (1992) Scanning force microscopy studies of the S-layers from Bacillus coagulans E38-66, *Bacillus sphaericus* CCM2177 and of an antibody binding process. *Ultramicroscopy*, **42–44** (B), 1236–1242.

Ohnesorge, F. M., Horber, J. K., Haberle, W., Czerny, C. P., Smith, D. P. and Binnig, G. (1997) AFM review study on pox viruses and living cells. *Biophys. J.*, **73**(4), 2183–2194.

Ottenbacher, D., Jähnig, F. and Göpel, W. (1993) *Sens. Actuat. B.*, **13–14**, 173.

Parpura, V. and Fernandez, J. M. (1996) Atomic force microscopy study of the secretory granule lumen. *Biophys. J.*, **71**(5), 2356–2366.

Pignataro, B., Steinem, C., Galla, H. J., Fuchs, H. and Janshoff, A. (2000) Specific adhesion of vesicles monitored by scanning force microscopy and quartz crystal microbalance. *Biophys. J.*, **78**(1), 487–498.

Plant, A. L., Gueguetchkeri, M. and Yap, W. (1994) Supported phospholipid/alkanethiol biomimetic membranes: insulating properties. *Biophys. J.*, **67**(3), 1126–1133.

Quist, A. P., Bergman, A. A., Reimann, C. T., Oscarsson, S. O. and Sundqvist, B. U. (1995) Imaging of single antigens, antibodies, and specific immunocomplex formation by scanning force microscopy. *Scanning Microsc.*, **9**(2), 395–400.

Radmacher, M., Cleveland, J. P., Fritz, M., Hansma, H. G. and Hansma, P. K. (1994) Mapping interaction forces with the atomic force microscope. *Biophys. J.*, **66**(6), 2159–2165.

Radmacher, M., Fritz, M., Kacher, C. M., Cleveland, J. P. and Hansma, P. K. (1996) Measuring the viscoelastic properties of human platelets with the atomic force microscope. *Biophys. J.*, **70**(1), 556–567.

Radmacher, M. (1997) Measuring the elastic properties of biological samples with the AFM. *IEEE Eng. Med. Biol. Mag.*, **16**(2), 47–57.

Razatos, A., Ong, Y.-L., Sharma, M. M. and Georgiou, G. (1998) Molecular determinants of bacterial adhesion monitored by atomic force microscopy. *Proc. Natl. Acad. Sci. USA*, **95**, 11059–11064.

Rief, M., Gautel, M., Schemmel, A. and Gaub, H. E. (1998) The mechanical stability of immunoglobulin and fibronectin III domains in the muscle protein titin measured by atomic force microscopy. *Biophys. J.*, **75**, 3008–3014.

Schabert, F. A. and Engel, A. (1994) Reproducible aquisition of E. coli porin surface topographs by atomic force microscopy. *Biophys. J.*, **67**, 2394–2403.

Schneider, S. W., Larmer, J., Henderson, R. M. and Oberleithner, H. (1998a) Molecular weights of individual proteins correlate with molecular volumes measured by atomic force microscopy. *Pflugers Arch.*, **435**(3), 362–367.

Schneider, S. W., Pagel, P., Storck, J., Yano, Y., Sumpio, B. E., Geibel, J. P. *et al.* (1998b) Atomic force microscopy on living cells: aldosterone-induced localized cell swelling. *Kidney Blood Press. Res.*, **21**(2–4), 256–258.

Schneider, S. W., Egan, M. E., Jena, B. P., Guggino, W. B., Oberleithner, H. and Geibel, J. P. (1999) Continuous detection of extracellular ATP on living cells by using atomic force microscopy. *PNAS*, **96**, 12180–12185.

Schneider, S. W., Pagel, P., Rotsch, C., Danker, T., Oberleithner, H., Radmacher, M. *et al.* (2000) Volume dynamics in migrating epithelial cells measured with atomic force microscopy. *Pflugers Arch.*, **439**(3), 297–303.

Schoenenberger, C.-A. and Hoh, J. H. (1994) Slow cellular dynamics in MDCK and R5 cells monitored by time-lapse atomic force microscopy. *Biophys. J.*, **67**, 929–936.

Shroff, S. G., Saner, D. R. and Lal, R. (1995) Dynamic micromechanical properties of cultured rat atrial myocytes measured by atomic force microscopy. *Am. J. Physiol.*, **269**, C286–C292.

Stephens, S. M. and Dluhy, R. A. (1996) In-situ and ex-situ structural analysis of phospholipid-supported planar bilayers using infrared spectroscopy and atomic force microscopy. *Thin Solid Films*, **284–285**, 381–386.

Thomson, N. H., Fritz, M., Radmacher, M., Cleveland, J. P., Schmidt, C. F. and Hansma, P. K. (1996) Protein tracking and detection of protein motion using atomic force microscopy. *Biophys. J.*, **70**(5), 2421–2431.

Thundat, T., Allison, D. P., Warmack, R. J. and Ferrell, T. L. (1992a) Imaging isolated strands of DNA molecules by atomic force microscopy. *Ultramicrosc.*, **42–44**, 1101–1106.

Thundat, T., Warmack, R. J., Allison, D. P., Bottomley, L. A., Lourenco, A. J. and Ferrell, T. L. (1992b) Atomic force microscopy of deoxyribonucleic acid strands adsorbed on mica: the effect of humidity on apparent width and image contrast. *J. Vac. Sci. Technol.*, **A10**, 630–635.

Thundat, T., Zheng, X. Y., Sharp, S. L., Allison, D. P., Warmack, R. J., Joy, D. C. *et al.* (1992c) Calibration of atomic force microscope tips using biomolecules. *Scanning Microsc.*, **6**, 903–910.

Thundat, T., Allison, D. P., Warmack, R. J., Brown, G. M., Jacobson, K. B., Schrick, J. J. *et al.* (1992d) Atomic force microscopy of DNA on mica and chemically modified mica. *Scanning Microsc.*, **6**(4), 911–918.

Tillmann, R. W., Hofmann, U. G. and Gaub, H. E. (1994) AFM-Investigation of the molecular structure of films from a polymerizable two-chain lipid. *Chem. Phys. Lipids*, **73**, 81–89.

Vikholm, I., Peltonen, J. and Teleman, O. (1995) Atomic force microscope images of lipid layers spread from vesicle suspensions. *BBA*, **1233**, 111–117.

Weisenhorn, A. L., Gaub, H. E., Hansma, H. G., Sinsheimer, R. L., Kelderman, G. L. and Hansma, P. K. (1990) Imaging single-stranded DNA, antigen-antibody reaction and polymerized Langmuir-Blodgett films with an atomic force microscope. *Scanning Microsc.*, **4**(3), 511–516.

Weisenhorn, A. L., Egger, M., Ohnesorge, F., Gould, S. A. C., Heyn, S.-P., Hansma, H. G., *et al.* (1991) Molecular-resolution images of Langmuir-Blodgett films and DNA by atomic force microscopy. *Langmuir*, **7**, 8–12.

Wiegräbe, W., Nonnenmacher, M., Guckenberger, R. and Wolter, O. (1991) Atomic force microscopy of a hydrated bacterial surface protein. *J. Microsc.*, **163**, 79–84.

Willemsen, O. H., Snel, M. M. E., Kuipers, L., Figdor, C. G., Greve, J. and de Grooth, B. G. (1999) A physical approach to reduce non-specific adhesion in molecular recognition atomic force microscopy. *Biophys. J.*, **76**, 716–724.

Wolfert, M. A. and Seymour, L. W. (1996) Atomic force microscopic analysis of the influence of the molecular weight of poly(L)lysine on the size of polyelectrolyte complexes formed with DNA. *Gene Ther.*, **3**, 269–273.

Wolfert, M. A., Schacht, E. H., Toncheva, V., Ulbrich, K., Nazarova, O. and Seymour, L. W. (1996) Characterization of vectors for gene therapy formed by self-assembly of DNA with synthetic block co-polymers. *Human Gene Ther.*, **7**, 2123–2133.

Wong, J., Chilkoti, A. and Moy, V. T. (1999) Direct force measurements of the streptavidin-biotin interaction. *Biomol. Eng.*, **16**(1–4), 45–55.

Wong, S. S., Joselevich, E., Woolley, A. T., Cheung, C. L. and Lieber, C. M. (1998) Covalently functionalized nanotubes as nanometre-sized probes in chemistry and biology. *Nature*, 2, **394**(6688), 52–55.

Woodward, J. T. and Zasadzinski, J. A. (1996) Amplitude, wave form, and temperature dependence of bilayer ripples in the $P_\beta$ phase. *Phys. Rev. E*, **53**, 3044–3047.

Yang, J., Tamm, L. K., Tillack, T. W. and Shao, Z. (1993) New approach for atomic force microscopy of membrane proteins. *J. Mol. Biol.*, **229**, 286–290.

# Chapter 24

# Fluorescence correlation spectroscopy

*Hanns Häberlein*

## INTRODUCTION

Fluorescence correlation spectroscopy (FCS) is a confocal fluorimetric method in which the fluorescence of a single dye-labeled molecule excited by a sharp focused laser beam is observed. With this technique, initially conceived in the early 1970s (Ehrenberg and Rigler, 1974; Elson and Madge, 1974; Koppel, 1974), the detection of single molecules in the millisecond range has been made possible due to improvements in the sensitivity of FCS (Eigen and Rigler, 1994; Rigler *et al.*, 1993; Rigler, 1995; Rigler and Metz, 1992; Rigler and Widengren, 1990). A disadvantage of conventional fluorescence spectroscopy is its limited resolution and sensitivity in recording fluorescence signals, owing to the high background noise. A relatively large volume element must be illuminated and the emitted fluorescence must be carefully separated from the following: scattered laser light, Raman scattering from the solvent, spurious fluorescence signals from the solvent and the optics and other luminescence.

The background emission is approximately proportional to the third power of the radius of the light cavity. In FCS, a laser beam can be focused into a volume element of one femtoliter, and even smaller, by using stable laser light sources in combination with confocal optics (Rigler, 1995). Thus, it is possible to discriminate the light quanta of a single dye-labeled molecule against a background with a signal to background ratio of approximately 1000:1 (Metz and Rigler, 1994; Rigler and Metz, 1992). Focusing the light into a volume element of one femtoliter means that in a given concentration of 1 nM there is theoretically only one molecule in the volume element at a given time. Thus, FCS is an extremely sensitive technique for studying molecular interactions on the single molecule level.

One requirement for FCS measurements is that the molecule under investigation is a fluorophore itself or that it is labeled with a fluorescent dye. Fluorescence is a phenomenon in which the excitation of a fluorescent molecule by light of an appropriate wavelength is followed by the emission of fluorescent light. The fluorophore molecule can be repeatedly excited unless it is irreversibly destroyed by photobleaching. For many fluorescent molecules it is believed that photobleaching originates from the triplet excited state (Widengren, 1996). Other processes, such as collisional quenching and fluorescence energy transfer, may also lead to a loss of fluorescence photons. Thus, not all fluorophore molecules initially excited by absorption return to the ground state. The ratio of the number of fluorescence

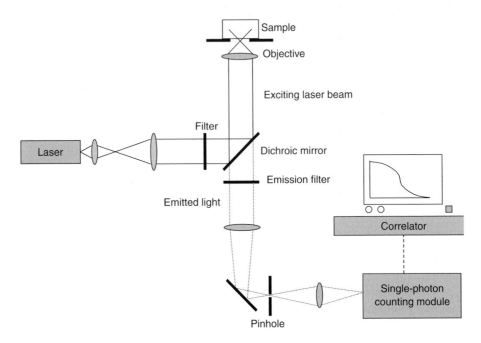

*Figure 24.1* Instrumental set-up of the first commercially available fluorescence correlation spectroscopy (FCS) instrument (ConfoCor®).

photons emitted and photons absorbed is called the quantum yield of the fluorophore. Also, environmental influences like temperature, ionic strength, pH value and covalent or non-covalent interactions of the fluorophore can affect its quantum efficiency (Widengren, 1996). The detection of emitted photons against a low background is possible by isolating them from the excitation photons using bandpass filters. In dilute solutions, the fluorescence intensity is linearly proportional to the molar excitation coefficient, the optical path length, the fluorophore concentration, the fluorescence quantum yield, the excitation intensity and the fluorescence collection efficiency of the instrument. Thus, when the illumination wavelength (energy) and the excitation intensity are held constant by the use of a laser light source, the fluorescence intensity is a linear function of the number of fluorophore molecules. At high concentrations of the fluorophore, the relationship becomes non-linear because other processes, like self-absorption, interfere with the measurement. Under high-intensity illumination conditions, photodestruction of the fluorophore (photobleaching) is a limiting factor (Widengren, 1996).

The set-up of the first commercially available FCS instrument (ConfoCor®) is given in Figure 24.1. In this instrument, based on a confocal microscope, a laser beam is guided by fiber optic coupling, then reflected by a dichroic mirror and focused through a microscope objective lens into a very small diffraction-limited spot with a diameter of approximately 0.4 µm. The emitted fluorescent light from the sample is captured by the microscope objective, then separated from the excitation laser light by an emission filter and guided through a confocal pinhole to hit a single-photon counting avalanche photodiode. By a motor-controlled, variable pinhole, an illuminated and detected volume element of about one femtoliter is determined. Fluctuations in the fluorescence intensity, caused by the diffusion of dye-labeled molecules through the volume element, are calculated. The sample can be situated on a variety of specimen carriers, from microscope slides, to micro-titer plates as well as capillaries. The precise positioning of the laser beam enables measurements on the surface or interior of cells.

The translational diffusion of fluorescent molecules through the illuminated volume element leads to fluctuations in the fluorescence intensity. Each individually emitted photon, resulting from a single molecule, can be registered in a time-resolved manner by high-sensitivity single-photon counting modules. The autocorrelation function for the Brownian motion of dye-labeled molecules in a 3D Gaussian volume element with half-axes 'w' and 'z' is described as:

$$G(\tau) = 1 + \frac{1}{N} \left( \frac{1}{1 + \frac{4D\tau}{\omega^2}} \right) \left( \frac{1}{1 + \frac{4D\tau}{z^2}} \right)^{\frac{1}{2}} \tag{24.1}$$

or

$$G(\tau) = 1 + \frac{1}{N} \left( \frac{1}{1 + \frac{4D\tau}{\omega^2}} \right) \left( \frac{1}{1 + \left(\frac{\omega}{z}\right)^2 \frac{\tau}{\tau_D}} \right)^{\frac{1}{2}}, \tag{24.2}$$

with

$$\tau_D = \frac{\omega^2}{4D}, \tag{24.3}$$

$\tau_D$ being the diffusion time and $D$ the diffusion coefficient.

For a dye-labeled ligand binding to a cell membrane its diffusion at the membrane takes place only at a 2D surface and $G(\tau)$ becomes:

$$G(\tau) = 1 + \frac{1}{N} \left( \frac{1}{1 + \frac{\tau}{\tau_D}} \right). \tag{24.4}$$

When the volume element with half-axes $\omega = 0.25\,\mu m$ and $z = 1.25\,\mu m$ is projected onto the cell surface, both bound dye-labeled ligand diffusing at the cell surface and unbound dye-labeled ligand diffusing above the cell surface will be seen. Then the autocorrelation function for 3D diffusion of the unbound dye-labeled ligand in solution above the cell surface and 2D diffusion of bound dye-labeled ligand to membranes on the cell surface is given by:

$$G(\tau) = 1 + \frac{1}{N}\left[ (1 - \Sigma y_i)Q_f^2 \left( \frac{1}{1 + \frac{\tau}{\tau_D^f}} \right)\left( \frac{1}{1 + \left(\frac{\omega}{z}\right)^2 \frac{\tau}{\tau_D^f}} \right)^{\frac{1}{2}} + \Sigma y_i Q_i^2 \left( \frac{1}{1 + \frac{\tau}{\tau_D^{bi}}} \right) \right], \quad (24.5)$$

where $y_i$ is the fraction of membrane bound dye-labeled ligand diffusing with diffusion time $\tau_D^{bi}$ and $(1 - \Sigma y_i)$ is the fraction of unbound dye-labeled ligand diffusion, with diffusion time $\tau_D^f$. For calculation of parameters of the autocorrelation function $G(\tau)$, nonlinear least square minimization was used (Marquardt, 1963). $Q_i$ is the quantum yield of the bound ligand represented by $y_i$. $Q_f$ is the quantum yield of the unbound ligand.

Taken together, the number of molecules in the volume element, their characteristic translational diffusion times and the ratio of rapidly to slowly diffusing components can be directly computed from the autocorrelation function. Thus, analyzing molecular kinetics of chemical or biological interactions, for example, between antigen and antibody, DNA-primer and DNA-targets or ligand and receptor, becomes possible. The basic steps of an FCS experiment are given in Figure 24.2.

## MATERIALS AND METHODS

### Equipment

FCS measurements were performed with confocal illumination of a volume element of 0.19 fl in a ConfoCor® instrument (*Zeiss*, Germany). For excitation the 488 or 514 nm line of an argon laser (LGK 7812 ML 2, *Lasos*) was focused through a water immersion objective (C-Apochromat, 63×, NA 1.2, *Zeiss*) into the sample (power density in the focal plane, measured before the objective: $p_{488}$ nm = 12.5 kW/cm²; $p_{514}$ nm = 14.2–109 kW/cm²). The emitted fluorescence was collected by the same objective and separated from the excitation light with a dichroic filter and a bandpass filter. The intensity fluctuations were detected by an avalanche photo diode (SPCM-AQ Series, *EG & G Optoelectronics*, Canada) and were correlated with a digital hardware correlator (ALV-5000, *ALV*, Germany).

### Cell measurements

For FCS experiments the coverslips were mounted to a coverslip carrier with an incubation volume of 300 μl. Focus positioning to the upper membrane of the cell soma was performed both by view in the *x-y* directions (resolution 1 μm) and by motor-aided scanning through the neuron in the *z*-direction (optoelectronical DC-servodrives, resolution 0.1 μm).

### Substances

The Alexa® 532-labeled Ro 7-1986/602 derivative (Ro-Ax) (Figure 24.3) was obtained by the labeling of Ro 7-1986/602 (*Novartis*, Basel) with Alexa Fluor® 532 carboxylic acid succinimidyl ester (*Molecular Probes*, Leiden, Netherlands). Midazolam was obtained from Sigma.

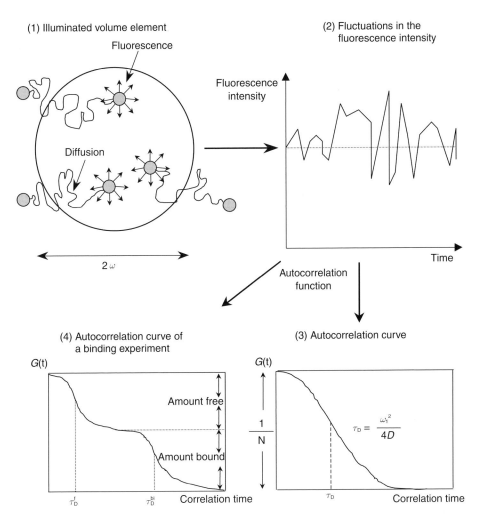

(1) Illuminated volume element

Fluorescence

Diffusion

$2\omega$

(2) Fluctuations in the fluorescence intensity

Fluorescence intensity

Time

Autocorrelation function

(4) Autocorrelation curve of a binding experiment

$G(t)$

Amount free

Amount bound

$\tau_D^f$  $\tau_D^{bi}$  Correlation time

(3) Autocorrelation curve

$G(t)$

$\dfrac{1}{N}$

$\tau_D = \dfrac{\omega_1^2}{4D}$

$\tau_D$  Correlation time

*Figure 24.2* Principle of fluorescence correlation spectroscopy (FCS) measurements. (1) Fluorescent molecules diffuse through the illuminated volume element, thereby emitting light quanta. (2) The fluctuation of the fluorescence intensity inside the volume element is recorded by a highly sensitive single-photon counting module. (3) The autocorrelation curve of a component with a diffusing time of $\tau_D$. The average number of molecules in the volume element is expressed by N. The amplitude of the correlation function increases as the number of molecules decreases. (4) Theoretical correlation curve of the interaction between a small labeled molecule and a large receptor. $\tau_D^f$ is the diffusion time of the free ligand and $\tau_D^{bi}$ is the diffusion time of the receptor–ligand complex. The binding properties of the ligand can be calculated directly from the diffusion times and the ratio of rapidly-to-slowly diffusing components.

## Cell culture

Hippocampal rat neurons were prepared on embryonic day 18 by microdissection. After trituation, cells were seeded at a density of approximately $2.5 \times 10^5$

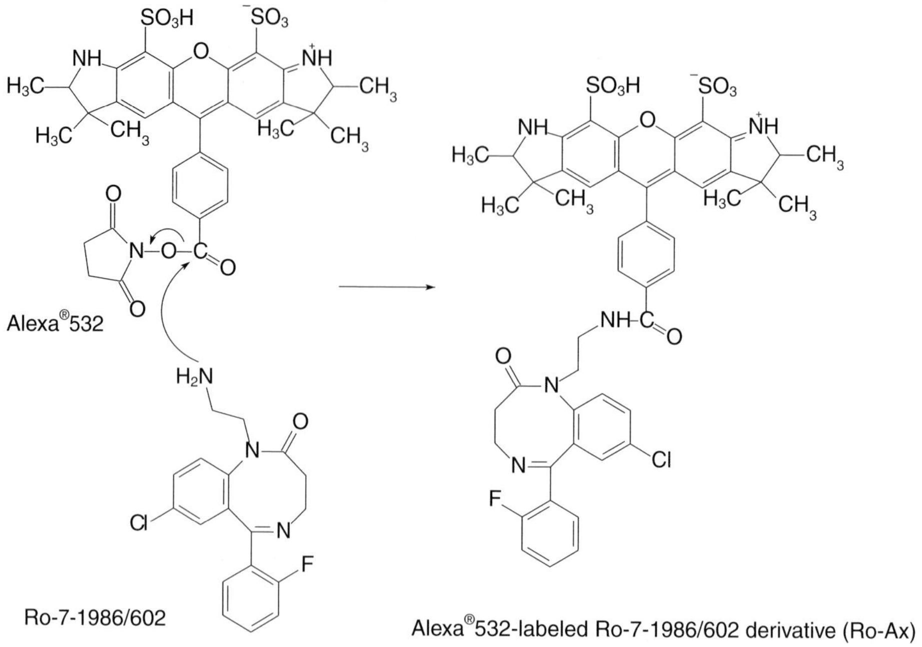

*Figure 24.3* Synthesis of a dye-labeled benzodiazepine derivative.

to 18 mm coverslips, coated with poly-D-lysin and placed in a nunclon Multiwell (12x). Cells were then cultured for ten days in Start-V medium (*Biochrome*, Berlin, Germany). Cells were used for experiments from day eight until day 14.

## Binding assay

Cells were washed three times with Locke's solution (HEPES: 5 mM; NaCl: 154 mM; KCl: 5.6 mM; $MgCl_2$: 1 mM; $Na_2CO_3$: 3.6 mM; glucose: 20 mM; $CaCl_2$: 2.3 mM; pH 7.4) at 37 °C before FCS measurements. In addition, incubation took place in Locke's solution that contained different concentrations of ligand (5, 10, 25, 50, 100 and 210 nM of Ro-Ax) for five minutes at room temperature. For the determination of non-specific binding the cells were incubated with 10 nM Ro-Ax after a preincubation with 10 µM of midazolam for 30 minutes. Displacement experiments were performed by a preincubation of 10 nM Ro-Ax and a secondary incubation of 10 µM midazolam.

If the free ligand is in equilibrium with the bound ligand, the concentration of bound ligand ($[L^*]_{bound}$) is given by the following equation:

$$[L^*]_{bound} = \left( \frac{[L^*]_0 + B_{max} + K_D \pm \sqrt{([L^*]_0 + B_{max} + K_D)^2 - 4 \times [L^*]_0 \times B_{max}}}{2} \right), \quad (24.6)$$

Where $[L^*]_0$: total amount of ligand in the sample; $B_{max}$: total number of binding sites in the sample; $K_D$: dissociation constant of the ligand–receptor complex.

From a plot of the bound ligand concentration versus the total amount of ligand, a dissociation constant ($K_D$) and the maximum number of binding sites ($B_{max}$) was obtained for each cell by non-linear curve fitting to Equation 24.6. Before averaging each concentration of bound ligand $[L^*]_{bound}$ was divided by the individual $B_{max}$ of the experiment so that a comparison of the different cell experiments was obtained. The averaged relative values of $[L^*]_{bound}/B_{max}$ were multiplied by the arithmetic average of $B_{max} = (37.8 \pm 7.7)\,nM$ ($n = 7$) and the resulting values of $[L^*]_{bound}$ versus $[L^*]_0$ were fitted by Equation 24.6 keeping $B_{max}$ constant.

## SPECIFIC APPLICATIONS

The interactions of benzodiazepines with specific receptors of the central nervous system (CNS) cause an amplification of the GABAergic synaptic inhibitory effect. GABAergic neurons are widely distributed in the CNS, with a high density in the frontal cortex, the cerebellum, and the hippocampus. This amplification of inhibitoric influence to several brain functions is the most important pharmacological therapy to treat, e.g. anxiety, insomnia and convulsions. Interactions between ligands and the benzodiazepine receptor have been investigated mainly by radioreceptor assays which require the use of radioactively labeled ligands and include several washing steps. Therefore, weak interactions between compounds and specific binding sites are often not found. In the case of detectable interactions, the binding characteristics are related to membrane fractions or receptor preparations and not to receptors embedded in their native environment in living cells. In the fully functional membrane, however, further characteristics can be expected. One relevant criteria for the regulation and functionality of membrane-standing receptors is their lateral mobility. Changes of this diffusion behavior in the membrane can give information about changes of receptor state. Thus, cell measurements with additional observation of the lateral mobility can provide more information about dynamic binding processes. FCS is well-suited to study the undisturbed interaction of ligands with binding sites on the molecular level in their native environment on cell surfaces. Because of the homogenous assay, parameters that describe both the kinetics of ligand–receptor interactions and the lateral movement of the ligand–receptor complex can be determined. Endogenous neurotransmitters and most of the exogenous ligands, like the benzodiazepines, are often structures of low molecular weight. Because of the need for fluorescently labeled ligands for FCS, the influence of the dye to the binding behavior – particularly of small ligands – has to be low.

The binding of the Ro-Ax to the chloride channel associated benzodiazepine binding site was analyzed at the membrane of prenatal rat hippocampal neurons. Ro-Ax was obtained by the labeling of Ro 7-1986/602 with an aminoreactive Alexa Fluor® 532 carboxylic acid succinimidyl ester (Figure 24.3).

After incubation of 5, 10, 25, 50, 100 and 210 nM of Ro-Ax with neurons, specific binding constants of Ro-Ax to the benzodiazepine binding site were

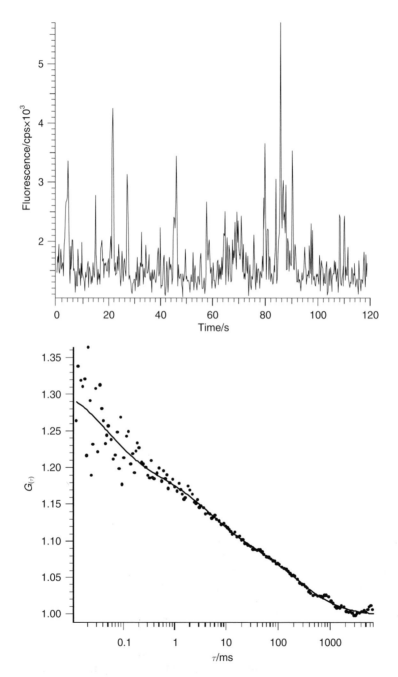

*Figure 24.4* Fluorescence time course and corresponding autocorrelation functions in presence of 10 nM Alexa® 532-labeled Ro 7-1986/602 derivative (Ro-Ax). The autocorrelation functions were fitted and yielded the following diffusion time constants for Ro-Ax: $\tau_{D1} = 487\,\mu s$ (41 per cent), $\tau_{D2} = 4.9\,ms$ (30 per cent) $\tau_{D3} = 250\,ms$ (29 per cent).

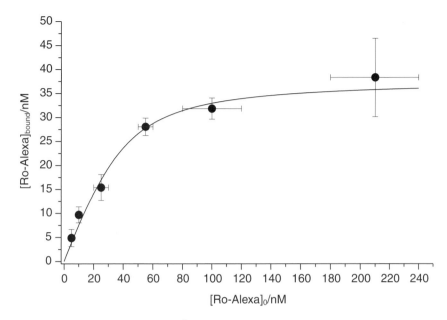

*Figure 24.5* Averaged bound Alexa® 532-labeled Ro 7-1986/602 derivative (Ro-Ax) concentration to neurons versus the total Ro-Ax concentration. The bound Ro-Ax fraction was determined from the autocorrelation function for different Ro-Ax concentrations. Each point is the average of seven experiments. Errors are standard error of the means (SEM).

determined. The diffusion time constants yielded by the autocorrelation analysis (Figure 24.4) were calculated according to Equation 24.3 into diffusion coefficients. The bound and the unbound ligand Ro-Ax leads to the following diffusion coefficients:

$$D_{\text{free}} = (220 \pm 3)\,\mu\text{m}^2/\text{s}\ (n=25),\ D_{\text{bound1}} = (1.32 \pm 0.26)\,\mu\text{m}^2/\text{s}\ (n=25),$$

$$\text{and } D_{\text{bound2}} = (2.63 \pm 0.63) \times 10^{-2}\,\mu\text{m}^2/\text{s}\ (n=22).$$

The fit to the average curve of a plot of bound Ro-Ax versus the total amount of Ro-Ax (Figure 24.5) yielded $K_D$-value of $9.9 \pm 1.9\,\text{nM}$ ($n=7$) and $B_{\text{max}} = 37.8 \pm 7.7\,\text{nM}$ ($n=7$). To test the specificity of the interaction of Ro-Ax with its receptor in the cell membrane, the bound ligand was replaced by an excess of non-labeled agonist. For these purposes, the benzodiazepine agonist midazolam was used in a concentration of $10\,\mu\text{M}$. After an application of midazolam the bound fraction of Ro-Ax was reduced in a time-dependent manner (Figure 24.6). The fluorescence time course was fitted by a monoexponential function which leads to a rate constant of ligand–receptor dissociation of $k_{\text{diss}} = (1.28 \pm 0.08) \times 10^{-3}/\text{s}$ ($n=5$).

The competitive displacement of the bound ligand took place with a maximum delay time of ten minutes. Seven–ten per cent of bound Ro-Ax could not be displaced by an excess of midazolam. This fraction corresponds to the non-specific

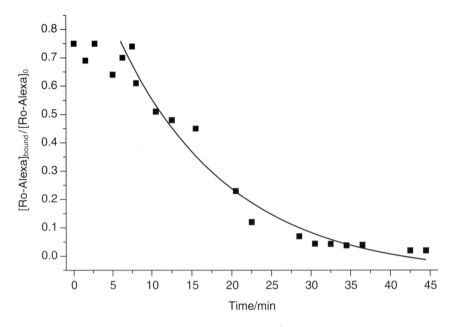

*Figure 24.6* Time course of displacement of 10 nM Alexa® 532-labeled Ro 7-1986/602 derivative (Ro-Ax) by unlabeled midazolam. The neuron was incubated with 10 nM Ro-Ax for two minutes. At $t=0$ a concentration of 10 μM midazolam was added to the extracellular solution. The bound Ro-Ax fraction was determined from autocorrelation functions recorded for two minutes. The line reflects a monoexponential fit through the measured points.

binding of Ro-Ax to the cell membrane. This value was independent from the illumination time of the sample. Thus, photobleaching does not change the binding fraction of bound ligand.

By dividing the complex dissociation constant ($K_D$) by the rate constant of ligand–receptor complex dissociation ($k_{diss}$), the rate constant of ligand–receptor complex association was calculated as: $k_{ass} = (1.30 \pm 0.26) \times 10^5$ l/mol/s ($n=5$).

The obtained binding constants give evidence that the affinity and the binding behavior of Ro-Ax to the benzodiazepine receptor is not essentially influenced by the bulky dye. Additionally, the lateral mobility of the benzodiazepine receptor on rat nerve cells was investigated. Two different populations were detected ($D_{bound1}$: $1.32 \pm 0.26$ μm$^2$/s, $D_{bound2}$: $2.63 \pm 0.63 \times 10^{-2}$ μm$^2$/s$^{-1}$), which correspond to earlier reported diffusion coefficients of freely lateral diffusing proteins in biomembranes and of immobilized proteins attached to anchor structures at their cytosolic domains.

## CONCLUDING REMARKS

With this work we are now able to perform a high-sensitive and non-invasive, homogenous assay to detect even weak ligand interactions to receptor sites. The great approach of this technique is the investigation of molecular interactions in native environments, e.g. on the membranes of living cells.

## REFERENCES

Ehrenberg, M. and Rigler, R. (1974) Rotational Brownian motion and fluorescence intensity fluctuation. *J. Chem. Phys.*, **4**, 390–401.

Eigen, M. and Rigler, R. (1994) Sorting single molecules: application to diagnostics and evolutionary biotechnology. *Proc. Natl. Acad. Sci. USA*, **91**, 5740–5747.

Elson, E. L. and Madge, D. (1974) Fluorescence correlation spectroscopy. I. Conceptional basis and theory. *Biopolymers*, **13**, 1–27.

Koppel, D. (1974) Statistical accuracy in fluorescence correlation spectroscopy. *Phys. Rev.*, **A10**, 1938–1945.

Marquardt, D. W. (1963) An algorithm for least-squares estimation of non-linear parameters. *J. Soc. Indust. Appl. Math.*, **11**, 431–441.

Metz, Ü. and Rigler, R. (1994) Submillisecond detection of single rhodamine molecules in water. *J. Fluor.*, **4**, 259–264.

Rigler, R. and Widengren, J. (1990) Ultrasensitive detection of single molecules by fluorescence correlation spectroscopy. *Bioscience*, **3**, 180–183.

Rigler, R., Metz, Ü., (1992) Diffusion of single molecules through a Gaussian laser beam. *SPIE*, **1921**, 239–248.

Rigler, R., Mets, Ü., Widengren, J. and Kask, P. (1993) Fluorescence correlation spectroscopy with high count rate and low background: analysis of translational diffusion. *Eur. Biophys. J.*, **22**, 169–175.

Rigler, R. (1995) Fluorescence correlations, single molecule detection and large number screening. Applications in biotechnology. *J. Biotechnol.*, **41**, 177–186

Widengren, J. (1996) *Fluorescence Correlation Spectroscopy, Photophysical Aspects and Applications*, Doctoral thesis. Stockholm: Repro Print AB.

# Index